HANDBOOK OF
REFRACTORY COMPOUNDS

IFI DATA BASE LIBRARY

COMPUTER TECHNOLOGY
Logic, Memory, and Microprocessors — A Bibliography
A. H. Agajanian

ACOUSTIC EMISSION
A Bibliography with Abstracts
Thomas F. Drouillard

MICROELECTRONIC PACKAGING
A Bibliography
A. H. Agajanian

SPEECH COMMUNICATION AND THEATER ARTS
A Classified Bibliography of Theses and Dissertations, 1973—1978
Merilyn D. Merenda and James W. Polichak

ATOMIC GAS LASER TRANSITION DATA
A Critical Evaluation
William Ralph Bennett, Jr.

BEHAVIORAL DEVELOPMENT OF NONHUMAN PRIMATES
An Abstracted Bibliography
Faren R. Akins, Gillian S. Mace, John W. Hubbard, and Dianna L. Akins

MASTER TABLES FOR ELECTROMAGNETIC DEPTH SOUNDING
INTERPRETATION
Rajni K. Verma

SPECTROSCOPIC REFERENCES TO POLYATOMIC MOLECULES
V. N. Verma

HANDBOOK OF REFRACTORY COMPOUNDS
G. V. Samsonov and I.M. Vinitskii

HANDBOOK OF
REFRACTORY COMPOUNDS

G. V. Samsonov   738·13
and
I. M. Vinitskii

Translated from Russian by
Kenneth Shaw

IFI/PLENUM · NEW YORK-WASHINGTON-LONDON

Library of Congress Cataloging in Publication Data

Samsonov, Grigorii Valentinovich.
  Handbook of refractory compounds.

  (IFI Data Base Library)
  Translation of Tugoplavkie soedineniia.
  Bibliography: p.
  Includes index.
  1. Heat resistant materials—Handbooks, manuals, etc. I. Vinitskii, Igor'Mar-
kovich, joint author. II. Title.
TN693.H4S313   620.1   79-17968
ISBN 0-306-65181-5

The original Russian Text, published by Metallurgiya in Moscow in 1976, has been
corrected by the author for the present edition. This translation is published under
an agreement with the Copyright Agency of the USSR (VAAP).

© 1980 IFI/Plenum Data Company
A Division of Plenum Publishing Corporation
227 West 17th Street, New York, N.Y. 10011

PREFACE

Scientific and technical progress in our country depends largely on supplying important sections of the national economy with modern materials. This may be done by improving traditional materials, as well as by developing new ones that may be used under severe temperature, stress, and velocity conditions and that have combinations of certain physical and chemical properties. Refractory, superhard, corrosion-resistant, semiconductor, dielectric, and other materials are thus being created that will permit the development of new, highly effective tool materials, the implementation of technological processes in plasmas, and the solution of some materials-related aerospace and nuclear power problems.

Refractory compounds play a vital role in the development of new materials and in the improvement of traditional materials. But information available on the properties of refractory compounds needed by scientists and engineers engaged in producing new materials for industry and technology has not yet been properly systematized.

A first attempt in 1963 at such systematization (the first edition of this book) played some part in expanding the development and use of refractory compounds, but the information has now become seriously outdated, especially since in the last decade the study of refractory compounds in the USSR and abroad has grown very rapidly. In 1964 the handbook was, with certain additions, translated and published in the USA, but that publication was not readily available to the Soviet reader.

A number of books of data on various classes of refractory materials have been published in the last decade, including R. B. Kotel'nikov et al., "Highly Refractory Elements and Compounds" (Metallurgiya, Moscow, 1969), V. S. Fomenko, "Emission Properties of Materials" (Naukova Dumka, Kiev, 1970), R. F. Voitovich, "Refractory Compounds (Thermodynamic Characteristics)" (Naukova Dumka, Kiev, 1971), A. S. Bolgar, A. G. Turchanin, and V. V. Fesenko, "Thermodynamic Properties of Carbides" (Naukova Dumka, Kiev, 1973). The most useful information in these books was taken into account in preparing the present book. We have also used new data from periodicals, as well as data obtained in the laboratories where the authors work and from neighboring laboratories and other institutes.

The handbook includes data on the physical, chemical, and technological properties of the most common and promising refractory compounds: borides, carbides, nitrides, beryllides, aluminides, silicides, phosphides, and sulfides. No information appears on that other important class of compounds, the oxides, since they have been covered in special publications, e.g., G. V. Samsonov's "The Physicochemical Properties of Oxides"* (Metallurgiya, Moscow, 1969).

While in the first edition there was an attempt at full coverage of the original literature, in this edition the literature references have had to be limited because there are now so many of them, and preference has been given to the review-type paper in which references to the various original publications can be found. References are given for the most important papers, especially when there was no adequate basis for selecting the most reliable data.

Chapters I–V cover the general character of refractory compounds: their crystal structures, density, and their thermochemical, thermal, electrical, magnetic, optical, mechanical, and refractory properties. In Chapters VI and VII an attempt is made to give a general account of the resistance of refractory compounds to the action of various chemical reagents and molten media, and also their resistance to oxidation. Chapter VIII contains abbreviated tables showing basic data on the actual and prospective areas of utilization of refractory compounds in various industries.

In each chapter the data are grouped according to the class of compound and are presented in the following sequence: beryllides, borides, carbides, nitrides, aluminides, silicides, phosphides, sulfides, and nonmetals. The presentation of the data within each class of compound is given in the following sequence: alkaline-earth compounds, rare-earth metals, actinides, transition metals of groups IV, V, and others. Within the framework of metal–nonmetal systems the sequence of phases consists of a transition to phases with the higher concentration of nonmetal, and in the case of beryllides and aluminides the atoms of beryllium and aluminum are tentatively taken to be nonmetallic.

The appendixes contain collections of currently known phase diagrams for binary systems that form refractory compounds; these diagrams will help the reader to compare the properties of phases with their positions in the appropriate systems.

The following observations must be made about the various tables.

The table "Ratio of the Radii of Some Atoms of Nonmetals and Metals" will help the reader compare the appropriate ratios R_X/R_{Me} with the condition established by Hägg for the formation of interstitial phases; as already mentioned, the atoms of beryllium and aluminum are tentatively taken to be nonmetallic.

The table "Crystal Structure" gives the main parameters for the structure of refractory compounds. It shows not only data for phases with structures that have been thoroughly studied, but also incomplete information, e.g., for some phases the data are limited to those on the type of structure or the space group, etc.

The section on density shows values determined by the pycnometric method and as calculated from x-ray data for refractory compounds. In a number of cases the density is calculated from parameters, and the respective literature references

---

*Published in English under the title "The Oxide Handbook," by IFI/Plenum, New York (1973).

then refer to the values of the periods in the crystal lattices of the phases. All the data is quoted for phases of the maximum composition relative to the content of nonmetal.

Chapter I concludes with data on the temperature stability ranges of refractory compounds, derived mainly from the phase diagrams of the corresponding binary systems.

Chapter II covers the main thermal and thermodynamic properties of refractory compounds.

The heats of formation are taken both from the original and from review works, and also from technical reports of the National Bureau of Standards (USA) [322-325].

The table "Entropy of Compounds" contains data on the absolute entropy, as well as the entropy of formation of refractory compounds from the elements, calculated by using the entropy of the elements. NBS data were also used in compiling this table [322-325].

The tables "Free Energy of Formation of Refractory Compounds" and "Specific Heat" were compiled from the most reliable data in original and review papers, and for the reader's convenience the specific heat values of the compounds are given at 20°C.

For convenience also, the table "Melting Points" gives the temperatures in degrees Celsius and Kelvin, and a note is made as to how fusion occurs, i.e., congruently or through a eutectic reaction.

In the table on boiling points, the values given are generally calculated values obtained by using the most reliable equations for the relationships between temperature and vapor pressures.

The values given in the table "Vapor Pressure and Evaporation Rates" are, strictly speaking, not physical constants, and essentially depend on the experimental conditions—the degree of roughness of the specimens, their porosity, etc.; however, they are important in practice for determining the possibility and duration of use of refractory compounds in vacuum at high temperatures.

The thermal conductivity values were obtained mainly by the stationary method; most of the data were collected from papers and refer to pore-free specimens, using justifiable extrapolation to zero porosity according to the Kingery equation, $\chi_{P=0} = \chi_P(1 - P)$, where P is the porosity expressed as a fraction of one. In spite of the fact that some papers contain no references to the state of the specimens and their porosity, it was decided that it was feasible to quote approximate thermal-conductivity data published in these reports both because of the practical importance of this property and the fact that such information is rather scarce for refractory compounds.

The thermal-expansion factors were determined on a dilatometer and also by x-ray structural methods.

The crystal lattice energy values were obtained chiefly by calculation using

É. S. Sarkisov's equation [351], which was originally used by Sarkisov to compute the values for ionic compounds. However, the essential features of this equation, which is based on assuming a collectivization of the electrons between the cores of atoms of two components, make it more suitable for calculating the lattice energy of metal-like compounds with a high proportion of metallic bond, as has been shown in the work of G. V. Samsonov [352] and O. I. Shulishova [353].

The characteristic temperatures were obtained partly from x-ray data and partly by calculation and measurements of the low-temperature specific heats.

The table "Parameters of Diffusion of Nonmetals into Metals" gives data on reaction-diffusion parameters, i.e., diffusion taking place together with the formation of the corresponding compounds, and also on ordinary heterodiffusion of nonmetals into transition metals. Data on spontaneous diffusion in refractory compounds are also presented.

Chapter III gives data on the electrical and magnetic properties of refractory compounds. It should be mentioned that the electrical properties are structurally sensitive and are critically dependent on the method of measurement, on the purity of the specimens, and on the extent of defects in them; furthermore, impurities may either reduce or increase the conductivity, a phenomenon especially marked in semiconductors. Therefore much of the data was made more precise by improving the methods of preparing the specimen. Nevertheless, it can be assumed that the values of the electrical resistance and other properties connected with the electrical conductivity of a range of compounds, at least those compounds with metallic conductivity, will change only very slightly with time, while this cannot, of course, be said about nonmetallic compounds, which are semiconductors. Most of the data given in the tables relate to polycrystals, so that the electrical conductivity may be affected by the grain boundaries, and the nature of this influence can, once again, be ambiguous. Recently it has been possible to obtain single crystals of certain refractory compounds, and data exist (though very scant) on their electrical and physical properties. In these cases the values are given for polycrystalline and for single-crystal specimens.

Information on the superconducting properties of refractory compounds is taken chiefly from References 245, 354, and 355. When superconductivity was not detected in the study of a compound, the table gives the lowest temperature down to which the compound was studied.

The thermal-emission properties of refractory compounds were taken from original work and were checked with recent handbook data of V. S. Fomenko [356]. Information presented in Chapter IV on the properties of light refers to powdered refractory compounds.

The emission coefficients were determined mainly by comparing the brilliance and true temperatures when powders were heated, using an ideal blackbody model.

Absorption spectra data in the infrared region are very limited; nevertheless, it is believed they should be presented in order to emphasize the importance of this relatively neglected method for providing important practical results for refractory compounds.

Chapter V gives the mechanical properties of refractory compounds. These data are less reliable and, since they are to a certain degree "semiqualitative" characteristics, it is hoped that researchers will pay more attention to developing a method of determining the mechanical properties and to making the appropriate measurements, especially for tensile, compressive, and bending strengths.

The authors decided to include data on the Rockwell and Vickers hardness, despite their poor reliability. This information is of interest, since in industry the hardness of fused-on and diffused coatings made from refractory compounds is usually determined by these techniques.

Data on the chemical (Chapter VI) and refractory (Chapter VII) properties are mainly of a qualitative and descriptive character. However, they are very important for the use of materials in some areas of new technology, and so are given as fully as possible; this may be of assistance to designers and technologists.

The data in Chapter VIII, "Examples of the Use of Refractory Compounds," do not pretend to be complete and cannot replace special monographs on the subject. This chapter is intended merely to familiarize readers with the most important examples of the use of refractory compounds in various areas of science and technology.

The authors would like to express their appreciation to Soviet researchers, especially P. V. Gel'd, E. M. Savitskii, R. A. Andrievskii, G. L. Gal'chenko, V. M. Gropyanov, S. S. Kiparisov, B. N. Oshcherin, D. N. Poluboyarinov, K. I. Portnoi, A. P. Brynza, V. S. Dergunova, E. N. Marmer, V. S. Neshpor, and also to the following foreign authors: F. Benesovsky, H. Nowotny (Austria), A. Pasteur, B. Bonier, P. Hagenmüller (France), A. Mersch (East Germany), E. Fischer (West Germany), M. Ristich (Yugoslavia), Y. S. Touloukian, A. Weber, B. Roberts (USA), G. S. Upadaya (India), P. Popper (England), and the staffs of the Max Planck Institute (West Germany) and the University of Uppsala (Sweden), who willingly placed numerous data, copies of articles, and monographs at our disposal.

The authors would be grateful for all comments on this handbook and hope that, in spite of the inevitable shortcomings, it will serve the further development of research on and the use of refractory compounds.

CONTENTS

INTRODUCTION

The term "refractory" has changed its meaning with time in that the temperature limit describing this concept has continuously increased. While in the literature the limiting temperatures frequently approach 3000°C, it is more proper to consider a temperature less than this as the point above which a "compound is refractory"; in all probability the limit should be taken as the melting point of iron (1535°C), and this is the basis for the traditional "nonrefractory" classification of numerous materials such as steels, cast irons, and other ferrocarbon alloys. Such a lower limit for the melting points of refractory compounds is tentative and corresponds to the level of technical development of the materials. Bearing in mind that the most refractory metal, tungsten, melts at 3340°C, and the most refractory nonmetal, carbon, at about 3700°C, it may be tentatively considered that the term refractory can be applied to metals, alloys, and compounds with melting points in the range 1600-4000°C, i.e., up to that level above which the melting points of substances under ordinary conditions are as yet unknown. Further upward movement of this limit, and any change in the meaning of refractory, is possible only under superhigh pressures (of the order of megabars) when the electron shells of atoms begin to be eliminated, and the electrons "collapse" into the nuclei via reactions with intranuclear protons, leading to the formation of superdense and superrefractory neutron matter.

Frequently, however, the term refractory is applied to substances with melting points below 1500-1600°C (so that traditionally they are not refractory) because they closely resemble, especially in regard to the chemical bond, substances that are normally called refractory. The classification of some materials as refractory because of the nature of their chemical bond, even though they would not be so classified on the basis of their melting point, is somewhat arbitrary but is of practical advantage. Thus, the list of refractory metals, alloys, and compounds will include substances melting at temperatures above 1500-1600°C, but in practice it will often include substances with lower melting points that have some similarity in the nature of the chemical bond and in certain other properties.

As is known, the elements of the Periodic Table fall into groups whose atoms have analogous electron shell structures:

1. s elements, having outer complete or incomplete s shells (alkali metals of the copper, beryllium, magnesium, and alkaline-earth metal subgroups, and metals of the zinc subgroup, i.e., all elements in groups I and II of the Periodic Table).

2. d and f elements having incomplete d and f shells (transition d metals, lanthanides, and actinides).

3. sp elements having valence s, p electrons (nonmetals, semimetals, and degenerate nonmetals).

According to this, refractory compounds should fall into the following groups:

1. Metal-like refractory compounds, i.e., compounds mainly of d and f transition metals (and also some s metals) with sp elements, i.e., nonmetals.

2. Metallic refractory compounds, i.e., compounds mainly of d and f transition metals with each other, but also with semimetals and degenerate metals from the sp elements.

3. Nonmetallic refractory compounds formed mainly by combinations of nonmetals, and also involving semimetals and degenerate metals.

The *metal-like* refractory compounds include borides, carbides, nitrides, silicides, phosphides, and sulfides of the transition metals, and also those s elements that have unfilled deep d and f levels in the isolated-atom state. In a combination of metal atoms and nonmetal atoms, both types of atoms tend to form, energywise, more stable electron configurations, corresponding to a minimum of the free energy [795].

In the formation of *borides*, the boron atom, which in the isolated state has the electron configuration s^2p, first tends to acquire the more stable configuration sp^2 because of the single-electron $s \to p$ transfer, and then to acquire the sp^3 configuration, i.e., it is mainly an acceptor of electrons. Its acceptor ability can be realized by the valence electrons of a partner in the combination, and also as a result of the formation of direct bonds B–B with a covalent nature. Hence, in borides there may be (for certain proportions of the metallic constituents) covalent bonds Me–Me between the metal atoms, predominantly metallic bonds Me–B, and covalent bonds B–B. In this case, the metal donors transfer their valence electrons to the boron atoms. When the donor capacity increases there is a greater probability of a complication of the structural elements because of the boron atoms in the lattice of the boride, a phenomenon typical of boride phases of most of the transition metals, which are electron donors. Thus, the high donor capacity of lanthanides (and actinides) is the reason for the formation of borides with complicated structural elements, e.g., compounds of the MeB_4, MeB_6, and MeB_{12} type; the more boron there is in a compound, the smaller the proportion of the covalent Me–Me bond, and the majority of the valence electrons will be "mobilized" to form complicated covalent-bonded structural elements made up of boron atoms. With a fall in the donor capacity of the transition-metal atoms, i.e., with the move from groups III to IV and then to groups V and VI, the Me–Me bonds are strengthened, and there is less chance of transferring the valence electrons of metal atoms to the boron atoms in the B–B bond; this leads to the formation of boride phases less enriched with boron. But since the Me–Me bonds are strengthened with an increase in the main quantum number of the valence electrons, a reduction in the tendency to form phases that are not so boron enriched is mainly observed on passing diagonally from titanium to tungsten in the Periodic Table. The number of phases in each Me–B system is also determined by the possible variants in the Me–Me, Me–B, and B–B bonds, and diminishes when some one of these bonds becomes dominant and uses a large portion of the valence electrons of the atoms of the metal and boron.

In the shift from groups IV to VII and especially to group VIII, the donor capacity of the metal atoms falls and begins to give way to an acceptor tendency, so that the transition metals of the subgroup VIIIc (Ni, Pd, Pt) are strong enough electron acceptors to reverse the acceptor tendency of boron atoms and to make them donors; this leads to the predominant formation of boride phases impoverished of boron, with boron atoms not combining directly with themselves, but separated by atoms of the metal, which form rather strong Me–Me bonds. Typical of these VIIIc metals is the formation of the strongest boride phases, the Me_3B, Me_2B, and other types.

Thus the borides generally form covalent-metallic bonds with a discrete-continuous change in the ratio of the shared constituents of these bonds, with the ratio a function of the donor-acceptor capacity of the metal atoms. This change in ratio provides an understanding of the nature of the formation of borides and the alteration in their physical and chemical properties, and on this basis the most probable numerical values for these properties can be selected from their entire range.

In the formation of *carbides*, the carbon atom, as a result of the $s \rightarrow p$ transfer, can acquire the energy-stable configuration sp^3, which, however, tends to share in equilibria of the type $sp^3 \rightleftarrows sp^2 + p$; as a result, displacement of this equilibrium in the sp^3 direction, i.e., stabilization of the sp^3 configuration, is possible when there is an excess of electrons transferred by the metallic partner—the transition metal. Therefore, with a strong electron donor such as titanium, the sp^3 configuration is stabilized (this gives rise, for example, to high hardness values for TiC) and a large proportion of the electrons do not participate in the chemical bond, thereby reducing the melting point of this carbide in comparison with carbides of transition metals in group V (niobium and tantalum), where there is a smaller stock of "unstable electrons" (lower hardness but a higher melting point for NbC as compared with TiC). With a transition to tungsten carbide, the chances of stabilization of the sp^3 configuration by the weak donor, tungsten, are so much reduced that this carbide acquires a certain plasticity as well as a thermodynamic instability.

In the case of carbides the tendency to form direct C–C bonds is more limited than the tendency to form the corresponding bonds in borides, and only metals that are very strong donors can form carbides of the Me_2C type and those having higher carbon contents; such strong donors include, for example, rare-earth metals and the alkaline and alkaline-earth metals.

Upsetting the probability of stabilizing the sp^3 configuration for acceptor atoms leads to the formation of very unstable carbides (e.g., carbides of iron, cobalt, nickel) or generally to the lack of formation of carbides, as is noted for platinoids.

Carbides typically have the covalent–metallic bond, changing to covalent–ionic for carbides of metals that are strong electron donors (the above-mentioned alkali, alkaline-earth, and rare-earth metals); therefore, carbides of the latter type are not metal-like, but salt-like (ionic). Nevertheless, in this book such carbides are classified as refractory for practical convenience, although they do not belong to the class of refractory metal-like compounds, either in terms of infusibility or in type of chemical bond.

In *nitrides*, the nitrogen atom (for which the valence electrons have the configuration s^2p^3 in the isolated state) may be a donor of electrons, $s^2p^3 \rightarrow sp^4 \rightarrow sp^3 + p$, as well as an acceptor, joining three electrons to itself and acquiring the configuration of an inert gas, s^2p^6. It is natural that the first possibility, donor, is manifest chiefly in the formation of nitrides of the transition metals, which act as electron acceptors, while the second possibility, acceptor, is chiefly manifest in forming nitrides of the metal donors. However, in most nitride phases nitrogen exhibits both tendencies, and only the relative amount of each varies from metal to metal. The strongest donor capacity of nitrogen atoms is apparently exhibited for transition metals of the iron triad, and the strongest acceptor capacity for transition metals of groups IV–VI of the Periodic Table.

Nitrides consequently have a typically complex covalent–metallic–ionic bond; but the covalent–metallic type of bond prevailing in most nitrides means we can classify them as metal-like refractory compounds. Clearly, they are less refractory than the borides and carbides. Furthermore, the nitrogen atom's tendency to form a thermodynamically strong molecule in the gaseous state leads to rather easy removal of nitrogen from nitrides, and this hinders their use as high-temperature materials.

Metal-like refractory compounds also include *silicides* and *phosphides*, which are essentially similar to carbides and nitrides but have special features because of the higher main quantum number of the valence electrons of silicon and phosphorus compared with carbon and nitrogen. Weakening of the corresponding sp^3 and other configurations means that the silicides and phosphides have a lower chemical strength and poorer properties than the carbides and nitrides. Furthermore, the weakening of the sp^3 configuration of the silicon atoms in silicides reinforces the role of the Me–Me bond, and this affects the formation of the structure and properties of silicides, which are the "most metallic" among the metal-like compounds except for certain silicide phases where semiconductor properties rather than metallic have begun to exhibit themselves.

Phosphides are very similar to nitrides but are chemically less strong, with a lower proportion of ionic bond than in the nitrides, i.e., they are more metal-like than nitrides, but less refractory. This is because the relatively larger size of the phosphorus atom leads to larger dimensions of the crystal lattice and a weakening of the bond forces.

Metal-like *sulfides* are formed mainly because of the high acceptor capacity of sulfur atoms, which, in the isolated state, have the valence-electron configuration s^2p^4, and this tends to build up to s^2p^6. This tendency determines the mixed ionic–metallic nature of the bond between the metal and sulfur atoms. As the sulfur content increases in the sulfide phases, the covalent bond appears and becomes stronger with increase in the localization of the valence electrons near the atoms of the transition metals, i.e., the lower the donor capacity of the metal; a change thus occurs from metallic to semiconductor conductivity and the habit of metal-like sulfide phases is lost.

Aluminides and beryllides, the *metallic, refractory compounds* presented in this book, are typical intermetallics, if we consider aluminum to be a degenerate *sp* element, i.e., a degenerate nonmetal, while beryllium is a rather typical *s* metal.

In the formation of *beryllides*, of considerable significance is the capacity of beryllium atoms to acquire, as a result of the single-electron $s \rightarrow p$ transfer, the *sp* configuration, which is a strong acceptor of electrons, and beryllium atoms tend thereby to acquire the sp^2 configuration, and then sp^3. Therefore, like borides, the beryllides form rather strong Be–Be bonds, becoming stronger with an increase in the donor capacity of the metal partner; these Be–Be bonds lead to complication of the crystal structures of beryllides.

Aluminides are more complex compounds since aluminum has donor properties, contributing to a metallic character, and acceptor properties, making it behave as a typical *sp* element. This gives rise to a greater variety of aluminide compositions than of beryllides, and a greater range of properties of these phases. In both types of compound a certain proportion of the bonds are covalent, but the metallic bonds prevail and become stronger with an increase in the donor capacity of the metal atoms.

Nonmetallic refractory compounds are compounds of *sp* elements; the covalent bond is typical of them, but it gradually shifts, and at some point these compounds join up with a series of compounds with an ionic bond. This book presents data on the most commonly used and most promising nonmetallic refractory compounds—carbides and nitrides of boron and silicon, and also aluminum nitride. In general, the atoms of all these compounds have a tendency to form sp^3 and s^2p^6 configurations, but the formation of less stable states is possible depending on external conditions. Boron nitride may be cited as an example, as it forms three modifications—graphitic, sphaleritic, and a type of wurtzite. In the first of these modifications sp^2 prevails, in the second sp^3, and in the third a mixed state prevails. The most widespread nonmetallic refractory compound, silicon carbide, is essentially a diamond crystal in which half of the carbon atoms are replaced by silicon atoms having less stable sp^3 configurations, making it possible to combine the functions of the

5

bond and to form a large number of α-SiC polytypes. All these compounds melt incongruently, decomposing during fusion; they are all dielectrics or semiconductors.

Many of the above groups have now been studied quite thoroughly, but the most accurate values need to be selected from a great variety of information. The authors have used the established relationships between properties and structure, on the one hand, and electron structure, on the other. Furthermore, the basis for selecting the most valuable data consisted of the reliability and up-to-dateness of the methods used to determine the properties, the purity and the condition of the specimens, and statistical indications. However, in some cases preference was given to old data if the new had been obtained more for advertisement than for science, or if the research had been done on compounds of an indefinite phase composition.

In choosing numerical values a certain role was played by the fundamental nature of the scientific school to which the investigator belongs, taking into account the possibility of an unwilling "adaptation" of numerical data to the particular theoretical conceptions of electronic structure.

In conclusion, we would like to emphasize once more that many properties of the refractory compounds have not been systematically studied, particularly the mechanical, electrical, and optical properties, a knowledge of which is quite essential for solving certain practical problems.

Chapter I

GENERAL DATA, STOICHIOMETRY, CRYSTAL-CHEMICAL PROPERTIES

ELECTRON STRUCTURE OF ISOLATED ATOMS

Atomic number	Element	K 1 s 0	L 2 s 0	L 2 p 1	M 3 s 0	M 3 p 1	M 3 d 2	N 4 s 0	N 4 p 1	N 4 d 2	N 4 f 3	O 5 s 0	O 5 p 1	O 5 d 2	O 5 f 3	P 6 s 0	P 6 p 1	P 6 d 2	P 6 f 3	Q 7 s 0	Q 7 p 1	Q 7 d 2	Q 7 f 3
1	H	1																					
2	He	2																					
3	Li	2	1																				
4	Be	2	2																				
5	B	2	2	1																			
6	C	2	2	2																			
7	N	2	2	3																			
8	O	2	2	4																			
9	F	2	2	5																			
10	Ne	2	2	6																			
11	Na	2	2	6	1																		
12	Mg	2	2	6	2																		
13	Al	2	2	6	2	1																	
14	Si	2	2	6	2	2																	
15	P	2	2	6	2	3																	
16	S	2	2	6	2	4																	
17	Cl	2	2	6	2	5																	

Shell / orbit

7

ELECTRON STRUCTURE OF ISOLATED ATOMS (continued)

Atomic number	Element	K 1	L 2		M 3			N 4				O 5				P 6				Q 7			
		s 0	s 0	p 1	s 0	p 1	d 2	s 0	p 1	d 2	f 3	s 0	p 1	d 2	f 3	s 0	p 1	d 2	f 3	s 0	p 1	d 2	f 3
18	Ar	2	2	6	2	6																	
19	K	2	2	6	2	6		1															
20	Ca	2	2	6	2	6		2															
21	Sc	2	2	6	2	6	1	2															
22	Ti	2	2	6	2	6	2	2															
23	V	2	2	6	2	6	3	2															
24	Cr	2	2	6	2	6	5	1															
25	Mn	2	2	6	2	6	5	2															
26	Fe	2	2	6	2	6	6	2															
27	Co	2	2	6	2	6	7	2															
28	Ni	2	2	6	2	6	8	2															
29	Cu	2	2	6	2	6	10	1															
30	Zn	2	2	6	2	6	10	2															
31	Ga	2	2	6	2	6	10	2	1														
32	Ge	2	2	6	2	6	10	2	2														
33	As	2	2	6	2	6	10	2	3														
34	Se	2	2	6	2	6	10	2	4														
35	Br	2	2	6	2	6	10	2	5														
36	Kr	2	2	6	2	6	10	2	6														
37	Rb	2	2	6	2	6	10	2	6			1											
38	Sr	2	2	6	2	6	10	2	6			2											
39	Y	2	2	6	2	6	10	2	6	1		2											
40	Zr	2	2	6	2	6	10	2	6	2		2											
41	Nb	2	2	6	2	6	10	2	6	4		1											
42	Mo	2	2	6	2	6	10	2	6	5		1											
43	Tc	2	2	6	2	6	10	2	6	6		1											
44	Ru	2	2	6	2	6	10	2	6	7		1											

Z		1s	2s	2p	3s	3p	3d	4s	4p	4d	4f	5s	5p	5d	5f	6s	6p
45	Rh	2	2	6	2	6	10	2	6	8	—	1	—	—	—	—	—
46	Pd	2	2	6	2	6	10	2	6	10	—	—	—	—	—	—	—
47	Ag	2	2	6	2	6	10	2	6	10	—	1	—	—	—	—	—
48	Cd	2	2	6	2	6	10	2	6	10	—	2	—	—	—	—	—
49	In	2	2	6	2	6	10	2	6	10	—	2	1	—	—	—	—
50	Sn	2	2	6	2	6	10	2	6	10	—	2	2	—	—	—	—
51	Sb	2	2	6	2	6	10	2	6	10	—	2	3	—	—	—	—
52	Te	2	2	6	2	6	10	2	6	10	—	2	4	—	—	—	—
53	I	2	2	6	2	6	10	2	6	10	—	2	5	—	—	—	—
54	Xe	2	2	6	2	6	10	2	6	10	—	2	6	—	—	—	—
55	Cs	2	2	6	2	6	10	2	6	10	—	2	6	—	—	1	—
56	Ba	2	2	6	2	6	10	2	6	10	—	2	6	—	—	2	—
57	La	2	2	6	2	6	10	2	6	10	—	2	6	1	—	2	—
58	Ce	2	2	6	2	6	10	2	6	10	2	2	6	—	—	2	—
59	Pr	2	2	6	2	6	10	2	6	10	3	2	6	—	—	2	—
60	Nd	2	2	6	2	6	10	2	6	10	4	2	6	—	—	2	—
61	Pm	2	2	6	2	6	10	2	6	10	5	2	6	—	—	2	—
62	Sm	2	2	6	2	6	10	2	6	10	6	2	6	—	—	2	—
63	Eu	2	2	6	2	6	10	2	6	10	7	2	6	—	—	2	—
64	Gd	2	2	6	2	6	10	2	6	10	7	2	6	1	—	2	—
65	Tb	2	2	6	2	6	10	2	6	10	8	2	6	1	—	2	—
66	Dy	2	2	6	2	6	10	2	6	10	10	2	6	—	—	2	—
67	Ho	2	2	6	2	6	10	2	6	10	11	2	6	—	—	2	—
68	Er	2	2	6	2	6	10	2	6	10	12	2	6	—	—	2	—
69	Tm	2	2	6	2	6	10	2	6	10	13	2	6	—	—	2	—
70	Yb	2	2	6	2	6	10	2	6	10	14	2	6	—	—	2	—
71	Lu	2	2	6	2	6	10	2	6	10	14	2	6	1	—	2	—
72	Hf	2	2	6	2	6	10	2	6	10	14	2	6	2	—	2	—
73	Ta	2	2	6	2	6	10	2	6	10	14	2	6	3	—	2	—
74	W	2	2	6	2	6	10	2	6	10	14	2	6	4	—	2	—
75	Re	2	2	6	2	6	10	2	6	10	14	2	6	5	—	2	—
76	Os	2	2	6	2	6	10	2	6	10	14	2	6	6	—	2	—
77	Ir	2	2	6	2	6	10	2	6	10	14	2	6	7	—	2	—
78	Pt	2	2	6	2	6	10	2	6	10	14	2	6	9	—	1	—
79	Au	2	2	6	2	6	10	2	6	10	14	2	6	10	—	1	—
80	Hg	2	2	6	2	6	10	2	6	10	14	2	6	10	—	2	—
81	Tl	2	2	6	2	6	10	2	6	10	14	2	6	10	—	2	1
82	Pb	2	2	6	2	6	10	2	6	10	14	2	6	10	—	2	2

ELECTRON STRUCTURE OF ISOLATED ATOMS (continued)

Atomic number	Element	K 1	L 2		M 3			N 4				O 5				P 6				Q 7			
		s 0	s 0	p 1	s 0	p 1	d 2	s 0	p 1	d 2	f 3	s 0	p 1	d 2	f 3	s 0	p 1	d 2	f 3	s 0	p 1	d 2	f 3
83	Bi	2	2	6	2	6	10	2	6	10	14	2	6	10	—	2	3	—	—				
84	Po	2	2	6	2	6	10	2	6	10	14	2	6	10	—	2	4	—	—				
85	At	2	2	6	2	6	10	2	6	10	14	2	6	10	—	2	5	—	—				
86	Rn	2	2	6	2	6	10	2	6	10	14	2	6	10	—	2	6	—	—				
87	Fr	2	2	6	2	6	10	2	6	10	14	2	6	10	—	2	6	—	—	1			
88	Ra	2	2	6	2	6	10	2	6	10	14	2	6	10	—	2	6	—	—	2			
89	Ac	2	2	6	2	6	10	2	6	10	14	2	6	10	—	2	6	1	—	2			
90	Th	2	2	6	2	6	10	2	6	10	14	2	6	10	—	2	6	2	—	2			
91	Pa	2	2	6	2	6	10	2	6	10	14	2	6	10	2	2	6	1	—	2			
92	U	2	2	6	2	6	10	2	6	10	14	2	6	10	3	2	6	1	—	2			
93	Np	2	2	6	2	6	10	2	6	10	14	2	6	10	4	2	6	1	—	2			
94	Pu	2	2	6	2	6	10	2	6	10	14	2	6	10	6	2	6	—	—	2			
95	Am	2	2	6	2	6	10	2	6	10	14	2	6	10	7	2	6	—	—	2			
96	Cm	2	2	6	2	6	10	2	6	10	14	2	6	10	7	2	6	1	—	2			
97	Bk	2	2	6	2	6	10	2	6	10	14	2	6	10	9	2	6	—	—	2			
98	Cf	2	2	6	2	6	10	2	6	10	14	2	6	10	10	2	6	—	—	2			
99	Es	2	2	6	2	6	10	2	6	10	14	2	6	10	11	2	6	—	—	2			
100	Fm	2	2	6	2	6	10	2	6	10	14	2	6	10	12	2	6	—	—	2			
101	Mv	2	2	6	2	6	10	2	6	10	14	2	6	10	13	2	6	—	—	2			
102	No	2	2	6	2	6	10	2	6	10	14	2	6	10	14	2	6	—	—	2			
103	Lw	2	2	6	2	6	10	2	6	10	14	2	6	10	14	2	6	1	—	2			
104	Ku	2	2	6	2	6	10	2	6	10	14	2	6	10	14	2	6	2	—	2			
105	—*	2	2	6	2	6	10	2	6	10	14	2	6	10	14	2	6	3	—	2			

*This element is tentatively called "Niels bohrium".

Atom	Electrons in outer shell							
	I	II	III	IV	V	VI	VII	VIII
	electron breakaway energy, eV							
1	2	3	4	5	6	7	8	9
H	13 54	—	—	—	—	—	—	—
He	24.54	54.16	—	—	—	—	—	—
Li	5.37	75.3	121.9	—	—	—	—	—
Be	9.30	18.12	153.1	216,6	—	—	—	—
B	8.28	24.99	37.70	258,0	338.5	—	—	—
C	11.24	24.28	47.55	64,1	390,1	487,4	—	—
N	14.51	29.41	47.36	77.0	97,3	549	663	—
O	13.57	34.75	54.8	77.5	113,3	137,3	735	867
F	17.46	34.71	62.3	87.3	114,8	156,5	184,2	949
Ne	21.47	40.67	63.2	97.1	127,0	159,1	206,6	237,9
Na	5.09	46.65	71.3	99.0	139.1	173,9	210,5	263,6
Mg	7.63	15.10	79.4	109.4	142.2	188,5	227,9	269,0
Al	5.94	18.85	28.35	119.6	154,9	192,7	245,1	289,2
Si	8.14	16.29	33,35	44.84	167.4	207,9	250,5	309,1
P	10.43	19.75	30,08	51.1	64,6	222,8	268,3	315.7
S	10,42	23.25	34.89	47,32	72.2	87,5	285.7	336.2
Cl	13.01	23.85	39.67	53,5	68.0	96.5	113,8	356,1
Ar	15.68	27.64	40.94	59,7	75.7	92.1	124,1	143.2
K	4.32	31.45	46.00	61,7	83.3	101,4	119,7	155.0
Ca	6.25	11.87	51.1	68.1	86.1	110,5	130.6	150.7
Sc	6.7	12.8	26,19	74,5	93,9	114,2	141,3	163.4
Ti	6.81	13.6	28.39	45.40	101,7	123,5	145.9	175.7
V	6,74	15.13	30.31	48,35	68,7	132.8	156.9	181.3
Cr	6.7	16.41	32.12	50.9	72,4	96.0	167,6	193.9
Mn	7.41	14.5	33.97	53.4	75,8	100,7	127,4	206.3
Fe	7.83	15.9	31,69	55.9	79.0	104.9	133,1	162.8
Co	7.8	17.47	33.77	53,2	82,2	108,9	138.2	169.6
Ni	7.6	18.88	35,92	56.0	79.1	112,9	143.1	175.7
Cu	7.67	20.33	37,93	58.9	82.7	109,3	148.0	181.5
Zn	9,37	18.04	40.00	61.6	86,3	113,7	148,8	187.5
Ga	5.97	20.39	30,66	64,3	89.8	118.3	149.2	182,7
Ge	8,10	15.95	33.68	45.51	93.3	122,6	154.7	189.1
As	10.05	18.88	28.30	49.25	62.6	127,0	159.9	195,7
Se	9.75	21.57	32,11	43.03	67.1	81,9	165.3	201,9

Atom	Electrons in outer shell							
	I	II	III	IV	V	VI	VII	VIII
	electron breakaway energy, eV							
1	2	3	4	5	6	7	8	9
Br	11,82	21,47	35,60	47.77	60,1	87,2	103,5	208,2
Kr	13.94	24.28	35,71	52,1	65,9	79,6	109,6	127,3
Rb	4,19	27,14	39.32	52,5	71,1	86,4	101,5	134,3
Sr	5.68	10.86	42,98	56.9	71.7	92,7	109,4	125.8
Y	6.6	12.3	20,46	61,5	77,1	93,5	116.7	134,8
Zr	6.92	13.97	22.64	34.83	82.6	99,9	117,9	143,2
Nb	6.88	13,48	24,7	37.7	51,9	106.3	125,3	144,7
Mo	7,2	15,17	27,00	40.53	55,6	71,7	132,7	153,2
Ru	7.5	16,37	28,62	46.52	62,9	80.6	99,6	119,3
Rh	7.7	18,07	31,03	45.63	66,7	85,2	105.0	125.8
Pd	8.30	19,85	33,36	48.77	65,6	89,9	110,5	132.2
Ag	7,58	21,50	35.79	51,8	69.6	88.7	116,2	138.7
Cd	8.94	16,80	38.00	55,0	73,4	93,5	114.7	145,5
In	5.76	18,76	27.85	57.8	77,4	98.2	120,5	143,7
Sn	7.54	14,56	30.45	40.72	80.9	103,1	126,1	150.5
Sb	8.35	17,01	25.22	44.02	55,4	107,3	132.0	157.2
Te	8,89	19.33	28,39	37.73	59.5	71,9	137,0	164,1
I	10.43	19,11	31,40	41.70	52,1	76,8	90,2	170,0
Xe	12,08	21,18	31.33	45.46	56.9	68.3	96,0	110.4
Cs	3.86	23,37	33,97	45.55	61,5	74,1	86,4	117.1
Ba	5.21	9.96	36,75	48.80	61,8	79.5	93,1	106.4
La	5.59	11,38	19,1	52,2	65,7	80,0	99,5	114.1
Ce	6.54	12,31	19,870	36.714	69.7	84.6	100,2	121,5
Pr	5,76	11,54	20.96	—	—	89.3	105,5	122.4
Nd	6.31	12.09	20.51	—	—	—	111,0	128.5
Pm	5.9	11.7	22,0	—	—	—	—	(135)
Sm	5.6	11,4	24.0	—	—	—	—	—
Eu	5.67	11,24	24,56	—	—	—	—	—
Gd	6,16	12	21.3	—	—	—	—	—
Tb	6.74	12.52	21.02	—	—	—	—	—
Dy	6.82	12,60	21,83	—	—	—	—	—
Ho	6.9	12.7	22.2	—	—	—	—	—
Er	6.7	12.5	22.4	—	—	—	—	—
Tm	6.6	12,4	23,9	—	—	—	—	—

IONIZATION POTENTIALS OF ATOMS (continued)

Atom	Electrons in outer shell							
	I	II	III	IV	V	VI	VII	VIII
	electron breakaway energy, eV							
1	2	3	4	5	6	7	8	9
Yb	6.22	12,10	25,61	—	—	—	—	—
Lu	6.15	12.0	23.7	—	—	—	—	—
Hf	5.5	14.9	(21)	(31)	—	—	—	—
Ta	7.7	16,2	22.27	33,08	(45)	—	—	—
W	7.98	17,7	24,08	35,36	(48)	(61)	—	—
Re	7.8	13.7	25,96	37.71	50.6	64,5	79,0	—
Os	8.7	17	(25)	(40)	(54)	(68)	(89)	(99)
Ir	9.2	17,0	(27)	(39)	(57)	(72)	(88)	(104)
Pt	8.8	17.37	28,55	41.13	54.8	75.3	91.9	109,3
Au	9,20	18,84	30.46	43.52	57.8	73,1	96,4	114.4
Hg	10,41	18.55	32,43	45.98	60,8	76,9	93.7	119,7
Tl	6,08	20,29	29,63	48,50	63,9	80.5	98,2	116,5
Pb	7.37	14,91	31.97	42,46	67,1	84.3	102,6	121,8
Bi	7.25	16,72	25.41	45,46	57,0	88.1	107.0	127,0
Po	8.2	19,4	27.3	(38)	(61)	(73)	(112)	(132)
At	9,2	20,1	29.3	(41)	(51)	(78)	(91)	(138)
Rn	10,69	20,02	29.78	43.78	55.1	66,8	96,7	111,2
Fr	3,98	22,5	33.5	(43)	(59)	(71)	(84)	(117)
Ra	5,21	10,19	34.26	46.41	58.5	76,0	89,3	102.8
Ac	5,7	11,5	18.9	(49)	(62)	(76)	(95)	(109)
Th	6.9	11.9	18.7	33.6	(65)	(80)	(94)	(115)
Pa	5,6	11,3	20.5	36.4	—	(84)	(100)	(115)
U	6.2	12.0	20.5	36,89	—	—	(104)	(121)
Np	5.9	11,7	22.0	38.1	—	—	—	—
Pu	5.5	11,2	23,5	39.5	—	—	—	—
Am	5,5	10,9	23,9	41,0	—	—	—	—
Cm	6.1	11.9	21,0	—	—	—	—	—
Bk	6,7	12,4	20,8	—	—	—	—	—
Cf	6,7	12.5	21.6	—	—	—	—	—
Es	6,8	12,6	22,1	—	—	—	—	—
Fm	6,7	12.5	22,5	—	—	—	—	—
Md	6.4	12.0	23,1	—	—	—	—	—
No	5,8	11,3	24,0	—	—	—	—	—
Lr	6,0	11.7	23,1	—	—	—	—	—

RATIO OF THE RADII OF SOME ATOMS OF NONMETALS AND METALS

Metal	Atomic radius R_{Me}, Å	Nonmetal (X)							
		N	C	B	S	P	Be	Si	Al
		atomic radius of nonmetal R_X, Å							
		0.70	0.77	0.80	1.04	1.10	1.12	1.17	1.43
		R_X/R_{Me}							
1	2	3	4	5	6	7	8	9	10
Mg	1.60	0.44	0.48	0.50	0.65	0.69	0.70	0.73	0.89
Ca	1.97	0.36	0.39	0.41	0.53	0.56	0.57	0.59	0.72
Sr	2.15	0.32	0.36	0.37	0.48	0.51	0.52	0.54	0.66
Ba	2.22	0.32	0.35	0.36	0.47	0.50	0.51	0.53	0.64
Sc	1.62	0.43	0.48	0.49	0.64	0.68	0.69	0.72	0.88
Y	1.80	0.39	0.43	0.44	0.58	0.61	0.62	0.65	0.79
La	1.87	0.37	0.41	0.43	0.56	0.59	0.60	0.62	0.76
Ce	1.82	0.38	0.42	0.44	0.57	0.60	0.61	0.64	0.78
Pr	1.82	0.38	0.42	0.44	0.57	0.60	0.61	0.64	0.78
Nd	1.82	0.38	0.42	0.44	0.57	0.60	0.61	0.64	0.78
Pm	1.81	0.39	0.42	0.44	0.57	0.61	0.62	0.65	0.79
Sm	1.85	0.38	0.42	0.43	0.56	0.59	0.60	0.63	0.77
Eu	2.08	0.34	0.37	0.38	0.50	0.53	0.54	0.56	0.69
Gd	1.80	0.39	0.43	0.44	0.58	0.61	0.62	0.65	0.79
Tb	1.77	0.40	0.44	0.45	0.59	0.62	0.63	0.66	0.81
Dy	1.77	0.40	0.44	0.45	0.59	0.62	0.63	0.66	0.81
Ho	1.76	0.40	0.44	0.45	0.59	0.62	0.63	0.66	0.81
Er	1.75	0.40	0.44	0.46	0.59	0.63	0.64	0.67	0.82
Tm	1.74	0.40	0.44	0.46	0.60	0.63	0.64	0.67	0.82
Yb	1.93	0.36	0.40	0.41	0.54	0.57	0.58	0.61	0.74
Lu	1.74	0.40	0.44	0.46	0.60	0.63	0.64	0.67	0.82
Ti	1.47	0.48	0.52	0.66	0.71	0.75	0.76	0.80	0.97
Zr	1.60	0.44	0.48	0.50	0.65	0.69	0.70	0.73	0.89
Hf	1.58	0.44	0.49	0.51	0.66	0.70	0.71	0.74	0.90
V	1.34	0.52	0.57	0.60	0.78	0.82	0.83	0.87	1.07
Nb	1.46	0.48	0.53	0.55	0.71	0.75	0.76	0.80	0.98
Ta	1.46	0.48	0.53	0.55	0.71	0.75	0.76	0.80	0.98
Cr	1.36	0.51	0.57	0.59	0.70	0.81	0.82	0.86	1.05
Mo	1.39	0.50	0.55	0.58	0.75	0.79	0.80	0.84	1.03
W	1.39	0.50	0.55	0.58	0.75	0.79	0.80	0.84	1.03
Mn	1.31	0.53	0.59	0.61	0.79	0.84	0.85	0.89	1.09
Tc	1.36	0.51	0.57	0.59	0.76	0.81	0.82	0.86	1.05
Re	1.37	0.51	0.56	0.58	0.76	0.80	0.81	0.85	1.04
Fe	1.26	0.56	0.61	0.63	0.82	0.87	0.88	0.93	1.13
Ru	1.34	0.52	0.57	0.60	0.78	0.82	0.83	0.87	1.07
Os	1.35	0.52	0.57	0.59	0.77	0.81	0.82	0.87	1.06
Co	1.25	0.56	0.62	0.64	0.83	0.88	0.89	0.94	1.14
Rh	1.34	0.52	0.57	0.60	0.78	0.82	0.83	0.87	1.07
Ir	1.36	0.51	0.57	0.59	0.76	0.81	0.82	0.86	1.05
Ni	1.24	0.56	0.62	0.64	0.84	0.89	0.90	0.94	1.15
Pd	1.37	0.51	0.56	0.58	0.76	0.80	0.81	0.85	1.04
Pt	1.38	0.51	0.56	0.58	0.75	0.80	0.81	0.85	1.03
Th	1.80	0.39	0.43	0.44	0.58	0.61	0.62	0.65	0.79
U	1.52	0.46	0.51	0.53	0.68	0.72	0.73	0.77	0.94

COMPOSITION OF REFRACTORY COMPOUNDS

Phase	Molecular weight	Content of nonmetal, %		Phase	Molecular weight	Content of nonmetal, %	
		atomic	by wt.			atomic	by wt.
1	2	3	4	1	2	3	4
$MgBe_{13}$	141,45	92,86	82,81	$TaBe_3$	207,98	75,00	13,00
$CaBe_{13}$	157,23	92,86	74,51	Ta_2Be_{17}	515,10	89,47	29,74
$SrBe_{13}$	204,78	92,86	57,21	$TaBe_{12}$	289,09	92,31	37,41
$ScBe_5$	90,01	83,33	50,06	$CrBe_2$	70,02	66,67	25,74
Sc_2Be_{17}	243,11	89,47	63,02	$CrBe_{12}$	160,14	92,31	67,53
$ScBe_{13}$	162,11	92,86	72,63	Mo_3Be	296,83	25,00	30,07
YBe_{13}	206,06	92,86	56,85	$MoBe_2$	113,96	66,67	15,81
$LaBe_{13}$	256,06	92,86	45,75	$MoBe_{12}$	204,09	92,31	52,99
$CeBe_{13}$	257,27	92,86	45,54	$MoBe_{22}$	294,21	95,65	67,39
$PrBe_{13}$	258,06	92,86	45,40	WBe_2	201,87	66,67	8,93
$NdBe_{13}$	261,39	92,86	44,83	WBe_{12}	292,00	92,31	37,04
$PmBe_{13}$	264,15	92,86	44,69	WBe_{22}	382,18	95,65	51,89
$SmBe_{13}$	267,51	92,86	43,80	$MnBe_2$	72,96	66,67	24,70
$EuBe_{13}$	269,12	92,86	43,53	$MnBe_5$	127,04	89,0	56,7
$GdBe_{13}$	274,41	92,86	42,69	$MnBe_{12}$	163,08	92,31	66,31
$TbBe_{13}$	276,08	92,86	42,44	$TcBe$	108,01	50,0	8,33
$DyBe_{13}$	279,66	92,86	41,89	$ReBe_2$	204,22	66,67	8,83
$HoBe_{13}$	282,09	92,86	41,53	$ReBe_{22}$	384,47	95,65	51,84
$ErBe_{13}$	284,42	92,86	41,19	$FeBe_2$	73,87	66,67	24,39
$TmBe_{13}$	286,09	92,86	69,35	$FeBe_5$	100,91	83,33	44,65
$YbBe_{13}$	290,20	92,86	40,37	$FeBe_{12}$	163,99	92,31	65,94
$LuBe_{13}$	292,13	92,86	40,10	$CoBe$	67,95	50,00	13,26
$TiBe$	56,91	50,00	15,8	$CoBe_5$	103,99	83,33	43,33
$TiBe_2$	65,92	66,67	27,34	$CoBe_{12}$	167,08	92,31	64,73
$TiBe_3$	74,94	75,00	36,08	Co_5Be_{21}	483,92	80,77	39,11
Ti_2Be_{17}	249,01	89,47	61,54	$NiBe$	67,72	50,00	13,40
$TiBe_{12}$	156,05	92,31	69,30	Ni_5Be_{21}	482,81	80,77	39,20
$ZrBe_2$	109,24	66,67	16,50	$RuBe_2$	119,09	66,67	15,1
$ZrBe_5$	136,28	83,33	33,06	Ru_3Be_{17}	456,42	85,00	33,57
Zr_2Be_{17}	335,65	89,47	45,64	$RhBe$	111,92	50,00	8,05
$ZrBe_{13}$	208,38	92,86	56,22	$RhBe_2$	120,93	66,67	14,9
$HfBe_2$	196,51	66,67	9,17	Rh_5Be_{22}	712,79	80,77	27,82
$HfBe_5$	223,55	83,33	20,16	$RhBe_{49}$	544,50	98,0	81,1
Hf_2Be_{17}	510,19	89,47	30,03	Pd_3Be	328,21	25,00	2,74
$HfBe_{13}$	295,65	92,86	39,63	Pd_2Be	222,81	33,33	4,04
VBe_2	68,97	66,67	26,13	Pd_3Be_2	337,22	40,00	5,35
VBe_{12}	159,09	92,31	67,98	Pd_4Be_3	452,64	42,90	5,97
Nb_3Be_2	296,74	40,00	6,07	$Pd_{13}Be_{12}$	1491,35	48,00	7,25
$NbBe_2$	110,93	66,67	16,25	$PdBe$	115,41	50,00	7,81
$NbBe_3$	119,94	75,00	22,54	$PdBe_5$	151,46	83,33	29,75
$NbBe_5$	137,97	83,33	32,66	$PdBe_{12}$	214,55	92,31	50,41
Nb_2Be_{17}	339,02	89,47	45,19	$OsBe_2$	208,22	66,67	8,65
$NbBe_{12}$	201,05	92,31	53,79	Os_3Be_{17}	723,81	85,00	21,17
Ta_2Be	370,91	33,33	2,43	$IrBe$	201,21	50,00	4,48
Ta_3Be_2	560,87	40,00	3,21	Ir_5Be_{22}	1159,27	80,77	17,10
$TaBe_2$	198,97	66,67	9,06	$Pt_{15}Be$	2935,36	6,25	0,307

15

Phase	Molecular weight	Content of nonmetal, %		Phase	Molecular weight	Content of nonmetal, %	
		atomic	by wt.			atomic	by wt.
1	2	3	4	1	2	3	4
$PtBe_5$	240.15	83,33	18.72	SmB_{100}	—	—	—
$PtBe_{12}$	303,24	92.31	35.60	EuB_6	216,92	85,71	29,92
$ThBe_{13}$	349,20	92,86	33,55	GdB_3	189,72	75,0	17,11
UBe_{13}	355.19	92.86	32.98	GdB_4	200.54	80,0	21,58
$NpBe_{13}$	354.0	92,86	33,08	GdB_6	222,18	85,71	29.22
$PuBe_{13}$	356,0	92,86	32,88	GdB_{100}	—	—	—
$AmBe_{13}$	360,0	92,86	32,53	TbB_4	202.21	80.0	21,40
Be_5B	55,89	16.67	19,36	TbB_6	223.85	85,71	29.00
Be_2B	28,84	33.33	37.5	TbB_{12}	288,66	92,35	44,94
BeB_2	30.65	66.67	70.61	TbB_{100}	—	—	—
BeB_4	46.86	80,0	83.09	DyB_2	184.12	66,67	11.74
BeB_6	73.94	85,71	87.81	DyB_4	205.79	80,0	21.03
BeB_9	106,39	90,0	91.53	DyB_6	227,43	85,71	28.55
BeB_{12}	138.85	92,28	93,51	DyB_{12}	292.23	92,35	44,39
MgB_2	45,96	66.67	47,00	DyB_{100}	—	—	—
MgB_4	67,55	80.0	64,03	HoB_4	208.22	80,0	20.79
MgB_6	89,24	85,68	72,8	HoB_6	229.86	85,71	23.24
MgB_{12}	154.16	92,31	84,5	HoB_{12}	294,66	92,35	44,03
CaB_6	105,00	85,71	61,83	HoB_{100}	—	—	—
SrB_6	152,55	85,71	42,56	ErB_4	210,55	80,0	20.56
BaB_6	202,28	85,71	32,09	ErB_6	332.19	85,71	27.96
AlB_2	48,62	66,67	44,51	ErB_{12}	296,99	92,35	43,68
AlB_4	70,26	80,00	61,60	ErB_{100}	—	—	—
AlB_{10}	135.18	90,91	80,05	TmB_4	212.22	80,0	20.39
AlB_{12}	156,82	92,31	82,80	TmB_6	233.86	85,71	27.76
ScB_2	66.74	66.67	32.42	TmB_{12}	298,67	92,35	43,44
ScB_4	88.20	80,00	49.1	TmB_{100}	—	—	—
ScB_6	110,02	85,71	59.00	YbB_3	205.50	75,0	15.80
ScB_{12}	174,688	92,31	74,3	YbB_4	216.32	80,0	20,00
YB_2	110,56	66.67	19,56	YbB_6	237.96	85,71	27,28
YB_3	121,38	75,0	26,74	YbB_{12}	302,77	92,35	42.85
YB_4	132,20	80,0	32,73	YbB_{100}	—	—	—
YB_6	153,84	85,71	42,19	LuB_2	196.59	66.67	11.00
YB_{12}	218.76	92,33	59.25	LuB_4	218,27	80,0	19.03
YB_{70}	—	—	89.49	LuB_6	239.91	85,71	27.06
LaB_3	171,38	75,0	18.94	LuB_{12}	304,70	92,35	42.58
LaB_4	182,20	80,0	23,75	LuB_{100}	—	—	—
LaB_6	203,84	85,80	31.80	ThB_2	232.05	66.67	9.32
CeB_4	183,41	80,0	23.59	ThB_4	275,33	80,0	15,72
CeB_6	205.05	85,71	31,66	ThB_6	296.97	85,71	21,86
PrB_3	173,38	75,00	18.72	ThB_{76}	—	—	77,98
PrB_4	184,20	80,00	23,49	UB	248.89	50.0	4,35
PrB_6	209,19	85,71	31.51	UB_2	259,71	66.67	8,33
NdB_4	187,55	80,0	23,1	UB_4	281,35	80,00	15,38
NdB_6	209.19	85,71	30.90	UB_{12}	367.91	92,35	35,29
PmB_6	209,87	85,71	30,91	PuB	252.82	50,0	4,28
SmB_4	193,63	80,0	22,35	PuB_2	263,64	66,67	8.21
SmB_6	215,27	85,71	30.16	PuB_4	285.28	80,0	15,17

Phase	Molecular weight	Content of nonmetal, %		Phase	Molecular weight	Content of nonmetal, %	
		atomic	by wt.			atomic	by wt.
1	2	3	4	1	2	3	4
PuB_6	306.42	85.71	21.18	W_2B_5	421.82	71.43	12.83
PuB_{70}	—	—	—	WB_4	227.12	80.0	19.06
Ti_2B	106.62	33.33	10.15	WB_{12}	313.58	92.31	41.24
TiB	58.72	50.0	18.43	Mn_4B	230.58	20.0	4.69
Ti_3B_4	186.94	57.2	23.2	Mn_2B	120.70	33.33	8.96
TiB_2	69.54	66.67	31.12	MnB	65.76	50.0	16.45
Ti_2B_5	149.90	71.43	36.09	Mn_3B_4	208.10	57.14	20.80
TiB_{12}	177.74	92.31	73.05	MnB_2	76.58	66.67	28.26
ZrB	102.04	50.0	10.60	MnB_4	98.18	80.0	44.04
ZrB_2	112.86	66.67	19.25	Tc_3B	307.8	25.0	3.51
ZrB_{12}	221.06	92.31	58.74	Tc_7B_3	725.4	30.0	4.47
HfB	189.32	50.0	5.71	TcB_2	120.6	66.67	17.9
HfB_2	200.14	66.67	10.81	Re_3B	569.75	25.0	1.90
V_3B_2	174.49	40.0	12.40	Re_7B_3	1336.63	30.0	2.43
VB	61.77	50.0	17.52	ReB_2	207.82	66.67	10.45
V_5B_6	319.58	54.5	20.29	Re_2B_5	426.72	71.43	12.65
V_3B_4	196.13	57.14	22.07	ReB_3	218.63	75.00	14.85
V_2B_3	134.32	60.00	24.14	Fe_3B	178.35	25.0	6.06
VB_2	72.13	66.67	29.8	Fe_2B	122.52	33.33	8.83
Nb_2B	196.64	33.33	5.50	FeB	66.67	50.0	16.23
Nb_3B_2	300.37	40.0	7.20	$FeB_{\sim19}$	—	—	>35
NbB	103.74	50.0	10.43	Co_3B	187.64	25.0	5.77
Nb_3B_4	322.01	57.14	13.44	Co_2B	128.70	33.33	8.4
NbB_2	114.55	66.67	18.89	CoB	69.76	50.0	15.51
Ta_2B	372.72	33.33	2.90	CoB_2	80.58	66.67	26.86
Ta_3B_2	564.49	40.0	3.83	Ni_3B	186.95	25.0	5.79
TaB	191.77	50.0	5.64	Ni_2B	128.24	33.33	8.44
Ta_3B_4	586.13	57.14	7.39	Ni_3B_2	197.71	40.0	10.95
TaB_2	202.59	66.67	10.68	Ni_4B_3	267.30	42.86	12.14
Cr_4B	218.86	20.0	4.94	NiB	69.53	50.00	15.56
Cr_2B	114.84	33.33	9.42	NiB_2	80.35	66.67	26.93
Cr_5B_3	292.51	37.50	11.10	NiB_{12}	188.44	92.31	68.9
CrB	62.83	50.0	17.24	Ru_7B_3	739.92	30.00	4.38
Cr_3B_4	199.31	57.14	21.72	$Ru_{11}B_8$	1198.26	42.1	7.22
CrB_2	73.65	66.67	29.38	$RuB_{\sim1.1}$	—	52.4	9.66
Cr_2B_5	158.12	71.46	34.22	Ru_2B_3	234.57	60.00	16.05
CrB_4	95.24	80.00	45.42	RuB_2	122.69	66.67	17.62
CrB_6	116.93	85.78	55.48	Ru_2B_5	256.20	71.5	21.1
Mo_2B	202.72	33.33	5.34	Rh_7B_3	752.77	30.00	4.31
Mo_3B_2	309.49	40.0	6.99	Rh_2B	216.62	33.33	4.99
MoB	106.77	50.0	10.13	$RhB_{\sim1.1}$	—	52.4	10.35
MoB_2	117.59	66.67	18.40	Pd_3B	330.01	25.00	3.27
Mo_2B_5	246.00	71.43	21.99	Pd_5B_2	553.62	28.60	3.90
MoB_4	139.29	80.0	31.12	Pd_3B_2	340.82	40.00	6.35
MoB_{12}	225.67	92.31	57.7	$OsB_{\sim1.2}$	—	54.5	5.38
W_2B	378.54	33.33	2.86	$OsB_{1.6}$	207.50	61.5	7.86
WB	194.62	50.0	5.56	OsB_2	211.82	66.67	10.21
WB_2	205.47	66.67	10.55	Os_2B_5	434.46	71.5	12.45

Phase	Molecular weight	Content of nonmetal, %		Phase	Molecular weight	Content of nonmetal, %	
		atomic	by wt.			atomic	by wt.
1	2	3	4	1	2	3	4
IrB~1.1	—	52.4	5.83	Ho C$_2$	188.94	66.67	12.7
IrB$_{1,35}$	206.79	57.4	7.59	Er$_3$C	513.6	25.00	2.34
Pt$_3$B	596.08	25.00	1.81	ErC$_2$	191.2	66.67	12.55
PtB~1.1	—	52.4	5.24	Tm$_3$C	520.2	25.00	2.31
Be$_2$C	30.04	33.33	39.98	TmC$_2$	192.94	66.67	12.45
BeC$_2$	33.04	66.67	72.71	Yb$_3$C	531.13	25.0	2.26
Mg$_2$C$_3$	84.67	60.0	42.56	YbC$_2$	197.04	66.67	12.18
MgC$_2$	48.34	66.67	49.69	Lu$_3$C	536.97	25.00	2.25
CaC$_2$	64.10	66.67	37.48	LuC$_2$	198.99	66.67	12.06
SrC$_2$	111.65	66.67	21.52	ThC	244.06	50.0	4.92
BaC$_2$	161.38	66.67	14.88	ThC$_2$	256.07	66.67	9.38
ScC	57.10	50.0	21.08	UC	250.08	50.0	4.80
Sc$_{15}$C$_{19}$	902.55	55.9	25.3	U$_2$C$_3$	512.17	60.0	7.04
Y$_3$C	278.76	25.0	4.34	UC$_2$	262.09	66.67	9.16
Y$_2$C	189.82	33.33	6.33	NpC$_2$	261.0	66.67	9.2
YC	100.92	50.0	11.89	PuC	254.01	50.0	4.72
Y$_5$C$_6$	516.59	54.5	13.95	Pu$_2$C$_3$	524.03	60.0	6.88
Y$_{15}$C$_{19}$	1561.79	55.9	14.6	PuC$_2$	268.0	66.67	8.97
Y$_2$C$_3$	213.84	60.0	16.80	TiC	59.91	50.0	20.05
YC$_2$	112.92	66.67	21.25	ZrC	103.23	50.0	11.64
La$_2$C$_3$	313.84	60.0	11.47	HfC	190.51	50.0	6.31
LaC$_2$	162.94	66.67	14.74	V$_2$C	113.91	33.33	10.54
CeC	152.13	50.0	7.9	VC	62.96	50.0	19.08
Ce$_2$C$_3$	316.24	60.0	11.39	Nb$_2$C	197.83	33.33	6.07
CeC$_2$	164.15	66.67	14.63	NbC	104.92	50.0	11.45
PrC	152.92	50.0	7.85	Ta$_2$C	373.91	33.33	3.21
Pr$_2$C$_3$	317.84	60.0	11.36	TaC	192.96	50.0	6.22
PrC$_2$	164.94	66.67	14.56	Cr$_{23}$C$_6$	1268.30	20.69	5.68
Nd$_2$C$_3$	324.54	60.0	11.09	Cr$_7$C$_3$	400.10	30.0	9.01
NdC$_2$	168.29	66.67	14.27	Cr$_3$C$_2$	180.05	40.0	13.34
Sm$_3$C	463.29	25.0	2.59	Mo$_2$C	203.91	33.33	5.89
Sm$_2$C$_3$	336.86	60.0	10.68	MoC	107.96	50.0	11.13
SmC$_2$	174.45	66.67	13.77	W$_2$C	397.73	33.33	3.16
Eu$_3$C	467.89	25.0	2.56	WC	195.87	50.0	6.13
EuC$_2$	175.98	66.67	13.65	Mn$_{23}$C$_6$	1335.69	20.69	5.4
Gd$_3$C	483.76	25.0	2.48	Mn$_3$C	176.83	25.0	6.79
Gd$_2$C$_3$	349.8	60.0	10.29	Mn$_7$C$_3$	298.72	28.57	8.04
GdC$_2$	180.9	66.67	13.3	Mn$_5$C$_2$	420.61	30.00	8.57
Tb$_3$C	488.78	25.0	2.46	Fe$_4$C	235.41	20.00	5.10
Tb$_2$C	329.86	33.33	3.64	Fe$_3$C	179.56	25.0	6.69
Tb$_2$C$_3$	355.86	60.0	10.13	Fe$_2$C	123.71	33.33	9.71
TbC$_2$	182.93	66.67	13.16	Co$_3$C	188.83	25.0	6.86
Dy$_3$C	499.38	25.00	2.41	Co$_2$C	129.89	33.33	9.25
Dy$_2$C$_3$	360.92	60.0	9.974	Ni$_3$C	188.14	25.0	6.38
DyC$_2$	186.46	66.67	12.87	Be$_3$N$_2$	55.06	40.0	50.89
Ho$_3$C	506.82	25.00	2.38	Mg$_3$N$_2$	100.98	40.0	27.75
Ho$_2$C	341.87	33.33	3.52	Ca$_3$N$_2$	148.26	40.0	18.89
Ho$_2$C$_3$	365.88	60.00	9.83	Sr$_3$N$_2$	290.91	40.0	9.63

COMPOSITION OF REFRACTORY COMPOUNDS (continued)

Phase	Molecular weight	Content of nonmetal, %		Phase	Molecular weight	Content of nonmetal, %	
		atomic	by wt.			atomic	by wt.
1	2	3	4	1	2	3	4
Ba_3N_2	440.10	40.0	6.37	TcN	113.0	50.0	12.4
ScN	58.97	50.0	23.76	Re_3N	572.61	25.0	2.45
YN	102.93	50.0	13.61	Re_2N	368.62	33.33	3.62
LaN	152.93	50.0	9.16	Fe_4N	237.40	20.0	5.90
CeN	154.14	50.0	9.09	Fe_3N	181.55	25.0	7.71
PrN	154.93	50.0	9.04	Fe_2N	125.70	33.33	11.14
NdN	158.28	50.0	8.85	Co_3N	190.82	25.0	7.34
SmN	164.44	50.0	8.52	Co_2N	131.88	33.33	10.62
EuN	166.01	50.0	8.44	Ni_3N	190.07	25.0	7.37
GdN	171.27	50.0	8.18	$CaAl_4$	148.01	80.0	72.92
TbN	172.94	50.0	8.10	$CaAl_2$	94.04	66.67	57.38
DyN	176.52	50.0	7.94	$SrAl_4$	195.55	80.0	55.19
HoN	178.95	50.0	7.83	SrAl	114.60	50.0	23.54
ErN	181.28	50.0	7.73	$BaAl_4$	245.27	80.0	44.00
TmN	182.95	50.0	7.66	$BaAl_2$	191.30	66.67	28.2
YbN	187.05	50.0	7.49	BaAl	164.32	50.0	16.42
LuN	189.00	50.0	7.41	ScAl	71.94	50.0	37.51
ThN	246.05	50.0	5.69	$ScAl_2$	98.92	66.67	54.55
Th_3N_4	752.18	57.14	7.42	$ScAl_3$	125.90	75.0	64.29
Th_2N_3	506.12	60.0	8.30	Y_2Al	204.79	33.33	13.17
UN	252.07	50.0	5.55	Y_3Al_2	320.68	40.0	16.83
U_2N_3	518.17	60.0	8.11	YAl	115.89	50.0	23.28
UN_2	266.07	66.67	10.52	YAl_2	142.87	66.67	37.77
NpN	251.00	50.0	5.58	Y_2Al_5	312.72	71.5	43.1
PuN	256.00	50.0	5.47	YAl_3	169.85	75.0	47.65
Ti_2N	109.81	33.33	12.75	YAl_4	196.83	80.0	54.83
TiN	61.90	50.0	22.63	La_3Al	443.71	25.0	6.08
ZrN	105.22	50.0	13.31	LaAl	165.89	50.0	16.26
HfN	192.60	50.0	7.28	$LaAl_2$	192.87	66.67	27.98
V_3N	166.85	25.0	8.39	$LaAl_3$	219.85	75.0	36.82
VN	64.95	50.0	21.56	$LaAl_4$	246.84	80.0	43.72
Nb_2N	199.82	33.33	7.01	Ce_3Al	447.34	25.0	6.03
$NbN_{0.75}$	103.41	42.85	10.13	CeAl	167.10	50.0	16.15
$NbNo_{0.98}$	106.62	49.50	12.89	$CeAl_2$	194.08	66.67	27.80
NbN	106.91	50.0	13.10	$CeAl_3$	221.06	75.0	36.61
Ta_2N	375.90	33.33	3.73	$CeAl_4$	248.05	80.0	43.51
TaN	194.95	50.0	7.19	Pr_3Al	449.70	25.0	6.00
Cr_2N	118.02	33.33	11.86	Pr_2Al	308.80	33.33	8.73
CrN	66.01	50.0	21.21	PrAl	167.89	50.0	16.07
Mo_3N	301.85	25.0	4.64	$PrAl_2$	194.87	66.67	27.69
Mo_2N	205.90	33.33	6.80	$PrAl_3$	221.85	75.0	36.05
MoN	109.95	50.0	12.73	$PrAl_4$	248.83	80.0	43.37
W_2N	318.72	33.33	4.39	Nd_3Al	459.70	25.0	5.87
WN	197.86	50.0	7.08	Nd_2Al	315.46	33.33	8.55
Mn_4N	233.76	20.0	6.00	NdAl	171.22	50.0	15.76
Mn_2N	123.88	33.33	11.30	$NdAl_2$	198.20	66.67	27.23
Mn_3N_2	192.82	40.0	14.52	$NdAl_3$	225.18	75.0	35.99
Mn_6N_5	399.66	45.5	17.5	$NdAl_4$	252.17	80.0	42.80

Phase	Molecular weight	Content of nonmetal, %		Phase	Molecular weight	Content of nonmetal, %	
		atomic	by wt.			atomic	by wt.
1	2	3	4	1	2	3	4
Sm$_3$Al	478,03	25.0	5.64	Zr$_2$Al	209.42	33.33	12.46
Sm$_2$Al	327.68	33.33	8.23	Zr$_5$Al$_3$	537.04	37.5	15.07
SmAl	177.33	50.0	15.21	Zr$_3$Al$_2$	327.62	40,0	16,47
SmAl$_2$	204,31	66.67	26,41	Zr$_4$Al$_3$	445.82	42.9	18.16
SmAl$_3$	231,29	75.0	34.99	ZrAl	118.20	50.0	22.83
SmAl$_4$	258.28	80.0	41,78	Zr$_2$Al$_3$	263.38	60.0	30.73
EuAl$_2$	205.92	66.67	26,20	ZrAl$_2$	145.18	66.67	37.17
EuAl$_4$	259.89	80.0	41.53	ZrAl$_3$	172.16	75.0	47,01
Gd$_2$Al	341.48	33.33	7.90	Hf$_2$Al	383.96	33.33	6.78
Gd$_3$Al$_2$	525.71	40.0	10.26	Hf$_3$Al$_2$	589.43	40,0	9.15
GdAl	184.23	50.0	14.64	Hf$_4$Al$_3$	794.90	42.9	10.18
GdAl$_2$	211,21	66,67	25.55	HfAl	205.47	50.0	13,13
GdAl$_3$	238.19	75.0	33.98	Hf$_2$Al$_3$	437,92	60.0	18:48
GdAl$_4$	265.18	80.0	40.70	HfAl$_2$	232,45	66,67	23.21
Tb$_3$Al$_2$	530.74	40.0	10,17	HfAl$_3$	259.43	75.0	31.20
TbAl	185.91	50.0	14.51	V$_3$Al	179.81	25.0	15.00
TbAl$_2$	212,89	66.67	25.35	V$_5$Al$_8$	470.56	61.5	45.87
TbAl$_3$	239.87	75,0	33.74	VAl$_3$	131.89	75.0	61.37
TbAl$_4$	266.85	80.0	40,44	V$_4$Al$_{23}$	824.34	85,2	75.28
Dy$_3$Al$_2$	541.46	40.0	9.97	VAl$_6$	212.83	85.8	76.06
DyAl	189.48	50.0	14.24	V$_7$Al$_{45}$	1570.76	86,6	77.30
DyAl$_2$	216.46	66.67	24.93	VAl$_{11}$	347.74	91,6	85,35
DyAl$_3$	243.44	75.0	33.25	Nb$_3$Al	305.70	25,0	8.83
Ho$_2$Al	356.84	33.33	7.57	Nb$_2$Al	212.79	33.33	12.27
Ho$_3$Al$_2$	548.75	40.0	9.83	NbAl$_3$	173.85	75.0	46,56
HoAl	191.91	50.0	14.06	Ta$_2$Al	388,88	33.33	6.94
HoAl$_2$	218.89	66.67	24.65	TaAl$_3$	261.89	75,0	30,91
HoAl$_3$	245.87	75.00	32.92	Cr$_2$Al	130.97	33,33	21,46
Er$_2$Al	361.50	33,33	7.21	Cr$_4$Al$_3$	288.93	42.9	28.0
Er$_3$Al$_2$	555.74	40,0	9.71	Cr$_5$Al$_8$	475.83	61.5	45.36
ErAl	194.24	50,0	13.89	Cr$_4$Al$_9$	450.82	69,2	53.9
ErAl$_2$	221,22	66.67	24.39	CrAl$_3$	132,94	75,0	60.9
ErAl$_3$	248.20	75.0	32.61	CrAl$_4$	159.92	80,0	67.5
TmAl	195.92	50.0	13.77	Cr$_2$Al$_{11}$	400.79	84.6	74.05
TmAl$_2$	222.90	66.67	24.21	CrAl$_7$	240.87	87.5	78.41
TmAl$_3$	249.88	75,0	32:39	Mo$_3$Al	314.80	25,0	8.57
Yb$_2$Al	373.06	33.33	7.25	MoAl$_2$	149.90	66.67	36,0
Yb$_3$Al$_2$	573.08	40,0	9.42	Mo$_3$Al$_8$	503.67	72,7	42.85
YbAl	200.02	50.0	13.5	MoAl$_3$	176.88	75.0	47.7
YbAl$_2$	227,00	66.67	23,77	MoAl$_4$	203.87	80.0	52.94
YbAl$_3$	253.98	75.0	31.87	MoAl$_5$	230.85	83.4	58.44
YbAl$_4$	280.97	80.0	38.4	MoAl$_7$	284.81	87.5	66.31
LuAl$_2$	228,93	66.67	23.57	MoAl$_{12}$	419.72	92.3	77,14
Ti$_3$Al	170,68	25.0	15.81	WAl$_2$	237.81	66,67	22,7
TiAl	74,88	50.0	36.03	W$_3$Al$_7$	740.42	70,0	25.5
TiAl$_2$	101,86	66.67	52.98	WAl$_3$	264.79	75.0	30.6
TiAl$_3$	128.84	75.0	62,82	WAl$_4$	291.78	80.0	36,99
Zr$_3$Al	300.64	25.0	8.97	WAl$_5$	318.76	83.4	42,32

Phase	Molecular weight	Content of nonmetal, %		Phase	Molecular weight	Content of nonmetal, %	
		atomic	by wt.			atomic	by wt.
1	2	3	4	1	2	3	4
WAl_{12}	507.63	92.3	63.78	$PtAl$	222.07	50.0	12,12
$MnAl$	81.92	50.0	32.94	Pt_2Al_3	471,12	60,0	17,14
Mn_4Al_{11}	516.55	73.3	57.46	$PtAl_2$	249.05	66.67	21,65
$MnAl_3$	135.88	75.0	62.6	$PtAl_3$	276.03	75.0	29.3
Mn_3Al_{10}	434.63	77.0	62.08	Pt_5Al_{21}	1542,06	80,8	36,74
$MnAl_4$	162.86	80.0	66.27	Th_3Al	723.10	25.0	3,73
$MnAl_6$	216.83	85.8	74.65	Th_2Al	491.06	33.33	5,30
$MnAl_{12}$	378.72	92.3	85.49	Th_3Al_2	750.08	40.0	7,19
Fe_3Al	194.52	25.0	13.87	$ThAl$	259.02	50.0	10,42
$FeAl$	82.83	50.0	32.57	Th_2Al_3	545.02	60.0	14,85
$FeAl_2$	109.81	66.67	49.1	$ThAl_2$	286.00	66.67	18,87
Fe_2Al_5	246.60	71.5	54.70	$ThAl_3$	312.98	75.0	25,86
$FeAl_{3,25}$	143.54	76.5	61,13	UAl_2	291.99	66.67	18,48
$FeAl_6$	217.74	85.8	74.35	UAl_3	318.97	75.0	25,38
$CoAl$	85.91	50.0	31,40	UAl_4	345.96	80.0	31,20
Co_2Al_5	252.77	71.5	53,37	$NpAl_2$	291.0	66.67	18,55
$CoAl_{3,25}$	146.62	76.5	59.81	$NpAl_3$	318.0	75.0	25,46
Co_2Al_9	360.70	81.8	67,32	$NpAl_4$	345.0	80.0	31,29
Ni_3Al	203.11	25.0	13.28	Pu_3Al	746.0	25.0	3.62
$NiAl$	85.69	50.0	31,49	$PuAl$	266.0	50.0	10.14
Ni_2Al_3	198.36	60.0	40,81	$PuAl_2$	293.0	66.67	18,41
$NiAl_3$	139.65	75.0	57,96	$PuAl_3$	320.0	75.0	25,29
Tc_2Al	224.98	33.33	11,60	$PuAl_4$	347.0	80.0	31,10
Tc_2Al_3	278.94	60,0	29,02	Mg_2Si	76,73	33,33	36,61
$TcAl_4$	206,93	80.0	52,16	Ca_2Si	108,25	33,33	25,95
$TcAl_6$	260,89	85.8	62,06	$CaSi$	68.17	50.00	41.20
$RuAl$	128.05	50.0	21,07	$CaSi_2$	96.26	66,67	58,36
Ru_2Al_3	283.08	60.0	28,59	Sr_2Si	203.33	33,33	13,81
$RuAl_2$	155.03	66.67	34.81	$SrSi$	115,72	50.0	24,28
$RuAl_3$	182,01	75.0	44.47	Sr_2Si_3	259,50	60.0	32,47
Rh_3Al	335.70	25.0	8,04	$SrSi_2$	143,81	66,67	39.07
$RhAl$	129.89	50.0	20.77	$BaSi$	165,45	50.0	16,98
Pd_2Al	239.78	33.33	10.88	$BaSi_2$	193,54	66,67	29,03
Pd_5Al_3	612.94	37.5	13,20	Sc_5Si_3	309.04	37,5	27,26
$PdAl$	133.38	50.0	20,23	$ScSi$	73.04	50.0	38,45
Pd_2Al_3	293.74	60.0	27,56	$ScSi_{1,66}$	91.58	62,5	51,01
$PdAl_3$	187,34	75.0	43,21	Y_5Si_3	528.87	37,52	15.97
$ReAl$	213.18	50.0	12,66	Y_5Si_4	556,87	44.5	20,17
$ReAl_2$	240,16	66,67	22.5	YSi	117,01	50.0	24,01
$ReAl_4$	294.13	80.0	36,69	$YSi_{1,66}$	135.53	62.5	34,40
$ReAl_6$	348.09	85.8	46,51	YSi_2	145,10	66,67	38,72
$ReAl_{12}$	509,98	92.4	63,49	La_5Si_3	778.81	37,5	10,82
$OsAl$	217,18	50,0	12,42	La_3Si_2	472,90	40.0	11,88
$IrAl$	219.18	50,0	12,31	La_5Si_4	806,89	44,5	13,92
$Pt_{13}Al_3$	2617,11	18.75	3,09	$LaSi$	167,00	50.0	16,82
Pt_3Al	612,25	25.0	4,41	$LaSi_2$	195,08	66.67	28,79
Pt_5Al_3	1056,39	37.5	7,64	Ce_5Si_3	784,86	37.5	10,73
Pt_3Al_2	639,23	40.0	8,45	Ce_3Si_2	476,53	40.0	11,79

Phase	Molecular weight	Content of nonmetal, %		Phase	Molecular weight	Content of nonmetal, %	
		atomic	by wt.			atomic	by wt.
1	2	3	4	1	2	3	4
Ce_5Si_4	812.94	44.5	13.82	LuSi	203.06	50.0	13.83
CeSi	168.22	50.0	16.70	$LuSi_{1.66}$	221.59	62.5	21.04
$CeSi_2$	196.29	66.67	28.62	$LuSi_2$	231.14	66.67	24.30
Pr_5Si_3	788.79	37.5	10.68	Th_3Si_2	752.33	40.0	7.47
Pr_3Si_2	478.89	40.0	11.73	ThSi	260.14	50.0	10.80
Pr_5Si_4	816.88	44.5	13.75	$ThSi_{1.67}$	278.66	62.5	16.81
PrSi	169.00	50.0	16.30	$ThSi_2$	288.23	66.67	19.49
$PrSi_2$	197.10	66.67	28.50	U_3Si	724.24	25.0	3.78
Nd_5Si_3	805.46	37.5	10.46	U_3Si_2	752.33	40.0	7.29
Nd_5Si_4	833.54	44.5	13.53	USi	266.16	50.0	10.55
NdSi	172.33	50.0	16.29	$USi_{1.67}$	284.65	62.5	16.5
$NdSi_2$	200.45	66.67	28.02	USi_2	294.25	66.67	19.09
Sm_5Si_3	836.01	37.5	10.07	USi_3	322.34	75.0	26.14
Sm_5Si_4	864.09	44.5	13.00	$NpSi_2$	293.18	66.67	19.16
SmSi	178.44	50.0	15.74	Pu_5Si_3	1280	37.5	6.58
$SmSi_2$	206.61	66.67	27.19	Pu_3Si_2	788	40.0	17.11
EuSi	180.05	50.0	15.60	PuSi	267	50.0	10.51
$EuSi_2$	208.18	66.67	26.99	$PuSi_{\sim1.5}$	—	$\sim$60.0	—
Gd_5Si_3	870.51	37.5	9.68	$PuSi_2$	298.18	66.67	18.84
Gd_5Si_4	898.59	44.5	12.50	Ti_3Si	171.79	25.0	16.35
GdSi	185.34	50.0	15.15	Ti_5Si_3	323.77	37.5	26.03
$GdSi_{1.66}$	203.87	62.5	22.87	Ti_5Si_4	351.84	44.5	31.9
$GdSi_2$	213.44	66.67	26.32	TiSi	75.99	50.0	36.97
Tb_5Si_3	878.88	37.5	9.59	$TiSi_2$	104.08	66.67	53.98
Tb_5Si_4	906.96	44.5	12.38	Zr_3Si	301.75	25.0	9.31
TbSi	187.01	50.0	15.02	Zr_2Si	210.53	33.33	13.34
$TbSi_2$	215.10	66.67	26.11	Zr_5Si_3	540.37	37.5	15.59
Dy_5Si_3	896.76	37.5	9.39	Zr_3Si_2	329.84	40.0	16.76
Dy_5Si_4	924.84	44.5	12.14	Zr_5Si_4	568.44	44.5	19.8
DySi	190.59	50.0	14.74	ZrSi	119.31	50.0	23.54
$DySi_{1.66}$	209.12	62.5	22.29	$ZrSi_2$	147.40	66.67	38.11
$DySi_2$	218.69	66.67	25.69	Hf_2Si	385.29	33.33	7.29
Ho_5Si_3	908.91	37.5	9.27	Hf_5Si_3	977.27	37.5	8.67
HoSi	193.02	50.0	14.55	Hf_3Si_2	591.64	40.0	9.5
$HoSi_2$	221.10	66.67	25.40	Hf_5Si_4	1004.79	44.5	10.7
Er_5Si_3	920.56	37.5	9.15	HfSi	206.69	50.0	13.59
Er_5Si_4	948.64	44.5	11.84	$HfSi_2$	234.78	66.67	23.93
ErSi	195.35	50.0	14.38	V_3Si	180.94	25.0	15.53
$ErSi_{1.66}$	213.88	62.5	21.80	V_5Si_3	339.02	37.5	24.86
$ErSi_2$	223.43	66.67	25.14	VSi_2	107.13	66.67	52.44
Tm_5Si_3	928.93	37.5	9.07	Nb_3Si	306.80	25.0	9.15
TmSi	197.02	50.0	14.25	Nb_5Si_3	548.82	37.5	15.36
$TmSi_2$	225.11	66.67	24.95	$NbSi_2$	149.09	66.67	37.68
Yb_5Si_3	949.46	37.5	8.87	Ta_3Si	570.93	25.0	4.92
YbSi	201.13	50.0	13.96	Ta_2Si	389.99	33.33	7.20
$YbSi_2$	229.21	66.67	24.51	Ta_5Si_3	989.02	37.5	8.52
Lu_5Si_3	959.11	37.5	8.78	$TaSi_2$	237.13	66.67	23.69
Lu_5Si_4	987.19	44.5	11.38	Cr_3Si	184.42	33.33	15.26

Phase	Molecular weight	Content of nonmetal, %		Phase	Molecular weight	Content of nonmetal, %	
		atomic	by wt.			atomic	by wt.
1	2	3	4	1	2	3	4
Cr_5Si_3	344.32	37.5	24.47	Rh_3Si_2	364.89	40,0	15.41
$CrSi$	80,10	50.0	35,07	$RhSi$	130.99	50.0	21,44
$CrSi_2$	108,19	66,67	51.93	Ir_3Si	604.69	25.0	4.64
Mo_3Si	315.94	25.0	8,89	Ir_2Si	412,49	33.33	6,81
Mo_5Si_3	564.02	37.5	14.94	Ir_3Si_2	632.77	40,0	8.88
$MoSi_2$	152.16	66,67	36.93	$IrSi$	220.29	50.0	12,74
W_5Si_3	1003.87	37.5	8,39	$IrSi_3$	276.46	75.0	30.5
WSi_2	240,10	66,67	23,40	Pd_3Si	347.29	25.0	8,09
Mn_3Si	192.91	25.0	14.55	Pd_2Si	240.89	33.33	11,66
Mn_5Si_2	330.86	28,6	17.0	$PdSi$	134.49	50.0	20,88
Mn_5Si_3	358.97	37,5	23,46	Pt_3Si	613.36	25,0	4.57
$MnSi$	83.03	50,0	33.85	$Pt_{12}Si_5$	2481.51	29,4	5.66
$MnSi_{1.727}$	—	—	—	Pt_2Si	418.27	33,33	6.71
$MnSi_{1.730}$	—	—	46.79	Pt_6Si_5	1310.97	45,5	10,71
$MnSi_{1.733}$	—	—	—	$PtSi$	223,18	50.0	12.55
$MnSi_{1.741}$	—	—	—	Be_3P_2	88,99	40,0	69.62
$MnSi_{1.750}$	—	—	—	Mg_3P_2	134,91	40.0	45.92
Tc_4Si	424,09	20,0	6.62	Ca_3P_2	182,19	40,0	34,00
Tc_3Si	325,09	25,0	8,63	Sr_3P_2	324,84	40,0	19.07
Tc_5Si_3	579.26	37.5	14,55	Ba_3P_2	474.03	40.0	13,07
$TcSi$	127.09	50,0	22,10	BaP_2	199.31	66,67	31,08
$TcSi_{1.750}$	148.15	63.6	33.18	AlP	57.96	50.0	53.45
Re_5Si_3	1015.82	37.5	8.30	ScP	75.93	50.0	40.79
$ReSi$	214.40	50.0	13,10	YP	119.88	50.0	25,84
$ReSi_2$	242,49	66,67	23,17	LaP	169,90	50.0	18,24
Fe_3Si	195,64	25,0	14.36	CeP	171,11	50.0	18,11
$Fe_{11}Si_5$	754.75	31,3	18,65	PrP	171,90	50.0	18,02
Fe_5Si_3	363,52	37,5	23,18	NdP	175.25	50.0	17,68
$FeSi$	83,94	50,0	33.46	SmP	181,41	50.0	17.08
$FeSi_2$	112,03	66,67	50,15	GdP	188,22	50.0	16,45
Co_2Si	145.97	33.33	19.24	TbP	189.90	50.0	16,31
$CoSi$	87,03	50.0	32.28	DyP	193.47	50.0	16,01
$CoSi_2$	115,12	66,67	48.80	HoP	195.90	50.0	15,81
Ni_3Si	204,16	25.0	13.76	ErP	198,23	50.0	15,62
Ni_5Si_2	349,63	28,6	16.07	TmP	199,91	50.0	15,49
Ni_2Si	145.47	33,33	19,31	YbP	204.01	50.0	15,18
Ni_3Si_2	232.25	40,0	24,19	$ThP_{0.75}$	255,28	42,86	9,10
$NiSi$	86.78	50.0	32,37	ThP	263.01	50.0	11,78
$NiSi_2$	114,88	66.67	48,89	Th_3P_4	820.05	57.14	15,11
Ru_2Si	230.23	33,33	12,20	UP	269.05	50.0	11,52
Ru_3Si_2	359.38	40,0	15,63	U_3P_4	838,11	57.14	14,78
$RuSi$	129,16	50.0	21,73	UP_2	299.98	66,67	20.65
$OsSi$	218,29	50.0	12.87	Np_3P_4	834.90	57.14	14,84
Os_2Si_3	464.66	60.0	18,15	PuP	272.98	50.0	11,35
$OsSi_{2.4}$	257,61	70,6	26.17	Ti_3P	174.68	25,0	17,74
Rh_2Si	233,90	33,33	12,01	Ti_2P	126.77	33,33	24,43
Rh_5Si_3	598,78	37,5	23,47	Ti_5P_3	332,42	37,5	27,9
$Rh_{20}Si_{13}$	2423,22	39,4	15.06	Ti_3P_2	205,65	40.0	30,12

Phase	Molecular weight	Content of nonmetal, %		Phase	Molecular weight	Content of nonmetal, %	
		atomic	by wt.			atomic	by wt.
1	2	3	4	1	2	3	4
Ti_4P_3	284,52	42,9	32,66	RuP	132,04	50,0	23,46
TiP	78,88	50.0	39,28	RuP_2	163,02	66.67	38,00
TiP_2	109,85	66.67	56,39	Rh_2P	236,78	33.33	13,08
Zr_3P	304,63	25,0	10,16	RhP_2	164,85	66,67	37,58
ZrP	122,20	50,0	25,35	RhP_3	195,83	75,0	47,45
ZrP_2	153,18	66,67	40,44	Pd_3P	350.17	25,0	8,84
HfP	209,46	50,0	14,79	Pd_7P_3	837,72	30,0	11,1
HfP_2	240,44	66,67	25,76	PdP_2	168,35	66,67	36,80
V_3P	183,83	25,0	16,85	PdP_3	199,32	75,0	46,62
VP	81,93	50,0	37,81	OsP_2	252,15	66,67	24,57
VP_2	112,89	66,67	54,87	Ir_2P	415.37	33,33	7,46
Nb_3P	309,69	25,0	10,00	IrP_2	254,15	66,67	24,37
NbP	123,89	50,0	25,01	IrP_3	285.12	75,0	32,59
NbP_2	154,85	66,67	40,00	PtP_2	257,04	66,67	24,05
Ta_3P	573,82	25,0	5,40	BeS	41,08	50,0	78,1
TaP	211,93	50,0	14,62	MgS	56,37	50,0	56,9
TaP_2	242,90	66,67	25,50	CaS	72.14	50,0	44,5
Cr_3P	187,01	25,0	16,57	SrS	119,68	50,0	26,8
Cr_2P	135,0	33,33	22,95	BaS	169,40	50,0	18,95
CrP	82,99	50,0	37,33	ScS	77,02	50,0	41,7
CrP_2	113,96	66,67	54,36	Sc_2S_3	186,12	60,0	51,69
Mo_3P	318,83	25,0	9,72	YS	120.99	50,0	26,51
MoP	126,93	50,0	24,41	Y_5S_7	669,06	58,33	33,55
MoP_2	157,90	66,67	39,23	Y_2S_3	274,04	60,0	35,11
W_2P	398,67	33,33	7,77	YS_2	153,05	66,67	41,90
WP	214,90	50,0	14,42	LaS	170.99	50,0	18,75
WP_2	245,87	66,67	25,20	La_3S_4	545.04	57.14	23,55
Mn_3P	195,80	25,0	15,82	La_2S_3	374.04	60,0	25,72
Mn_2P	140,86	33,33	21,99	La_5S_7	919,00	58,4	24,4
Mn_3P_2	226,77	40,00	27,32	LaS_2	203,05	66,67	31,53
MnP	85,92	50,00	36,06	CeS	172,20	50,0	18,62
Re_2P	403,60	33,33	7,68	Ce_3S_4	548.65	57.14	23,38
ReP	217,29	50,00	14,26	Ce_2S_3	376,46	60,00	25,55
ReP_2	248,26	66,67	24,95	Ce_5S_7	925,05	58,4	24,3
ReP_3	279,23	75,00	33,28	CeS_2	204,26	66,67	31,40
Fe_3P	198,53	25,00	15,61	PrS	172,99	50,0	18,54
Fe_2P	142,68	33,33	21,71	Pr_3S_4	551.02	57,14	23,28
FeP	86,83	50,00	35,68	Pr_2S_3	378,05	60,0	25,40
FeP_2	117,80	66,67	52,59	Pr_5S_7	928,98	37,5	24,20
Co_2P	143,86	33,33	20,81	PrS_2	205,4	66,67	31,2
CoP	89,92	50,0	34,45	NdS	176,34	50,0	18,18
CoP_3	151,86	75,0	61,19	Nd_3S_4	561,07	57,14	22,86
Ni_3P	207,05	25,0	14,96	Nd_2S_3	384,74	60,0	25,0
$Ni_{12}P_5$	859,16	29,41	18,03	Nd_5S_7	945,65	58,4	23,7
Ni_2P	148,36	33,33	20,88	Nd_4S_7	801,41	63,6	28,0
NiP_2	120,66	66,67	51,34	NdS_2	208,37	66,67	30,8
NiP_3	151,62	75,0	61,29	SmS	182,50	50,0	17,57
Ru_2P	233.11	33,33	13.29	Sm_3S_4	579,55	57,14	22,13

COMPOSITION OF REFRACTORY COMPOUNDS (continued)

Phase	Molecular weight	Content of nonmetal, %		Phase	Molecular weight	Content of nonmetal, %	
		atomic	by wt.			atomic	by wt.
1	2	3	4	1	2	3	4
Sm_2S_3	397,06	60,0	24.23	Np_3S_5	871.32	62,5	18,4
Sm_5S_7	976.20	58,4	23,0	Np_2S_5	634,32	71,5	25,3
EuS	184.07	50,02	17,42	NpS_3	333,19	75,0	28,9
Eu_3S_4	584,28	57,14	22.0	PuS	274,07	50,0	11,70
$Eu_2S_{3,81}$	426.10	65,50	28,65	Pu_3S_4	860,26	57,2	14.9
GdS	189.33	50,0	16,94	Pu_2S_3	580,20	60,0	16.58
Gd_2S_3	410.72	60,0	23,42	PuS_2	308,13	66,67	20.8
GdS_2	221,39	66,67	28.97	Am_2S_3	582,20	60,0	16,52
TbS	190,99	50.0	16,8	Bk_2S_3	590.19	60,0	16.3
Tb_2S_3	414,04	60,0	23,2	Cf_2S_3	600.19	60,0	16.0
DyS	194,56	50,0	16,5	Ti_6S	319,46	14,3	10.65
Dy_5S_7	1037,01	58,33	21.65	Ti_3S	175,76	25,0	18.25
Dy_2S_3	421.22	60,00	22,84	Ti_2S	127,86	33.33	25,1
DyS_2	226,64	66,67	28.30	TiS	79,96	50,0	40,1
HoS	196.99	50,0	16,25	Ti_4S_5	351,92	55,6	45,6
Ho_2S_3	426,05	60,0	22,6	Ti_3S_4	271,96	57,2	47.2
ErS	199,27	50,0	16.05	Ti_2S_3	191,99	60,0	50,1
Er_5S_7	1060,81	58,33	21.16	TiS_2	112,03	66,67	57,3
Er_2S_3	430,74	60,0	22.33	TiS_3	144.09	75,0	66,7
TmS	201,00	50,0	15,95	Zr_2S	214,50	33.33	14.95
Tm_2S_3	434,06	60,0	22,18	ZrS	123,28	50,0	26,0
YbS	205,10	50.0	15,63	Zr_2S_3	278,63	60,0	34.5
Yb_3S_4	647,36	57,14	19.80	ZrS_2	155,35	66,67	41,3
Yb_2S_3	442,28	60,0	21,75	ZrS_3	187,41	75,0	51,3
LuS	207.03	50,0	15.5	Hf_2S	389,04	33,33	8,24
Lu_2S_3	446.13	60,0	21.55	HfS	210,55	50,0	15.24
Ac_2S_3	550,20	60,0	17,49	Hf_2S_3	453,17	60,0	21,2
ThS	264,12	50,0	12,14	HfS_2	242,62	66,67	26,45
Th_2S_3	560,30	60,00	17.17	HfS_3	274,68	75,0	35,0
Th_4S_7	1152,66	63,60	19,47	V_3S	184,89	25,0	17,35
ThS_2	296,18	66,67	21,65	VS	83.01	50,0	38,6
Th_2S_5	624.40	71,5	25,7	V_4S_5	364,09	55,6	44.00
US	270,14	50,0	11,87	V_3S_4	281,08	57,2	45,6
U_2S_3	572,34	60,0	16,81	V_2S_3	198,08	60,0	48,5
U_3S_5	874,54	62.5	18,33	VS_2	115,07	66,67	55,7
US_2	302.20	66,67	21,22	VS_4	179,20	80,0	71.6
US_3	334,22	75.0	28.8	NbS	124.97	50,0	25,6
NpS	269,0	50.0	11.9	Nb_2S_3	282,00	60,0	34,1
Np_2S_3	570,20	60,0	16.87	Nb_3S_4	406,97	57,2	31,5

Phase	Molecular weight	Content of nonmetal, %		Phase	Molecular weight	Content of nonmetal, %	
		atomic	by wt.			atomic	by wt.
1	2	3	4	1	2	3	4
NbS_2	157.03	66.67	40.9	CoS_2	123.06	66.67	52.1
NbS_3	189.10	75.0	50.9	Ni_2S	149.48	33.33	21.4
NbS_4	221.16	80.0	58.0	Ni_3S_2	240.26	40.0	26.7
Ta_6S	1117.75	14.3	2.87	Ni_6S_5	512.58	45.5	31.3
Ta_2S	393.96	33.33	8.14	Ni_7S_6	603.35	46.1	31.9
TaS_2	245.08	66.67	26.2	NiS	90.77	50.0	35.3
TaS_3	277.14	75.0	34.7	Ni_3S_4	304.39	57.2	42.1
CrS	84.06	50.0	38.1	NiS_2	122.84	66.67	52.2
Cr_7S_8	620.48	53.4	41.4	Ru_2S_3	298.33	60.0	32.2
Cr_5S_6	452.36	54.5	42.5	RuS_2	165.20	66.67	38.8
Cr_3S_4	284.24	57.2	45.1	Rh_9S_8	1182.66	47.1	21.7
Cr_2S_3	200.18	60.0	48.0	Rh_3S_4	436.97	57.2	29.4
Cr_5S_8	516.49	61.6	49.7	Rh_2S_3	302.00	60.0	31.9
Mo_2S_3	288.07	60.0	33.4	RhS_2	167.03	66.67	38.4
MoS_2	160.07	66.67	40.1	Rh_2S_5	366.13	71.5	43.8
Mo_2S_5	352.20	71.5	45.5	Pd_4S	457.66	20.0	7.00
MoS_3	192.13	75.0	50.1	Pd_2S	244.86	33.33	13.1
WS_2	247.98	66.67	25.9	PdS	138.46	50.0	23.1
WS_3	280.04	75.0	34.4	PdS_2	170.53	66.67	37.6
MnS	87.00	50.0	36.9	OsS_2	254.33	66.67	25.2
MnS_2	119.07	66.67	53.9	IrS	224.26	50.0	14.3
TcS_2	163.13	66.67	39.3	Ir_2S_3	480.59	60.0	20.0
Tc_2S_7	422.45	77.8	53.1	IrS_2	256.33	66.67	25.0
ReS_2	250.33	66.67	25.6	Ir_3S_8	833.11	72.7	30.8
Re_2S_5	532.72	71.5	30.1	IrS_3	288.39	75.0	33.3
Re_2S_7	596.85	77.8	37.6	PtS	227.15	50.0	14.1
FeS	87.91	50.0	36.5	PtS_2	259.22	66.67	24.7
Fe_3S_4	295.80	57.2	43.4	B_4C *	55.29	20.0	21.72
FeS_2	119.98	66.67	53.5	SiC *	40.10	50.0	29.95
Co_4S_3	331.92	42.8	29.0	BN **	24.82	50.0	56.44
Co_9S_8	786.91	47.0	32.6	Si_3N_4 **	140.30	57.14	39.94
CoS	91.00	50.0	35.2	AlN **	40.99	50.0	34.18
Co_3S_4	305.06	57.2	42.0				

*Content of C.
**Content of N.

Phase	Content of nonmetal, %		Temp., °C	Source	Year	Notes
	atomic	by wt.				
1	2	3	4	5	6	7
$ScBe_{13}$ YBe_{13} $LaBe_{13}$ $CeBe_{13}$	—	—	—	[1]	1962	Narrow
$CrBe_2$	70.9—77.5	29.7—37.4	—	[2]	1966	
$CrBe_{12}$	—	—	—	[2]	1966	Narrow
$MnBe_8$	75—92.8	32.9—67.8	1100	[267]	1970	
$ReBe_2$	60—70	6.76—10.15	—	[267]	1970	
Be_4B	20—25	23,1—28.6	—	[267]	1970	
MgB_2	64.0—66.6	44—47	—	[3]	1975	
MgB_6	83.4—88.3	69—77	—	[3]	1975	
MgB_{12}	91.1—91.9	82—83.5	—	[3]	1975	
LaB_6	85.8—86.8	31.8—33.6	—	[5]	1967	
SmB_4 SmB_6	—	—	—	[267]	1970	Narrow
GdB_4	—	—	—	[6]	1971	Narrow
GdB_6	—	—	—	[267]	1970	Very narrow
DyB_2 DyB_4 DyB_6 DyB_{12}	—	—	—	[267]	1970	Very narrow
ThB_4	—	—	—	[3]	1975	Very narrow
ThB_6	85.8—88.5	21,9—26.4	—	[4]	1971	
UB_2	—	—	—	[3]	1975	Narrow
TiB_2	66.6—66.8	31.0—31.3	530	[320]	1969	
	66.2—66.8	30.6—31.3	930	[320]	1969	
	65.6—66.9	30.1—31.4	1330	[320]	1969	
	65.0—67.4	29.5—31.8	1730	[320]	1969	
ZrB_2	66.6—66.8	19,1—19.25	530	[320]	1969	
	66.4—66.8	19.0—19,25	930	[320]	1969	
	66.0—67.1	18.7—19.45	1330	[320]	1969	
	65.5—67.9	18,35—20.05	1730	[320]	1969	
HfB_2	66.6—66.8	10.8—10.9	530	[320]	1969	
	66.4—66.8	10,7—10.9	930	[320]	1969	
	65.9—67.0	10.45—10.95	1330	[320]	1969	
	65.5—67.7	10.3—11.25	1730	[320]	1969	
VB_2	66—68	29.2—31.1	—	[10]	1973	
NbB_2	66.4—66.8	18.7—19.0	530	[320]	1969	
	65.3—66.8	18,0—19,0	930	[320]	1969	
	64.0—67.0	17,15—19.15	1330	[320]	1969	
	62.5—67.5	16.25—19.5	1730	[320]	1969	
TaB_2	66,0—66,8	10,4—10.75	530	[320]	1969	
	64.6—66.8	9,85—10.75	930	[320]	1969	
	62.9—66,9	9,2—10.8	1330	[320]	1969	
	58.9—67.3	7.88—10.95	1730	[320]	1969	

Phase	Content of nonmetal, %		Temp., °C	Source	Year	Notes
	atomic	by wt.				
1	2	3	4	5	6	7
Cr_2B	32.5—34	9.09—9.66	—	[268]	1973	
Cr_5B_3	36—37.5	10.45—11.1	—	[268]	1973	
CrB	48—50	16.1—17.2	—	[268]	1973	
Cr_3B_4	58—62.5	22.3—25.7	—	[268]	1973	
CrB_2	66—70	28.8—32.7	—	[268]	1973	
Mo_2B	—	—	—	[7]	1973	Not present
Mo_3B_2	—	—	—	[7]	1973	
α-MoB	48—51	9.7—10.0	—	[7]	1973	
β-MoB	51—52	10.0—10.7	—	[7]	1973	
Mo_2B_5	70—71.4	19.5—20.8	—	[7]	1973	
W_2B	—	—	—	[7]	1973	Very narrow
α-WB	48—51	5.2—5.8	—	[7]	1973	
W_2B_5	68—71.4	11.1—12.8	—	[7]	1973	
WB_4	78.3—81.1	17.5—20.1	—	[8]	1974	
MnB_2	66.7—69	28.2—30.4	—	[9]	1967	
ReB_2	67—75	10.5—14.8	—	[3]	1975	
Co_3B	—	—	—	[267]	1970	Narrow
Ni_3B	—	—	—	[268]	1973	"
Ni_2B	—	—	—			
Pd_3B	—	—	—	[267]	1970	"
Pd_5B_2	—	—	—			
OsB_2	—	—	—	[268]	1973	"
Y_3C	20—28.5	3.27—5.11	—	[267]	1970	
La_2C_3	55—60	9.55—11.58	806	[267]	1970	
ThC	38—50	3.07—4.92	1500	[268]	1973	See [12]
ThC_2	62.2—67.2	8—9.56	—	[11]	1968	
NpC	45—49	3.99—4.64	—	[268]	1973	
Pu_3C_2	—	—	—	[268]	1973	Very narrow
PuC	48.5—49.8	4.43—4.65	1500	[268]	1973	
UC	48.5—50	4.52—4.80	2000	[320]	1969	
UC_2	—	—	—	[320]	1969	Narrow
TiC	33—50	11.0—20.0	1750	[320]	1969	
ZrC	35—50	6.62—11.6	1400	[320]	1969	
HfC	37.5—50	3.88—6.3	1030—1530	[320]	1968	
V_2C	31—33.3	9.9—10.5	—	[11]	1968	
VC	42.1—46.5	14.7—17.0	1030—1330	[320]	1969	
Nb_2C	31—33.3	5.8—6.1	—	[11]	1968	
NbC	41.9—50.0	8.6—11.5	—	[320]	1969	
Ta_2C	32.3—33.0	3.07—3.19	1230—2430	[320]	1969	
TaC	40.5—50	4.62—6.23	1230—2430	[320]	1969	
Mo_2C	31.2—33.3	5.4—5.9	—	[11]	1968	
MoC	39—43	7.4—8.6	2550	[17]	1974	
W_2C	28.5—33.3	2.54—3.16	830—2030	[320]	1969	

Phase	Content of nonmetal, %		Temp., °C	Source	Year	Notes
	atomic	by wt.				
1	2	3	4	5	6	7
α-WC	—	—	—	[13]	1970	Very narrow
β-WC	37—48	3.68—5.69	—	[17]	1974	
Fe₃C	—	—	—	[268]	1973	Narrow
ThN ⎫	—	—	—	[320]	1969	"
UN ⎬						
Ti₂N ⎭	—	—	—	[14]	1969	Very narrow
TiN	37.5—50	14.9—22.6	—	[14]	1969	
ZrN	35—50	7.6—13.3	1500	[267]	1970	See [14]
HfN	42—52	5.40—7.85	—	[320]	1969	
V₃N	25—33	8.4—11.9	—	[14]	1969	
VN	41—50	16.0—21.6	—	[14]	1969	
Nb₂N	28.5—33.8	5.7—7.1	—	[15]	1972	β; hex.
NbN₀.₇₅	42.9—44.0	10.2—10.6	—	[15]	1972	γ; tetr.
NbN₀.₉₈	46.8—49.5	11.55—12.85	—	[15]	1972	δ; cub.
NbN	49.2—49.5	12.75—12.85	—	[15]	1972	δ'; hex. B8₁
NbN	50.0—50.6	13.1—13.3	—	[15]	1972	ε; hex. Bi
Ta₂N	28.6—31	3.0—3.4	—	[16]	1969	γ; hex.
TaN	—	—	—	[17]	1974	ε; hex. CoSn type absent
TaN	44.5—47.3	5.8—6.5	—	[16]	1969	0; hex. WC type
Ta₃N₅	60.9—62.5	10.75—11.4	—	[16]	1969	
Cr₂N	32—33.3	11.3—11.8	—	[14]	1969	
CrN	—	—	—	[10]	1973	Narrow
Mo₂N	32—33	6.4—6.7	—	[14]	1969	
Mn₄N	18—20	5.8—6.1	<400	[14]	1969	
Mn₂N	28.3—34.7	9.14—11.9	<400	[14]	1969	
Mn₃N₂	38.2—41.0	13.1—15.1	—	[14]	1969	
Mn₆N₅	45.7—47.9	17.7—19.0	—	[14]	1969	
Re₂N	30—33.3	10—11.2	—	[14]	1969	
Co₃N	26—26.7	7.7—8	—	[14]	1969	
SrAl₄	74.3—85.0	47.09—63.57	—	[18]	1965	
TiAl	46—53	35.5—44.5	—	[10]	1973	
Zr₂Al	—	—	—	[267]	1970	Narrow
Hf₂Al	—	—	—	[19]	1961	"
V₅Al₈	59.7—63.5	44—48	—	[20]	1964	
Nb₂Al	26—36	9.25—14.0	1400	[268]	1973	
Cr₂Al₁₁	83.5—85.9	72.5—76	—	[18]	1965	
CrAl₇	86.5—87.6	76.75—78.4	—	[18]	1965	
CoAl	45.7—57.6	27—34	—	[10]	1973	
Ni₃Al	21.5—28	11.2—15.15	1150	[267]	1970	See [18]
β-NiAl	40—55	23.5—36.0	—	[21]	1971	
RhAl	47.8—54.8	19.35—24.15	—	[268]	1973	
Pd₂Al	33.3—34	11.25—11.55	—	[268]	1973	
PdAl	47—51.5	18.35—21.2	<540	[268]	1973	
	43—55	16.1—23.7	940	[268]	1973	

Phase	Content of nonmetal, %		Temp., °C	Source	Year	Notes
	atomic	by wt.				
1	2	3	4	5	6	7
Pd_2Al_3	58.5—61.5	26.4—28.8	<600	[268]	1973	
Ti_3Si	—	—	—	[22]	1971	Narrow
α-Nb_5Si_3	36.7—39.8	14.9—16.6	1250	[268]	1973	
$NbSi_2$	64.9—68.8	35.8—40.0	1250	[268]	1973	
Cr_3Si	22.4—30.3	13.5—19	1680	[304]	1964	
Cr_5Si_3 ⎫	—	—	—	[304]	1964	Very
$CrSi$ ⎭						narrow
$CrSi_2$	66.4—67.0	51.7—52.4	—	[305]	1962	
$MoSi_2$	65.8—66.7	36.0—36.9	—	[10]	1973	
Fe_5Si_3	37—40	22.8—25.1	—	[23]	1966	
$CoSi$	49.0—50.6	31.4—32.8	800—1000	[268]	1973	
	48.5—50.9	31.0—33.1	1200	[268]	1973	
Ni_3Si	—	—	—	[303]	1971	Narrow
Ni_5Si_2	—	—	—	[25]	1961	"
α-Ni_2Si ⎫						
Ni_3Si_2 ⎬	—	—	—	[303]	1971	"
$NiSi_2$ ⎭						
Ru_2Si_3	—	—	—	[268]	1973	Very narrow
Pd_3Si	—	—	—	[267]	1970	
$Pt_{12}Si_5$	—	—	—	[26]	1969	Narrow
PrP	—	—	1000	[27]	1973	Narrow; increases as temperature rises
ThP	35.5—49.0	6.84—11.4	—	[268]	1973	
Ti_2P	—	—	—	[28]	1968	Narrow
TiP	48.0—48.5	37.4—37.9	—	[29]	1961	
α-ZrP	—	—	—	[29]	1961	Very narrow
β-ZrP	48.0—48.5	23.9—24.2	—	[29]	1961	
NbP	44.7—50	21.2—25.0	—	[313]	1963	
Fe_2P	—	—	—	[268]	1973	Narrow
$Pd_{4.8}P$	—	—	—	[30]	1966	"
Pd_3P	20—25	~6.8—8.85	740	[267]	1970	
PdP_2	—	—	—	[267]	1970	Absent
YS	43—50	21.4—26.5	—	[31]	1965	
LaS	43—50	14.85—18.75	—	[31]	1965	
CeS	43—50	14.75—18.65	—	[31]	1965	
PrS	43—50	14.65—18.55	—	[31]	1965	
Pr_5S_7	58.4—59.2	24.2—24.8	1100	[32]	1968	
NdS	43—50	14.35—18.20	—	[31]	1965	
α-Nd_2S_3	59—60	24.2—25.0	—	[33]	1972	
β-Nd_2S_3	57—60	22.8—25.0	—	[33]	1972	
Nd_4S_7	62.7—63.5	27.2—27.9	—	[33]	1972	
NdS_2	64.0—69.0	28.3—33.2	—	[33]	1972	

Phase	Content of nonmetal, %		Temp., °C	Source	Year	Notes
	atomic	by wt.				
1	2	3	4	5	6	7
Eu_3S_4	—	—	—	[267]	1970	Absent
GdS	43—50	13.3—16.9	—	[31]	1965	
TbS	43—50	13.2—16.8	—	[31]	1965	
DyS	43—50	13.0—16.5	—	[31]	1965	
HoS	43—50	12.8—16.3	—	[31]	1965	
ErS	43—50	12.6—16.1	—	[31]	1965	
Yb_3S_4	57.08—59.35	19.77—21.29	—	[33]	1972	
YbS	50—53.5	15.6—17.6	—	[31]	1965	
TmS	43—50	12.4—15.9	—	[31]	1965	
LuS	43—55.5	12.15—18.6	—	[31]	1965	
Th_2S_3	55—60.8	14.45—17.65	—	[33]	1972	
TiS	43.5—54	34.0—44.0	—	[33]	1972	
Ti_2S_3	57.5—59.1	47.5—49.1	—	[34]	1971	$4H$
Ti_2S_3	59.1—62.3	49.1—52.5	—	[34]	1971	$12H$
TiS_{2-x}	64.3—65.8	54.7—56.25	—	[17]	1974	
TiS_2	56.5—66.4	46.5—57.0	—	[33]	1972	
Zr_2S_3	47.4—60	24.0—34.5	—	[33]	1972	
ZrS_2	65.6—66.6	40.1—41.3	500—900	[267]	1970	
Hf_2S	31.5—33.7	7.63—8.37	—	[33]	1972	
VS	51.2—55.0	39.8—43.5	900	[267]	1970	
V_2S_3	57.5—58.4	46.0—47.0	900	[267]	1970	
V_4S_5	54.5—57.2	43.0—45.7	>600	[33]	1972	
NbS	50—54	25.7—28.9	—	[35]	1963	
NbS	47.367—54.54	23.6—29.3	—	[36]	1965	$P6_3/mmc$
Nb_2S_3	57.16—64.22	31.5—38.2	800	[36]	1965	} R_{3m}
Nb_2S_3	61.5—64.22	35.5—38.2	1100	[36]	1965	
Nb_2S_3	58.3—66.67	32.5—40.8	—	[36]	1965	$P6_3/mmc$
$2s\text{-}Ta_{1+x}S_2$	60—62.5	21.0—22.8	—	[36]	1965	Hex.
TaS_2	58—66.7	19.65—26.2	—	[37]	1971	
CrS	—	—	—	[38]	1967	Narrow
Cr_7S_8	53.1—53.5	41.1—41.5	—	[38]	1967	
Cr_5S_6	54.1—54.6	42.1—42.6	—	[38]	1967	
Cr_3S_4	55.9—56.8	43.8—44.7	—	[38]	1967	
Cr_2S_3	—	—	—	[38]	1967	Narrow
WS_2	65—69.7	24.5—28.6	—	[33]	1972	
FeS	50—53.2	36.5—39.5	—	[33]	1972	
Co_4S_3	41.5—44.5	27.8—30.4	785	[33]	1972	
Co_9S_8	—	—	—	[33]	1972	Very narrow
CoS	50.86—54	36.0—39.0	600	[33]	1972	
Co_3S_4	—	—	—	[33]	1972	Very narrow
CoS_2						
Ni_3S_2	35.2—42.9	23.2—29.4	640	[39]	1967	
PtS	—	—	—	[33]	1972	Narrow
B_4C	17.6—29.5	19.2—31.7	—	[40]	1960	
BN	—	—	—	[10]	1973	Narrow; see [268]
AlN	—	—	—	[10]	1973	Narrow
$\alpha\text{-}Si_3N_4$	—	—	—	[10]	1973	"

CRYSTAL STRUCTURE

Phase	Unit cell	Space group	Structural type	a, Å	b, Å	c, Å	α	c/a	Source	Year	Notes
1	2	3	4	5	6	7	8	9	10	11	12
$MgBe_{13}$	Cub.	O_h^6-Fm3c	$NaZn_{13}$	10.16	—	—	—	—	[41]	1962	
$CaBe_{13}$	"	O_h^6-Fm3c	$NaZn_{13}$	10.31	—	—	—	—	[41]	1962	
$SrBe_{13}$	"	O_h^6-Fm3c	$NaZn_{13}$	10,45	—	—	—	—	[42]	1969	
$ScBe_5$	Hex.	$D_{6h}'-P6/mmm$	$CaCu_5$	4,55	—	3,50	—	0,77	[42]	1969	
Sc_2Be_{17}	"	D_{3d}^5-R3m	Th_2Zn_{17}	7,61	—	11.25	—	1.47	[42]	1969	
$ScBe_{13}$	Cub.	O_h^6-Fm3c	$NaZn_{13}$	10,082	—	—	—	—	[1]	1962	
YBe_{13}	"	O_h^6-Fm3c	$NaZn_{13}$	10,230	—	—	—	—	[1]	1962	
$LaBe_{13}$	"	O_h^6-Fm3c	$NaZn_{13}$	10.45	—	—	—	—	[43]	1963	
$CeBe_{13}$	"	O_h^6-Fm3c	$NaZn_{13}$	10.378	—	—	—	—	[43]	1963	
$PrBe_{13}$	Cub.	O_h^6-Fm3c	$NaZn_{13}$	10.367	—	—	—	—	[43]	1963	C
$NdBe_{13}$	"	O_h^6-Fm3c	$NaZn_{13}$	10.356	—	—	—	—	[43]	1963	See [44]
$PmBe_{13}$	"	O_h^6-Fm3c	$NaZn_{13}$	10.33	—	—	—	—	[43]	1963	
$SmBe_{13}$	"	O_h^6-Fm3c	$NaZn_{13}$	10.28	—	—	—	—	[43]	1963	See [44]
$EuBe_{13}$	"	O_h^6-Fm3c	$NaZn_{13}$	10.288	—	—	—	—	[43]	1963	See [44]
$GdBe_{13}$	"	O_h^6-Fm3c	$NaZn_{13}$	10.27	—	—	—	—	[43]	1963	
$TbBe_{13}$	"	O_h^6-Fm3c	$NaZn_{13}$	10.251	—	—	—	—	[43]	1963	See [44]
$DyBe_{13}$	"	O_h^6-Fm3c	$NaZn_{13}$	10,240	—	—	—	—	[43]	1963	See [44]
$HoBe_{13}$	"	$O_h^{61}-Fm3c$	$NaZn_{13}$	10,220	—	—	—	—	[43]	1963	See [44]

Compound	System	Space group	Structure type	a		c	angle	c/a	Ref.	Year	
$ErBe_{13}$	"	O_h^6—$Fm3c$	$NaZn_{13}$	10.215	—	—	—	—	[43]	1963	See [44]
$TmBe_{13}$	"	O_h^6—$Fm3c$	$NaZn_{13}$	10.192	—	—	—	—	[43]	1963	See [44]
$YbBe_{13}$	"	O_h^6—$Fm3c$	$NaZn_{13}$	10.19	—	—	—	—	[43]	1963	See [44]
$LuBe_{13}$	"	O_h^6—$Fm3c$	$NaZn_{13}$	10.177	—	—	—	—	[43]	1963	See [44]
$TiBe$	Hex.	—	—	3.203	—	3.693	—	1.15	[42]	1969	
$TiBe_2$	Cub.	O_h^7—$Fd3m$	$MgCu_2$	6.451	—	—	—	—	[42]	1969	
$TiBe_3$	Hex.	D_{3d}^5—$R\bar3m$	$NbBe_3$	4.49	—	21.32	—	4.74	[42]	1969	
Ti_2Be_{17}	"	D_{3d}^5—$R3m$	Th_2Zn_{17}	7.35	—	7.26	—	0.984	[42]	1969	
$TiBe_{12}$	Tetr.	D_{4h}^{17}—$I4/mmm$	$ThMn_{12}$	7.36	—	4.195	—	0.571	[42]	1969	
$ZrBe_2$	Hex.	D_{6h}^1—$P6/mmm$	AlB_2	3.82	—	3.25	—	0.85	[42]	1969	
$ZrBe_5$	"	D_{6h}^1—$P6/mmm$	$CaCu_5$	4,564	—	3.485	—	0.764	[126]	1959	
Zr_2Be_{17}	"	D_{3d}^5—$R3m$	Th_2Zn_{17}	7.548	—	10.997	—	1.457	[126]	1959	
$ZrBe_{13}$	Cub.	O_h^6—$Fm3c$	$NaZn_{13}$	10.067	—	—	—	—	[42]	1969	
$HfBe_2$	Hex.	D_{6h}^1—$P6/mmm$	AlB_2	3.788	—	3.168	—	0.836	[42]	1969	
$HfBe_5$	"	D_{6h}^1—$P6/mmm$	$CaCu_5$	4.534	—	3.471	—	0.765	[42]	1969	
Hf_2Be_{17}	"	D_{3d}^5—$R3m$	Th_2Zn_{17}	7.499	—	21,905	—	2.921	[42]	1969	
$HfBe_{13}$	Cub.	O_h^6—$Fm3c$	$NaZn_{13}$	10.010	—	—	—	—	[45]	1961	
VBe_2	Hex.	D_{6h}^4—$P6_3/mmc$	$MgZn_2$	4.394	—	7.144	—	1.639	[42]	1969	
VBe_{12}	Tetr.	D_{4h}^{17}—$I4/mmm$	$ThMn_{12}$	7.278	—	4.212	—	0.579	[130]	1957	
Nb_3Be_2	"	D_{4h}^5—$P4/mbm$	U_3Si_2	—	—	—	—	—	[42]	1969	
$NbBe_2$	Cub.	O_h^7—$Fd3m$	$MgCu_2$	6.535	—	—	—	—	[127]	1959	
$NbBe_3$	Rhombohedral, Hex.	—	$NbBe_3$	7.495	—	21.05	35.4°	4.61	[127]	1959	
$NbBe_5$	Hex.	D_{3d}^5—$R\bar3m$	$NbBe_3$	4.561	—	—	—	—	[127]	1959	
Nb_2Be_{17}	" Rhombohedral	—	Th_2Zn_{17}	7.409 / 5,599	—	10.84	82.4°	1.46	[2] / [42]	1966 / 1969	

CRYSTAL STRUCTURE (continued)

Phase	Unit cell	Space group	Structural type	a, Å	b, Å	c, Å	α	c/a	Source	Year	Notes
1	2	3	4	5	6	7	8	9	10	11	12
$NbBe_{12}$	Tetr.	$D_{4h}^{17}—I4/mmm$	$ThMn_{12}$	7.376	—	4.258	—	0.577	[42]	1969	
Ta_2Be	"	$D_{4h}^{18}—I4/mcm$	$CuAl_2$	5.997	—	4.892	—	0.816	[42]	1969	
Ta_3Be_2	"	$D_{4h}^{5}—P4/mbm$	U_3Si_2	6,50	—	3.32	—	0.51	[128]	1961	
$TaBe_2$	Cub.	$O_h^{7}—Fd3m$	$MgCu_2$	6.51	—	—	—	—	[128]	1961	
$TaBe_3$	Hex.	$D_{3d}^{5}—R\bar{3}m$	$NbBe_3$	4.53	—	20.95	—	4.62	[128]	1961	
Ta_2Be_{17}	"	$D_{3d}^{5}—R3m$	Th_2Zn_{17}	7.388	—	10.74	—	1.45	[128]	1961	
$TaBe_{12}$	Tetr.	$D_{4h}^{17}—I4/mmm$	$ThMn_{12}$	7.334	—	4.267	—	0.582	[130]	1957	
$CrBe_2$	Hex.	$D_{6h}^{4}—P6_3/mmc$	$MgZn_2$	4.259	—	6.975	—	1,638	[42]	1969	
$CrBe_{12}$	Tetr.	$D_{4h}^{17}—I4/mmm$	$ThMn_{12}$	7.234	—	4.176	—	0.577	[42]	1969	
Mo_3Be	Cub.	$O_h^{3}—Pm3n$	Cr_3Si	4 89	—	—	—	—	[129]	1960	
$MoBe_2$	Hex.	$D_{6h}^{4}—P6_3/mmc$	$MgZn_2$	4.442	—	7.356	—	1,658	[42]	1969	
$MoBe_{12}$	Tetr.	$D_{4h}^{17}—I4/mmm$	$ThMn_{12}$	7.271	—	4.234	—	0.585	[42]	1969	
$MoBe_{22}$	Cub.	$O_h^{7}—Fd3m$	$ReBe_{22}$	11.64	—	—	—	—	[129]	1960	
WBe_2	Hex.	$D_{6h}^{4}—P6_3mmc$	$MgZn_2$	4.446	—	7.289	—	1.63	[42]	1969	
WBe_{12}	Tetr.	$D_{4h}^{17}—I4/mmm$	$ThMn_{12}$	7.362	—	4.216	—	0.573	[42]	1969	
WBe_{22}	Cub.	$O_h^{7}—Fd3m$	$ReBe_{22}$	11.631	—	—	—	—	[42]	1969	
$MnBe_2$	Hex.	$D_{6h}^{4}—P6_3/mmc$	$MgZn_2$	4.231	—	6,909	—	1,632	[42]	1969	
$MnBe_3$	Cub.	—	—	5.931	—	—	—	—	[42]	1969	
$MnBe_{12}$	Tetr.	$D_{4h}^{17}—I4/mmm$	$ThMn_{12}$	7.276	—	4.256	—	0.585	[130]	1957	

				a	b	c		c/a		Ref.	Year
$ReBe_2$	Hex.	D_{6h}^4—$P6_3/mmc$	$MgZn_2$	4.354	—	7.101	—	1.631	—	[42]	1969
$ReBe_{22}$	Cub.	O_h^7—$Fd3m$	$ReBe_{22}$	11.54	—	—	—	—	—	[129]	1960
$FeBe_2$	Hex.	D_{6h}^4—$P6_3/mmc$	$MgZn_2$	4.212	—	6.853	—	1.626	—	[42]	1969
$FeBe_5$	Cub.	O_h^7—$Fd3m$	UNi_5	5.884	—	—	—	—	—	[42]	1969
$FeBe_{11}$	Hex.	—	—	—	—	—	—	—	—	[47]	1971
$FeBe_{12}$	Tetr.	D_{4h}^{17}—$I4/mmm$	$ThMn_{12}$	7.253	—	4.232	—	0.583	—	[130]	1957
$CoBe$	Cub.	O_h^1—$Pm3m$	$CsCl$	2.615	—	—	—	—	—	[42]	1969
$CoBe_5$	"	O_h^7—$Fd3m$	UNi_5	5.852	—	—	—	—	—	[130]	1957
$CoBe_{12}$	Tetr.	D_{4h}^{17}—$I4/mmm$	$ThMn_{12}$	7.237	—	4.249	—	0.587	—	[130]	1957
$NiBe$	Cub.	O_h^1—$Pm3m$	$CsCl$	2.621	—	—	—	—	—	[42]	1969
Ni_5Be_{21}	"	T_d^3—$I/43m$	Ni_5Zn_{21}	7.625	—	—	—	—	—	[42]	1969
Ru_2Be_{17}	Hex.	$P6/mmm$	—	4.203	—	10.90	—	2.59	—	[49]	1971
Ru_5Be_{17}	Cub.	T_h^5—$Im3$	Ru_3Be_{17}	11.337	—	—	—	—	—	[42]	1969
$RhBe_{3,4}$	"	—	—	13.122	—	—	—	—	—	[50]	1971
$RhBe$	Rhomb.	O_h^1—$Pm3m$	$CsCl$	2.739	11.609	2.803	—	—	—	[42]	1969
$RhBe_2$	Cub.	—	—	3.642	—	—	—	—	—	[48]	1971
Rh_5Be_{22}	Hex.	T^2—$F23$	Pb_5Li_{22}	15.81	—	—	—	—	—	[42]	1969
Rh_2Be_{17}	Cub.	$P6/mmm$	—	4.190	—	10.88	—	2.60	—	[49]	1971
$RhBe_{49}$	"	—	—	11.84	—	—	—	—	—	[42]	1969
$PdBe$	"	O_h^1—$Pm3m$	$CsCl$	2.819	—	—	—	—	—	[42]	1969
$PdBe_5$	"	O_h^7—$Fd3m$	$PdBe_5$	5.969	—	—	—	—	—	[131]	1959
$PdBe_{12}$	Tetr.	D_{4h}^{17}—$I4/mmm$	$ThMn_{12}$	7.271	—	4.251	—	0.584	—	[131]	1959
Ir_5Be_{22}	Cub.	T^2—$F23$	Pb_5Li_{22}	15.87	—	—	—	—	—	[50]	1971
Ir_2Be_{17}	Hex.	$P6/mmm$	—	4.193	—	10.89	—	2.60	—	[49]	1971
Os_5Be_{17}	Cub.	T_h^5—$Im3$	Ru_3Be_{17}	11.342	—	10.95	—	2.60	—	[42]	1969
Os_2Be_{17}	Hex.	$P6/mmm$	—	4.221	—	—	—	—	—	[49]	1971
$Pt_{15}Be$	"	—	—	22.13	—	27.12	—	1.225	—	[42]	1969

35

CRYSTAL STRUCTURE (continued)

Phase	Unit cell	Space group	Structural type	a, Å	b, Å	c, Å	α	c/a	Source	Year	Notes
1	2	3	4	5	6	7	8	9	10	11	12
PtBe$_5$	Cub.	O_h^7—$Fd3m$	PdBe$_5$	—	—	—	—	—	[47]	1971	
PtBe$_{12}$	Tetr.	D_{4h}^{17}—$I4/mmm$	ThMn$_{12}$	7.237	—	4.252	—	0.587	[131]	1959	
ThBe$_{13}$	Cub.	O_h^6—$Fm3c$	NaZn$_{13}$	10.395	—	—	—	—	[42]	1969	
UBe$_{13}$	"	O_h^6—$Fm3c$	NaZn$_{13}$	10.256	—	—	—	—	[42]	1969	
NpBe$_{13}$	"	O_h^6—$Fm3c$	NaZn$_{13}$	10.266	—	—	—	—	[42]	1969	
PuBe$_{13}$	"	O_h^6—$Fm3c$	NaZn$_{13}$	10.278	—	—	—	—	[42]	1969	
AmBe$_{13}$	"	O_h^6—$Fm3c$	NaZn$_{13}$	10.283	—	—	—	—	[42]	1969	
Be$_5$B	Tetr.	D_{4h}^7—$P4/nmm$	—	3.369	—	7.050	—	2.093	[51]	1973	
Be$_2$B	Cub.	O_h^5—$Fm3m$	CaF$_2$	4.661	—	—	—	—	[51]	1973	
BeB$_2$	Hex.	D_{6h}^1—$C6/mmm$	—	9.79	—	9.55	—	0.98	[51]	1973	
BeB$_4$	Tetr.	D_{4h}^5—$P4/mbm$	UB$_4$	—	—	—	—	—	[51]	1973	
BeB$_6$	"	D_{4h}^{12}—$C4/nnm$	—	10.16	—	14.28	—	1.41	[51]	1973	Existence doubtful; see [52]
BeB$_{12}$	Hex.	—	AlB$_{12}$	5.08	—	8.80	—	1.73	[51]	1973	
BeB$_9$	"	D_4C	—	5.46	—	12.42	—	2.27	[51]	1973	
MgB$_2$	"	D_{6h}^1—$P6/mmm$	AlB$_2$	3.0846	—	3.5224	—	1.142	[53]	1973	
MgB$_4$	Rhomb.	—$P\,nam$	MgB$_4$	5.464	7.472	4.428	—	—	[53]	1973	
MgB$_6$	Tetr.	—	—	7.07	—	6.45	—	0.912	[3]	1975	
MgB$_{12}$		Not identified									

36

Compound	Crystal system	Space group	Structural type	a		c		c/a	Ref.	Year	Notes
CaB$_6$	Cub.	O_h^1–$Pm3m$	CaB$_6$	4.144	—	—	—	—	[54]	1961	
SrB$_6$	"	O_h^1–$Pm3m$	CaB$_6$	4.195	—	—	—	—	[54]	1961	
BaB$_6$	"	O_h^1–$Pm3m$	CaB$_6$	4.268	—	—	—	—	[54]	1961	
ScB$_2$	Hex.	D_{6h}^1–$P6/mmm$	AlB$_2$	3.146	—	3.517	—	1.118	[55]	1958	
ScB$_4$	Tetr.	D_{4h}^5–$P4/mbm$	UB$_4$	7.7	—	3.64	—	0.47	[42]	1969	In [56] presence not confirmed
ScB$_6$	Cub.	O_h^1–$Pm3m$	CaB$_6$	4.435	—	—	—	—	[42]	1969	In [56] presence not confirmed
ScB$_{12}$	"	O_h^5–$Fm3m$	UB$_{12}$	7.422	—	—	—	—	[42]	1969	In [56] presence not confirmed
ScB$_{12}$	Tetr.	$I4/mmm$	—	5.22	—	7.35	—	1.408	[56]	1970	
YB$_2$	Hex.	D_{6h}^1–$P6/mmm$	AlB$_2$	3.78	—	4.40	—	1.16	[42]	1969	
YB$_3$	Tetr.	—	—	3.78	—	3.55	—	0.94	[42]	1969	
YB$_4$	"	D_{4h}^5–$P4/mbm$	UB$_4$	7.12	—	4.04	—	0.57	[6]	1966	
YB$_6$	Cub.	O_h^1–$Pm3m$	CaB$_6$	4.102	—	—	—	—	[6]	1966	
YB$_{12}$	"	O_h^5–$Fm3m$	UB$_{12}$	7.506	—	—	—	—	[6]	1966	
YB$_{70}$	Tetr.	—	—	11.75	—	12.62	—	1.07	[3]	1975	
YB$_{66}$	Cub.	O_h^6–$Fm3c$	—	23.44	—	—	—	—	[57]	1968	
LaB$_3$	Tetr.	—	—	3.82	—	3.96	—	1.04	[42]	1969	
LaB$_4$	"	D_{4h}^5–$P4/mbm$	UB$_4$	7.3240	—	4.1811	—	0.57	[42]	1969	
LaB$_6$	Cub.	O_h^1–$Pm3m$	CaB$_6$	4.1561	—	—	—	—	[42]	1969	
CeB$_4$	Tetr.	D_{4h}^5–$P4/mbm$	UB$_4$	7.205	—	4.090	—	0.558	[42]	1969	
CeB$_6$	Cub.	O_h^1–$Pm3m$	CaB$_6$	4.138	—	—	—	—	[54]	1961	
PrB$_3$	Pseudo-cub.	—	—	3.81	—	—	—	—	[42]	1969	
PrB$_4$	Tetr.	D_{4h}^5–$P4/mbm$	UB$_4$	7.20	—	4.11	—	0.571	[42]	1969	
PrB$_6$	Cub.	O_h^1–$Pm3m$	CaB$_6$	4.131	—	—	—	—	[54]	1961	
NdB$_4$	Tetr.	D_{4h}^5–$P4/mbm$	UB$_4$	7.219	—	4.102	—	0.568	[132]	1959	

CRYSTAL STRUCTURE (continued)

Phase	Unit cell	Space group	Structural type	a, Å	b, Å	c, Å	α	c/a	Source	Year	Notes
1	2	3	4	5	6	7	8	9	10	11	12
NdB_6	Cub.	O_h^1—$Pm3m$	CaB_6	4.126	—	—	—	—	[132]	1959	
PmB_6	"	O_h^1—$Pm3m$	CaB_6	4.128	—	—	—	—	[42]	1969	
SmB_4	Tetr.	D_{4h}^5—$P4/mbm$	UB_4	7.174	—	4.0696	—	0.557	[132]	1959	
SmB_6	Cub.	O_h^1—$Pm3m$	CaB_6	4.1333	—	—	—	—	[132]	1959	
SmB_{100}	Cub.	—	—	23.487	—	—	—	—	[3]	1975	
EuB_6	"	O_h^1—$Pm3m$	CaB_6	4.182	—	—	—	—	[54]	1961	
GdB_3	Tetr.	—	—	3.79	—	3.63	—	0.958	[42]	1969	
GdB_4	"	D_{4h}^5—$P4/mbm$	UB_4	7.120	—	4.048	—	0.567	[6]	1966	
GdB_6	Cub.	O_h^1—$Pm3m$	CaB_6	4.1144	—	—	—	—	[6]	1966	
GdB_{100}	"	—	—	23.474	—	—	—	—	[3]	1975	
TbB_4	Tetr.	D_{4h}^5—$P4/mbm$	UB_4	7.118	—	4.0286	—	0.566	[132]	1959	
TbB_6	Cub.	O_h^1—$Pm3m$	CaB_6	4.1020	—	—	—	—	[132]	1959	
TbB_{12}	"	O_h^5—$Fm3m$	UB_{12}	7.504	—	—	—	—	[42]	1969	
TbB_{100}	"	—	—	23.457	—	—	—	—	[3]	1975	
DyB_2	Hex.	—	—	3.83	—	4.45	—	1.16	[3]	1975	
DyB_4	Tetr.	D_{4h}^5—$P4/mbm$	UB_4	7.101	—	4.0174	—	0.573	[132]	1959	
DyB_6	Cub.	O_h^1—$Pm3m$	CaB_6	4.0976	—	—	—	—	[132]	1959	
DyB_{12}	Cub.	O_h^5—$Fm3m$	UB_{12}	7.501	—	—	—	—	[42]	1969	
DyB_{100}	"	—	—	23.441	—	—	—	—	[3]	1975	
HoB_4	Tetr.	D_{4h}^5—$P4/mbm$	UB_4	7,086	—	4.008	—	0.566	[132]	1959	

Compound	System	Space group	Structure type	a	c	c/a	Ref.	Year	Notes
HoB$_6$	Cub.	O_h^1–$Pm3m$	CaB$_6$	4.096	—	—	[132]	1959	
HoB$_{12}$	"	O_h^5–$Fm3m$	UB$_{12}$	7.492	—	—	[42]	1969	
HoB$_{100}$	Tetr.	D_{4h}^5–$P4/mbm$	—	23.441	3,9972	0.565	[3]	1975	
ErB$_4$	Tetr.	D_{4h}^5–$P4/mbm$	UB$_4$	7.071	—	—	[132]	1969	
ErB$_6$	Cub.	O_h^1–$Pm3m$	CaB$_6$	4.101	—	—	[42]	1959	
ErB$_{12}$	"	O_h^5–$Fm3m$	UB$_{12}$	7.484	—	—	[42]	1969	
ErB$_{100}$	"	—	—	23.428	3.99	0.57	[3]	1975	
TmB$_4$	Tetr.	D_{4h}^5–$P4/mbm$	UB$_4$	7.06	—	—	[58]	1961	
TmB$_6$	Cub.	O_h^1–$Pm3m$	CaB$_6$	4.110	—	—	[58]	1961	
TmB$_{12}$	"	O_h^5–$Fm3m$	UB$_{12}$	7.476	—	—	[42]	1969	
TmB$_{100}$	Tetr.	—	—	23.433	3.56	0.94	[3]	1975	
YbB$_3$	Tetr.	—	—	3.77	4.00	0.57	[42]	1969	
YbB$_4$	Tetr.	D_{4h}^5–$P4/mbm$	UB$_4$	7.01	—	—	[42]	1969	
YbB$_6$	Cub.	O_h^1–$Pm3m$	CaB$_6$	4.1468	—	—	[42]	1969	
YbB$_{12}$	"	O_h^5–$Fm3m$	UB$_{12}$	7.469	—	—	[59]	1971	
YbB$_{100}$	"	—	—	23.422	—	—	[3]	1975	
LuB$_2$	Hex.	D_{6h}^1–$P6/mmm$	AlB$_2$	3.246	3.704	1.141	[42]	1969	
LuB$_4$	Tetr.	D_{4h}^5–$P4/mbm$	UB$_4$	6.983	3.930	0.562	[60]	1958	
LuB$_6$	Cub.	O_h^1–$Pm3m$	CaB$_6$	4.11	—	—	[61]	1958	
LuB$_{12}$	"	O_h^5–$Fm3m$	UB$_{12}$	7.464	—	—	[42]	1969	
LuB$_{100}$	"	—	—	23.412	4.113	0.567	[3]	1975	
ThB$_4$	Tetr.	D_{4h}^5–$P4/mbm$	UB$_4$	7.256	—	—	[42]	1969	
ThB$_6$	Cub.	O_h^1–$Pm3m$	CaB$_6$	4.1105	—	—	[57]	1968	
ThB$_{76}$	"	—	—	23.518	—	—	[62]	1972	
UB	"	O_h^5–$Fm3m$	NaCl	4.88	—	—	[42]	1969	See [64]
UB$_2$	Hex.	D_{6h}^1–$P6/mmm$	AlB$_2$	3.136	3.988	1.272	[42]	1969	
UB$_4$	Tetr.	D_{4h}^5–$P4/mbm$	UB$_4$	7.080	3.978	0.562	[42]	1969	

CRYSTAL STRUCTURE (continued)

Phase	Unit cell	Space group	Structural type	a, Å	b, Å	c, Å	α	c/a	Source	Year	Notes
1	2	3	4	5	6	7	8	9	10	11	12
UB$_{12}$	Cub.	O_h^5—$Fm3m$	UB$_{12}$	7.472	—	—	—	—	[63]	1961	
PuB	"	O_h^5—$Fm3m$	NaCl	4.92	—	—	—	—	[133]	1960	
PuB$_2$	Hex.	D_{6h}^1—$P6/mmm$	AlB$_2$	3.18	—	3.90	—	1.23	[133]	1960	
PuB$_4$	Tetr.	D_{4h}^5—$P4/mbm$	UB$_4$	7.10	—	4.014	—	0.57	[133]	1960	
PuB$_6$	Cub.	O_h^1—$Pm3m$	CaB$_6$	4.115	—	—	—	—	[133]	1960	
TiB	Rhomb.	D_{2h}^{16}—$Pbnm$	FeB	6,12	3.06	4.56	—	—	[42]	1969	
Ti$_3$B$_4$	"	D_{2h}^{25}—$Immm$	Ta$_3$B$_4$	3,259	13.73	3,0424	—	—	[42]	1969	
TiB$_2$	Hex.	D_{6h}^1—$P6/mmm$	AlB$_2$	3.026	—	3.213	—	1.062	[42]	1969	
Ti$_2$B$_5$	"	D_{6h}^4—$C6/mmc$	W$_2$B$_5$	2,98	—	13.98	—	4.70	[42]	1969	
ZrB	Cub.	O_h^5—$Fm3m$	NaCl	4.65	—	—	—	—	[65]	1970	
ZrB$_2$	Hex.	D_{6h}^1—$P6/mmm$	AlB$_2$	3,168	—	3.528	—	1.114	[65]	1970	
ZrB$_{12}$	Cub.	O_h^5—$Fm3m$	UB$_{12}$	7.408	—	—	—	—	[65]	1970	
HfB	Cub.	O_h^5—$Fm3m$	NaCl	4.62	—	—	—	—	[42]	1969	
HfB$_2$	Hex.	D_{6h}^1—$P6/mmm$	AlB$_2$	3,141	—	3.470	—	1,105	[42]	1969	See [66]
V$_3$B$_2$	Tetr.	D_{4h}^5—$P4/mbm$	U$_3$Si$_2$	5,746	—	3,032	—	0.528	[67]	1963	
V$_5$B$_6$	Rhomb.	$Ammm$	—	3,058	21.25	2,974	—	—	[3]	1975	
VB	"	D_{2h}^{17}—$Cmcm$	TaB	3,058	8,043	2.966	—	—	[67]	1963	
V$_3$B$_4$	"	D_{2h}^{25}—$Immm$	Ta$_3$B$_4$	3,030	13,18	2.986	—	—	[67]	1963	
V$_2$B$_3$	"	D_{2h}^{17}—$Cmcm$	—	3.061	18.40	2.984	—	—	[3]	1975	

Compound	System	Space group	Structure type	a	b	c		c/a	Ref.	Year
VB_2	Hex.	D_{6h}^{1}—P6/mmm	AlB_2	3.001		3.061	—	1.020	[67]	1963
Nb_3B_2	Tetr.	D_{4h}^{5}—P4/mbm	U_3Si_2	6.185		3.281	—	0.530	[67]	1963
NbB	Rhomb.	D_{2h}^{17}—Cmcm	TaB	3.297	8.72	3.166	—	—	[67]	1963
Nb_3B_4	»	D_{2h}^{25}—Immm	Ta_3B_4	3.312	14.11	3.143	—	—	[67]	1963
NbB_2	Hex.	D_{6h}^{1}—P6/mmm	AlB_2	3.089		3.303	—	1.06	[67]	1963
Nb_2B	Tetr.	D_{4h}^{18}—I4/mcm	$CuAl_2$	5.778		4.864	—	0.841	[42]	1969
Ta_2B	Tetr.	D_{4h}^{18}—I4/mcm	$CuAl_2$	—		—	—	—	[67]	1963
Ta_3B_2	»	D_{4h}^{5}—P4/mbm	U_3Si_2	6.184		3.187	—	0.523	[57]	1963
TaB	Rhomb.	D_{2h}^{17}—Cmcm	TaB	3.276	8.669	3.157	—	—	[67]	1963
Ta_3B_4	»	D_{2h}^{25}—Immm	Ta_3B_4	3.29	14.0	3.13	—	—	[57]	1963
TaB_2	Hex.	D_{6h}^{1}—P6/mmm	AlB_2	3.078		3.265	—	1.050	[57]	1963
Cr_4B	Rhomb.	D_{2h}^{24}—Fddd	Mn_4B	4.26	7.38	14.71	—	—	[42]	1969
Cr_2B	Tetr.	D_{4h}^{18}—I4/mcm	$CuAl_2$	5.180		4.316	—	0.832	[42]	1969
Cr_5B_3	»	D_{4h}^{18}—I4/mcm	Cr_5B_3	5.46		10.64	—	1.945	[42]	1969
CrB	Rhomb.	D_{2h}^{17}—Cmcm	TaB	2.969	7.858	3.002	—	—	[38]	1961
Cr_3B_4	»	D_{2h}^{25}—Immm	Ta_3B_4	2.986	13.02	2.952	—	—	[42]	1969
CrB_2	Hex.	D_{6h}^{1}—P6/mmm	AlB_2	2.970		3.074	—	1.035	[39]	1968
CrB_4	Rhomb.	D_{2h}^{25}—Immm	—	4.744	5.477	2.866	—	—	[42]	1969
CrB_6	»	—	—	5.468		7.152	—	1.313	[39]	1968
Mo_2B	Tetr.	D_{4h}^{18}—I4/mcm	$CuAl_2$	5.543		4.735	—	0.854	[57]	1963
Mo_3B_2	»	D_{4h}^{5}—P4/mbm	U_3Si_2	6.00		3.15	—	0.524	[57]	1963
$\alpha\text{-}MoB$	»	D_{4h}^{19}—I4/amd	MoB	3.110		16.97	—	5.45	[57]	1963
$\beta\text{-}MoB$	Rhomb.	D_{2h}^{17}—Cmcm	TaB	3.16	8.61	3.08	—	—	[57]	1963
MoB_2	Hex.	D_{6h}^{1}—P6/mmm	AlB_2	3.05		3.113	—	1.01	[67]	1963
Mo_2B_5	Rhombo-hed.	D_{3d}^{5}—R3m	Mo_2B_5	3.011		20.93	—	6.95	[67]	1963

CRYSTAL STRUCTURE (continued)

Phase	Unit cell	Space group	Structural type	a, Å	b, Å	c, Å	α	c/a	Source	Year	Notes
1	2	3	4	5	6	7	8	9	10	11	12
MoB_4	Tetr.	D_{4h}^5—$P4/mbm$	UB_4	6.34	—	4.50	—	0.710	[67]	1963	
MoB_{12}	Hex.	—	—	3.004	—	3.174	—	1.055	[67]	1963	
W_2B	Tetr.	D_{4h}^{18}—$I4/mcm$	$CuAl_2$	5.564	—	4.740	—	0,852	[67]	1963	See [70]
α-WB	"	D_{4h}^{19}—$I4/amd$	MoB	3.115	—	16.93	—	5.42	[67]	1963	See [70]
β-WB	Rhomb.	D_{2h}^{17}—$Cmcm$	TaB	3.19	8.46	3.07	—	—	[67]	1963	
WB_2	Hex.	D_{6h}^4—$P6_3/mmc$	—	2.9831	—	13.879	—	4.65	[70]	1968	
WB_2	"	—	—	6.363	—	16.433	—	2.582	[124]	1962	
W_2B_5	"	D_{6h}^4—$C6/mmc$	—	2.982	—	13.87	—	4.65	[67]	1963	See [70]
WB_4	Hex.	D_{6h}^4—$P6_3/mmc$	W_2B_5	5.202	—	6.333	—	0.794	[8]	1974	See [70]
WB_4	Tetr.	—	WB_4	6.34	—	4.50	—	0.71	[67]	1963	
WB_{12}	Hex.	—	—	3.994	—	3.174	—	0.795	[67]	1963	
Mn_4B	Rhomb.	D_{2h}^{24}—$Fddd$	Mn_4B	14.53	7.293	4.209	—	—	[42]	1969	
Mn_2B	Tetr.	D_{4h}^{18}—$I4/mcm$	$CuAl_2$	5.148	—	4.208	—	0.82	[42]	1969	
MnB	Rhomb.	D_{2h}^{16}—$Pbnm$	FeB	4.145	5.560	2.977	—	—	[42]	1969	
Mn_3B_4	"	D_{2h}^{25}—$Immm$	Ta_3B_4	3.302	12.86	2.960	—	—	[42]	1969	
MnB_2	Hex.	D_{6h}^4—$P6/mmm$	AlB_2	3.009	—	3.039	—	1.01	[71]	1960	
MnB_4	Monocl.	$C2/m$	—	5.503	5.367	2.949	122.71°	—	[72]	1970	
Tc_3B	Rhomb.	D_{2h}^{17}—$Cmcm$	Re_3B	2.891	9.161	7.246	—	—	[73]	1964	
Tc_7B_3	Hex.	C_{6v}^4—$P6_3/mc$	Th_7Fe_3	7.417	—	4.777	—	0.644	[73]	1964	
TcB_2	"	D_{6h}^4—$P6/mmc$	ReB_2	2.892	—	7.453	—	2.577	[73]	1964	

42

Compound	System	Space group	Structure	a	b	c	angle	ratio	Ref.	Year	Notes
Re$_3$B	Rhomb.	D_{2h}^{17}—$Cmcm$	Re$_3$B	2.890	9.313	7.258	—	—	[74]	1960	
Re$_7$B$_3$	Hex.	C_{6v}^4—$C6mc$	Cr$_7$C$_3$	7.504	—	4.772	—	0.651	[75]	1960	See [76]
Re$_2$B$_5$	Hex.	D_{6h}^4—$C6/mmc$	W$_2$B$_5$	2.97	—	13.8	—	4.65	[77]	1958	
ReB$_2$	"	D_{6h}^4—$P6_3/mmc$	ReB$_2$	2.900	—	7.478	—	2.579	[42]	1969	
ReB$_3$	"	D_{6h}^4—$C6/mmc$	ReB$_3$	2.900	—	7.475	—	2.578	[75]	1960	
Fe$_3$B	Rhomb.	D_{2h}^{16}—$Pbnm$	Fe$_3$C	5.4052	6.6685	4.450	—	—	[78]	1962	
Fe$_2$B	Tetr.	D_{4h}^{18}—$I4/mcm$	CuAl$_2$	5.1087	—	4.2497	—	0.842	[79]	1968	
FeB	Rhomb.	D_{2h}^{16}—$Pnma$	FeB	4.061	5.506	2.952	—	—	[42]	1969	
FeB~$_{19}$	Pseudo-cub.	—	—	—	—	—	—	—	[3]	1975	
Co$_3$B	Rhomb.	D_{2h}^{16}—$Pbnm$	Fe$_3$C	4.411	5.235	6.635	—	—	[80]	1958	
Co$_2$B	Tetr.	D_{4h}^{18}—$I4/mcm$	CuAl$_2$	5.016	—	4.220	—	0.841	[42]	1969	
CoB	Rhomb.	D_{2h}^{16}—$Pbnm$	FeB	3.956	5.253	3.043	—	—	[42]	1969	
Ni$_3$B	"	D_{2h}^{16}—$Pbnm$	Fe$_3$C	5.211	6.619	4.389	—	—	[80]	1958	
Ni$_2$B	Tetr.	D_{4h}^{18}—$I4/mcm$	CuAl$_2$	4.993	—	4.249	—	0.851	[32]	1959	
Ni$_4$B$_3$	Rhomb.	D_{2h}^{16}—$Pbnm$	o-Ni$_4$B$_3$	11.953	2.981	6.569	—	—	[31]	1967	
Ni$_4$B$_3$	Monocl.	C_{2h}^6—$C2/c$	m-Ni$_4$B$_3$	6.430	4.882	7.818	103°18'	—	[81]	1967	
NiB	Rhomb.	D_{2h}^{16}—$Pbnm$	FeB	2.925	7.396	2.966	—	—	[82]	1959	
NiB$_{12}$	Cub.			7.385	—	—	—	—	[42]	1969	
Ru$_7$B$_3$	Hex.	C_{6v}^4—$P6_3mc$	Th$_7$Fe$_3$	7.469	—	4.713	—	0.631	[42]	1969	See [76]
Ru$_{11}$B$_8$	Rhomb.		Ru$_{11}$B$_8$	11.609	11.342	2.836	—	—	[83]	1960	
RuB~$_{1.1}$	Hex.	D_{3h}^1—$P\bar{6}m2$	WC	2.852	—	2.855	—	1.001	[84]	1962	
Ru$_2$B$_3$	"	D_{6h}^4—$P6_3/mmc$	—	2.9051	—	12.8125	—	4.43	[70]	1968	
Ru$_4$B$_5$	"	D_{6h}^4—$P6_3/mmc$	W$_2$B$_5$	2.89	—	12.81	—	4.43	[84]	1962	
RuB$_2$	Rhomb.	D_{2h}^{13}—$Pmmn$	RuB$_2$	4.6443	2.8668	4.0449	—	—	[85]	1963	
Rh$_7$B$_3$	Hex.	C_{6v}^4—$P6_3mc$	Th$_7$Fe$_3$	7.471	—	4.777	—	0.6394	[75]	1960	

CRYSTAL STRUCTURE (continued)

Phase	Unit cell	Space group	Structural type	a, Å	b, Å	c, Å	α	c/a	Source	Year	Notes
1	2	3	4	5	6	7	8	9	10	11	12
Rh_2B	Rhomb.	$D_{2h}^{16}-Pnam$	Rh_2Ge	5,42	3.98	7 44	—	—	[42]	1969	See [86]
$RhB\sim1,1$	Hex.	$D_{6h}^4-P6_3/mmc$	NiAs	3.309	—	4.224	—	1.277	[75]	1960	
Pd_3B	Rhomb.	$D_{6h}^{16}-Pbnm$	Fe_3C	5.463	7,567	4.852	—	—	[78]	1962	
Pd_5B_2	Monocl.	$C_{2h}^6-C^2/C$	Pd_5B_2	12,786	4.955	5.472	97° 2'	0.528	[78]	1962	
Pd_3B_2	Hex.	—	—	6.49	—	3.43	—	0.528	[42]	1969	
$OsB\sim1,2$	"	$D_{3h}^1-P\bar{6}m2$	WC	2.876	—	2.871	—	0.998	[84]	1962	
Os_2B_5	"	$D_{6h}^4-P6_3/mmc$	W_2B_5	2.91	—	12.91	—	4.44	[84]	1962	
OsB_2	Rhomb.	$D_{2h}^{13}-Pmmn$	RuB_2	4.684	4.076	2.872	—	—	[85]	1962	
$IrB_{0,9}$	"	$C_{2v}^{12}-Cmc2_1$	—	2,771	7.578	7.314	—	—	[87]	1971	Low temp.
	Hex.	D_{3h}^1-P6m2	WC	2,815	—	2.823	—	1.005	[87]	1971	High temp.
$IrB\sim1,1$	Tetr.	$D_{4h}^{19}-I4/amd$	α-$ThSi_2$	2,81	—	10.26	—	3.652	[84]	1962	
$IrB_{1,35}$	Monocl.	C_{2h}^3-C2/m ·	$IrB_{1,35}$	10,525	2.910	6.099	91° 4'	—	[85]	1962	
PtB	Hex.	$D_{6h}^4-P6_3/mmc$	NiAs	3.358	—	4.058	—	1.208	[75]	1960	
Pt_3P	Tetr.	—	—	2,78	—	2.96	—	1,06	[124]	1962	
	"	—	—	2,63	—	3.83	—	1.46	[3]	1975	
Be_2C	Cub.	O_h^5-Fm3m	CaF_2	4,34	—	—	—	—	[88]	1972	
Mg_2C_3	Tetr.	—	—	7,45	—	10.61	—	1,424	[42]	1969	
MgC_2	Tetr.	$D_{4h}^{17}-I4/mmm$	CaC_2	5,55	—	5.03	—	0.906	[125]	1962	

Compound	System	Space group	Structure type	a, Å	b, Å	c, Å	Angles	c/a	Ref.	Year	Notes
CaC_2	•	D_{4h}^{17}—$I4/mmm$	CaC_2	3.80	—	6.38	—	1.64	[11]	1968	298—720 K
CaC_2	Cub.	T_h^6—$Pa3$	FeS_2	5.88	—	—	—	—	[11]	1968	<298
	Monocl.	—	—	—	—	—	$\beta = 93.4°$	—	[11]	1968	—
	Tricl.	—	—	8.42	3.94	11.84	$\beta = 92.5°$ $\gamma = 89.9°$	—	[11]	1968	>720 K
SrC_2	Tetr.	D_{4h}^{17}—$I4/mmm$	CaC_2	4.11	—	6.68	—	1.63	[125]	1962	
BaC_2	•	D_{4h}^{17}—$I4/mmm$	CaC_2	4.40	—	7.06	—	1.60	[124]	1962	<423 K
	Cub.	T_h^6—$Pa3$	FeS_2	6.55	—	—	—	—	[125]	1962	High temp.
$Sc_{2-3}C$	•	O_h^5—$Fm3m$	$NaCl$	4,67—4,72	—	—	—	—	[89]	1967	
ScC	•	O_h^5—$Fm3m$	$NaCl$	4,539	—	—	—	—	[91]	1962	
	Hex.	—	—	5.46	—	10.24	—	1.88	[95]	1959	
$Sc_{15}C_{19}$	Tetr.	D_{2d}—$P\bar{4}2_1C$	$Sc_{15}C_{19}$	7,50	—	15.00	—	2.00	[90]	1971	
Sc_2C_3	Cub.	T_d^6—$I\bar{4}3d$	Pu_2C_3	7.205	—	—	—	—	[89]	1967	
Y_2C	Rhombo-hed.	D_{3d}^5—$R\bar{3}m$	$CdCl_2$	3.167	—	17.96	—	4.695	[42]	1969	
YC	Cub.	O_h^5—$Fm3m$	$NaCl$	4.99	—	—	—	—	[92]	1962	
Y_2C_3	•	T_d^6—$I\bar{4}3d$	Pu_2C_3	8.2142	—	—	—	—	[94]	1958	See [93]
$Y_{15}C_{19}$	Tetr.	D_{2d}—$P\bar{4}2_1C$	$Sc_{15}C_{19}$	7.94	—	15.00	—	1.89	[42]	1969	
YC_2	•	D_{4h}^{17}—$I4/mmm$	CaC_2	3.664	—	6.169	—	1.684	[94]	1958	
La_2C_3	Cub.	T_d^6—$I\bar{4}3d$	Pu_2C_3	8.8034	—	—	—	1.67	[94]	1958	See [93]
LaC_2	Tetr.	D_{4h}^{17}—$I4/mmm$	CaC_2	3.92	—	6.56	—	—	[94]	1958	
	Cub.	T_h^8—$Pa3$	FeS_2	6.0	—	—	—	—	[42]	1969	High temp.
CeC	•	O_h^5—$Fm3m$	$NaCl$	5.13	—	—	—	—	[125]	1962	

CRYSTAL STRUCTURE (continued)

Phase	Unit cell	Space group	Structural type	a, Å	b, Å	c, Å	α	c/a	Source	Year	Notes
1	2	3	4	5	6	7	8	9	10	11	12
Ce_2C_3	Cub.	D_{4h}^{17}—I4/mmm	—	5.22	—	6,21	—	1.19	[95]	1959	
	″	T_d^6—$I\bar{4}3d$	Pu_2C_3	8.4476	—	—	—	—	[94]	1958	
CeC_2	Tett.	D_{4h}^{17}—I4/mmm	CaC_2	5,49	—	6.49	—	1,18	[95]	1959	
Pr_2C_3	Cub.	T_d^6—$I\bar{4}3d$	Pu_2C_3	8,5731	—	—	—	—	[94]	1958	
PrC_2	Tett.	D_{4h}^{17}—I4/mmm	CaC_2	3.5	—	6.42	—	1,67	[94]	1958	
Nd_2C_3	Cub.	T_d^6—$I\bar{4}3d$	Pu_2C_3	8.5207	—	—	—	—	[94]	1958	
NdC_2	Tett.	D_{4h}^{17}—I4/mmm	CaC_2	3,2	—	6.37	—	1.67	[94]	1958	
Sm_3C	Cub.	O_h^1—Pm3m	Fe_4N	5,172	—	—	—	—	[94]	1958	
Sm_2C_3	″	T_d^6—$I\bar{4}3d$	Pu_2C_3	8.3989	—	—	—	—	[94]	1958	
SmC_2	Tett.	D_{4h}^{17}—I4/mmm	CaC_2	3.76	—	6,29	—	1.676	[94]	1958	
Eu_3C	Cub.	O_h^1—Pm3m	Fe_4N	5.14	—	—	—	—	[42]	1969	
EuC_2	Tett.			4.215	—	5.96	—	1.414	[42]	1969	
Gd_3C	Cub.	O_h^1—Pm3m	Fe_4N	5.126	—	—	—	—	[94]	1958	
Gd_2C_3	″	T_d^6—$I\bar{4}3d$	Pu_2C_3	8.3221	—	—	—	—	[94]	1958	
GdC_2	Tett.	D_{4h}^{17}—I4/mmm	CaC_2	3.71	—	6.275	—	1.688	[94]	1958	
	″	D_{4h}^{17}—I4/mmm		5.25	—	6.27	—	1,19	[95]	1959	
Tb_3C	Cub.	O_h^1—Pm3m	Fe_4N	5.107	—	—	—	—	[94]	1958	
Tb_2C	Rhombo-hed.	D_{3d}^5—$R\bar{3}m$	$CdCl_2$	3.595	—	18.19	—	5.060	[42]	1969	
Tb_2C_3	Cub.	T_d^6—$I\bar{4}3d$	Pu_2C_3	8.2434	—	—	—	—	[94]	1958	
TbC_2	Tett.	D_{4h}^{17}—I4/mmm	CaC_2	3.690	—	6.217	—	1.685	[94]	1958	

Carbide	System	Space group	Structure type	a, Å	b, Å	c, Å	α	c/a	References	Year	Notes
Dy₃C	Cub.	O_h^1-Pm3m	Fe₄N	5.079	—	—	—	—	[94]	1958	
Dy₂C₃	»	$T_d^6-I\bar43d$	Pu₂C₃	8.198	—	—	—	—	[94]	1958	
DyC₂	Tetr.	$D_{4h}^{17}-I4/mmm$	CaC₂	3.669	—	6.178	—	1.383	[94]	1958	
	»	$D_{4h}^{17}-I4/mmm$	—	5,15	—	6.15	—	1.19	[95]	1959	
Ho₃C	Cub.	O_h^1-Pm3m	Fe₄N	5.061	—	—	—	—	[94]	1958	
Ho₂C	Rhombohed.	$D_{3d}^5-R\bar3m$	CdCl₂	6.248	—	—	33°04'	—	[95]	1966	
Ho₂C₃	»	$T_d^6-I\bar43d$	Pu₂C₃	5.0729	—	—	89°30'	—	[97]	1964	
HoC₂	Tetr.	$D_{4h}^{17}-I4/mmm$	CaC₂	3.643	—	6.139	—	1.685	[94]	1958	
	»	$D_{4h}^{17}-I4/mmm$	—	5.13	—	6.11	—	1.19	[95]	1958	
Er₃C	Cub.	O_h^1-Pm3m	Fe₄N	5.034	—	—	—	—	[94]	1958	
Er₁₅C₁₉	Tetr.	$D_{2d}^4-P\bar42_1c$	Sc₁₅C₁₉	7.989	—	15.794	—	1.977	[98]	1974	
ErC₂	»	$D_{4h}^{17}-I4/mmm$	CaC₂	3.620	—	6.094	—	1.687	[94]	1958	
	»	$D_{4h}^{17}-I4/mmm$	—	5.11	—	6.11	—	1.20	[95]	1959	
Tm₃C	Cub.	O_h^1-Pm3m	Fe₄N	5.016	—	—	—	—	[94]	1958	
TmC₂	Tetr.	$D_{4h}^{17}-I4/mmm$	CaC₂	3.600	—	6.047	—	1.690	[94]	1958	
Yb₃C	Cub.	O_h^1-Pm3m	Fe₄N	4.993	—	—	—	—	[94]	1958	
YbC₂	Tetr.	$D_{4h}^{17}-I4/mmm$	CaC₂	2.637	—	6.109	—	1.680	[94]	1958	
	»	$D_{4h}^{17}-I4/mmm$	—	5.12	—	6.09	—	1.19	[95]	1959	
Lu₃C	Cub.	O_h^1-Pm3m	Fe₄N	4.965	—	—	—	—	[94]	1958	
LuC₂	Tetr.	$D_{4h}^{17}-I4/mmm$	CaC₂	3.563	—	5.964	—	1.674	[94]	1958	
ThC	Cub.	O_h^5-Fm3m	NaCl	5.338	—	—	—	—	[42]	1969	With the composition $ThC_{0.76}$ superstructure [99]

CRYSTAL STRUCTURE (continued)

Phase	Unit cell	Space group	Structural type	a, Å	b, Å	c, Å	α	c/a	Source	Year	Notes
1	2	3	4	5	6	7	8	9	10	11	12
ThC_2	Monocl.	C_{2h}^6—$B2/b$	ThC_2	6.691	4.231	6.744	103° 50'	—	[42]	1969	
UC	Cub.	O_h^5—$Fm3m$	NaCl	4,961	—	—	—	—	[100]	1963	See [101]
U_2C_3	"	T_d^6—$I\bar{4}3d$	Pu_2C_3	8.089	—	—	—	—	[101]	1968	
α-UC_2	Tetr.	D_{4h}^{17}—$I4/mmm$	CaC_2	3,517	—	5.987	—	1.7	[42]	1969	
β-UC_2	Cub.	O_h^5—$Fm3m$	CaF_2	5,46	—	—	—	—	[42]	1969	
PuC	"	O_h^5—$Fm3m$	NaCl	4,989	—	—	—	—	[101]	1968	
Pu_2C_3	"	T_d^6—$I\bar{4}3d$	Pu_2C_3	8.1260	—	—	—	—	[101]	1968	
PuC_2	Tetr.	D_{4h}^{17}—$I4/mmm$	CaC_2	3,63	—	6.094	—	1.679	[42]	1969	
NpC	Cub.	O_h^5—$Fm3m$	NaCl	5,0026	—	—	—	—	[101]	1968	
Np_2C_3	"	T_d^6—$I\bar{4}3d$	Pu_2C_3	8.1023	—	—	—	—	[101]	1968	
TiC	"	O_h^5—$Fm3m$	NaCl	4,3178	—	—	—	—	[11]	1968	
ZrC	"	O_h^5—$Fm3m$	NaCl	4,6828	—	—	—	—	[11]	1968	
HfC	Cub.	O_h^5—$Fm3m$	NaCl	4,6395	—	—	—	—	[11]	1968	
V_2C	Hex.	D_{3d}^3—$C\bar{3}m$	Mo_2C	2.884	—	4.587	—	1.584	[11]	1968	See [102, 104]
VC	Rhomb.	$P3_12$	ζ-Fe_2N	11,49	10.06	4.55	—	—	[42]	1969	
	—	—	—	4.555	5.745	5.037	—	—	[103]	1966	
	Cub.	O_h^5—$Fm3m$	NaCl	4.118	—	—	—	—	[102]	1968	For compositions V_8C_7 and V_6C_5

Compound	System	Space group	Structure type	a		c		ratio	Ref.	Year	super-structures [104, 105]
Nb_2C	Hex.	D_{6h}^4—$P6_3/mmc$	Mo_2C	3.126	—	4.965	—	1.588	[102]	1968	
»		$P3_12$	ζ-Fe_2N	5.408	—	4.965	—	0.915	[102]	1968	
NbC	Cub.	O_h^5—$Fm3m$	$NaCl$	4.433	—	—	—	—	[102]	1968	
Ta_2C	Hex.	$Pbcm$	Mo_2C	3.101	—	4.933	—	1.591	[102]	1968	
	Rhomb.	$P3_12$	ζ-Fe_2N	4.933	9.866	8.534	—	—	[102]	1968	
	Rhombohed.	D_{3d}^5—$R\bar{3}m$	Sn_4P_3	—	—	—	—	—	[106]	1971	
TaC	Cub.	O_h^5—$Fm3m$	$NaCl$	4.410	—	—	—	—	[102]	1968	Above 2200° C
(CrC)	»	O_h^5—$Fm3m$	$NaCl$	3.62	—	—	—	—	[102]	1968	
$Cr_{23}C_6$	»	O_h^5—$Fm3m$	$Cr_{23}C_6$	10.656	—	—	—	—	[76]	1967	See [107]
Cr_7C_3	Rhomb.	D_{2h}^{16}—$Pnma$	—	4.526	7.010	12.142	—	—	[108]	1969	
Cr_3C_2	»	D_{2h}^{16}—$Pnma$	—	5.5329	2.829	11.4719	—	—	[109]	1969	
α-MoC	Cub.	O_h^5—$Fm3m$	$NaCl$	4.321	—	—	—	—	[102]	1968	Above 2200°C
γ-MoC	Hex.	$P\bar{6}m2$	WC	2.898	—	2.809	—	0,969	[102]	1968	
Mo_2C	»	D_{6h}^4—$P6_3/mmc$	Mo_2C	2.997	—	4.727	—	1,54	[102]	1968	
Mo_2C	Rhomb.	D_{2h}^{14}—$Pbcn$	ζ-Fe_2N	4.727	9.454	8.178	—	—	[102]	1968	
W_2C	Hex.	—	ε-Fe_2N	5.814	—	4.721	—	0.911	[110]	1968	
β-W_2C	Cub.	O_h^5—$Fm3m$	$NaCl$	4.222	—	—	—	—	[102]	1968	
WC	Hex.	$P\bar{6}m2$	WC	2.906	—	2.837	—	0.975	[102]	1968	
α-W_2C	»	D_{6h}^4—$P6_3/mmc$	Mo_2C	3.001	—	4.736	—	1.578	[102]	1968	
	Rhomb.	D_{2h}^{14}—$Pbcn$	ζ-Fe_2N	4.736	9.472	8.193	—	—	[102]	1968	
$Mn_{23}C_6$	Cub.	O_h^5—$Fm3m$	$Cr_{23}C_6$	10.598	—	—	—	—	[76]	1967	See [111]

CRYSTAL STRUCTURE (continued)

Phase	Unit cell	Space group	Structural type	a, Å	b, Å	c, Å	α	c/a	Source	Year	Notes
1	2	3	4	5	6	7	8	9	10	11	12
Mn4C	Tetr.	—	—	7.66	—	10.57	—	1.38	[111]	1967	
Mn15C4	Hex.	—	—	7.4920	—	12.070	—	1.61	[76]	1967	See [112]
Mn3C	Rhomb.	D_{2h}^{16}–$Pbmn$	Fe3C	4.545	5.103	6.787	—	—	[76]	1967	See [111]
Mn8C3	Triclin.	—	—	—	—	—	—	—	[111]	1967	
	Rhomb.	—	—	13.9	—	4.56	—	0.327	[111]	1967	
	Hex.	—	—	—	—	—	—	—	[111]	1967	
Mn5C2	Monoclin.	C_{2h}^{6}–$C2/C$	Pd5B2	11.673	4.583	5.094	97° 41'	—	[76]	1967	See [111]
Mn7C3	Rhomb.	D_{2h}^{16}–$Pnma$	—	4.546	6.959	11.976	—	—	[76]	1967	See [107]
TcC	Cub.	—	—	3.982	—	—	—	—	[268]	1973	
Fe3C	Rhomb.	D_{2h}^{16}–$Pbmn$	Fe3C	4.5235	5.089	6.7353	—	—	[112]	1964	
Fe20C9	Monoclin.	—	Pd5B2	11.562	4.57	5.06	97.74°	—	[42]	1969	
Fe2C	Hex.	D_{6h}^{4}–$P6_3/mmc$	Mo2C	2.755	—	4.349	—	1.579	[125]	1962	
	Monoclin.	—	—	2.794	2.794	4.360	120.92°	—	[42]	1969	
Fe5C2	Monoclin.	C_{2h}^{6}–$C2/C$	Pd5B2	11.563	4.573	5.058	97° 45'	—	[76]	1967	
Fe7C3	Rhomb.	D_{2h}^{16}–$Pnma$	—	4.537	6.892	11.913	—	—	[76]	1967	
Co3C	Rhomb.	D_{2h}^{16}–$Pbmn$	Fe3C	4.53	5.09	6.74	—	—	[112]	1964	
Co2C	Rhomb.	—	Co2C	2.891	4.463	4.369	—	—	[125]	1962	
Ni3C	Hex.	—	—	2.631	—	4.314	—	1.640	[125]	1962	
ReC	Hex.	D_{6h}^{4}–$P6_3/mmc$	γ'-MoC	2.840	—	9.85	—	3.47	[113]	1971	Synthesized when p ≥ 60 kbar

Compound	System	Space group	Structure type	a	b	c	angle	c/a	Ref.	Year	Temp.
α-Be$_3$N$_2$	Cub.	T_h^7—$Ia3$	Mn$_2$O$_3$	8.15	—	—	—	—	[88]	1972	<1400° C
β-Be$_3$N$_2$	Hex.	—	Mn$_2$O$_3$	2.8413	—	9.693	—	3.42	[88]	1972	>1400° C
Mg$_3$N$_2$	Cub.	T_h^7—$Ia3$	CdCl$_2$	9.97	—	—	—	—	[14]	1969	
Ca$_2$N	Rhombohed.	D_{3d}^5—$R\bar3m$	—	—	—	—	—	—	[115]	1972	
α-Ca$_3$N$_2$	Pseudotetra.	—	Mn$_2$O$_3$	3.56	—	4.12	—	1.157	[14]	1969	<700° C
β-Ca$_3$N$_2$	Cub.	T_h^7—$Ia3$	—	11.42	—	—	—	—	[14]	1969	700—1050° C
γ-Ca$_3$N$_2$	Rhomb.	—	CdCl$_2$	—	—	—	30.83°	—	[14]	1969	>1050° C
Sr$_2$N	Rhombohed.	D_{3d}^5—$R\bar3m$	—	7.246	—	—	—	—	[115]	1972	
Sr$_3$N$_2$	Pseudotetra.	—	—	—	—	—	—	—	[14]	1969	
SrN	Cub.	O_h^5—$Fm3m$	NaCl	11.82	11.47	6.08	—	—	[115]	1972	
SrN$_6$	Rhomb.	—	—	—	—	—	—	—	[14]	1969	
Ba$_3$N$_2$	Pseudohex.	—	—	—	—	—	—	—	[14]	1969	
BaN$_6$	Monoclin.	—	—	6.22	22.99	7.02	105° 14'	—	[14]	1969	
ScN	Cub.	O_h^5—$Fm3m$	NaCl	4.45	—	—	—	—	[42]	1969	
YN	"	O_h^5—$Fm3m$	NaCl	4.877	—	—	—	—	[42]	1969	
LaN	Cub.	O_h^5—$Fm3m$	NaCl	5.30	—	—	—	—	[134]	1956	
CeN	"	O_h^5—$Fm3m$	NaCl	5.02	—	—	—	—	[42]	1969	
PrN	"	O_h^5—$Fm3m$	NaCl	5.16	—	—	—	—	[42]	1969	
NdN	"	O_h^5—$Fm3m$	NaCl	5.15	—	—	—	—	[42]	1969	
SmN	"	O_h^5—$Fm3m$	NaCl	5.046	—	—	—	—	[134]	1956	
EuN	"	O_h^5—$Fm3m$	NaCl	5.014	—	—	—	—	[134]	1956	
GdN	"	O_h^5—$Fm3m$	NaCl	4.999	—	—	—	—	[134]	1956	
TbN	"	O_h^5—$Fm3m$	NaCl	4.933	—	—	—	—	[134]	1956	

CRYSTAL STRUCTURE (continued)

Phase	Unit cell	Space group	Structural type	a, Å	b, Å	c, Å	α	c/a	Source	Year	Notes
1	2	3	4	5	6	7	8	9	10	11	12
DyN	"	O_h^5–$Fm3m$	NaCl	4.905	—	—	—	—	[134]	1956	
HoN	"	O_h^5–$Fm3m$	NaCl	4.874	—	—	—	—	[134]	1956	
ErN	"	O_h^5–$Fm3m$	NaCl	4.839	—	—	—	—	[134]	1956	
TmN	"	O_h^5–$Fm3m$	NaCl	4.809	—	—	—	—	[134]	1956	
YbN	"	O_h^5–$Fm3m$	NaCl	4.786	—	—	—	—	[134]	1956	
LuN	Cub.	O_h^5–$Fm3m$	NaCl	4.766	—	—	—	—	[134]	1956	
ThN	Cub.	O_h^5–$Fm3m$	NaCl	5.21	—	—	—	—	[125]	1962	
Th_3N_4	Hex.	—	—	3.871	—	27.385	—	7.074	[42]	1969	
	Rhombohed.	—	—	9.398	—	—	23.78°	—	[42]	196	
	Cub.	—	—	4.55	—	—	—	—	[125]	1962	
Th_2N_3	Hex.	D_{3d}^3–$P3m$	La_2O_3	3.883	—	6.187	—	1,593	[124]	1962	
UN	Cub.	O_h^5–$Fm3m$	NaCl	4.890	—	—	—	—	[101]	1968	
α-U_2N_3	"	T_h^7–$Ia3$	Mn_2O_3	10.678	—	—	—	—	[100]	1963	
β-U_2N_3	Hex.	—	—	3.69	—	5.83	—	1.58	[42]	1969	
UN_0	Cub.	O_h^5–$Fm3m$	CaF_2	5.31	—	—	—	—	[100]	1963	
NpN	"	O_h^5–$Fm3m$	NaCl	4.8979	—	—	—	—	[101]	1968	
PuN	"	O_h^5–$Fm3m$	NaCl	4.906	—	—	—	—	[101]	1968	
β-Ti_2N	Tett.	D_{4h}^{14}–$P4_2/mnm$	TiO_2	4.946	—	3.030	—	0.613	[116]	1969	
α-Ti_2N	"	$I4_1/amd$	—	4.140	—	8.805	—	2.13	[116]	1969	
TiN	Cub.	O_h^5–$Fm3m$	NaCl	4.249	—	—	—	—	[125]	1962	

Formula	System	Space group	Structure type	a	b	c	Angle	c/a	Ref.	Year	Remarks
ZrN	—	O_h^5—$Fm3m$	NaCl	4.537	—	—	—	—	[125]	1962	
Hf_3N	Rhombohed.	D_{3d}^5 or $R\bar{3}m$	—	7.972	—	—	23° 12′	—	[117]	1973	
HfN	Cub.	O_h^5—$Fm3m$	NaCl	4.50	—	—	—	—	[117]	1973	β
V_3N	Hex.	—	—	2.835	—	4.541	—	1.60	[42]	1969	
VN	Cub.	O_h^5—$Fm3m$	NaCl	4.136	—	—	—	—	[125]	1962	
Nb_2N	Hex.	C_{6v}^4—$P6_3mc$	ZnS	3.056	—	4.995	—	1.637	[16]	1969	γ
$NbN_{0.79}$	Tetr.	—	—	4.386	—	4.335	—	0.989	[16]	1969	
$NbN_{0.98}$	Cub.	O_h^5—$Fm3m$	NaCl	4.392	—	—	—	—	[16]	1969	δ
$NbN_{0.95}$	Hex.	D_{3h}^1—$C\bar{6}m2$	$B8_1$	2.968	—	5.549	—	1.865	[16]	1969	δ
NbN	"	—	γ-MoC	2.959	—	11.271	—	3.804	[16]	1969	ε
Ta_2N	"	C_{6v}^4—$P6_3mc$	ZnS	3.042	—	4.909	—	1.61	[15]	1972	γ
TaN	Cub.	O_h^5—$Fm3m$	NaCl	4.344	—	—	—	—	[15]	1972	δ
TaN	Hex.	D_{3h}^1—$P6m2$	WC	2.933	—	2.864	—	0.98	[15]	1972	Formed under pressure
TaN	Hex.	D_{6h}^1—$P6/mmm$	CoSn	5.183	—	2.907	—	0.56	[15]	1972	Stabilized with oxygen, see [16]
Ta_5N_6	"	D_{6h}^3—$P6_3/mcm$	—	5.213	—	10.426	—	2.0	[15]	1972	ε
Ta_4N_5	Tetr.	C_{4h}^5—$I4/m$	—	6.840	—	4.241	—	0.62	[15]	1972	See [16], [118]
Ta_3N_5	"	—	—	10.252	—	3.896	—	0.38	[15]	1972	See [16], [118]
Cr_2N	Hex.	—	—	4.806	—	4.479	—	0.928	[42]	1969	
CrN	Cub.	O_h^5—$Fm3m$	NaCl	4.148	—	—	—	—	[42]	1969	
CrN	Hex.	—	—	2.756	—	4.424	—	1.605	[124]	1962	

CRYSTAL STRUCTURE (continued)

Phase	Unit cell	Space group	Structural type	a, Å	b, Å	c, Å	α	c/a	Source	Year	Notes
1	2	3	4	5	6	7	8	9	10	11	12
Mo_3N	Tetr.	—	—	4.188	—	4.024	—	0,961	[125]	1962	
Mo_2N	Cub.	$O_h^5 - Fm3m$	NaCl	4.168	—	—	—	—	[119]	1961	
$Mo_{16}N_7$	Tetr.	—	—	8.41	—	8.05	—	0,955	[120]	1970	
MoN	Hex.	$D_{3h}^3 - C\bar{6}m2$	γ-MoC	5.725	—	5.608	—	0,980	[42]	1969	
W_2N	Cub.	$O_h^5 - Fm3m$	NaCl	4.128	—	—	—	—	[124]	1962	
WN	Hex.	$D_{3h}^1 - P\bar{6}m2$	WC	2.893	—	2.826	—	0,977	[42]	1969	
Mn_4N	Cub.	$O_h^1 - Pm3m$	Fe_4N	3.860	—	—	—	—	[42]	1969	
Mn_2N	Hex.	—	—	2.773	—	4.520	—	1,630	[42]	1969	
Mn_3N_2	Tetr.	—	—	4,20	—	4.03	—	0,97	[42]	1969	
Mn_6N_5	"	—	—	4.214	—	4.148	—	0,984	[42]	1969	
MnN	Cub.	$O_h^5 - Fm3m$	NaCl	4.435	—	—	—	—	[125]	1962	
TcN	"	—	—	3.980	—	—	—	—	[73]	1964	
Re_2N	Cub.	$O_h^5 - Fm3m$	NaCl	3.92	—	—	—	—	[121]	1964	
Fe_4N	"	$O_h^1 - Pm3m$	Fe_4N	3.790	—	—	—	—	[111]	1967	
Fe_3N	Hex.	$D_6^6 - P6_222$	ε-Fe_3N	2.660	—	4.343	—	1,633	[125]	1962	
Fe_2N	Rhomb.	$D_6^6 - P6_222$	Co_2N	2.764	4.425	4.829	—	—	[125]	1962	
Co_3N	Hex.	$D_6^6 - P6_222$	ε-Fe_3N	2.663	—	4.360	—	1,637	[125]	1962	
Co_2N	Rhomb.	$D_6^6 - P6_222$	Co_2N	2.848	4.636	4.339	—	—	[125]	1962	
Ni_3N	Hex.	$D_6^6 - P6_222$	ε-Fe_3N	2.667	—	4.312	—	1,616	[125]	1962	
Os_2N	Cub.	—	—	3.45	—	—	—	—	[267]	1970	
$CaAl_4$	Tetr.	$D_{4h}^{17} - I4/mmm$	$BaAl_4$	4.358	—	11.092	—	2,545	[124]	1962	

				a	b	c					
CaAl$_2$	Cub.	O_h^7 — $Fd3m$	Cu$_2$Mg	8,038	—	—	—	—	[125]	1962	
SrAl$_4$	Tetr.	D_{4h}^{17} — $I4/mmm$	BaAl$_4$	4.459	—	11.07	—	2.482	[125]	1962	
SrAl	Cub.	O_h^1 — $Pm3m$	CsCl	15.83	—	—	—	—	[124]	1962	
BaAl$_4$	Tetr.	D_{4h}^{17} — $I4/mmm$	BaAl$_4$	4.540	—	11.16	—	2.451	[124]	1962	
BaAl	Hex.	—	—	6.01	—	17.78	—	2.958	[125]	1962	
ScAl$_2$	Cub.	O_h^7 — $Fd3m$	Cu$_2$Mg	7.58	—	—	—	—	[122]	1964	
ScAl$_3$	"	O_h^1 — $Pm3m$	Cu$_3$Au	4.10	—	—	—	—	[122]	1964	
Y$_2$Al	Rhomb.		Er$_2$Al	7.62	9.22	11.14	—	—	[135]	1965	
Y$_3$Al$_2$	Tetr.	D_{4h}^4 — $P4_2/mmm$	Zr$_3$Al$_2$	8,239	—	7.648	—	0.928	[135]	1965	
YAl	Rhomb.	D_{2h}^{17} — $Cmcm$	TlI	11.52	3.88	4.38	—	—	[42]	1969	
YAl$_2$	Cub.	O_h^7 — $Fd3m$	Cu$_2$Mg	7.843	—	—	—	3.41	[42]	1969	
YAl$_3$	Hex.	—	BaPb$_3$	6.194	—	21.138	46° 8'	—	[42]	1969	
	Rhombohed.	—	BaPb$_3$	7,901	—	—	—	—	[136]	1964	
YAl$_4$	Hex.	D_{6h}^4 — $P6_3/mmc$	Mg$_3$Cd	6,276	—	4.582	—	0.730	[42]	1969	
La$_3$Al	Tetr.	D_{4h}^{17} — $I4/mmm$	BaAl$_4$	4.17	—	9.77	—	2.34	[42]	1969	
	Hex.	D_{6h}^4 — $P6_3/mmc$	CdMg$_3$	7.195	—	5.503	—	0.765	[137]	1965	
	Cub.	O_h^1 — $Pm3m$	Cu$_3$Au	5.093	—	—	—	—	[42]	1969	
LaAl	Rhomb.	D_{2h}^{17} — $Cmcm$	CeAl	9.531	7.734	5.809	—	—	[137]	1965	
LaAl$_2$	Cub.	O_h^7 — $Fd3m$	Cu$_2$Mg	8.153	—	—	—	—	[137]	1965	
LaAl$_{2,4}$	Hex.	D_{6h}^1 — $P6/mmm$	AlB$_2$	4.478	—	4.347	—	0.97	[137]	1965	
LaAl$_4$	Tetr.	D_{4h}^{17} — $I4/mmm$	BaAl$_4$	4.431	—	10.23	—	2.309	[136]	1964	>1086 K
	Rhomb.		α-La$_3$Al$_{11}$	4.431	13.14	10.132	—	—	[137]	1965	<1086 K
LaAl$_3$	Hex.	D_{6h}^4 — $P6_3/mmc$	Mg$_3$Cd	6.662	—	4.609	—	0.691	[137]	1965	<928 K
Ce$_3$Al	"		Ni$_3$Sn	7,043	—	5.451	—	0.774	[42]	1969	>503 K
	Cub.	O_h^1 — $Pm3m$	Cu$_3$Au	4.985	—	—	—	—	[42]	1969	
CeAl	Rhomb.	D_{2h}^{17} — $Cmcm$	CeAl	9.270	7.680	5.760	—	—	[123]	1967	<1118 K

CRYSTAL STRUCTURE (continued)

Phase	Unit cell	Space group	Structural type	a, Å	b, Å	c, Å	α	c/a	Source	Year	Notes
1	2	3	4	5	6	7	8	9	10	11	12
$CeAl$	Cub.	O_h^1—$Pm3m$	CsCl	3.86	—	—	—	—	[42]	1969	
$CeAl_2$	"	O_h^7—$Fd3m$	Cu_2Mg	8.059	—	—	—	—	[138]	1960	
$CeAl_3$	Hex.	D_{6h}^4—$P6_3/mmc$	Mg_3Cd	6.541	—	4.610	—	0.705	[42]	1969	
$CeAl_4$	Tetr.	D_{4h}^{17}—$I4/mmm$	$BaAl_4$	4.374	—	10.12	—	2.31	[136]	1964	1293—1508 K
	Rhomb.	—	α-La_3Al_{11}	4.395	13.025	10.092	—	—	[139]	1966	
Pr_3Al	Hex.	D_{6h}^4—$P6_3/mmc$	$CdMg_3$	7.02	—	5.43	—	0.773	[164]	1969	<903 K
Pr_3Al	Cub.	O_h^1—$Pm3m$	Cu_3Au	4.950	—	—	—	—	[42]	1969	
Pr_2Al	Rhomb.	D_{2h}^{11}—$Pbcm$	Er_2Al	7.872	9.461	11.45	—	—	[139]	1966	<1008 K
$PrAl$	"	D_{2h}^{17}—$Cmcm$	DyAl	5.97	11.77	5.73	—	—	[123]	1967	<1178 K
	"	—	CeAl	9.22	7.63	5.71	—	—	[123]	1967	
$PrAl_2$	Cub.	O_h^7—$Fd3m$	Cu_2Mg	8.035	—	—	—	—	[138]	1960	
$PrAl_3$	Hex.	D_{6h}^4—$P6_3/mmc$	Mg_3Cd	6.504	—	4.604	—	0.7078	[42]	1969	
$PrAl_4$	Tetr.	D_{4h}^{17}—$I4/mmm$	$BaAl_4$	4.360	—	10.010	—	2.29	[136]	1964	<1238 K
	Rhomb.	—	α-La_3Al_{11}	4.446	12.949	10.005	—	—	[139]	1966	1238—1513 K
Nd_3Al	Hex.	—	Ni_3Sn	6.968	—	5.407	—	0.776	[141]	1965	
Nd_2Al	Rhomb.	—	Er_2Al	7.848	9.395	11.38	—	—	[141]	1965	
$NdAl$		D_{2h}^{11}—$Pbcm$	DyAl	5.940	11.728	5.729	—	—	[141]	1965	
	Cub.	O_h^1—$Pm3m$	CsCl	3.738	—	—	—	—	[124]	1962	
$NdAl_2$	"	O_h^7—$Fd3m$	Cu_2Mg	8.000	—	—	—	—	[138]	1960	

Compound	System	Space group	Structure type	a	b	c	angle	c/a	Ref.	Year	Notes
$NdAl_3$	Hex.	D_{6h}^{4}—$P6_3/mmc$	Mg_3Cd	6.472	—	4.606	—	0.7117	[141]	1965	
$NdAl_4$	Tetr.	D_{4h}^{17}—$I4/mmm$	$BaAl_4$	4.338	—	9.996	—	2.304	[144]	1961	<1083 K
Sm_3Al	Rhomb.	—	$\alpha\text{-}La_3Al_{11}$	4.359	12.924	10.017	—	—	[141]	1965	
Sm_2Al	Cub.	O_h^{1}—$Pm3m$	Cu_3Au	4.901	—	—	—	—	[145]	1966	
	Rhomb.	—	Er_2Al	7.782	9.302	11.21	—	—	[42]	1969	
$SmAl$	″	D_{2h}^{11}—$Pbcm$	$DyAl$	5.678	11.622	5.899	—	—	[140]	1965	See [123]
	Cub.	O_h^{1}—$Pm3m$	$CsCl$	3.739	—	—	—	—	[42]	1969	
$SmAl_2$	″	O_h^{7}—$Fd3m$	Cu_2Mg	7.940	—	—	—	—	[42]	1969	
$SmAl_3$	Hex.	D_{6h}^{4}—$P6_3/mmc$	Mg_3Cd	6.380	—	4.597	—	0.720	[42]	1969	
$SmAl_4$	Tetr.	D_{4h}^{17}—$I4/mmm$	$BaAl_4$	4.287	—	9.905	—	2.310	[42]	1969	
Gd_2Al	Rhomb.	—	Er_2Al	7.69	9.24	11.21	—	—	[141]	1965	
Gd_3Al_2	Tetr.	D_{4h}^{4}—$P4_2/mnm$	Zr_3Al_2	8.329	—	7.578	—	0.909	[141]	1965	
$GdAl$	Rhomb.	D_{2h}^{11}—$Pbcm$	$DyAl$	5.888	11.527	5.656	—	—	[141]	1965	
$GdAl$	″	D_{2h}^{17}—$Cmcm$	$CeAl$	9.274	7.679	5.584	—	—	[42]	1969	
	Cub.	O_h^{1}—$Pm3m$	$CsCl$	3.721	—	—	—	—	[42]	1969	
$GdAl_2$	″	O_h^{7}—$Fd3m$	Cu_2Mg	7.900	—	—	—	—	[136]	1969	
$GdAl_3$	Hex.	D_{6h}^{4}—$P6_3/mmc$	Mg_3Cd	6.320	—	4.592	—	0.726	[141]	1964	
$GdAl_4$	Rhomb.	D_{2h}^{28}—$Imma$	UAl_4	4.442	6.316	13.739	—	—	[140]	1965	
Tb_3Al_2	Tetr.	D_{4h}^{4}—$P4_2/mnm$	Zr_3Al_2	8.255	—	7.568	—	0.916	[140]	1965	
$TbAl$	Rhomb.	D_{2h}^{11}—$Pbcm$	$DyAl$	5.834	11.370	5.621	—	—	[138]	1965	
$TbAl_2$	Cub.	O_h^{7}—$Fd3m$	Cu_2Mg	7.867	—	—	—	—	[136]	1960	
$TbAl_3$	Hex.	D_{6h}^{4}—$P6_3/mmc$	$BaPb_3$	6.175	—	21.18	—	3.43	[136]	1964	
	Rhombohed.	—	$BaPb_3$	7.910	—	—	46°	—	[136]	1964	
$TbAl_4$	Rhomb.	D_{2h}^{28}—$Imma$	UAl_4	4.430	6.261	13.706	—	—	[136]	1964	
Dy_3Al_2	Tetr.	D_{4h}^{4}—$P4_2/mnm$	Zr_3Al_2	8.170	—	7.523	—	0,921	[140]	1965	
$DyAl$	Rhomb.	D_{2h}^{11}—$Pbcm$	$DyAl$	5.80	11.44	5.62	—	—	[123]	1967	

CRYSTAL STRUCTURE (continued)

Phase	Unit cell	Space group	Structural type	a, Å	b, Å	c, Å	α	c/a	Source	Year	Notes
1	2	3	4	5	6	7	8	9	10	11	12
$DyAl$	Cub.	O_h^1—$Pm3m$	$CsCl$	3.6826	—	—	—	—	[42]	1969	
$DyAl_2$	"	O_h^7—$Fd3m$	Cu_2Mg	7.840	—	—	—	—	[138]	1960	
$DyAl_3$	Hex.	D_{6h}^4—$P6_3/mmc$	Ni_3Ti	6.097	—	9.534	—	1.564	[42]	1969	
	Rhombohed.	—	$HoAl_3$	6.080	—	35.940	—	5.91	[136]	1964	
		—	$HoAl_3$	12.483	—	—	28° 11′	—	[136]	1964	
Ho_2Al	Rhomb.	—	Er_2Al	7.59	9,14	11.00	—	—	[42]	1969	
Ho_3Al_2	Tett.	D_{4h}^4—$P4_2/mnm$	Zr_3Al_2	8.182	—	7.525	—	0.919	[140]	1965	
$HoAl$	Rhomb.	D_{2h}^{11}—$Pbcm$	$DyAl$	5.801	11.339	5.621	—	—	[140]	1965	
$HoAl_2$	Cub.	O_h^7—$Fd3m$	Cu_2Mg	7.813	—	—	—	—	[138]	1960	
$HoAl_3$	"	O_h^1—$Pm3m$	Cu_3Au	4.248	—	—	—	—	[42]	1969	
	Hex.	—	$HoAl_3$	6.052	—	35.93	—	5.93	[136]	1964	
	Rhombohed.	—	$HoAl_3$	12.476	—	—	28° 4′	—	[136]	1964	
Er_2Al	Rhomb.	—	Er_2Al	7.523	9.056	10.925	—	—	[135]	1965	
Er_3Al_2	Tetr.	D_{4h}^4—$P4_2/mnm$	Zr_3Al_2	8.123	—	7.620	—	0.938	[140]	1965	
$ErAl$	Rhomb.	D_{2h}^{11}—$Pbcm$	$DyAl$	5.801	11.272	5.570	—	—	[140]	1965	
$ErAl_2$	Cub.	O_h^7—$Fd3m$	Cu_2Mg	7.795	—	—	—	—	[135]	1965	
$ErAl_3$	Cub.	O_h^1—$Pm3m$	Cu_3Au	4.216	—	—	—	—	[42]	1969	
$EuAl_2$	"	O_h^7—$Fd3m$	Cu_2Mg	8.121	—	—	—	—	[42]	1969	
$EuAl_4$	Tetr.	D_{4h}^{17}—$I4/mmm$	$BaAl_4$	4.396	—	11.17	—	2.54	[42]	1969	
Yb_3Al_2	"	—	—	8.237	—	7.646	—	0.929	[18]	1965	

Compound	System	Space group	Structure type	a	b	c	angle	c/a	Ref.	Year	Temp.
YbAl$_2$	Cub.	O_h^7—$Fd3m$	Cu$_2$Mg	7.877	—	—	—	—	[142]	1960	—
YbAl$_3$	"	O_h^1—$Pm3m$	Cu$_3$Au	4.202	—	—	—	—	[136]	1964	—
TmAl$_3$	"	O_h^1—$Pm3m$	Cu$_3$Au	4.200	—	—	—	—	[42]	1969	—
TmAl	Rhomb.	D_{2h}^{11}—$Pbcm$	DyAl	5,77	5.56	11.24	—	—	[140]	1965	—
TmAl$_2$	Cub.	O_h^7—$Fd3m$	Cu$_2$Mg	7.770	—	—	—	—	[142]	1960	—
LuAl$_2$	"	O_h^7—$Fd3m$	Cu$_2$Mg	7.742	—	—	—	—	[142]	1960	—
Th$_3$Al	Hex.	—	—	6.500	—	4.626	—	0.711	[18]	1965	—
Th$_2$Al	Tetr.	D_{4h}^{18}—$I4/mcm$	CuAl$_2$	7.614	—	5.857	—	0.769	[42]	1969	—
Th$_3$Al$_2$	"	D_{4h}^{5}—$P4/mbm$	U$_3$Si$_2$	8.125	—	4,217	—	0.519	[42]	1969	—
ThAl	Rhomb.	D_{2h}^{17}—$Cmcm$	TlI	11.45	4.42	4.19	—	—	[42]	1969	—
ThAl$_2$	Hex.	D_{6h}^{1}—$P6/mmm$	AlB$_2$	4.393	—	4.164	—	0.947	[42]	1969	—
ThAl$_3$	"	D_{6h}^{4}—$P6_3/mmc$	Mg$_3$Cd	6.499	—	4.626	—	0.712	[42]	1969	—
UAl$_2$	Cub.	O_h^7—$Fd3m$	Cu$_2$Mg	7.811	—	—	—	—	[124]	1962	—
UAl$_3$	"	O_h^1—$Pm3m$	Cu$_3$Au	4.287	—	—	—	—	[143]	1962	—
UAl$_4$	Rhomb.	D_{2h}^{28}—$Imma$	UAl$_4$	4.41	6.27	13,71	—	—	[143]	1962	—
NpAl$_2$	Cub.	O_h^7—$Fd3m$	Cu$_2$Mg	7.785	—	—	—	—	[125]	1962	—
NpAl$_3$	"	O_h^1—$Pm3m$	Cu$_3$Au	4.262	—	—	—	—	[145]	1966	—
NpAl$_4$	Rhomb.	D_{2h}^{28}—$Imma$	UAl$_4$	4.42	6.26	13.71	—	—	[125]	1962	—
Pu$_3$Al	Tetr.	—	SrPb$_3$	4.499	—	4.536	—	1.008	[143]	1962	—
PuAl	Cub.	—	—	10.76	—	—	—	—	[143]	1962	—
PuAl$_2$	"	O_h^7—$Fd3m$	Cu$_2$Mg	7.848	—	—	—	3.443	[143]	1962	—
PuAl$_3$	Hex.	D_{6h}^{4}—$P6_3/mmc$	PuAl$_3$	6.150	—	21,175	—	—	[146]	1965	<1188 K
	Rhombohed.	—	—	7.879	—	—	45.94°	—	[42]	1969	1188—1300 K
PuAl$_3$	Hex.	—	—	6.084	—	14.427	—	2.371	[42]	1969	1300—1483 K
PuAl$_3$	Cub.	O_h^1—$Pm3m$	Cu$_3$Au	4.262	—	—	—	—	[146]	1965	<1483 K

CRYSTAL STRUCTURE (continued)

Phase	Unit cell	Space group	Structural type	a, Å	b, Å	c, Å	α	c/a	Source	Year	Notes
1	2	3	4	5	6	7	8	9	10	11	12
$PuAl_4$	Rhomb.	D_{2h}^{28}—$Imma$	UAl_4	4.387	6.262	13.714	—	—	[143]	1962	
Ti_3Al	Hex.	—	Ni_3Sn	11.543	—	4.659	—	0.404	[42]	1969	
$TiAl$	Tetr.	D_{4h}^{1}—$P4/mmm$	$CuAu$	4.105	—	4.0625	—	0.989	[42]	1969	
$TiAl_2$	"	D_{4h}^{19}—$I4_1/amd$	$HfGa_2$	3.976	—	24.36	—	6.127	[42]	1969	
$TiAl_3$	"	D_{4h}^{17}—$I4/mmm$	$TiAl_3$	5.435	—	8.591	—	1,580	[42]	1969	
$TiAl_3$	"	—	—	3.850	—	8.60	—	2.234	[145]	1966	
Zr_3Al	Cub.	O_h^{1}—$Pm3n$	Cu_3Au	4.372	—	—	—	—	[147]	1962	
Zr_2Al	Hex.	D_{6h}^{4}—$P6_3/mmc$	Ni_2In	4.894	—	5.928	—	1.211	[147]	1962	
Zr_5Al_3	Tetr.	D_{4h}^{18}—$I4/mcm$	$CuAl_2$	6.854	—	5.501	—	0.803	[19]	1961	
Zr_5Al_3	Hex.	D_{6h}^{3}—$P6_3/mcm$	Mn_5Si_3	8.184	—	5.702	—	0,697	[42]	1969	
Zr_5Al_3	Tetr.	D_{4h}^{18}—$I4/mcm$	W_5Si_3	11.049	—	5.396	—	0.488	[42]	1969	
Zr_3Al_2	"	D_{4h}^{4}—$P4_2/mnm$	Zr_3Al_2	7.630	—	6.998	—	0.917	[42]	1969	
Zr_4Al_3	Hex.	C_{3h}^{1}—$P\bar{6}$	Zr_4Al_3	5.430	—	5.389	—	0.992	[42]	1969	
$ZrAl$	Rhomb.	D_{2h}^{17}—$Cmcm$	TlI	10.887	3.359	4.274	—	—	[42]	1969	
Zr_2Al_3	"	C_{2v}^{19}—$Fdd2$	Zr_2Al_3	9.601	13.906	5.574	—	—	[42]	1969	
$ZrAl_2$	Hex.	D_{6h}^{4}—$P6_3/mmc$	Zn_2Mg	5.282	—	8.748	—	1.656	[42]	1969	
$ZrAl_3$	Tetr.	D_{4h}^{17}—$I4/mmm$	$ZrAl_3$	4.306	—	16.90	—	3.925	[147]	1962	
Hf_2Al	"	D_{4h}^{18}—$I4/mcm$	$CuAl_2$	6.785	—	5.385	—	0.794	[19]	1961	
Hf_3Al_2	"	D_{4h}^{4}—$P4_2/mnm$	Zr_3Al_2	7,535	—	6.906	—	0.916	[147]	1962	

Compound	System	Space group	Structure type	a	b	c	angle	c/a	Ref.	Year	Notes
Hf_4Al_3	Hex.	C_{3h}^1–$P\bar{6}$	Zr_4Al_3	5.343	—	5.422	—	1.015	[147]	1962	See [150]
$HfAl$	Rhomb.	D_{2h}^{17}–$Cmcm$	TlI	10.831	3.253	4.282	—	—	[42]	1969	
Hf_2Al_3	"	C_{2v}^{19}–$Fdd2$	Zr_2Al_3	9.523	13.763	5.522	—	—	[147]	1962	
$HfAl_2$	Hex.	D_{6h}^4–$P6_3/mmc$	Zn_2Mg	5.230	—	8.651	—	1.654	[147]	1962	
$HfAl_3$	Tetr.	D_{4h}^{17}–$I4/mmm$	$ZrAl_3$	3.989	—	17.156	—	4.301	[42]	1969	
$HfAl_3$	"	D_{4h}^{17}–$I4/mmm$	$TiAl_3$	5.555	—	8.887	—	1.600	[42]	1969	
V_3Al	Cub.	O_h^3–$Pm3n$	Cr_3Si	4.926	—	—	—	—	[148]	1963	
V_5Al_8	Cub.	T_d^3–$I43m$	γ-brass	9.207	—	—	—	—	[148]	1963	
VAl_3	Tetr.	D_{4h}^{17}–$I4/mmm$	$TiAl_3$	5.343	—	8.325	—	1.558	[149]	1964	
VAl_3	"	—	—	3.780	—	8.322	—	2.202	[145]	1966	
V_4Al_{23}	Hex.	D_{6h}^4–$P6_3/mmc$	V_4Al_{23}	7.6928	—	17.04	—	2.215	[42]	1969	
VAl_6	"	D_{6h}^4–$P6_3/mmc$	—	7.718	—	17.15	—	2.222	[124]	1962	
V_7Al_{45}	Monoclin.	C_{2h}^3–$C2/m$	V_7Al_{45}	25.604	7.6213	11.081	28° 5'	—	[42]	1969	
VAl_{10}	Cub.	—	—	14.492	—	—	—	—	[42]	1969	
VAl_{11}	"	—	—	14.465	—	—	—	—	[124]	1962	
Nb_3Al	"	O_h^3–$Pm3n$	Cr_3Si	5.187	—	—	—	—	[42]	1969	
Nb_2Al	Tetr.	D_{4h}^{14}–$P4_2/mnm$	Nb_2Al	9.957	—	5.167	—	0.519	[42]	1969	
$NbAl_3$	"	D_{4h}^{17}–$I4/mmm$	$TiAl_3$	5.438	—	8.601	—	1.582	[42]	1969	
$NbAl_3$	"	—	—	3.845	—	8.601	—	2.237	[145]	1966	
Ta_2Al	Tetr.	D_{4h}^{14}–$P4_2/mnm$	Nb_2Al	9.828	—	5.232	—	0.532	[42]	1969	
$TaAl_3$	"	D_{4h}^{17}–$I4/mmm$	$TiAl_3$	5.433	—	8.553	—	1.574	[125]	1962	
Cr_2Al	"	D_{4h}^{17}–$I4/mmm$	$MoSi_2$	3.001	—	8.637	—	2.878	[42]	1969	
Cr_5Al_8	Rhombohed.	C_{3v}^5–$R3m$	Cr_5Al_8	9.051	—	—	89° 16.4'	—	[125]	1962	
Cr_5Al_8				7.805	—	—	109° 7.6'	—	[124]	1962	
Cr_2Al_{11}	Rhomb.	—	—	24.85	24.75	30.26	—	—	[124]	1962	
$CrAl_7$	Monoclin.	—	—	20.47	7.64	25.36	155° 10'	—	[125]	1962	
	Rhomb.	—	—	11.01	34.58	12.49	—	—	[124]	1962	

CRYSTAL STRUCTURE (continued)

Phase	Unit cell	Space group	Structural type	a, Å	b, Å	c, Å	α	c/a	Source	Year	Notes
1	2	3	4	5	6	7	8	9	10	11	12
Mo_3Al	Cub.	$O_h^3 - Pm3n$	Cr_3Si	4.950	—	—	—	—	[42]	1969	
Mo_3Al_8	Monoclin.	$C_{2h}^3 - B2/m$	Mo_3Al_8	9.208	3,638	10,065	100° 4'	1,588	[42]	1969	>1403 K
Mo_3Al_8	Tetr.	—	—	6.297	—	10,000	—	—	[42]	1969	>1073 K
$MoAl_4$	Monoclin.	$C_s^3 - Bm$	WAl_4	5.255	17.768	5.225	103° 5'	—	[42]	1969	
$MoAl_5$	Hex.	$C_6^6 - P6_3$	WAl_5	4.98	—	8.964	—	1.80	[42]	1969	
"	Monoclin.	—	—	4.937	—	13.07	—	2,779	[42]	1969	
$MoAl_7$	Monoclin.	—	—	5,12	13.0	13.5	95°	—	[42]	1969	
$MoAl_{12}$	Cub.	$T_h^5 - Im3$	WAl_{12}	7,575	—	—	—	—	[42]	1969	
WAl_4	Monoclin.	$C_s^3 - Bm$	WAl_4	5.272	17.77	5.218	100°	—	[42]	1969	
WAl_5	Hex.	$C_6^6 - P6_3$	WAl_5	4.902	—	8.857	—	1.807	[124]	1962	
WAl_{12}	Cub.	$T_h^5 - Im3$	WAl_{12}	7.580	—	—	—	1.278	[42]	1969	
$Mn_{1.11}Al_{0.89}$	Tetr.	—	—	2.77	—	3.54	—		[42]	1969	
$MnAl$	Cub.	$O_h^1 - Pm3m$	$CsCl$	2.98	—	—	—	—	[42]	1969	
Mn_4Al_{11}	Tricl.	$C_i^1 - P\bar{1}$	Mn_4Al_{11}	5.092	8.862	5,047	85° 1' $\beta = 105°\ 2'$ $= \gamma = 100°\ 2'$	—	[42]	1969	
$MnAl_3$	Rhomb.	—	—	14,79	12,42	12.59	—	—	[42]	1969	
	"	—	—	14,82	12,63	12,46	—	—	[151]	1960	
Mn_3Al_{10}	Hex.	$D_{6h}^4 - P6_3/mmc$	Mn_3Al_{10}	7.543	—	7.898	—	1.047	[42]	1969	

62

Compound	System	Space group	Structure type	a	b	c	Angle	c/a	Ref.	Year
$MnAl_4$	•	—	—	28.41	—	12.38	—	0.436	[151]	1960
$MnAl_6$	Rhomb.	$D_{2h}^{17}-Ccmm$	$MnAl_6$	6.498	7.552	8.870	—	—	[125]	1962
$MnAl_{12}$	Monoclin.	—	—	13.28	8.870	4.981	98° 34′	—	[42]	1969
Fe_3Al	•	O_h^5-Fm3m	BiF_3	5.792	—	—	—	—	[125]	1962
$FeAl$		O_h^1-Pm3m	$CsCl$	2.909	—	—	—	—	[124]	1962
Fe_2Al_5	Rhomb.	—	—	7.675	6.403	4.203	—	—	[124]	1962
$FeAl_3$	Monoclin.	—	$FeAl_3$	15.489	8.083	12.476	107° 4′	—	[125]	1962
$FeAl_3$	Rhomb.	—	—	47.52	15.491	8.096	—	—	[42]	1969
$FeAl_6$	•	$D_{2h}^{17}-Ccmm$	$MnAl_6$	6.464	7,440	8,779	—	—	[124]	1962
$CoAl$	Cub.	O_h^1-Pm3m	$CsCl$	2.8619	—	—	—	—	[42]	1969
Co_2Al_5	Hex.	$D_{6h}^4-P6_3/mmc$	Co_2Al_5	7.671	—	7.608	—	0.992	[42]	1969
$CoAl_3$	Monoclin.	—	$FeAl_3$	15.183	8.122	12.34	107° 5′	—	[124]	1962
Co_2Al_9	″	$P2_1/a$	—	8.5738	6.303	6.225	94.76°	—	[42]	1969
Ni_3Al	Cub.	O_h^1-Pm3m	Cu_3Au	3.589	—	—	—	—	[124]	1962
$NiAl$	″	O_h^1-Pm3m	$CsCl$	2.887	—	—	—	—	[125]	1962
Ni_2Al_3	Hex.	D_{3d}^3-P3m1	Ni_2Al_3	4.036	—	4.900	—	1.214	[42]	1969
$NiAl_3$	Rhomb.	$D_{2h}^{16}-Pnma$	$NiAl_3$	6.6115	7.3664	4.8118	—	—	[125]	1962
Tc_2Al	Tetr.	$D_{4h}^{17}-I4/mmm$	$MoSi_2$	2.977	—	9.476	—	3.183	[78]	1962
Tc_2Al_3	Hex.	D_{3d}^3-P3m1	Ni_2Al_3	4.16	—	5.13	—	1.233	[42]	1969
$TcAl_4$	Monoclin.	C_s^3-Bm	WAl_4	5.11	17.0	5.11	100°	—	[42]	1969
$TcAl_6$	Rhomb.	$D_{2h}^{17}-Ccmm$	$MnAl_6$	6.5944	7.629	9.0011	—	—	[42]	1969
$RuAl$	Cub.	O_h^1-Pm3m	$CsCl$	2.95	—	—	—	—	[42]	1969
			—	3.03	—	—	—	—	[152]	1963
$RuAl_2$	Rhomb.	$D_{2h}^{24}-Fddd$	$TiSi_2$	8.015	4,715	8.780	—	—	[42]	1969
	Tetr.	—	—	4.40	—	6.38	—	1.45	[152]	1963
$RuAl_3$	Hex.	$D_{6h}^4-P6_3/mmc$	Ni_3Ti	4.81	—	7.84	—	1.63	[152]	1963

CRYSTAL STRUCTURE (continued)

Phase	Unit cell	Space group	Structural type	a, Å	b, Å	c, Å	α	c/a	Source	Year	Notes
1	2	3	4	5	6	7	8	9	10	11	12
Rh_3Al	Cub.	O_h^1—$Pm\bar{3}m$	Cu_3Au	3.808	—	—	—	—	[125]	1962	
RhAl	"	O_h^1—$Pm\bar{3}m$	CsCl	2.986	—	—	—	—	[145]	1966	
Pd_2Al	Rhomb.	D_{2h}^{16}—$Pnam$	Rh_2Ge	7.761	5.415	4.057	—	—	[42]	1969	
Pd_5Al_3	Hex.	D_{2h}^9—$Pbam$	Rh_5Ge_3	10.41	5.35	4.028	—	—	[42]	1969	
PdAl	Hex.	—	—	3.959	—	5.614	—	1.420	[42]	1969	
PdAl	Cub.	O_h^1—$Pm\bar{3}m$	CsCl	3.049	—	—	—	—	[18]	1965	>1128 K
Pd_2Al_3	Hex.	D_{3d}^3—$P\bar{3}m1$	Ni_2Al_3	4.216	—	5.166	—	1.225	[125]	1962	
$PdAl_3$	Rhomb.	—	$PdAl_3$	7.085	7.531	5.087	—	—	[42]	1969	
ReAl	Cub.	O_h^1—$Pm\bar{3}m$	CsCl	2.880	—	—	—	—	[42]	1969	
$ReAl_4$	Triclin.	—	—	9.13	13.8	5.16	99.5°, $\beta = 94°$, $\gamma = 103.5°$	—	[42]	1969	
$ReAl_6$	Rhomb.	D_{2h}^{17}—$Ccmm$	$MnAl_6$	6.6117	7.6091	9.023	—	—	[42]	1969	
$ReAl_{12}$	Cub.	T_h^5—$Im\bar{3}$	WAl_{12}	7.527	—	—	—	—	[42]	1969	
OsAl	"	O_h^1—$Pm\bar{3}m$	CsCl	3.005	—	—	—	—	[145]	1966	
IrAl	Cub.	O_h^1—$Pm\bar{3}m$	CsCl	2.983	—	—	—	—	[145]	1966	
$Pt_{13}Al_3$	Tetr.	—	—	7.698	—	7.844	—	1.019	[153]	1964	
Pt_3Al	Cub.	O_h^1—$Pm\bar{3}m$	Cu_3Au	3.876	—	—	—	—	[153]	1964	
Pt_5Al_3	Rhomb.	D_{2h}^9—$Pbam$	Rh_5Ge_3	10.70	5.41	3.95	—	—	[42]	1969	
PtAl	Cub.	T^4—$P2_13$	FeSi	4.866	—	—	—	—	[42]	1969	<1827 K

Compound	Symmetry	Space group	Structure type	a	b	c	angle	c/a	Ref.	Year	Notes
Pt_2Al_3	•	O_h^1–$Pm3m$	CsCl	3.03	—	—	—	—	[18]	1965	
	Hex.	D_{3d}^3–$P3m1$	Ni_2Al_3	4.208	—	5.172	—	1.229	[42]	1969	<1794 K
	"			4.209	—	10.35	—	2.459	[153]	1964	
$PtAl_2$	Cub.	O_h^5–$Fm3m$	CaF_2	5.922	—	—	—	—	[42]	1969	
Pt_5Al_{21}	•	T_d^3–$I43m$	γ-brass	19.23	—	—	—	—	[153]	1964	
Mg_2Si	•	O_h^5–$Fm3m$	CaF_2	6.351	—	—	—	—	[42]	1969	
Ca_2Si	Rhomb.	D_{2h}^{16}–$Pbnm$	$PbCl_2$	9.002	7.667	4.799	—	—	[42]	1969	
$Ca_{2\pm x}Si$	Cub.		—	~9.45	—	—	—	—	[154]	1971	See [154]
CaSi	Rhomb.	D_{2h}^{17}–$Cmcm$	TlI	4.45	10.73	3.890	—	—	[86]	1961	
$CaSi_2$	Rhombohed.	D_{3d}^5–$R3m$	$CaSi_2$	10.42	—	—	21°30'	—	[125]	1962	
$CaSi_2$	Hex.		$CaSi_2$	3.89	—	30.5	—	7.84	[86]	1961	
SrSi	Rhomb.	D_{2h}^{17}–$Cmcm$	TlI	4.826	11.287	4.042	—	—	[42]	1969	
Sr_2Si_3	Tetr.	D_{4h}^{19}–$I4/amd$	α-$ThSi_2$	4.40	—	13.92	—	3.161	[154]	1971	
$SrSi_2$	Cub.	O^7–$P4_132$	$SrSi_2$	6.535	—	—	—	—	[154]	1971	
BaSi	Rhomb.	D_{2h}^{17}–$Cmcm$	TlI	5.028	11.929	4.131	—	—	[42]	1969	
$BaSi_2$	Hex.	D_{6h}^1–$P6/mmm$	AlB_2	4.39	—	4.83	—	1.10	[154]	1971	
$BaSi_2$	Rhomb.	D_{2h}^{16}–$Pnma$	$BaSi_2$	8.92	6.80	11.58	—	—	[154]	1971	
Sc_5Si_3	Hex.	D_{6h}^3–$P6_3/mcm$	Mn_5Si_3	7.861	—	5.812	—	0.739	[155]	1967	See [157]
ScSi	Rhomb.	D_{2h}^{17}–$Cmcm$	TlI	3.988	9.882	3.659	—	—	[155]	1967	
$ScSi_{1.66}$	Hex.	D_{6h}^1–$P6/mmm$	AlB_2	3.66	—	3.87	—	1.06	[154]	1971	
Y_5Si_3	Hex.	D_{6h}^3–$P6_3/mcm$	Mn_5Si_3	8.403	—	6.303	—	0.750	[155]	1967	
Y_5Si_4	Rhomb.	D_{2h}^{16}–$Pnma$	Sm_5Ge_4	7.39	4.52	7.64	—	—	[158]	1967	
YSi	"	D_{2h}^{17}–$Cmcm$	TlI	4.257	10.527	3.839	—	—	[159]	1966	
YSi_2	Tetr.	D_{4h}^{19}–$I4/amd$	α-$ThSi_2$	4.04	—	13.42	—	3.322	[86]	1961	
YSi_2	Rhomb.	D_{2h}^{28}–$Imma$	α-$GdSi_2$	4.052	3.954	13.36	—	—	[155]	1967	
$YSi_{1.66}$	Hex.	D_{6h}^1–$P6/mmm$	AlB_2	3.836	—	4.139	—	1.079	[155]	1967	

CRYSTAL STRUCTURE (continued)

Phase	Unit cell	Space group	Structural type	a, Å	b, Å	c, Å	α	c/a	Source	Year	Notes
1	2	3	4	5	6	7	8	9	10	11	12
La_5Si_3	Tetr.	D_{4h}^{18}—$I4/mcm$	Cr_5B_3	7.95	—	14.04	—	1.766	[160]	1965	
La_5Si_4	"	D_4^4—$P4_12_12$	Zr_5Si_4	8.04	—	15.43	—	1.919	[158]	1967	
La_3Si_2	"	D_{4h}^5—$P4/mbm$	U_3Si_2	7.87	—	4.50	—	0.572	[156]	1965	
$LaSi$	Rhomb.	D_{2h}^{16}—$Pnma$	FeB	8.404	4.010	6,059	—	—	[159]	1966	
$LaSi_2$	Tetr.	D_{4h}^{19}—$I4/amd$	α-$ThSi_2$	4.322	—	13.86	—	3.207	[154]	1971	
$LaSi_2$	Rhomb.	D_{2h}^{28}—$Imma$	α-$GdSi_2$	4.272	4.184	14.02	—	—	[156]	1965	
Ce_5Si_3	Tetr.	D_{4h}^{18}—$I4/mcm$	Cr_5B_3	7,89	—	13.77	—	1.745	[160]	1965	
Ce_5Si_4	"	D_4^4—$P4_12_12$	Zr_5Si_4	7.93	—	15.04	—	1.897	[158]	1967	
Ce_3Si_2	"	D_{4h}^5—$P4/mbm$	U_3Si_2	7.79	—	4.36	—	0.560	[154]	1971	
$CeSi$	Rhomb.	D_{2h}^{16}—$Pnma$	FeB	8.306	3.967	5.978	—	—	[159]	1966	
$CeSi_2$	Tetr.	D_{4h}^{19}—$I4/amd$	α-$ThSi_2$	4.175	—	13.848	—	3.317	[154]	1971	
$CeSi_2$	Rhomb.	D_{2h}^{28}—$Imma$	α-$GdSi_2$	4.19	4.13	13.92	—	—	[154]	1971	
Pr_5Si_3	Tetr.	D_{4h}^{18}—$I4/mcm$	Cr_5B_3	7.93	—	13.97	—	1.762	[160]	1965	
Pr_5Si_4	"	D_4^4—$P4_12_12$	Zr_5Si_4	7.90	—	14.91	—	1.887	[158]	1967	
Pr_3Si_2	"	D_{4h}^5—$P4/mbm$	U_3Si_2	7.75	—	4.38	—	0.565	[154]	1971	
$PrSi$	Rhomb.	D_{2h}^{16}—$Pnma$	FeB	8.240	3.941	5.920	—	—	[159]	1966	
$PrSi_2$	Tetr.	D_{4h}^{19}—$I4/amd$	α-$ThSi_2$	4,148	—	13.67	—	3.295	[125]	1962	
$PrSi_2$	Rhomb.	D_{2h}^{28}—$Imma$	α-$GdSi_2$	4.17	4.11	13.85	—	—	[154]	1971	
Nd_5Si_3	Tetr.	D_{4h}^{18}—$I4/mcm$	Cr_5B_3	7.81	—	13.91	—	1.781	[160]	1965	

Nd_5Si_4	Hex.	D_{6h}^3—$P6_3/mcm$	Mn_5Si_3	8,66	—	6,53	—	0.754	[160]	1965
NdSi	Tett.	D_4^4—$P4_12_12$	Zr_5Si_4	7.87	—	14.78	—	1.878	[158]	1967
NdSi	Rhomb.	D_{2h}^{16}—$Pnma$	FeB	8.156	3.920	5,881	—	—	[159]	1966
$NdSi_2$	"	D_{2h}^{28}—$Imma$	α-$GdSi_2$	4.180	4.150	13.56	—	3.30	[155]	1967
	Tett.	D_{4h}^{19}—$I4/amd$	α-$ThSi_2$	4.111	—	13.56	—	0.754	[155]	1967
Sm_5Si_3	Hex.	D_{6h}^3—$P6_3/mcm$	Mn_5Si_3	8.56	—	6,45	—	—	[160]	1965
Sm_5Si_4	Rhomb.	D_{2h}^{16}—$Pnma$	Sm_5Ge_4	7.57	14.88	7,78	—	—	[158]	1967
SmSi	"	D_{2h}^{16}—$Pnma$	FeB	8,055	3.888	5,804	—	—	[159]	1966
$SmSi_2$	"	D_{2h}^{28}—$Imma$	α-$GdSi_2$	4.105	4.035	13.46	—	3.286	[155]	1967
$SmSi_2$	Tett.	D_{4h}^{19}—$I4/amd$	α-$ThSi_2$	4.08	—	13.41	—	—	[86]	1961
EuSi	Rhomb.	D_{2h}^{17}—$Cmcm$	TlI	4.72	11.15	3.99	—	—	[154]	1971
$EuSi_2$	Tett.	D_{4h}^{19}—$I4/amd$	α-$ThSi_2$	4.29	—	13.66	—	3,18	[155]	1967
Gd_5Si_3	Hex.	D_{6h}^3—$P6_3/mcm$	Mn_5Si_3	8.51	—	6.39	—	0.751	[154]	1971
Gd_5Si_4	Rhomb.	D_{2h}^{16}—$Pnma$	Sm_5Ge_4	7.45	14.67	7.73	—	—	[158]	1967
GdSi	"	D_{2h}^{16}—$Pnma$	FeB	7.996	3.859	5,724	—	—	[159]	1966
$GdSi_2$	Tett.	D_{4h}^{19}—$I4/amd$	α-$ThSi_2$	4.10	—	13.61	—	3.319	[86]	1961
$GdSi_2$	Rhomb.	D_{2h}^{28}—$Imma$	α-$GdSi_2$	4.09	4.01	13.44	—	—	[155]	1967
$GdSi_{1.66}$	Hex.	D_{6h}^1—$P6/mmm$	AlB_2	3.877	—	4.172	—	1.076	[155]	1967
Tb_5Si_3	Hex.	D_{6h}^3—$P6_3/mcm$	Mn_5Si_3	8.42	—	6,29	—	0.747	[154]	1971
Tb_5Si_4	Rhomb.	D_{2h}^{16}—$Pnma$	Sm_5Ge_4	7.41	14.58	7,69	—	—	[158]	1967
TbSi	"	D_{2h}^{16}—$Pnma$	FeB	7.919	3.833	5,703	—	—	[159]	1966
$TbSi_{1.66}$	Hex.	D_{6h}^1—$P6/mmm$	AlB_2	3.847	—	4.146	—	1.078	[154]	1971
$TbSi_2$	Rhomb.	D_{2h}^{28}—$Imma$	α-$GdSi_2$	4.07	3.98	13.37	—	—	[154]	1971
Dy_5Si_3	Hex.	D_{6h}^3—$P6_3/mcm$	Mn_5Si_3	8.37	—	6.26	—	0.748	[154]	1971
Dy_5Si_4	Rhomb.	D_{2h}^{16}—$Pnma$	Sm_5Ge_4	7.36	14.48	7.65	—	—	[158]	1967

CRYSTAL STRUCTURE (continued)

Phase	Unit cell	Space group	Structural type	a, Å	b, Å	c, Å	α	c/a	Source	Year	Notes
1	2	3	4	5	6	7	8	9	10	11	12
DySi	Rhomb.	D_{2h}^{17}–$Cmcm$	TlI	4.237	10.494	3.818	—	—	[159]	1966	
"	"	D_{2h}^{16}–$Pnma$	FeB	7.844	3.820	5.668	—	—	[159]	1966	High temp.
DySi$_{1.66}$	Hex.	D_{6h}^{1}–$P6/mmm$	AlB$_2$	3.831	—	4.121	—	1.076	[154]	1971	
DySi$_2$	Tetr.	D_{4h}^{19}–$I4/amd$	α-ThSi$_2$	4.03	—	13.38	—	3.32	[86]	1961	
DySi$_2$	Rhomb.	D_{2h}^{28}–$Imma$	α-GdSi$_2$	4.03	13.33	3.94	—	0.746	[155]	1967	
Ho$_5$Si$_3$	Hex.	D_{6h}^{3}–$P6_3/mcm$	Mn$_5$Si$_3$	8.34	—	6.22	—	0.746	[154]	1971	
HoSi	Rhomb.	D_{2h}^{17}–$Cmcm$	TlI	4.228	10.429	3.801	—	—	[159]	1966	
"	"	D_{2h}^{16}–$Pnma$	FeB	7.808	3.801	5.633	—	—	[159]	1966	High temp.
HoSi$_{1.66}$	Hex.	D_{6h}^{1}–$P6/mmm$	AlB$_2$	3.816	—	4.107	—	1.076	[154]	1971	
HoSi$_2$	Rhomb.	D_{2h}^{28}–$Imma$	α-GdSi$_2$	4.03	13.29	3.92	—	—	[154]	1971	
Er$_5$Si$_3$	Hex.	D_{6h}^{3}–$P6_3/mcm$	Mn$_5$Si$_3$	8.293	—	6.207	—	0.749	[154]	1971	
Er$_5$Si$_4$	Rhomb.	D_{2h}^{16}–$Pnma$	Sm$_5$Ge$_4$	7.27	14.32	7.58	—	—	[158]	1967	
ErSi	"	D_{2h}^{17}–$Cmcm$	TlI	4.197	10.382	3.791	—	—	[159]	1966	
"	"	D_{2h}^{16}–$Pnma$	FeB	7.772	3.785	5.599	—	—	[159]	1966	High temp.
ErSi$_{1.66}$	Hex.	D_{6h}^{1}–$P6/mmm$	AlB$_2$	3.799	—	4.090	—	1.076	[155]	1967	
ErSi$_2$	"	D_{6h}^{1}–$P6/mmm$	AlB$_2$	3.799	—	4.089	—	1.076	[154]	1971	
Tm$_5$Si$_3$	"	D_{6h}^{3}–$P6_3/mcm$	Mn$_5$Si$_3$	8.25	—	6.18	—	0.749	[154]	1971	

Compound	System	Space group	Structure type	a	b	c	β	c/a	Ref.	Year	Remarks
TmSi	Rhomb.	D_{2h}^{17}—Cmcm	TlI	4.18	10.35	3.78	—	—	[154]	1971	
TmSi$_2$	Hex.	D_{6h}^{1}—P6/mmm	AlB$_2$	3.773	—	4.070	—	1.079	[155]	1967	
Yb$_5$Si$_3$	〃	D_{6h}^{3}—P6$_3$/mcm	Mn$_5$Si$_3$	8.23	—	6.19	—	0.752	[154]	1971	
YbSi	Rhomb.	D_{2h}^{17}—Cmcm	TlI	4.19	10.35	3.77	—	—	[154]	1971	
YbSi$_2$	Hex.	D_{6h}^{1}—P6/mmm	AlB$_2$	3.771	—	4.098	—	1.087	[155]	1967	
Lu$_5$Si$_3$	〃	D_{6h}^{3}—P6$_3$/mcm	Mn$_5$Si$_3$	8.21	—	6.14	—	0.748	[154]	1971	
Lu$_5$Si$_4$	Monocl.	$P2_1/a$	—	7.20	14.11	7.46	92.1°	—	[158]	1967	
LuSi	Rhomb.	D_{2h}^{17}—Cmcm	TlI	4.15	10.24	3.75	—	—	[154]	1971	
LuSi$_{1.66}$	Hex.	D_{6h}^{1}—P6/mmm	AlB$_2$	3.745	—	4.042	—	1.079	[155]	1967	
LuSi$_2$	〃	D_{6h}^{1}—P6/mmm	AlB$_2$	3.745	—	4.050	—	1.081	[154]	1971	
Th$_3$Si$_2$	Tetr.	D_{4h}^{5}—P4/mbm	U$_3$Si$_2$	7.835	—	4.154	—	0.530	[143]	1962	
ThSi	Rhomb.	D_{2h}^{16}—Pnma	FeB	5.896	7.88	4.148	—	—	[154]	1971	
ThSi$_{1.66}$	Hex.	D_{6h}^{1}—P6/mmm	AlB$_2$	3.986	—	4.228	—	1.061	[154]	1971	
ThSi$_{1.83}$	Tetr.	Similar to α-ThSi$_2$		4.01	—	13.89	—	3.463	[154]	1971	
ThSi$_2$	Tetr.	D_{4h}^{19}—I4/amd	α-ThSi$_2$	4.125	—	14.221	—	3.447	[154]	1971	
	Hex.	D_{6h}^{1}—P6/mmm	AlB$_2$	4.136	—	4.126	—	0.997	[154]	1971	High temp.
δ-U$_3$Si	Tetr.	D_{4h}^{18}—I4/mcm	U$_3$Si	6.030	—	8.696	—	1.442	[161]	1965	<765° C
δ'-U$_3$Si	Cub.	O_h^1—Pm3m	Cu$_3$AuI	4.346	—	—	—	—	[161]	1965	>765° C
U$_3$Si$_2$	Tetr.	D_{4h}^{5}—P4/mbm	U$_3$Si$_2$	7.3298	—	3.900	—	0.532	[86]	1961	
USi	Tetr.	—	—	10.61	—	24.42	—	2.3	[162]	1971	See [162]
USi	Rhomb.	D_{2h}^{16}—Pnma	FeB	5.56	7.67	3.91	—	—	[100]	1963	
USi$_{1.5}$	Hex.	D_{6h}^{1}—P6/mmm	AlB$_2$	4.028	—	3.852	—	0.956	[86]	1961	
USi$_2$	Tetr.	D_{4h}^{19}—I4/amd	α-ThSi$_2$	3.98	—	13.74	—	3.45	[143]	1962	
USi$_3$	Cub.	O_h^1—Pm3m	Cu$_3$Au	4.035	—	—	—	—	[154]	1971	
NpSi$_2$	Tetr.	D_{4h}^{19}—I4/amd	d-ThSi$_2$	3.968	—	13.697	—	3.452	[154]	1971	

CRYSTAL STRUCTURE (continued)

Phase	Unit cell	Space group	Structural type	a, Å	b, Å	c, Å	α	c/a	Source	Year	Notes
1	2	3	4	5	6	7	8	9	10	11	12
Pu_5Si_3	Tett.	D_{4h}^{18}—$I4/mcm$	W_5Si_3	11.409	—	5.448	—	0.477	[154]	1971	
$PuSi$	Rhomb.	D_{2h}^{16}—$Pnma$	FeB	5.727	7.933	3.847	—	—	[143]	1962	
$PuSi_{1.5}$	Hex.	D_{6h}^{1}—$P6/mmm$	AlB_2	3.884	—	4.082	—	0.951	[154]	1971	
$PuSi_2$	Tett.	D_{4h}^{19}—$I4/amd$	α-$ThSi_2$	3.967	—	13.72	—	3.458	[86]	1961	
Ti_3Si	Tett.	C_{4h}^{4}—$P4_2/n$	Ti_3P	10.39	—	5.17	—	0.497	[154]	1971	
Ti_5Si_3	Hex.	D_{6h}^{3}—$P6_3/mcm$	Mn_5Si_3	7.465	—	5.162	—	0.691	[125]	1962	
Ti_5Si_4	Tett.	D_{4h}^{4}—$P4_12_12$	Zr_5Si_4	7.133	—	12.997	—	1.82	[163]	1970	
$TiSi$	Rhomb.	D_{2c}^{16}—$Pbnm$	Ti_5Si_4	6,645	6,506	12.69	—	—	[164]	1970	
$TiSi$	"	D_{2h}^{16}—$Pnma$	FeB	3.638	4.997	6,544	—	—	[159]	1966	
$TiSi_2$	"	D_{2h}^{24}—$Fddd$	$TiSi_2$	8.279	4.819	8,568	—	—	[42]	1969	
	"	D_{2h}^{17}—$Cmcm$	$ZrSi_2$	3.62	13.76	3.65	—	—	[42]	1969	
Zr_3Si	Tett.	C_{4h}^{4}—$P4_2/n$	Ti_3P	11.01	—	5.45	—	0.495	[154]	1971	
Zr_2Si	"	D_{4h}^{18}—$I4/mcm$	$CuAl_2$	6.612	—	5,294	—	0.8007	[154]	1971	
Zr_5Si_3	Hex.	D_{6h}^{3}—$P6_3/mcm$	Mn_5Si_3	7.886	—	5.558	—	0.7048	[154]	1971	
Zr_3Si_2	Tett.	D_{4h}^{5}—$P4/mbm$	U_3Si_2	7.082	—	3.314	—	0.524	[154]	1971	
Zr_5Si_4	"	D_{4h}^{4}—$P4_12_12$	Zr_5Si_4	7,123	—	13.002	—	1.825	[154]	1971	
$ZrSi$	Rhomb.	D_{2h}^{16}—$Pnma$	FeB	6.995	3.786	5.296	—	—	[169]	1965	
	"	D_{2h}^{17}—$Cmcm$	TlI	3,762	9.912	3.754	—	—	[154]	1971	
$ZrSi_2$	"	D_{2h}^{1}—$Cmcm$	$ZrSi_2$	3.72	14.76	3.67	—	—	[86]	1961	

Compound	System	Space group	Structure type	a	b	c		c/a	Ref.	Year	Notes
Hf_2Si	Tetr.	$D_{4h}^{18}-I4/mcm$	$CuAl_2$	6.48	—	5.21	—	0.804	[119]	1961	
Hf_5Si_3	Hex.	$D_{6h}^{3}-P6_3/mcm$	Mn_5Si_3	7.89	—	5.55	—	0.703	[119]	1961	
Hf_3Si_2	Tetr.	$D_{4h}^{5}-P4/mbm$	U_3Si_2	7.000	—	3.714	—	0.530	[154]	1971	
Hf_5Si_4	"	$D_{4h}^{4}-P4_12_12$	Zr_5Si_4	7.04	—	12.83	—	1.82	[154]	1971	See [166]
$HfSi$	Rhomb.	$D_{2h}^{16}-Pnma$	FeB	6.855	3.753	5.191	—	—	[119]	1961	See [167]
$HfSi_2$	"	$D_{2h}^{17}-Cmcm$	$ZrSi_2$	3.69	14.46	3.64	—	—	[119]	1961	
V_3Si	Cub.	$O_h^{3}-Pm3n$	Cr_3Si	4.726	—	—	—	—	[165]	1970	
V_5Si_3	Hex.	$D_{6h}^{3}-P6_3/mcm$	Mn_5Si_3	7.121	—	4.832	—	0.679	[165]	1970	
V_5Si_3	Tetr.	$D_{4h}^{18}-I4/mcm$	W_5Si_3	9.410	—	4.747	—	0.5045	[165]	1970	
V_6Si_5	Rhomb.	$D_{2h}^{26}-Ibam$	—	15.966	7.501	4.858	—	—	[165]	1970	
VSi_2	Hex.	$D_6^{4}-P6_222$	$CrSi_2$	6.374	—	4,572	—	0.718	[167]	1971	
Nb_3Si	Cub.	$O_h^{1}-Pm3m$	Cu_3Au	4.211	—	—	—	0.51	[154]	1971	
Nb_3Si	Tetr.	$C_{4h}^{4}-P4_2/n$	Ti_3P	10.23	—	5.19	—	0.5063	[154]	1971	
Nb_5Si_3	"	$D_{4h}^{18}-I4/mcm$	W_5Si_3	10.018	—	5.072	—	1.8083	[168]	1957	
	"	$D_{4h}^{18}-I4/mcm$	Cr_5B_3	6.570	—	11.884	—	1.375	[154]	1971	
$NbSi_2$	Hex.	$D_6^{4}-P6_222$	$CrSi_2$	4.803	—	6.604	—	0.51	[154]	1971	
Ta_3Si	Tetr.	$C_{4h}^{4}-P4_2/n$	Ti_3P	10.19	—	5.17	—	0.818	[154]	1971	
Ta_2Si	"	$D_{4h}^{18}-I4/mcm$	$CuAl_2$	6.157	—	5.039	—	0.512	[168]	1957	
Ta_5Si_3	"	$D_{4h}^{18}-I4/mcm$	W_5Si_3	9.88	—	5.06	—	1.822	[168]	1957	
Ta_5Si_3	"	$D_{4h}^{18}-I4/mcm$	Cr_5B_3	6.516	—	11.873	—	1.372	[86]	1961	
$TaSi_2$	Hex.	$D_6^{4}-P6_222$	$CrSi_2$	4.778	—	6.558	—	1.372	[154]	1971	
Cr_3Si	Cub.	$O_h^{3}-Pm3n$	Cr_3Si	4.564	—	—	—	—	[168]	1957	
Cr_5Si_3	Tetr.	$D_{4h}^{18}-I4/mcm$	W_5Si_3	9.170	—	4.636	—	0.505	[168]	1957	
$CrSi$	Cub.	$T^{4}-P2_13$	$FeSi$	4,629	—	—	—	—	[86]	1961	
$CrSi_2$	Hex.	$D_6^{4}-P6_222$	$CrSi_2$	4.431	—	6.364	—	1.436	[170]	1963	

CRYSTAL STRUCTURE (continued)

Phase	Unit cell	Space group	Structural type	a, Å	b, Å	c, Å	α	c/a	Source	Year	Notes
1	2	3	4	5	6	7	8	9	10	11	12
Mo_3Si	Cub.	O_h^3-Pm3n	Cr_3Si	4.890	—	—	—	—	[168]	1957	
Mo_5Si_3	Tetr.	$D_{4h}^{18}-I4/mcm$	W_5Si_3	9.62	—	4.90	—	0.5087	[172]	1955	
$MoSi_2$	"	$D_{4h}^{17}-I4/mmm$	$MoSi_2$	3.203	—	7.855	—	2.452	[171]	1964	
$MoSi_2$	Hex.	$D_6^4-P6_222$	$CrSi_2$	4.605	—	6.56	—	1.425	[173]	1968	Low temp.
W_5Si_3	Tetr.	$D_{4h}^{18}-I4/mcm$	W_5Si_3	9.645	—	4.97	—	0.516	[172]	1955	
WSi_2	"	$D_{4h}^{17}-I4/mmm$	$MoSi_2$	3.212	—	7.835	—	2.439	[171]	1964	
Mn_3Si	Cub.	O_h^9-Im3m	α-Fe	2.857	—	—	—	—	[71]	1960	
	"	O_h^5-Fm3m	BiF_3	5.722	—	—	—	—	[71]	1960	Super-structure ($a = 2\times$ $\times$ 2,861 Å)
Mn_5Si_2	Tetr.	—	—	8.910	—	8.716	—	0.978	[154]	1971	
Mn_5Si_3	Hex.	$D_{6h}^3-P6_3/mcm$	Mn_5Si_3	6.910	—	4.814	—	0.696	[71]	1960	
$MnSi$	Cub.	T^4-P2_13	$FeSi$	4.557	—	—	—	—	[119]	1961	
$MnSi_{1.727}$	Tetr.	$D_{2d}^8-\bar{P}4n2$	$Mn_{11}Si_{19}$	5.518	—	48,136	—	8.723	[174]	1970	
$MnSi_{1.730}$	"	—	$Mn_{26}Si_{45}$	5.515	—	113.36	—	20.554	[174]	1970	
$MnSi_{1.733}$	"	$D_{2d}^{12}-\bar{I}42d$	$Mn_{15}Si_{26}$	5.525	—	65.55	—	10.808	[174]	1970	
$MnSi_{1.741}$	Tetr.	—	$Mn_{27}Si_{47}$	5.530	—	117.94	—	21.327	[174]	1970	

Compound	System	Space group	Structure type	a	b	c		c/a	Ref	Year	Remarks
MnSi$_{1.750}$	Tetr.	D_{2d}–$P\bar{4}c2$	Mn$_4$Si$_7$	5.534	—	17.550	—	3.367	[154]	1971	
Tc$_4$Si	Cub.	O_h^9–$Im3m$	α-Fe	3.009	—	—	—	—	[154]	1971	
Tc$_3$Si	»	T_d^3–$I\bar{4}3m$	Fe$_3$Zn$_{10}$	9.014	—	—	—	—	[154]	1971	
Tc$_5$Si$_3$	Tetr.	D_{4h}^{18}–$I4/mcm$	W$_5$Si$_3$	9.403	—	4.849	—	0.516	[154]	1971	
TcSi	Cub.	T^4–$P2_13$	FeSi	4.755	—	—	—	—	[154]	1971	
TcSi$_{1.750}$	Tetr.	D_{2d}–$P\bar{4}c2$	Mn$_4$Si$_7$	5.737	—	18.099	—	3.1548	[174]	1970	
Re$_5$Si$_3$	»	D_{4h}^{18}–$I4/mcm$	W$_5$Si$_3$	9.53	—	4.81	—	0.505	[154]	1971	
ReSi	Cub.	T^4–$P2_13$	FeSi	4.775	—	—	—	—	[154]	1971	
ReSi$_2$	Tetr.	D_{4h}^{17}–$I4/mmm$	MoSi$_2$	3.129	—	7.674	—	2.452	[86]	1961	High temp.
Fe$_3$Si	Cub.	O_h^5–$Fm3m$	BiF$_3$	5.651	—	—	—	—	[175]	1968	
Fe$_{11}$Si$_5$	»	T_h^1–$Pm3$	Fe$_{11}$Si$_5$	5.629	—	—	—	—	[176]	1974	
Fe$_2$Si	»	O_h^1–$Pm3m$	—	2.81	—	—	—	—	[176]	1974	
Fe$_5$Si$_3$	Hex.	D_{6h}^3–$P6_3/mcm$	Mn$_5$Si$_3$	6.7552	—	4.7174	—	0.6983	[154]	1971	
FeSi	Cub.	T^4–$P2_13$	FeSi	4.489	—	—	—	—	[154]	1971	
Fe$_{0.87}$Si	Tetr.	D_{4h}^1–$P4/mmm$	Lebeauite	2.69	—	5.13	—	1.91	[177]	1972	α-FeSi$_2$
FeSi$_2$	Rhomb.	D_{2h}^{18}–$Cmca$	FeSi$_2$	9.863	7.791	7.833	—	—	[177]	1972	β-FeSi$_2$ See [178]
Co$_3$Si	Hex.	—	Ni$_3$Sn	—	—	—	—	—	[179]	1973	1193—1214° C
Co$_2$Si	Rhomb.	D_{2h}^{16}–$Pnma$	δ-Ni$_2$Si	4.918	3.738	7.109	—	—	[154]	1971	
CoSi	Cub.	T^4–$P2_13$	FeSi	4.4445	—	—	—	—	[154]	1971	
CoSi$_2$	»	O_h^5–$Fm3m$	CaF$_2$	5.3627	—	—	—	—	[154]	1971	
Ni$_3$Si	»	O_h^1–$Pm3m$	Cu$_3$Au	3.504	—	—	—	—	[154]	1971	
Ni$_{31}$Si$_{12}$	Hex.	D_3^2–$P321$	—	6.671	—	12.288	—	1.842	[180]	1971	
Ni$_5$Si$_2$	Trigon.	—	Ni$_5$Si$_2$	6.670	—	12.332—12,267	—	1.833	[25]	1961	

CRYSTAL STRUCTURE (continued)

Phase	Unit cell	Space group	Structural type	a, Å	b, Å	c, Å	α	c/a	Source	Year	Notes
1	2	3	4	5	6	7	8	9	10	11	12
θ-Ni$_2$Si	Hex.	D_{6h}^4–$P6_3/mmc$	Ni$_2$In	3.805	—	4.890	—	1.285	[154]	1971	
δ-Ni$_2$Si	Rhomb.	D_{2h}^{16}–$Pnma$	δ-Ni$_2$Si	4.99	3.72	7.03	—	—	[154]	1971	
Ni$_3$Si$_2$	"	C_{2v}^2–$Cmc2_1$	Ni$_3$Si$_2$	12.229	10.805	6.924	—	—	[25]	1961	
NiSi	Rhomb.	D_{2h}^{16}–$Pnma$	MnP	5.18	3.34	5.62	—	—	[154]	1971	
ζ-NiSi$_2$	Rhombohed.	—	—	8.881	—	—	90° 23′	—	[181]	1970	
α-NiSi$_2$	Cub.	O_h^5–$Fm3m$	CaF$_2$	5.406	—	—	—	—	[181]	1970	
Ru$_2$Si	Rhomb.	D_{2h}^{16}–$Pnma$	δ-Ni$_2$Si	5.279	4.005	7.418	—	—	[184]	1961	
Ru$_5$Si$_3$	"	D_{2h}^9–$Pbam$	Rh$_5$Ge$_3$	5.246	9.815	4.023	—	—	[182]	1970	
Ru$_4$Si$_3$	"	$Pnma$	Ru$_4$Si$_3$	5.1936	4.0216	17.134	—	—	[182]	1970	
RuSi	Cub.	O_h^1–$Pm3m$	CsCl	2.909	—	—	—	—	[183]	1962	
	"	T^4–$P2_13$	FeSi	4.703	—	—	—	—	[183]	1962	
RuSi$\sim$1.5	Tetr.		RuSi$\sim$1.5	11.075	—	8.954	—	0.813	[182]	1970	
OsSi	Cub.	T^4–$P2_13$	FeSi	4.729	—	—	—	—	[183]	1962	
	"	O_h^1–$Pm3m$	CsCl	2.963	—	—	—	—	[183]	1962	
OsSi$\sim$1.5	Tetr.		RuSi$\sim$1.5	11.158	—	8.962	—	0.803	[182]	1970	
OsSi$_2$	Rhomb.	D_{2h}^{18}–$Cmca$	β-FeSi$_2$	10.150	8.117	8.223	—	—	[182]	1970	
OsSi$_{2.4}$	Monocl.	C_{2h}^3–$C2/m$	OsGe$_2$	8.77	3.00	7.38	118° 30′	—	[183]	1962	
Rh$_2$Si	Rhomb.	D_{2h}^{16}–$Pnma$	δ-Ni$_2$Si	5.408	3.930	7.383	—	—	[185]	1963	
Rh$_5$Si$_3$	"	D_{2h}^9–$Pbam$	Rh$_5$Ge$_3$	5.317	10.131	3.895	—	—	[185]	1963	

Compound	System	Space group	Structure type	a	b	c	angle	c/a	Ref.	Year	Remarks
$Rh_{20}Si_{13}$	Hex.	C_{6h}^2–$P6_3/m$	$Rh_{20}Si_{13}$	11.851	—	3.623	—	0.305	[182]	1970	
Rh_3Si_2	"	D_{6h}^4–$P6_3/mmc$	NiAs	3.949	—	5.047	—	1.278	[183]	1962	
RhSi	Cub.	T^4–$P2_13$	FeSi	4.674	—	—	—	—	[186]	1965	
	Rhomb.	D_{2h}^{16}–$Pnma$	MnP	6.362	3.063	5.531	—	—	[183]	1962	
	Cub.	O_h^1–$Pm3m$	CsCl	2.963	—	—	—	—	[133]	1962	
Rh_4Si_5	Monocl.	$P2_1/m$	Rh_4Si_5	12.335	3.508	5.924	100,18°	—	[132]	1970	
Rh_3Si_4	Rhomb.	D_{2h}^{16}–$Pnma$	Rh_3Si_4	18.810	3.614	5.813	—	—	[132]	1970	
Ir_3Si	Tetr.	D_{4h}^{18}–$I4/mcm$	U_3Si	5.222	—	7.954	—	1.523	[182]	1970	
Ir_2Si	Rhomb.	D_{2h}^{16}–$Pnma$	δ-Ni_2Si	5.284	3.989	7.615	—	—	[132]	1970	
Ir_3Si_2	Hex.	D_{6h}^4–$P6_3/mmc$	NiAs	3.963	—	5.126	—	1.293	[183]	1962	
IrSi	Rhomb.	D_{2h}^{16}–$Pnma$	MnP	5.558	3.211	6.273	—	—	[182]	1970	
Ir_4Si_5	Monocl.	$P2_1/m$	Rh_4Si_5	12.354	3.618	5.881	100,14°	—	[182]	1970	
Ir_3Si_4	Rhomb.	D_{2h}^{16}–$Pnma$	Rh_3Si_4	18.870	3.679	5.774	—	—	[182]	1970	
$IrSi_{\sim 1.5}$	Monocl.	—	—	5.542	14.166	12.426	120,61°	—	[182]	1970	
$IrSi_3$	Hex.	D_{6h}^4–$P6_3/mmc$	Na_3As	4.350	—	6.610	—	1.519	[183]	1962	See [182]
$Pd_{4,5}Si$	Rhomb.	—	—	7.418	9.396	9.048	—	—	[182]	1970	
Pd_3Si	"	D_{2h}^{16}–$Pnma$	Fe_3C	5.735	7.555	5.260	—	—	[187]	1960	
Pd_2Si	Hex.	D_{3h}^3–$P\bar{6}2m$	Fe_2P	6.528	—	3.437	—	0.526	[187]	1960	
Pd_2Si	"	—	—	13.055	—	27.490	—	2.105	[182]	1970	Super-structure Fe_2P, see [182]
PdSi	Rhomb.	D_{2h}^{16}–$Pnma$	MnP	5.599	3.381	6.133	—	—	[187]	1960	
Pt_3Si	Tetr.	D_{4h}^{18}–$I4/mcm$	U_3Si	5.46	—	7.86	—	1.439	[182]	1970	
Pt_2Si	Monocl.	C_{2h}^3–$C2m$	Pt_3Ge	7.072	7.765	5.558	135,79°	—	[182]	1970	
$Pt_{12}Si_5$	Tetr.	C_{4h}^3–$P4/n$	$Pt_{12}Si_5$	13.395	—	5.54	—	0.414	[183]	1964	See [26], [182]

CRYSTAL STRUCTURE (continued)

Phase	Unit cell	Space group	Structural type	a, Å	b, Å	c, Å	α	c/a	Source	Year	Notes
1	2	3	4	5	6	7	8	9	10	11	12
Pt_2Si	Hex.	$D_{3h}^3-P\bar{6}2m$	Fe_2P	6.440	—	3.573	—	0.554	[182]	1970	See [188]
	Cub.	O_h^5-Fm3m	CaF_2	5.63	—	—	—	—	[188]	1964	
Pt_6Si_5	Tetr.	$D_{4h}^{13}-I4/mmm$	Pt_2Si	3.933	—	5.910	—	1,503	[188]	1964	
PtSi	Monocl.	$C_{2h}^2-P2_1/m$	Pt_6Si_5	15.308	3.48	6.120	86.6°	—	[188]	1964	See [182]
PtSi	Rhomb.	$D_{2h}^{16}-Pnma$	MnP	5.595	3.603	5.932	—	—	[189]	1950	See [182]
ScP	Cub.	O_h^5-Fm3m	NaCl	5.312	—	—	—	—	[42]	1969	
YP	"	O_h^5-Fm3m	NaCl	5.661	—	—	—	—	[42]	1969	
LaP	"	O_h^5-Fm3m	NaCl	6.025	—	—	—	—	[190]	1962	
CeP	"	O_h^5-Fm3m	NaCl	5.909	—	—	—	—	[190]	1962	
PrP	"	O_h^5-Fm3m	NaCl	5.872	—	—	—	—	[190]	1962	
NdP	"	O_h^5-Fm3m	NaCl	5.838	—	—	—	—	[190]	1962	
SmP	Cub.	O_h^5-Fm3m	NaCl	5.780	—	—	—	—	[191]	1965	See [190]
GdP	"	O_h^5-Fm3m	NaCl	5.723	—	—	—	—	[42]	1969	
TbP	"	O_h^5-Fm3m	NaCl	5.686	—	—	—	—	[42]	1969	
DyP	"	O_h^5-Fm3m	NaCl	5.654	—	—	—	—	[42]	1969	
HoP	"	O_h^5-Fm3m	NaCl	5.626	—	—	—	—	[42]	1969	
ErP	"	O_h^5-Fm3m	NaCl	5.606	—	—	—	—	[42]	1969	
TmP	"	O_h^5-Fm3m	NaCl	5.573	—	—	—	—	[42]	1969	
YbP	"	O_h^5-Fm3m	NaCl	5.554	—	—	—	—	[42]	1969	

Compound	System	Space group	Structure type	a	b	c		c/a		Ref.	Year	Remarks
ThP	Cub.	O_h^5—$Fm3m$	NaCl	5.830	—	—	—	—	—	[190]	1962	$ThP_{0.7}$
Th₃P₄	Cub.	T_d^6—$I\bar{4}3d$	Th₃P₄	8.618	—	—	—	—	—	[190]	1962	
UP	Cub.	O_h^5—$Fm3m$	NaCl	5.601	—	—	—	—	—	[190]	1962	
U₃P₄	Cub.	T_d^6—$I\bar{4}3d$	Th₃P₄	8.214	—	—	—	—	—	[190]	1962	
UP₂	Tetr.	D_{4h}^7—$P4/nmm$	Fe₂As	3.800	—	7.762	—	2.043	—	[190]	1962	
PuP	Cub.	O_h^5—$Fm3m$	NaCl	5.644	—	—	—	—	—	[190]	1962	
Ti₃P	Tetr.	C_{4h}^4—$P4_2/n$	Ti₃P	9.9592	—	4.9869	—	0.500	—	[192]	1967	
Ti₂P	Hex.	—	—	11.5314	—	3.4575	—	0.328	—	[192]	1967	
Ti~1.7P	Hex.	D_{3h}^3—$P\bar{6}2m$	Fe₂P	6.715	—	3.462	—	0.515	—	[193]	1965	
Ti₅P₃	Rhomb.	—	—	7.438	9.752	6.505	—	—	—	[28]	1968	
Ti₄P₃	Hex.	D_{6h}^3—$P6_3/mcm$	Mn₅Si₃	7.2226	—	5.0936	—	0.706	—	[192]	1967	See [192]
Ti₃P₂	Cub.	$I\bar{4}3d$	Th₃P₄	7.425	—	—	—	—	—	[193]	1965	See [192]
TiP	Tetr.	—	—	7.483	—	10.495	—	1.402	—	[193]	1965	See [195]
TiP₂	Hex.	D_{6h}^4—$P6_3/mmc$	TiP	3.493	—	11.67	—	3.341	—	[193]	1965	
Zr₃P	Rhomb.	D_{2h}^{16}—$Pbnm$	ZrAs₂	6.181	8.256	3.346	—	—	—	[42]	1969	
ZrP_x	Tetr.	C_{4h}^4—$P4_2/n$	Ti₃P	10.799	—	5.354	—	0.4958	—	[194]	1966	
ZrP	Rhomb.	$Pnnm$ or $Pnn2$	—	16.715	27.572	3.6742	—	—	—	[195]	1966	$0.5 < x < 1$
ZrP₂	Hex.	D_{6h}^4—$P6_3/mmc$	TiP	3.684	—	12.554	—	3.408	—	[42]	1969	
Hf₂P	Rhomb.	D_{2h}^{16}—$Pbnm$	ZrAs₂	6.494	8.744	3.513	—	—	—	[195]	1966	
HfP	Rhomb.	$Pnnm$	Ta₂P	15.031	12.258	3.5738	—	—	—	[196]	1968	
Hf₃P₂	Hex.	D_{6h}^4—$P6_3/mmc$	TiP	3.650	—	12.375	—	3.396	—	[197]	1962	
HfP₂	Rhomb.	D_{2h}^{16}—$Pnma$	Sb₂S₃	10.138	3.578	9.881	—	—	—	[198]	1968	
V₃P	Rhomb.	D_{2h}^{16}—$Pbnm$	ZrAs₂	6.467	8.646	3.497	—	—	—	[42]	1969	
V₃P	Tetr.	C_{4h}^4—$P4_2/n$	Ti₃P	9.387	—	4.756	—	0.5067	—	[199]	1971	
VP	Hex.	D_{6h}^4—$P6_3/mmc$	NiAs	3.18	—	6.22	—	1.96	—	[190]	1962	

CRYSTAL STRUCTURE (continued)

Phase	Unit cell	Space group	Structural type	a, Å	b, Å	c, Å	α	c/a	Source	Year	Notes
1	2	3	4	5	6	7	8	9	10	11	12
$V_{12}P_7$	Hex.	C_{6h}^2—$P6_3/m$	—	9.299	—	3.279	—	0.353	[200]	1970	·
VP_2	Monocl.	—	$NbAs_2$	8.466	3.106	7.170	119° 16′	—	[42]	1969	
Nb_3P	Tetr.	C_{4h}^4—$P4_2/n$	Ti_3P	10.128	—	5.091	—	0.5027	[201]	1966	See [194]
Nb_7P_4	Monocl.	$C2/m$	—	14.950	3.440	13.848	104.74°	—	[202]	1966	
NbP	Tetr.	C_{4v}^{11}—$I4_1md$	$NbAs$	3.334	—	11.378	—	3.412	[203]	1963	
NbP_2	Monocl.	—	$NbAs_2$	8,8715	3.2663	7.9514	119.097°	—	[201]	1966	
Ta_3P	Tetr.	C_{4h}^4—$P4_2/n$	Ti_3P	10.154	—	5.012	—	0.4936	[201]	1966	
Ta_2P	Rhomb.	$Pnnm$	Ta_2P	14.420	11.547	3.400	—	—	[201]	1966	
TaP	Tetr.	C_{4v}^{11}—$I4_1md$	$NbAs$	3.319	—	11.341	—	3.40	[201]	1966	
TaP_2	Monocl.	—	$NbAs_2$	8.870	3.267	7.497	119° 24′	—	[42]	1969	See [201]
Cr_3P	Tetr.	—	Fe_3P	9.186	—	4.558	—	0.496	[190]	1962	
CrP	Rhomb.	D_{2h}^{16}—$Pcmn$	MnP	6.018	5.362	3.113	—	—	[204]	1962	
Mo_3P	Tetr.	$\bar{I}42m$	α-V_3S	9.794	—	4.827	—	0.493	[205]	1963	
$Mo_{\sim1.7}P$	Hex.	—	—	16.673	—	3.3644	—	0.202	[207]	1972	
Mo_8P_5	Monocl.	Pm	—	9.399	3.209	6.537	109.59°	—	[207]	1972	
Mo_4P_3	Rhomb.	$Pnma$	—	12.428	3.158	20.440	—	—	[206]	1965	
MoP	Hex.	D_{3h}^1—$P\bar{6}m2$	WC	3.230	—	3.207	—	—	[42]	1969	See [205]
MoP_2	Rhomb.	C_{2v}^{12}—$Cmc2_1$	MoP_2	3.145	11.184	4.984	—	0.993	[205]	1963	
W_3P	Tetr.	$\bar{I}42m$	α-V_3S	9.858	—	4.800	—	0.486	[201]	1966	

Compound	System	Space group	Structure type	a	b	c	angle	ratio	Ref.	Year
W$_2$P	Hex.	—	—	6.191	—	6.794	—	1.097	[42]	1969
WP	Rhomb.	D_{2h}^{16}–$Pcmn$	MnP	6,223	5.726	3.238	—	—	[124]	1962
α-WP$_2$	Monocl.	C_{2v}^{12}–$Cmc2_1$	NbAs$_2$	8.5022	3.1695	7.466	119.367°	—	[201]	1966
β-WP$_2$	Rhomb.	—	MoP$_2$	3.166	11.161	4.973	—	0.497	[201]	1966
Mn$_3$P	Tetr.	—	Fe$_3$P	9.181	—	4,568	—	0.569	[209]	1962
Mn$_2$P	Hex.	D_{3h}^{3}–$P\bar{6}2m$	Fe$_2$P	6,081	—	3,460	—	—	[208]	1962
MnP	Rhomb.	D_{2h}^{16}–$Pnma$	MnP	5.918	5.258	3.172	—	—	[204]	1962
Re$_2$P	•	D_{2h}^{16}–$Pcmn$	PbCl$_2$	10.040	5.540	2.939	—	—	[210]	1961
Re$_3$P$_4$	Monocl.	$C2/m$	Fe$_3$Se$_4$	12.179	3.012	6.042	114.07°	0.4891	[211]	1966
Fe$_3$P	Tetr.	S_4^{2}–$I\bar{4}$	Fe$_3$P	9.107	—	4.460	—	0.590	[190]	1962
Fe$_2$P	Hex.	D_{3h}^{3}–$P\bar{6}2m$	Fe$_2$P	5.864	—	3.456	—	—	[190]	1962
FeP	Rhomb.	D_{2h}^{12}–$Pnnm$	MnP	5.792	5.191	3.099	—	—	[204]	1962
FeP$_2$	•	D_{2h}^{16}–$Pbmm$	FeS$_2$	2.730	4.985	5.668	—	—	[42]	1969
Co$_2$P	•	D_{2h}^{16}–$Pcmn$	PbCl$_2$	6.608	5.646	3.513	—	—	[190]	1962
CoP	•	D_{2h}^{16}–$Pcmn$	MnP	5.599	5.076	3.281	—	—	[125]	1962
CoP$_3$	Cub.	T_h^{5}–$Im3$	CoAs$_3$	7.7073	—	—	—	0.49	[213]	1968
Ni$_3$P	Tetr.	S_4^{2}–$I\bar{4}$	Fe$_3$P	8.954	—	4.386	—	3.72	[214]	1965
Ni$_5$P$_2$	Hex.	$P\bar{3}$ or $P3$	—	6.608	—	24.634	—	0.587	[214]	1965
Ni$_{12}$P$_5$	Tetr.	C_{4h}^{5}–$I4/m$	Ni$_{12}$P$_5$	8.646	—	5.070	—	0.577	[214]	1965
Ni$_2$P	Hex.	D_{3h}^{3}–$P\bar{6}2m$	Fe$_2$P	5.865	—	3.387	—	1.62	[214]	1965
Ni$_5$P$_4$	•	$P6_3mc$	—	6.789	—	10.986	—	—	[214]	1965
NiP	Rhomb.	$Pbca$	NiP	6.050	4,881	6.890	—	—	[214]	1965
NiP$_2$	Monocl.	Cc or $C2/c$	NiP$_2$	6.366	5.615	6.072	126,22°	—	[214]	1965
NiP$_3$	Cub.	T_h^{5}–$Im3$	CoAs$_3$	7.8192	—	—	—	—	[213]	1968
Ru$_2$P	Rhomb.	D_{2h}^{16}–$Pnma$	PbCl$_2$	6.896	5.902	3.859	—	—	[190]	1962

See [204]

CRYSTAL STRUCTURE (continued)

Phase	Unit cell	Space group	Structural type	a, Å	b, Å	c, Å	α	c/a	Source	Year	Notes
1	2	3	4	5	6	7	8	9	10	11	12
RuP	Rhomb.	$D_{2h}^{12}-Pcmn$	MnP	6.120	5.520	3.168·	—	—	[204]	1962	
RuP$_2$	″	$D_{2h}^{12}-Pnnm$	FeS$_2$	5.115	5.888	2.870	—	—	[190]	1962	
Rh$_2$P	Cub.	O_h^5-Fm3m	CaF$_2$	5.516	—	—	—	—	[125]	1962	
Rh$_4$P$_3$	Rhomb.	—	Rh$_4$P$_3$	11.662	3.317	9,994	—	—	[190]	1962	
RhP$_2$	Monocl.	—	RhSb$_2$	5.743	5.794	5.837	112.92°	—	[215]	1961	
RhP$_3$	Cub.	T_n^5-Im3	CoAs$_3$	7.9951	—	—	—	—	[213]	1968	
Pd$_{4.8}$P	Monocl.	$P2_1$	—	5.004	7.606	8.416	95.63°	—	[30]	1966	
Pd$_5$P	Rhomb.	$D_{2h}^{16}-Pbnm$	Fe$_3$C	5.890	7.440	5.164	—	—	[216]	1960	
Pd$_7$P$_3$	Rhombohed.	—	—	7.28	—	—	110,12°	—	[217]	1963	
PdP$_2$	Monocl.	—	PdP$_2$	6.207	5.857	5.874	111.8°	—	[218]	1963	
PdP$_3$	Monocl.	—	—	6.777	5.856	6.206	126° 43'	—	[190]	1962	
OsP$_2$	Rhomb.	$D_{2h}^{12}-Pnnm$	FeS$_2$	7.705	5.898	2.918	—	—	[190]	1962	
Ir$_2$P	Cub.	O_h^5-Fm3m	CaF$_2$	5.098	—	—	—	—	[190]	1962	
IrP$_2$	Monocl.	—	RhSb$_2$	5.543	5.790	5.85	111.6°	—	[190]	1962	
IrP$_3$	Cub.	T_h^5-Im3	CoAs$_3$	5.746	—	—	—	—	[215]	1961	
Pt$_5$P$_2$	Monocl.	C2/c	—	8.0151	5,385	7.438	99.17°	—	[213]	1968	
PtP$_2$	Cub.	—	—	10.764	—	—	—	—	[219]	1967	
BeS	Cub.	T_h^6-Pa3	FeS$_2$	5.6956	—	—	—	—	[212]	1969	
	″	$T_d^2-F\bar{4}3m$	Sphalerite	4.87	—	—	—	—	[33]	1972	

80

Compound	System	Space group	Structure type	a	b	c	β	c/a	Ref.	Year	Note
MgS	Cub.	O_h^5—$Fm3m$	NaCl	5.1913					[33]	1972	
CaS	"	O_h^5—$Fm3m$	NaCl	5.6836					[33]	1972	
SrS	"	O_h^5—$Fm3m$	NaCl	6.008					[33]	1972	
BaS	"	O_h^5—$Fm3m$	NaCl	6.37					[33]	1972	
ScS	"	O_h^5—$Fm3m$	NaCl	5.19					[33]	1972	
Sc_2S_3	Rhomb.	—	Sc_2S_3	10.41	7.38	22.05			[33]	1972	
YS	Cub.	O_h^5—$Fm3m$	NaCl	5.466					[33]	1972	See [221]
Y_5S_7	Monocl.	$C2/m$	—	12.768	3.803	11.545	104.82°		[220]	1964	
Y_2S_3	"	—	—	17.520	4.019	10.17	98.64°		[33]	1972	
YS_2	Tetr.	—	—	7.71		7.89		1.02	[221]	1956	
LaS	Cub.	O_h^5—$Fm3m$	NaCl	5.860					[222]	1964	
La_3S_4	"	T_d^6—$I\bar{4}3d$	Th_3P_4	8.748					[221]	1956	
α-La_2S_3	Rhomb.	$Pnma$	—	7.66	4.22	15.95			[224]	1968	
β-La_2S_3	Cub.	—	—	20.46					[226]	1971	See [223]
γ-La_2S_3	"	T_d^6—$I\bar{4}3d$	Th_3P_4	8.723					[226]	1971	See [225]
La_5S_7	Tetr.	—$I4_1/acd$	—	15.61		20.58		1.318	[32]	1968	See [223]
LaS_2	"	—	—	4.147		8.176		1.96	[226]	1971	
CeS	Cub.	O_h^5—$Fm3m$	NaCl	5.790					[222]	1964	
α-Ce_2S_3	Rhomb.	$Pnma$	—	7.84	8.56	15.45			[227]	1966	
β-Ce_2S_3	"	—	—	7.55	4.14	15.79			[224]	1968	
Ce_3S_4	"	—	—	7.28	10.65	13.17			[227]	1966	
γ-Ce_2S_3	Cub.	T_d^6—$I\bar{4}3d$	Th_3P_4	8.630					[223]	1969	
Ce_3S_7	"	T_d^6—$I\bar{4}3d$	Th_3P_4	8.630					[223]	1969	
CeS_2	Tetr.	$I4_1/acd$	—	15.19		20.19		1.329	[32]	1968	
	Tetr.	—	—	8.115		8.115		~1	[227]	1966	

CRYSTAL STRUCTURE (continued)

Phase	Unit cell	Space group	Structural type	a, Å	b, Å	c, Å	α	c/a	Source	Year	Notes
1	2	3	4	5	6	7	8	9	10	11	12
CeS_2	Cub.	—	—	8.12	—	—	—	—	[221]	1956	See [227]
PrS	"	O_h^5-Fm3m	NaCl	5.763	—	—	—	—	[222]	1964	
$\alpha\text{-}Pr_2S_3$	Rhomb.	$Pnma$	—	7,49	4.10	15.69	—	—	[224]	1968	
Pr_3S_4	Cub.	$T_d^6-I\bar{4}3d$	Th_3P_4	8.611	—	—	—	—	[221]	1956	
$\gamma\text{-}Pr_2S_3$	"	$T_d^6-I\bar{4}3d$	Th_3P_4	8,611	—	—	—	—	[221]	1956	See [223]
Pr_5S_7	Tetr.	$I4_1/acd$	—	15.10	—	20.05	—	1.328	[32]	1968	
NdS	Cub.	O_h^5-Fm3m	NaCl	5.715	—	—	—	—	[222]	1964	
Nd_3S_4	"	$T_d^6-I\bar{4}3d$	Th_3P_4	8,524	—	—	—	—	[226]	1971	
$\alpha\text{-}Nd_2S_3$	Rhomb.	$Pnma$	—	7.42	4.05	15.57	—	—	[224]	1968	
$\beta\text{-}Nd_2S_3$	Cub.	—	—	19.92	—	—	—	—	[226]	1971	
$\gamma\text{-}Nd_2S_3$	"	$T_d^6-I\bar{4}3d$	Th_3P_4	8.527	—	—	—	1.327	[226]	1971	
Nd_5S_7	Tetr.	$I4_1/acd$	—	14.94	—	19.82	—	1.327	[32]	1968	
Nd_4S_7	Tetr.	—	—	7,915	—	8,029	—	1,015	[226]	1971	
NdS_2	"	—	—	4,022	—	8.031	—	1.995	[226]	1971	
SmS	Cub.	O_h^5-Fm3m	NaCl	5.863	—	—	—	—	[33]	1972	
$\alpha\text{-}Sm_2S_3$	Rhomb.	$Pnma$	—	7.33	4.00	15,46	—	—	[224]	1968	
Sm_3S_4	Cub.	$T_d^6-I\bar{4}3d$	Th_3P_4	8.563	—	—	—	—	[221]	1956	
$\beta\text{-}Sm_2S_3$	Rhomb.	—	—	—	—	—	—	—	[228]	1966	Isotope $\beta\text{-}Ce_2S_3$

Formula	System	Space group	Type	a	b	c	Angle	Ratio	Ref.	Year
γ-Sm_2S_3	Cub.	T_d^6—$I\overline{4}3d$	Th_3P_4	8.465	—	—	—	—	[221]	1956
Sm_5S_7	Tetr.	$I4_1/acd$	—	14.88	—	19.76	—	1.328	[32]	1968
SmS_2	Cub.	—	—	—	—	—	—	—	[47]	1971
EuS	"	O_h^5—$Fm3m$	NaCl	5.970	—	—	—	—	[33]	1972
Eu_3S_4	"	T_d^6—$I\overline{4}3d$	Th_3P_4	8.537	—	—	—	—	[33]	1972
$Eu_2S_{3.81}$	Tetr.	—	—	7.86	—	8.03	—	1.02	[33]	1972
GdS	Cub.	O_h^5—$Fm3m$	NaCl	—	—	—	—	—	[33]	1972
α-Gd_2S_3	Rhomb.	$Pnma$	—	7.23	15.30	3.93	—	—	[224]	1968
γ-Gd_2S_3	Cub.	T_d^6—$I\overline{4}3d$	Th_3P_4	8.387	—	—	—	—	[33]	1972
GdS_2	Tetr.	—	—	7.85	—	7.96	—	1.01	[33]	1972
TbS	Cub.	—	—	5.517	—	—	—	—	[33]	1972
α-Tb_2S_3	Rhomb.	—	—	8.32	—	—	—	—	[229]	1965
γ-Tb_2S_3	Cub.	T_d^6—$I\overline{4}3d$	Th_3P_4	—	—	—	—	—	[229]	1965
DyS	"	O_h^5—$Fm3m$	NaCl	—	—	—	—	—	[33]	1972
Dy_5S_7	Monocl.	$C2/m$	—	12.785	3.813	11.565	104.85°	—	[220]	1964
γ-Dy_2S_3	Cub.	T_d^6—$I\overline{4}3d$	Th_3P_4	8.292	—	—	—	—	[33]	1972
δ-Dy_2S_3	Monocl.	$C2/m$	—	10.17	4.02	17.57	—	—	[33]	1972
DyS_2	Tetr.	—	—	7.69	—	7.85	—	1.02	[33]	1972
HoS	Cub.	O_h^5—$Fm3m$	NaCl	5.457	—	—	—	—	[33]	1972
Ho_5S_7	Monocl.	$C2/m$	—	12.729	3.796	11.515	104,83°	—	[220]	1964
Ho_2S_3	"	—	—	17.50	4.002	10,15	99° 4′	—	[33]	1972
ErS	Cub.	O_h^5—$Fm3m$	NaCl	5.624	—	—	—	—	[33]	1972
Er_5S_7	Monocl.	$C2/m$	—	12.671	3.775	11.480	104,75°	—	[220]	1964
Er_2S_3	"	—	—	10.07	4.00	17.33	—	—	[33]	1972
TmS	Cub.	O_h^5—$Fm3m$	NaCl	5.412	—	—	—	—	[33]	1972

CRYSTAL STRUCTURE (continued)

Phase	Unit cell	Space group	Structural type	a, Å	b, Å	c, Å	α	c/a	Source	Year	Notes
1	2	3	4	5	6	7	8	9	10	11	12
Tm_5S_7	Monocl.	$C2/m$	—	12,628	3.761	11.462	104.82°	—	[220]	1964	—
Tm_2S_3	Cub.	—	—	10.51	—	—	—	—	[230]	1969	Low temp.
YbS	Monocl	—	—	—	—	—	—	—	[230]	1969	—
	Cub.	O_h^5—$Fm3m$	NaCl	5.673	—	—	—	—	[33]	1972	For $YbS_{1.13}$
Yb_3S_4	Rhomb.	$Pnma$	—	12.81	12.97	3.84	—	—	[33]	1972	See [231]
Yb_2S_3	Hex.	—	Al_2O_3	6.772	—	18.28	—	2.699	[232]	1964	—
	Rhombohed.	—	Al_2O_3	7.240	—	—	55°46'	—	[232]	1964	—
	Rhomb.	—	—	6.78	9.95	3.61	—	—	[33]	1972	—
	Cub.	—	Tl_2O_3	12.47	—	—	—	—	[230]	1969	Low temp.
LuS	"	—	—	5.323	—	—	—	—	[33]	1972	—
Lu_2S_3	Hex.	—	Al_2O_3	6.73	—	18.21	—	2.706	[232]	1964	—
Lu_2S_3	Rhombohed.	—	Al_2O_3	7.207	—	—	55°40'	—	[232]	1964	—
Ac_2S_3	Cub.	T_d^6—$I\bar{4}3d$	Th_3P_4	8.99	—	—	—	—	[33]	1972	—
ThS	"	O_h^5—$Fm3m$	NaCl	5.682	—	—	—	—	[33]	1972	—
Th_2S_3	Rhomb.	D_{2h}^{16}—$Pbnm$	Sb_2S_3	10.99	10.85	3.96	—	—	[33]	1972	—
Th_4S_7	Hex.	C_{6h}^2—$C6_3/m$	Th_4S_7	11.041	—	3.983	—	0.36	[33]	1972	See [200]
ThS_2	Rhomb.	D_{2h}^{16}—$Pbnm$	$PbCl_2$	4.268	7.264	8.617	—	—	[33]	1972	—

Compound	System	Space group	Structure type	a	b	c	Angle	c/a	Ref.	Year	Notes
Th_2S_5	Tetr.	—	—	7.64	—	10.20	—	1.335	[233]	1967	
$ThS_{2.5}$	"	—	—	5.43	—	10.15	—	1.87	[33]	1972	
US	Cub.	O_h^5—$Fm3m$	NaCl	5.489	—	—	—	—	[234]	1964	
U_2S_3	Rhomb.	D_{2h}^{16}—$Pbnm$	Sb_2S_3	10.37	10.58	3.88	—	—	[234]	1964	
U_3S_5	"	$Pcmn$	—	7.39	8.10	11.75	—	—	[234]	1964	
α-US_2	Tetr.	—	—	10.25	—	6.41	—	0.625	[234]	1964	
β-US_2	Rhomb.	—	—	4.124	7.117	8.479	—	—	[235]	1968	
γ-US_2	Hex.	—	—	7.238	—	4.059	—	0.56	[33]	1972	
U_2S_5	Tetr.	—	—	7.47	—	9.88	—	1.32	[233]	1967	
US_3	Monocl.	—	—	5.45	3.92	18.22	80° 30′	—	[234]	1964	
				5.37	3.96	9.06	97.20°	—	[235]	1968	
NpS	Cub.	O_h^5—$Fm3m$	NaCl	5.532	—	—	—	—	[233]	1967	
Np_3S_4	Rhomb.	T_d^6—$I\bar{4}3d$	Th_3P_4	8.440	—	—	—	—	[233]	1967	
α-Np_2S_3	Rhomb.	—	—	—	—	—	—	—	[233]	1967	Isotope α-Pu_2S_3
β-Np_2S_3	"	—	—	—	—	—	—	—	[233]	1967	Isotope β-Pu_2S_3
γ-Np_2S_3	Cub.	—	—	—	—	—	—	—	[233]	1967	Isotope γ-Pu_2S_3
η-Np_2S_3	Rhomb.	D_{2h}^{16}—$Pbnm$	Sb_2S_3	10.32	10.62	3.86	—	—	[233]	1967	
Np_2S_5	Tetr.	—	—	7.40	—	9.84	—	1.33	[233]	1967	
Np_3S_5	Rhomb.	—	—	7.42	8.07	11.71	—	—	[233]	1967	
NpS_3	Monocl.	—	—	5.36	3.87	18.10	99° 30′	—	[233]	1967	
PuS	Cub.	O_h^5—$Fm3m$	NaCl	5.536	—	—	—	—	[228]	1966	
Pu_3S_4	"	T_d^6—$I\bar{4}3d$	Th_3P_4	8.4155	—	—	—	—	[33]	1972	
α-Pu_2S_3	Rhomb.	—	—	7.69	8.41	15.15	—	—	[227]	1966	See [227]

CRYSTAL STRUCTURE (continued)

Phase	Unit cell	Space group	Structural type	a, Å	b, Å	c, Å	α	c/a	Source	Year	Notes
1	2	3	4	5	6	7	8	9	10	11	12
β-Pu_2S_3	Rhomb.	—	—	7.18	10.50	12.98	—	—	[227]	1966	
γ-Pu_2S_3	Cub.	—	—	8.4546	—	—	—	—	[227]	1966	
PuS_2	Tetr.	—	—	7.962	—	7.962	—	~1	[227]	1966	
	Monocl.	—	—	7.962	3.981	7.962	90°	—	[33]	1972	
PuS_{2-x}	Tetr.	—	Fe_2As	3.943	—	7.962	—	2.02	[228]	1966	$0,1 < x < 0,2$
Am_2S_3	Cub.	$T_d^6 - \bar{I}43d$	Th_3P_4	8.445	—	—	—	—	[33]	1972	
Cf_2S_3	"	—	Ce_2S_3	8.388	—	—	—	—	[33]	1972	
Bk_2S_3	"	—	—	8.44	—	—	—	—	[33]	1972	
Ti_6S	Hex.	—	—	2.966	—	14.4495	—	3×1.629	[35]	1963	
Ti_2S	Rhomb.	$Pnnm$	Ta_2P	11.35	14.06	3.32	—	—	[33]	1972	
TiS	Hex.	—	—	3.41	—	26.4	—	7.7	[33]	1972	
Ti_4S_5	"	$P6_3/mmc$	—	3.439	—	28.93	—	8.413	[236]	1970	
Ti_3S_4	"	—	—	3.42	—	11.42	—	3.33	[33]	1972	
Ti_2S_3	Hex.	$P6_3mc$	—	3.42	—	4×2.86	—	0.836	[34]	1971	
	Hex.	—	—	3.42	—	10×2.86	—	0.836	[34]	1971	Poly-
	"	$R\bar{3}m$	—	3.42	—	12×2.86	—	0.836	[34]	1971	typism
	"	—	—	3.42	—	24×2.87	—	0.837	[34]	1971	
	"	—	—	3.42	—	40×2.87	—	0.837	[34]	1971	
TiS_2	Trigon.	—	CdI_2	3.405	—	5.687	—	1.67	[237]	1967	

Formula	System	Space group	Structure type	a	b	c	β	ratio	Ref.	Year	
TiS₃	Hex.	—	—	3.40	—	11,38	—	3.34	[33]	1972	
ZrS	Monocl.	—	—	4.98	3.37	17,6	—	—	[33]	1972	
Zr₂S	Cub.	—	—	10,24	14.85	—	—	—	[33]	1972	
Zr₂S₃	Rhomb.	Pnnm	Ta₂P	12.46	—	3.33	—	—	[33]	1972	
ZrS₂	Hex.	—	—	3.436	—	3.435	—	~1.00	[55]	1963	
ZrS₂	Trig.	—	CdI₂	3.67	—	5,84	—	1,59	[228]	1973	
ZrS₃	Monocl.	—	—	5.04	3.60	8,89	97,1°	3,49	[33]	1972	
Hf₂S	Hex.	—	—	3.3736	—	11,7882	—	1,02	[33]	1972	
HfS	Hex. "	—	—	3.3748	—	3,4351	—	—	[33]	1972	
HfS₂	Trigon.	—	CdI₂	3.62	—	5,82	—	1,61	[238]	1973	
HfS₃	Monocl.	—	—	5.08	3.58	8.96	98,4°	—	[33]	1972	
α-V₃S	Tetra.	I̅42m	α-V₃S	9.470	—	4.589	—	0,483	[33]	1972	
β-V₃S	Tetr.	—	—	9.381	—	4,663	—	0,493	[33]	1972	
VS	Hex.	—	—	8.987	—	3,224	—	0,358	[33]	1972	
V₁₋ₓS	Hex. "	—	—	3.36	—	5,81	—	1,73	[33]	1972	
V₁₊ₓS₂	Trigon.	—	—	6.73	—	5,82	—	0,858	[33]	1972	
V₇S₈	Monocl.	—	—	3.29	—	5,66	—	1,72	[33]	1972	
V₃S₄				11.671	6.646	23,118	90.99°	—	[240]	1972	
V₂S₃	Hex.	C2, Cm or C2/m	—	5,86	3.28	11,37	92° 31'	0,338	[239]	1965	
V₅S₈	Monocl.	—	—	6.59	—	2,213	—	—	[33]	1972	
VS₄		C2, Cm or C2/m	—	11.37	6.65	11,29	91° 31'	—	[241]	1964	
	Hex. "	—	—	12.67	10.41	12.11	148.37°	—	[33]	1972	
NbS	Hex.	—	NiAs	3,33	—	6,39	—	1,919	[35]	1963	
	Rhomb.	—	—	3.34	5,86	6.37	—	—	[242]	1969	
	Hex.	P6₃/mmc	—	3,32	—	12,52	—	2×1,946	[35]	1963	
Nb₂S₃	Hex. "	P6₃/mmc	—	3.31	—	12,65	—	2×1,911	[36]	1965	See [243]

CRYSTAL STRUCTURE (continued)

Phase	Unit cell	Space group	Structural type	a, Å	b, Å	c, Å	α	c/a	Source	Year	Notes
1	2	3	4	5	6	7	8	9	10	11	12
NbS_2	Hex.	$P2_1/m$	—	3.33	—	17.81	—	3×1.783	[35]	1963	
"	"	$R3m$	—	3.31	—	11.89	—	2×1.796	[35]	1963	
NbS_3	Monocl.	—	—	4.94	6.74	18.1	97.5°	—	[35]	1963	
Nb_3S_4	Hex.	—	—	9.583	—	3.3755	—	0.3522	[33]	1972	
$Nb_{2.1}S_8$	Tetr.	—	—	16.794	—	3.359	—	0.202	[33]	1972	
Ta_6S	Monocl.	—	—	14.158	5.284	14.789	108.01°	—	[33]	1972	
Ta_2S	Rhomb.	—	—	7.381	5.574	15.195	—	—	[33]	1972	
TaS_2	Trigon.	—	—	3.346	—	5.860	—	1.751	[35]	1963	
	Hex.	—	—	3.31	—	12.10	—	2×1.825	[35]	1963	
	Hex.	—	—	3.32	—	17.9	—	3×1.797	[35]	1963	
	"	—	—	3.33	—	35.85	—	6×1.792	[35]	1963	
	Hex.	—	—	1.92	—	5.99	—	$\sqrt{3}\times \times1.801$	[35]	1963	
TaS_3	Rhomb.	—	—	36.804	3.34	15.173	—	—	[33]	1972	
TaS_3	Monocl.	—	—	—	—	—	—	—	[36]	1965	
CrS		—	—	3.826	5.913	6.089	101° 36'	—	[33]	1972	
CrS	Trigon.	—	—	3.476	—	5.816	—	1.673	[244]	1967	
Cr_7S_8	"	—	—	3.459	—	5.761	—	1.666	[33]	1972	
Cr_5S_6	"	—	—	5.982	—	11.509	—	$\frac{2}{\sqrt{3}}\times$	[33]	1972	

Compound	System	Space group	Structure type	a	b	c	angle	ratio	Ref.	Year	Notes
Cr_3S_4	Monocl.	$C2/m$	Fe_3Se_4	5.960	3.427	11.253	91° 38′	×1.666			See [239]
Cr_2S_3	Rhombohed.	—	—	5.939	—	16.65	—	$\dfrac{-}{\sqrt{3}\times}$ ×1.619	[33]	1972	
Cr_2S_3	Trigon.	—	—	5,943	—	11.171	—	$\dfrac{2}{\sqrt{3}}\times$ ×1.628	[33]	1972	
Cr_5S_8	Monocl.	—	—	11.783	6.786	11.063	90° 82′	—	[33]	1972	
Mo_2S_3	″	—	—	8.634	3.208	6.092	102.43°	—	[33]	1972	
MoS_2	Hex.	—	—	3.16	—	12.32	—	3.89	[33]	1972	
WS_2	Rhombohed.	—	—	3.162	—	18.50	—	3×1.950	[33]	1972	
$\alpha\text{-MnS}$	Cub.	$O_h^5 - Fm3m$	NaCl	5.22	—	—	—	—	[33]	1972	
$\beta\text{-MnS}$	″	$T_d^2 - F\bar{4}3m$	ZnS	5.61	—	—	—	—	[33]	1972	
$\gamma\text{-MnS}$	Hex.	$C_{6v}^4 - P6_3mc$	Wurtzite	3.98	—	6.44	—	1.618	[33]	1972	
MnS_2	Cub.	$T_h^6 - Pa3$	FeS_2	6.108	—	—	—	—	[33]	1972	
TcS_2	Triclin.	—	—	6.465	6.375	6.659	103.61°, β = 62.97°, γ = 118.96°	—	[33]	1971	
ReS_2	Rhombohed.	—	—	3.18	—	18.2	—	3×1.91	[35]	1963	
ReS_2	Hex.	—	—	3.14	—	12.20	—	2×1.943	[35]	1963	
ReS_2	Triclin.	—	—	6.455	6.362	6.401	105.04°, β = 91.60°, γ = 118.97°	—	[33]	1971	
Re_2S_7	Tetr.	—	—	13.7	—	10.24	—	0.746	[33]	1972	
$\alpha\text{-FeS}$	Hex.	$D_{6h}^4 - P6_3/mmc$	NiAs	3.436	—	5.858	—	1.705	[33]	1972	

CRYSTAL STRUCTURE (continued)

Phase	Unit cell	Spatial group	Structural type	a, Å	b, Å	c, Å	α	c/a	Source	Year	Notes
1	2	3	4	5	6	7	8	9	10	11	12
α'-FeS	Hex.	D_{6h}^4—P6₃/mmc	NiAs	5.952	—	11.72	—	1.969	[33]	1972	
α''-FeS	"	D_{6h}^4—P6₃/mmc	NiAs	3.442	—	5.774	—	1.678	[33]	1972	
β-FeS	"	D_{6h}^4—P6₃/mmc	NiAs	3.461	—	5.789	—	1.673	[33]	1972	180° C
δ-FeS	"	D_{6h}^4—P6₃/mmc	NiAs	3.533	—	5.797	—	1.641	[33]	1972	
Fe_3S_4	Rhombohed.	—	—	—	—	—	—	—	[33]	1972	
FeS_2	Cub.	T_h^6—$Pa3$	Pyrites	5.4189	—	—	—	—	[33]	1972	
FeS_2	Rhomb.	D_{2h}^{12}—$Pnnm$	Marcasite	—	—	—	—	—	[33]	1972	
Co_9S_8	Cub.	—	Pentlandite	9.932	—	—	—	—	[33]	1972	
CoS	Hex.	D_{6h}^4—P6₃/mmc	NiAs	3.383	—	5.197	—	1.536	[33]	1972	
Co_3S_4	Cub.	—	—	9.40	—	—	—	—	[33]	1972	
CoS_2	"	—	—	5.5359	—	—	—	—	[33]	1972	
Ni_3S_2	Rhombohed.	—	—	4.08	—	—	89° 25′	—	[33]	1972	
Ni_6S_5	Rhomb.	—	—	11.22	16.56	3.27	—	—	[33]	1972	
β-NiS	Hex.	—	—	9.62	—	3.39	—	0.353	[33]	1972	
γ-NiS	"	—	—	3.43	—	5.34	—	1.56	[33]	1972	
Ni_3S_4	Cub.	—	—	9.46	—	—	—	—	[33]	1972	
NiS_2	"	—	—	5.68	—	—	—	—	[33]	1972	
RuS_2	"	T_h^6—$Pa3$	FeS_2	5.6095	—	—	—	—	[33]	1972	

Formula	System	Space group	Structure type	a	b	c	angle	c/a	Ref	Year	Notes
Rh_2S_3	Rhomb.	—	—	8.462	—	6.138	—	—	[33]	1972	
$Rh_{17}S_{15}$	Cub.	—	—	9.911	5.985	—	—	—	[245]	1969	
Pd_4S	Tetr.	—	—	5.1147	—	5.5903	—	1.095	[246]	1965	
$Pd_{2.2}S$	Cub.	—	—	8.93	—	—	—	—	[217]	1963	
PdS	Tetr.	—	—	6.43	5.54	6.63	—	1.031	[33]	1972	
PdS_2	Rhomb.	—	—	5.460	—	7.531	—	—	[33]	1972	
OsS_2	Cub.	T_h^6—$Pa3$	FeS_2	5.6196	—	—	—	—	[33]	1972	
IrS_2	"	T_h^6—$Pa3$	FeS_2	5.68	—	—	—	—	[33]	1972	For $IrS_{1.0}$
PtS	Tetr.	D_{3d}^3—$P\bar{3}m1$	—	3.4701	—	6.1092	—	1.765	[33]	1972	
PtS_2	Hex.	D_{3d}^5—$R\bar{3}m$	CdI_2	3.5432	—	5.0388	—	1.425	[33]	1972	
B_4C	Rhombohed.	D_{3d}^5—$R\bar{3}m$	B_4C	5.598	—	12.12	—	2.165	[11]	1968	
α-SiC_I	"	C_{3v}^5—$R3m$		12.73	—	—	13° 55'	—	[247]	1954	
α-SiC_{II}	Hex.	C_{6v}^4—$P6mc$		3.080	—	15.098	—	4.90	[247]	1954	
α-SiC_{III}	"	C_{6v}^4—$P6mc$		3.080	—	10.081	—	3.27	[247]	1954	Polytypism
α-SiC_{IV}	Rhombohed.	C_{3v}^5—$R3m$		17.718	—	—	9° 58'	—	[247]	1954	
α-SiC_V	"	C_{3v}^5—$R3m$		42.84	—	—	4° 07'	—	[247]	1954	
α-SiC_{VI}	"	C_{3v}^5—$R3m$		27.759	—	—	6° 21.5'	—	[247]	1954	
α-SiC_{VII}	"	C_{3v}^5—$R3m$		73.053	—	—	2° 25'	—	[247]	1954	
β-SiC	Cub.	T_d^2—$F\bar{4}3m$	ZnS	4.358	—	—	—	—	[247]	1954	
α-BN	Hex.	D_{3h}^1—$P\bar{6}m2$	BN	2.504	—	6.661	—	2.662	[248]	1969	
β-BN	Cub.	T_d^2—$F\bar{4}3m$	ZnS	3.615	—	—	—	—	[248]	1969	
γ-BN	Rhombohed.	—		2.504	—	10.01	—	4.0	[248]	1969	
AlN	Hex.	C_{6v}^4—$C6mc$	ZnS	3.111	—	4.978	—	1.600	[249]	1964	
α-Si_3N_4	"	C_{3v}^4—$P31c$	Si_3N_4	7.765	—	5.622	—	0.724	[250]	1969	
β-Si_3N_4	"	D_3^2—$P6_3/m$	—	7.606	—	2.909	—	0.3825	[42]	1969	

91

DENSITY

Phase	Density, g/cm^3		Source	Year	Notes
	pycnometric	x-ray			
1	2	3	4	5	6
MgBe$_{13}$	—	1,788	[41]	1962	
CaBe$_{13}$	—	1,904	[41]	1962	
SrBe$_{13}$	—	2,378	[42]	1969	
ScBe$_5$	—	2.381	[42]	1969	
Sc$_2$Be$_{17}$	—	2.146	[42]	1969	
ScBe$_{13}$	—	2.088	[1]	1962	
YBe$_{13}$	2.32	2.550	[42]	1969	
LaBe$_{13}$	—	2,980	[43]	1963	
CeBe$_{13}$	—	3.041	[43]	1963	
PrBe$_{13}$	—	3,076	[43]	1963	
NdBe$_{13}$	—	3,126	[43]	1963	
SmBe$_{13}$	—	3,227	[44]	1963	
EuBe$_{13}$	—	3.282	[43]	1963	
GdBe$_{13}$	—	3.363	[43]	1963	
TbBe$_{13}$	—	3,404	[43]	1963	
DyBe$_{13}$	—	3,459	[44]	1963	
HoBe$_{13}$	—	3.510	[43]	1963	
ErBe$_{13}$	—	3,543	[43]	1963	
TmBe$_{13}$	—	3.589	[43]	1963	
YbBe$_{13}$	—	3,651	[43]	1963	
LuBe$_{13}$	—	3,681	[43]	1963	
TiBe$_2$	3.23	3.262	[42]	1969	
TiBe$_3$	—	3.008	[128]	1961	
Ti$_2$Be$_{17}$	—	2.429	[42]	1969	
TiBe$_{12}$	—	2,288	[42]	1969	
ZrBe$_2$	—	4.429	[42]	1969	
ZrBe$_5$	—	3.598	[126]	1959	
Zr$_2$Be$_{17}$	—	3,081	[126]	1959	
ZrBe$_{13}$	—	2,712	[42]	1969	
HfBe$_2$	—	8.322	[42]	1969	
HfBe$_5$	—	6.006	[42]	1969	
Hf$_2$Be$_{17}$	—	4.763	[42]	1969	
HfBe$_{13}$	—	3.920	[45]	1961	
VBe$_2$	—	3.834	[42]	1969	
VBe$_{12}$	—	2,367	[130]	1957	
NbBe$_2$	—	5.279	[127]	1959	
NbBe$_3$	—	4,726	[127]	1959	
Nb$_2$Be$_{17}$	3,31	3,276	[42]	1969	
NbBe$_{12}$	—	2,881	[130]	1957	
Ta$_3$Be$_2$	—	13.270	[128]	1961	
TaBe$_2$	9.79	9,577	[42]	1969	
TaBe$_3$	—	8,345	[128]	1961	
Ta$_2$Be$_{17}$	—	5.041	[128]	1961	
TaBe$_{12}$	—	4,182	[130]	1957	
CrBe$_2$	—	4.245	[42]	1969	
CrBe$_{12}$	—	2,437	[42]	1969	
Mo$_3$Be	—	8,425	[129]	1960	
MoBe$_2$	5.92	6,058	[42]	1969	
MoBe$_{12}$	3,13	3,027	[42]	1969	

DENSITY (continued)

Phase	Density, g/cm^3		Source	Year	Notes
	pycnometric	x-ray			
1	2	3	4	5	6
$MoBe_{22}$	2.51	2.485	[42]	1969	
WBe_2	—	10.744	[42]	1969	
WBe_{12}	—	4.243	[42]	1969	
WBe_{22}	—	3,225	[42]	1969	
$MnBe_2$	—	4.494	[42]	1969	
$MnBe_{12}$	—	2,403	[130]	1957	
$ReBe_2$	—	11,632	[42]	1969	
$ReBe_{22}$	—	3,304	[129]	1960	
$FeBe_2$	—	4,632	[42]	1969	
$FeBe_5$	3,17	3.289	[42]	1969	
$FeBe_{12}$	~2.50	2.446	[130]	1957	
$CoBe$	—	6.307	[42]	1969	
$CoBe_5$	—	3,446	[130]	1957	
$CoBe_{12}$	—	2,493	[130]	1957	
$NiBe$	6,01	6.252	[42]	1969	
Ni_5Be_{21}	—	3.688	[42]	1969	
Ru_3Be_{17}	—	4,158	[42]	1969	
Ru_2Be_{17}	3.40	3.54	[49]	1971	
$RhBe$	—	9,036	[2]	1966	
$RhBe_2$	6.82	6.775	[48]	1971	
$RhBe_{3,4}$	—	5.314	[50]	1971	
Rh_5Be_{22}	4,66	4,781	[2]	1966	
Rh_2Be_{17}	3,63	3.60	[49]	1971	
$PdBe$	—	8.540	[42]	1969	
$PdBe_5$	4,56	4.670	[42]	1969	
$PdBe_{12}$	—	3.169	[131]	1959	
Os_2Be_{17}	5,20	5.27	[49]	1971	
Os_3Be_{17}	—	6.586	[42]	1969	
Ir_2Be_{17}	—	5.38	[49]	1971	
$PtBe_{12}$	—	4.521	[131]	1959	
$ThBe_{13}$	—	4.169	[42]	1969	
UBe_{13}	4,420	4.371	[143]	1962	
$NpBe_{13}$	—	4.347	[42]	1969	
$PuBe_{13}$	4.35—4.36	4.357	[42]	1969	
$AmBe_{13}$	—	4.399	[42]	1969	
Be_5B	2.06—2,14	—	[42]	1969	
Be_4B	2,1	—	[51]	1973	
Be_2B	2.01—2.2	1.9	[51]	1973	
BeB_2	2.42	—	[51]	1973	
BeB_6	2,35	2,33	[51]	1973	
BeB_{12}	2,36	2.342	[51]	1973	
MgB_2	2.48—2,67	2.63	[42]	1969	
MgB_4	2,495	—	[251]	1972	
MgB_6	2,45—2,47	—	[42]	1969	
MgB_{12}	2,44	—	[42]	1969	
CaB_6	2.49	2.44	[54]	1961	
SrB_6	3,39	3.42	[54]	1961	
BaB_6	4,26	4,25	[54]	1961	
ScB_2	3,65	3.667	[55]	1958	See [56]
ScB_4	—	2.713	[42]	1969	

DENSITY (continued)

Phase	Density, g/cm³		Source	Year	Notes
	pycnometric	x-ray			
1	2	3	4	5	6
ScB_6	—	2,090	[42]	1969	
ScB_{12}	—	2,836	[42]	1969	Cub.
ScB_{12}	2,883	2,890	[56]	1970	Tetr.
YB_2	—	3,370	[42]	1969	
YB_3	—	3,97	[42]	1969	
YB_4	3.98—4,20	4,356	[42]	1969	
YB_6	3,72	3.723	[54]	1961	
YB_{12}	—	3.43	[6]	1966	
LaB_3	—	4,92	[42]	1969	
LaB_4	—	5,392	[42]	1969	
LaB_6	4,76	4.722	[54]	1961	
CeB_4	—	5.734	[42]	1969	
CeB_6	4.87	4,796	[54]	1961	
PrB_3	—	5,20	[42]	1969	
PrB_4	—	5.739	[42]	1969	
PrB_6	4,53	4,84	[54]	1961	
NdB_4	—	5,824	[132]	1959	
NdB_6	4.86	4,94	[42]	1969	
SmB_4	—	6,138	[132]	1959	
SmB_6	—	5,090	[132]	1959	
EuB_6	4.88	4,91	[252]	1973	
GdB_3	—	6.03	[42]	1969	
GdB_4	—	6.444	[132]	1959	See [6]
GdB_6	5.0	5.30	[42]	1969	See [6]
TbB_4	—	6.576	[132]	1959	
TbB_6	—	5.422	[132]	1959	
TbB_{12}	—	4.536	[42]	1969	
DyB_4	—	6.744	[132]	1959	
DyB_6	—	5.486	[132]	1959	
DyB_{12}	—	4.598	[42]	1969	
HoB_4	—	6.925	[132]	1959	
HoB_6	—	5,551	[132]	1959	
HoB_{12}	—	4.652	[42]	1969	
ErB_4	—	6.994	[132]	1959	
ErB_6	5.58	5,58	[42]	1969	
ErB_{12}	—	4.704	[42]	1969	
TmB_4	—	7.084	[58]	1961	
TmB_6	5,55	5.59	[58]	1961	
TmB_{12}	—	4.746	[42]	1969	
YbB_3	—	6.74	[42]	1969	
YbB_4	—	7.306	[42]	1969	
YbB_6	5.50	5.54	[252]	1973	
YbB_{12}	—	4,825	[59]	1971	
LuB_2	—	9,656	[42]	1969	
LuB_4	—	7.515	[60]	1958	
LuB_6	—	5,701	[61]	1958	
LuB_{12}	—	4.866	[42]	1969	
ThB_4	8.4	8,44	[42]	1969	
ThB_6	6.99	7.10	[57]	1968	
ThB_{18}	4.42	—	[57]	1968	

Phase	Density, g/cm^3		Source	Year	Notes
	pycnometric	x-ray			
1	2	3	4	5	6
ThB_{66}	2,98	2,92	[57]	1968	
UB	—	14,219	[42]	1969	
UB_2	—	12.692	[42]	1969	
UB_4	9,32	9.378	[42]	1969	
UB_{12}	5.65	5.87	[42]	1969	
PuB	—	13.935	[133]	1960	
PuB_2	—	12,674	[133]	1960	
PuB_4	—	9,265	[133]	1960	
PuB_6	—	7.566	[133]	1960	
TiB	5.09	4,565	[42]	1969	
Ti_3B_4	—	4.558	[42]	1969	
TiB_2	4,38	4,530	[42]	1969	
Ti_2B_5	—	4,627	[42]	1969	
ZrB	5.7	6,48	[42]	1969	
ZrB_2	6.17	6,09	[42]	1969	
ZrB_{12}	3,70	3.609	[42]	1969	
HfB	—	12.405	[42]	1969	
HfB_2	10,5	11,2	[42]	1969	
V_3B_2	—	5,787	[67]	1963	
VB	—	5.620	[67]	1963	
V_3B_4	—	5.427	[67]	1963	
VB_2	5,06—5,28	5,062	[254]	1960	
Nb_3B_2	—	7.906	[67]	1963	
NbB	—	7.561	[67]	1963	
Nb_3B_4	—	7.277	[67]	1963	
NbB_2	6.97	7.00	[42]	1969	
Ta_2B	—	15,240	[67]	1963	
Ta_3B_2	—	14,923	[67]	1963	
TaB	14.0	14.204	[42]	1969	
Ta_3B_4	13,50	13,60	[42]	1969	
TaB_2	12,38	12,62	[42]	1969	
Cr_4B	—	6,279	[42]	1969	
Cr_2B	6,11	6,57	[42]	1969	
Cr_5B_3	6,10	6,12	[42]	1969	
CrB	6.05	6,11	[42]	1969	
Cr_3B_4	—	5.763	[42]	1969	
CrB_2	5,22	5.60	[42]	1969	
CrB_6	—	3,60	[42]	1969	
Mo_2B	9.1	9,254	[42]	1969	
Mo_3B_2	—	9,071	[42]	1969	
MoB	8,3	8.665	[42]	1969	
MoB_2	8.01	7,988	[42]	1969	
Mo_2B_5	7.01	7.48	[42]	1969	
MoB_4	4,8	4,96	[42]	1969	
W_2B	16,0	17,13	[42]	1969	
WB	15,1—15.4	15,734	[42]	1969	
WB_2	9.6	—	[124]	1962	
W_2B_5	11,0	13,111	[42]	1969	
WB_4	8,3	8,40	[42]	1969	
Mn_4B	6,60	6,865	[42]	1969	

Phase	Density, g/cm^3		Source	Year	Notes
	pycnometric	x-ray			
1	2	3	4	5	6
Mn_2B	7.20	7.186	[42]	1969	
MnB	6.45	6.320	[42]	1969	
Mn_3B_4	5.90	5.983	[42]	1969	
MnB_2	—	5.344	[71]	1960	
Tc_3B	—	10.651	[73]	1964	
Tc_7B_3	—	10.581	[73]	1964	
TcB_2	—	7.417	[73]	1964	
Re_3B	—	19.355	[74]	1960	
Re_7B_3	—	18.629	[75]	1960	
Re_2B_5	—	13.433	[77]	1958	
ReB_2	—	12.669	[42]	1969	
ReB_3	—	11.66	[75]	1960	
Fe_3B	—	7.383	[78]	1962	
Fe_2B	—	7.336	[42]	1969	
FeB	7.15	6,706	[42]	1969	
Co_3B	—	8.131	[80]	1958	
Co_2B	7,9—8.3	8,048	[42]	1969	
CoB	7.25	7,323	[42]	1969	
Ni_3B	8.17	8,19	[42]	1969	
Ni_2B	7.90	8,030	[42]	1969	
Ni_3B_2	7.5	—	[42]	1969	
Ni_4B_3	—	7.56	[42]	1969	
NiB	7.13	7,195	[42]	1969	
NiB_{12}	—	3,107	[42]	1969	
Ru_7B_3	—	10.789	[42]	1969	
Ru_2B_5	—	9,181	[84]	1962	
RuB_2	—	7.567	[42]	1969	
Rh_7B_3	—	10.823	[75]	1960	
Rh_2B	10,5	8.963	[42]	1969	
$RhB_{\sim1,1}$	—	9.514	[75]	1960	
Pd_3B	—	10,925	[78]	1962	
Pd_5B_2	—	10.684	[78]	1962	
Pd_3B_2	—	11,819	[42]	1969	
OsB_2	—	12.828	[85]	1962	
Os_2B_5	—	15.234	[84]	1962	
$IrB_{0.9}$	—	17,3	[87]	1971	
$IrB_{\sim1.1}$	—	16,730	[84]	1962	
$IrB_{1,35}$	—	14,704	[84]	1962	
PtB	—	17,253	[75]	1960	
Be_2C	2.26	2.44	[11]	1968	See [88]
Mg_2C_3	—	2.204	[42]	1969	
MgC_2	—	2.073	[125]	1962	
CaC_2	2.1	2,204	[42]	1969	Tetr.
CaC_2	2.2	—	[42]	1969	Triclinic
SrC_2	3.19	3.285	[125]	1962	
BaC_2	3,75	3,895	[124]	1962	Tetr.
ScC	3,06	4.045	[42]	1969	Cub. see [89]
ScC	3.60	3.577	[42]	1969	Hex.
YC	—	5.38	[11]	1968	

Phase	Density, g/cm^3		Source	Year	Notes
	pycnometric	x-ray			
1	2	3	4	5	6
Y_2C_3	3.66	—	[92]	1962	
YC_2	4.13	4.528	[42]	1969	
La_2C_3	5.992	6.078	[42]	1969	
α-LaC_2	5.29	5.35	[11]	1968	Tetr.
β-LaC_2	—	5.0	[11]	1968	Cub.
CeC	—	7.483	[125]	1962	
Ce_2C_3	—	6.948	[94]	1958	
CeC_2	5.47	5.60	[93]	1964	
Pr_2C_3	—	6.620	[94]	1958	
PrC_2	5.67	5.74	[93]	1964	
Nd_2C_3	—	6.902	[94]	1958	
NdC_2	5.93	5.97	[11]	1968	
Sm_3C	—	8.139	[11]	1968	
Sm_2C_3	—	7.477	[94]	1958	
SmC_2	6.421	6.50	[11]	1968	
Eu_4C	—	7 570	[42]	1969	
Gd_3C	—	8.701	[11]	1968	
Gd_2C_3	—	8.024	[94]	1958	
GdC_2	6.93	6.939	[11]	1968	
Tb_3C	—	8.882	[11]	1968	
Tb_2C_3	—	8.335	[11]	1968	
TbC_2	7.09	7.176	[11]	1968	
Dy_3C	—	9.211	[11]	1968	
Dy_2C_3	—	8.703	[94]	1958	
DyC_2	7.38	7.45	[11]	1968	
Ho_3C	—	9.434	[11]	1968	
Ho_2C_3	—	8.892	[94]	1958	
HoC_2	7.76	7.701	[11]	1968	
Er_3C	—	8.915	[94]	1958	
ErC_2	7.70	7.954	[11]	1968	
Tm_3C	—	9.901	[11]	1968	
TmC_2	—	8.175	[94]	1958	
Yb_3C	—	10.26	[11]	1968	
YbC_2	7.97	8.097	[11]	1968	
Lu_3C	—	10.54	[11]	1968	
LuC_2	—	8.728	[94]	1958	
ThC	—	10.605	[42]	1969	
ThC_2	9.5	9.736	[42]	1969	
UC	12.97	13.63	[100]	1963	
U_2C_3	12.7	12.88	[11]	1968	
α-UC_2	11.28	11.79	[11]	1968	
PuC	13.52	13 486	[42]	1969	
Pu_2C_3	—	12.73	[42]	1969	
PuC_2	—	10.881	[42]	1969	
Np_2C_3	—	12.785	[301]	1968	
TiC	4.93	4.92	[11]	1968	
ZrC	6.73	6.66	[11]	1968	
HfC	12.60	12.67	[11]	1968	
V_2C	—	5.75	[11]	1968	
VC	5.36	5.48	[11]	1968	

Phase	Density, g/cm^3		Source	Year	Notes
	pycnometric	x-ray			
1	2	3	4	5	6
Nb$_2$C	7,86	7,85	[11]	1968	
NbC	7,56	7,82	[11]	1968	
Ta$_2$C	14.8	15,04	[11]	1968	
TaC	14.3	14,4	[11]	1968	
Cr$_{23}$C$_6$	6.97	6,98	[11]	1968	
Cr$_7$C$_3$	6,92	6,97	[11]	1968	
Cr$_3$C$_2$	6,68	6.74	[11]	1968	
Mo$_2$C	9.04	9,18	[42]	1969	
MoC	8,4	8,88	[102]	1968	
W$_2$C	17,2	17,34	[11]	1968	
WC	15,5—15.7	15,77	[11]	1968	
Mn$_{23}$C$_6$	—	7.478	[125]	1962	
Mn$_{15}$C$_4$	—	7,30	[76]	1967	
Mn$_3$C	6,89	7.53	[256]	1957	
Mn$_5$C$_2$	—	7.36	[257]	1974	
Mn$_7$C$_3$	—	7,356	[257]	1974	
Fe$_3$C	7,67	7.69	[256]	1957	
Fe$_2$C	—	7.185	[125]	1962	
Co$_3$C	—	8.098	[112]	1964	
Co$_2$C	—	7.651	[125]	1962	
Ni$_3$C	7,96	7.88	[11]	1968	
ReC	17,9	19.5	[113]	1971	Synthesized at $p \geqslant$ 60 kbar
α-Be$_3$N$_2$	2,709	—	[88]	1972	<1400° C
β-Be$_3$N$_2$	2,696	—	[88]	1972	>1400° C
Mg$_3$N$_2$	2.71—2.74	2.704	[42]	1969	
α-Ca$_3$N$_2$	—	2.67	[258]	1964	
β-Ca$_3$N$_2$	—	2.64	[258]	1964	
γ-Ca$_3$N$_2$	—	2.63	[258]	1964	
Sr$_2$N	3,56	3.547	[115]	1972	
ScN	4,2	4.298	[42]	1969	
YN	5,60	5,887	[42]	1969	
LaN	6,73	6,813	[42]	1969	
CeN	7,89	8.076	[42]	1969	
PrN	—	7.465	[42]	1969	
NdN	—	7.693	[42]	1969	
SmN	—	8.494	[134]	1956	
EuN	—	8.743	[134]	1956	
GdN	—	9,103	[134]	1956	
TbN	—	9,566	[134]	1956	
DyN	—	9,932	[134]	1956	
HoN	—	10.572	[134]	1956	
ErN	—	10.621	[134]	1956	
TmN	—	10,923	[134]	1956	
YbN	—	11.328	[134]	1956	
LuN	—	11,592	[134]	1956	
ThN	—	11.554	[125]	1962	
Th$_3$N$_4$	—	10.55	[42]	1969	Hex.
Th$_2$N$_3$	—	10.399	[124]	1962	
UN	—	14,402	[42]	1969	

Phase	Density, g/cm^3		Source	Year	Notes
	pycnometric	x-ray			
1	2	3	4	5	6
U_2N_3	—	11.302	[100]	1963	Cub.
UN_2	11,3	11,799	[100]	1963	
NpN	—	14,191	[124]	1962	
PuN	—	14.151	[42]	1969	
Ti_2N	4,86	4,913	[42]	1969	
TiN	5,43	5,44	[259]	1959	
ZrN	7,09	7,35	[259]	1959	
HfN	11.696	13.386	[14]	1969	
V_3N	5,967	5,987	[14]	1969	
VN	6,040	6,102	[14]	1969	
Nb_2N	8.08	8.231	[100]	1963	
γ-NbN$_{0.75}$	8,32	—	[14]	1969	
ε-NbN	8,40	8,305	[42]	1969	
δ-NbN	8,30	—	[14]	1969	
Ta_2N	15.42—15,46	15,78—15,86	[14]	1969	
TaN	13.80	14,36	[14]	1969	
Cr_2N	—	6.51	[14]	1969	
CrN	6,1	6,141	[42]	1969	Cub. see [14]
Mo_2N	8.04	9.441	[42]	1969	
MoN	8.60	9,18	[14]	1969	
W_2N	—	17.893	[42]	1969	See [14]
WN	12,12	15,94	[42]	1969	
Mn_4N	—	6.789	[125]	1962	
Mn_5N_2	—	6,21	[124]	1962	
Mn_2N	6,2—6,6	6,572	[42]	1969	
Mn_3N_2	6,6	8.68	[124]	1962	
MnN	—	5.247	[125]	1962	
Re_2N	—	21,188	[125]	1962	
Fe_4N	6,57	7.21	[42]	1969	
Fe_3N	—	7.353	[125]	1962	
Fe_2N	6,35	7,08	[42]	1969	
Co_3N	7,1	7,884	[125]	1962	
Co_2N	6,4—6.5	7,66	[124]	1962	
Ni_3N	7,66	7.916	[124]	1962	
$CaAl_4$	2.33	2.33	[124]	1962	
$CaAl_2$	2.35	2,405	[125]	1962	
$SrAl_4$	2.98	2.948	[125]	1962	
SrAl	2,8	2.782	[124]	1962	
$BaAl_4$	3.52	3,54	[124]	1962	
BaAl	—	3,924	[125]	1962	
$ScAl_2$	—	3,017	[122]	1964	
$ScAl_3$	—	3,032	[122]	1964	
Y_2Al	4.3	4,344	[135]	1965	
Y_3Al_2	—	4.101	[135]	1965	
YAl	—	3,921	[42]	1969	
YAl_2	—	3,933	[42]	1969	
YAl_3	3,6	3,613	[42]	1969	
YAl_4	3.94	3.847	[42]	1969	
La_3Al	—	5,971	[137]	1965	Hex.
La_3Al	—	5.576	[42	1969	Cub.

Phase	Density, g/cm^3		Source	Year	Notes
	pycnometric	x-ray			
1	2	3	4	5	6
LaAl	—	5.145	[137]	1965	
LaAl$_2$	4.79	4.740	[137]	1965	
LaAl$_{\sim 2.4}$	4.1	—	[137]	1965	
LaAl$_3$	—	4.120	[137]	1965	
LaAl$_4$	—	4.167	[137]	1965	Rhomb.
LaAl$_4$	3.86	4.080	[136]	1964	Tetr.
Ce$_3$Al	—	6.342	[42]	1969	Hex.
CeAl	—	5.411	[42]	1969	Rhomb.
CeAl$_2$	4.97	4.924	[138]	1960	
CeAl$_3$	—	4.297	[42]	1969	
CeAl$_4$	4.02	4.253	[136]	1964	Tetr.
CeAl$_4$	4.05	—	[139]	1966	Rhomb.
Pr$_3$Al	—	6.586	[42]	1969	Hex.
Pr$_2$Al	—	6.011	[139]	1966	
PrAl	—	5.525	[140]	1965	
PrAl$_2$	—	5.007	[138]	1960	
PrAl$_3$	—	4.36	[42]	1969	
PrAl$_4$	—	4.342	[136]	1964	Tetr.
PrAl$_4$	4.05	—	[139]	1966	Rhomb.
Nd$_3$Al	—	6.713	[141]	1965	
Nd$_2$Al	—	6.241	[141]	1965	
NdAl	—	5.697	[141]	1965	Rhomb.
NdAl	6.05	5.442	[124]	1962	Cub.
NdAl$_2$	—	5.141	[138]	1960	
NdAl$_3$	—	4.475	[141]	1965	
NdAl$_4$	—	4.451	[144]	1961	Tetr.
Sm$_3$Al	—	6.741	[145]	1966	
Sm$_2$Al	—	6.703	[42]	1969	
SmAl	—	6.049	[140]	1965	Rhomb.
SmAl	—	5.632	[42]	1969	Cub.
SmAl$_2$	—	5.689	[138]	1960	
SmAl$_3$	—	4.739	[42]	1969	
SmAl$_4$	—	4.710	[42]	1969	
Gd$_2$Al	—	7.116	[141]	1965	
Gd$_3$Al$_2$	—	6.549	[141]	1965	
GdAl	—	6.374	[141]	1965	Rhomb.
GdAl	—	5.937	[42]	1969	Cub.
GdAl$_2$	—	5.689	[42]	1969	
GdAl$_3$	—	4.979	[136]	1964	
GdAl$_4$	—	4.568	[141]	1965	
Tb$_3$Al$_2$	—	6.833	[140]	1965	
TbAl	—	6.621	[140]	1965	
TbAl$_2$	—	5.807	[138]	1960	
TbAl$_3$	—	5.124	[136]	1964	Hex.
TbAl$_4$	—	4.661	[136]	1964	
Dy$_3$Al$_2$	—	7.135	[140]	1965	
DyAl	—	6.783	[140]	1965	Rhomb.
DyAl	—	6.298	[42]	1969	Cub.
DyAl$_2$	5.65	5.965	[138]	1960	
DyAl$_3$	5.3	5.267	[136]	1964	

Phase	Density, g/cm³		Source	Year	Notes
	pycnometric	x-ray			
1	2	3	4	5	6
Ho_3Al_2	—	7.232	[140]	1965	
HoAl	—	6.892	[140]	1965	
$HoAl_2$	—	6,095	[138]	1960	
$HoAl_3$	—	5,324	[42]	1969	Cub.
$HoAl_3$	5.4	5.372	[136]	1964	Hex.
Er_2Al	8.0	8.062	[135]	1965	
Er_3Al_2	—	7.339	[140]	1965	
ErAl	—	7.082	[140]	1965	
$ErAl_2$	—	6,203	[135]	1965	
$ErAl_3$	5.30	5,498	[42]	1969	
$EuAl_2$	—	5.098	[42]	1969	
$EuAl_4$	—	3.997	[42]	1969	
$YbAl_2$	—	6.168	[142]	1960	
$YbAl_3$	—	5.683	[136]	1964	
TmAl	—	7.215	[140]	1965	
$TmAl_2$	—	6.286	[142]	1960	
$TmAl_3$	—	5,598	[42]	1969	
$LuAl_2$	—	6,552	[142]	1960	
Th_3Al	—	14,183	[18]	1965	
Th_2Al	9,7	9.604	[42]	1969	
Th_3Al_2	8.98	8.945	[42]	1969	
ThAl	8.10	8,111	[42]	1969	
$ThAl_2$	6.84	6.822	[42]	1969	
$ThAl_3$	—	6.141	[42]	1969	
UAl_2	8.3	8,137	[124]	1962	
UAl_3	6.8	6,72	[143]	1962	
UAl_4	5.7	6.06	[143]	1962	
$NpAl_2$	—	8,189	[125]	1962	
$NpAl_3$	—	6,817	[145]	1966	
$NpAl_4$	—	6.037	[125]	1962	
Pu_3Al	—	13,459	[143]	1962	
$PuAl_2$	—	8.089	[143]	1962	
$PuAl_3$	—	6,894	[146]	1965	Hex., <1188 K
$PuAl_3$	6,4	—	[42]	1969	Hex., 1300 – 1483 K
$PuAl_3$	—	6.862	[146]	1965	Cub., >1483 K
$PuAl_4$	—	6,117	[143]	1962	
Ti_3Al	—	4.216	[42]	1969	
TiAl	3.84	3.63	[42]	1969	
$TiAl_2$	—	3.512	[42]	1969	
$TiAl_3$	3.31	3.371	[42]	1969	
Zr_3Al	—	5.971	[147]	1962	
Zr_2Al	5.78	5.654	[42]	1969	
Zr_5Al_3	5.61	5,392	[42]	1969	
Zr_3Al_2	5.35	5.34	[42]	1969	
Zr_4Al_3	5.28	5,372	[42]	1969	
ZrAl	4.8	5.022	[42]	1969	
Zr_2Al_3	4.79	4.70	[42]	1969	
$ZrAl_2$	4.42	4.561	[42]	1969	
$ZrAl_3$	4,11	4.098	[42]	1969	

Phase	Density, g/cm³		Source	Year	Notes
	pycnometric	x-ray			
1	2	3	4	5	6
Hf_2Al	—	10,283	[19]	1961	
Hf_3Al_2	—	9.941	[147]	1962	
Hf_4Al_3	—	9.844	[147]	1962	
HfAl	8.97	9.042	[42]	1969	
Hf_2Al_3	8.00	8.036	[42]	1969	
$HfAl_2$	—	7.532	[147]	1962	
$HfAl_3$	6,37	6.31	[42]	1969	
V_3Al	—	4,995	[148]	1963	
V_5Al_8	4.0	4,004	[42]	1969	
VAl_3	3.57	3.685	[149]	1964	
V_4Al_{23}	3.089	3.134	[42]	1969	
VAl_6	3.1	3.195	[124]	1962	
V_7Al_{45}	3,10	3,099	[42]	1969	
VAl_{10}	2,79	—	[42]	1969	
VAl_{11}	2.9	3.052	[124]	1962	
Nb_3Al	—	7,272	[42]	1969	
Nb_2Al	6.87	6.895	[42]	1969	
$NbAl_3$	4.52	4.538	[42]	1969	
Ta_2Al	—	12.774	[42]	1969	
$TaAl_3$	6,73	6.889	[125]	1962	
Cr_2Al	—	5.633	[125]	1962	
Cr_5Al_8	—	4.261	[125]	1962	
$CrAl_7$	—	2.881	[125]	1962	Monocl.
$CrAl_7$	3.18	3.027	[124]	1962	Rhomb.
Mo_3Al	—	8.619	[42]	1969	
Mo_3Al_8	5.28	5.049	[42]	1962	Monocl. >1403 K
Mo_3Al_8	4.60	4,443	[42]	1969	Tetr.
$MoAl_4$	4.35	4,287	[42]	1969	
$MoAl_5$	3.90	3.981	[42]	1969	
$MoAl_7$	—	3.169	[42]	1969	
$MoAl_{12}$	3.28	—	[42]	1969	
WAl_4	6.6	6.040	[42]	1969	
WAl_5	5.5	5.742	[124]	1962	
WAl_{12}	—	3.869	[42]	1969	
$Mn_{1,11}Al_{0.89}$	5.16	5.195	[42]	1969	
Mn_4Al_{11}	4.41	—	[42]	1969	
$MnAl_3$	3.90	—	[42]	1969	
$MnAl_3$	—	3.896	[151]	1960	
Mn_3Al_{10}	3.65	3.708	[42]	1969	
$MnAl_4$	3.45	3,624	[151]	1960	
$MnAl_6$	3,27	3.307	[125]	1962	
$MnAl_{12}$	3:0	3.221	[125]	1962	
Fe_3Al	—	6.648	[124]	1962	
FeAl	—	5.585	[124]	1962	
Fe_2Al_5	—	3.963	[125]	1962	
$FeAl_3$	3,77	3.844	[42]	1969	$FeAl_{3.25}$
$FeAl_6$	3.45	3,407	[42]	1969	
CoAl	6,07	6:084	[42]	1969	
Co_2Al_5	4.19	4,329	[124]	1962	
$CoAl_3$	3,81	4,034	[42]	1969	$CoAl_{3.25}$

DENSITY (continued)

Phase	Density, g/cm^3		Source	Year	Notes
	pycnometric	x-ray			
1	2	3	4	5	6
Co_2Al_9	3,60	3.572	[124]	1962	
Ni_3Al	—	7.293	[125]	1962	
$NiAl$	—	5.910	[42]	1969	$NiAl_{1,008}$
Ni_2Al_3	—	4.787	[125]	1962	
$NiAl_3$	3.81	3.957	[78]	1962	
Tc_2Al	—	8.895	[42]	1969	
Tc_2Al_3	—	6.022	[42]	1969	
$TcAl_4$	—	4.723	[42]	1969	
$TcAl_6$	—	3.83	[42]	1969	
$RuAl$	—	7.641	[152]	1963	
$RuAl_2$	—	6.206	[42]	1969	
Rh_3Al	—	10.089	[125]	1962	
$RhAl$	—	8.10	[145]	1966	
Pd_2Al	—	9.339	[42]	1969	
Pd_5Al_3	—	9.073	[42]	1969	
$PdAl$	5.96	8.717	[42]	1969	Hex.
$PdAl$	—	7.812	[18[	1965	Cub. >1128 K
Pd_2Al_3	—	6.132	[125]	1962	
$PdAl_3$	4.59	4.583	[42]	1969	
$ReAl$	—	14.812	[42]	1969	
$ReAl_4$	6.4	6.485	[42]	1969	
$ReAl_6$	—	5,10	[42]	1969	
$ReAl_{12}$	3.90	3.97	[42]	1969	
$OsAl$	—	13.288	[145]	1966	
$IrAl$	—	13.709	[145]	1966	
Pt_3Al	17,40	17.453	[153]	1964	
Pt_5Al_3	15,28	15.342	[42]	1969	
$PtAl$	—	12.80	[42]	1969	
Pt_2Al_3	—	9.862	[42]	1969	<1794 K
Pt_2Al_3	9.65	—	[153]	1964	
$PtAl_2$	—	7.963	[42]	1969	
Pt_5Al_{21}	5,65	5.70	[153]	1964	
Mg_2Si	—	1.988	[42]	1969	
Ca_2Si	—	1.570	[42]	1969	
$CaSi$	—	2.386	[86]	1961	
$CaSi_2$	—	2.399	[125]	1962	
$SrSi$	3,39	3.490	[42]	1969	
Sr_2Si_3	—	3.197	[42]	1969	
$SrSi_2$	3.35	3.401	[42]	1969	
$BaSi$	—	4.433	[42]	1969	
$BaSi_2$	—	3.985	[42]	1969	
Sc_5Si_3	—	3.299	[42]	1969	
$ScSi$	—	3.363	[42]	1969	
$ScSi_{1,66}$	—	3.393	[42]	1969	
Y_5Si_3	4,36	4.514	[42]	1969	
Y_5Si_4	—	4.510	[158]	1967	
YSi	4.33	4.515	[159]	1966	
YSi_2	4.5	4.501	[42]	1969	Rhomb.
YSi_2	—	4.265	[42]	1969	Hex.
La_5Si_3	—	5.828	[160]	1965	

DENSITY (continued)

Phase	Density, g/cm^3		Source	Year	Notes
	pycnometric	x-ray			
1	2	3	4	5	6
La$_5$Si$_4$	—	5,372	[158]	1967	
La$_3$Si$_2$	5.59	5,633	[156]	1965	
LaSi	5,38	5,43	[159]	1966	
LaSi$_2$	5.0	5.14	[42]	1969	
Ce$_5$Si$_3$	—	6.079	[160]	1965	
Ce$_5$Si$_4$	—	5,707	[158]	1967	
Ce$_3$Si$_2$	—	5,979	[42]	1969	
CeSi	—	5.694	[159]	1966	
CeSi$_2$	5.31	5.40	[42]	1969	
Pr$_5$Si$_3$	—	5.962	[160]	1965	
Pr$_5$Si$_4$	—	5,829	[158]	1967	
Pr$_3$Si$_2$	—	6,044	[42]	1969	
PrSi	—	5,837	[159]	1966	
PrSi$_2$	5.46	5,564	[125]	1962	
Nd$_5$Si$_3$	—	6.303	[160]	1965	
Nd$_5$Si$_4$	—	6.046	[158]	1967	
NdSi	—	6.086	[159]	1966	
NdSi$_2$	4.7	5,657	[42]	1969	
Sm$_5$Si$_3$	—	6,781	[160]	1965	
Sm$_5$Si$_4$	—	6,547	[158]	1967	
SmSi	—	6.518	[159]	1966	
SmSi$_2$	5.14	6.15	[42]	1969	
EuSi	—	5,693	[42]	1969	
EuSi$_2$	—	5.497	[42]	1969	
Gd$_5$Si$_3$	—	7,212	[42]	1969	
Gd$_5$Si$_4$	—	7.062	[158]	1967	
GdSi	—	6.976	[159]	1966	
GdSi$_2$	5.94	6.429	[42]	1969	
Tb$_5$Si$_3$	—	7.556	[42]	1969	
Tb$_5$Si$_4$	—	7,247	[158]	1967	
TbSi	—	7.173	[159]	1966	
TbSi$_2$	—	6.719	[42]	1969	
Dy$_5$Si$_3$	—	7,839	[42]	1969	
Dy$_5$Si$_4$	—	7,533	[158]	1967	
DySi	—	7.454	[159]	1966	
DySi$_2$	5.2	6.861	[42]	1969	
Ho$_5$Si$_3$	—	8.05	[42]	1969	
HoSi	—	7.647	[159]	1966	
HoSi$_2$	—	7.087	[42]	1969	
Er$_5$Si$_3$	8.28	8.267	[42]	1969	
Er$_5$Si$_4$	—	7,983	[158]	1967	
ErSi	7.73	7.852	[42]	1969	Type TlI
ErSi	—	7.874	[159]	1966	Type FeB
ErSi$_2$	—	7,257	[42]	1969	
Tm$_5$Si$_3$	—	8,466	[42]	1969	
TmSi	—	8.00	[42]	1969	
TmSi$_2$	—	7.448	[42]	1969	
Yb$_5$Si$_3$	—	8.682	[42]	1969	
YbSi	—	8,169	[42]	1969	
YbSi$_2$	—	7,539	[42]	1969	

DENSITY (continued)

Phase	Density, g/cm³		Source	Year	Notes
	pycnometric	x-ray			
1	2	3	4	5	6
Lu_5Si_3	—	8.884	[42]	1969	
Lu_5Si_4	—	8,658	[158]	1967	
$LuSi$	—	8,461	[42]	1969	
$LuSi_2$	—	7,800	[42]	1969	
Th_3Si_2	—	9.79	[143]	1962	
$ThSi$	—	8,962	[42]	1969	
$ThSi_{1,67}$	—	7.960	[42]	1969	
$ThSi_2$	—	7.908	[42]	1969	Tetr.
$ThSi_2$	—	7.828	[42]	1969	Hex.
U_3Si	—	15,590	[86]	1961	
U_3Si_2	—	12.204	[86]	1961	
USi	10,40	10,598	[100]	1963	
α-USi_2	9,0	8,98	[42]	1969	
β-USi_2	9.2	9,25	[42]	1969	
USi_3	—	8,12	[42]	1969	
$NpSi_2$	—	9.027	[42]	1969	
Pu_5Si_3	12.0	11,982	[42]	1969	
$PuSi$	—	10.152	[143]	1962	
$PuSi_2$	9,08	9,08	[86]	1961	
Ti_5Si_3	—	4,315	[125]	1962	
$TiSi$	4.21	4.32	[42]	1969	
$TiSi_2$	4,02	4,043	[42]	1969	
Zr_2Si	5.99	6,04	[260]	1959	
Zr_5Si_3	5.90	6.04	[260]	1959	
Zr_5Si_4	5,674	5,682	[42]	1969	
$ZrSi$	5,56	5.654	[169]	1965	
$ZrSi_2$	4.88	4.86	[260]	1959	
Hf_2Si	—	11,685	[119]	1961	
Hf_5Si_3	—	10.837	[119]	1961	
$HfSi$	—	10,270	[119]	1961	
$HfSi_2$	7.2	7,981	[42]	1969	
V_3Si	5.67	5,706	[42]	1969	
V_5Si_3	4.80	5.321	[168]	1957	
VSi_2	4.34	4.627	[42]	1969	
Nb_4Si	8.01	—	[260]	1959	
α-Nb_5Si_3	6,56	7.13	[260]	1959	
β-Nb_5Si_3	7.34	7,19	[260]	1959	
$NbSi_2$	5.45	5,659	[261]	1969	
$Ta_{4,5}Si$	12,7	12.86	[260]	1959	
Ta_2Si	12.4	13.554	[168]	1957	
Ta_5Si_3	11 6	13.401	[168]	1957	
$TaSi_2$	8.83	9.1	[260]	1959	
Cr_3Si	—	6.429	[168]	1957	
Cr_5Si_3	5.6	5,864	[168]	1957	
$CrSi$	5,18	5 36	[86]	1961	
$CrSi_2$	4.91	4.978	[170]	1963	
Mo_3Si	—	8.968	[168]	1957	
Mo_5Si_3	7.4	8.24	[260]	1959	
$MoSi_2$	5.9—6.3	6.24	[260]	1959	
W_5Si_3	14.43	14,523	[42]	1969	

DENSITY (continued)

Phase	Density, g/cm³ pycnometric	Density, g/cm³ x-ray	Source	Year	Notes
1	2	3	4	5	6
WSi_2	9.25	9.857	[171]	1964	
α-Mn_3Si	—	6.71	[42]	1969	
β-Mn_3Si	—	6.60	[42]	1969	
Mn_5Si_3	—	5.985	[247]	1954	
$MnSi$	—	5.826	[119]	1961	
Tc_3Si	—	9.822	[42]	1969	
Tc_5Si_3	—	8.968	[42]	1969	
$TcSi$	—	7.848	[42]	1969	
$TcSi_{1.750}$	—	6.605	[42]	1969	
Re_5Si_3	—	15.433	[42]	1969	
$ReSi$	—	13.077	[42]	1969	
$ReSi_2$	—	10.694	[86]	1961	
Fe_3Si	—	7.181	[175]	1968	
Fe_5Si_3	—	6.474	[42]	1969	
$FeSi$	—	6.162	[42]	1969	
α-$FeSi_2$	4.69	—	[42]	1969	
β-$FeSi_2$	4.93	4.94	[178]	1971	
Co_2Si	—	7.46	[247]	1954	
$CoSi$	—	6.582	[42]	1969	
$CoSi_2$	4.94	4.95	[42]	1969	
Ni_3Si	—	7.857	[261]	1958	
Ni_2Si	—	7.405	[42]	1969	
$NiSi$	—	5.927	[42]	1969	
$NiSi_2$	—	4.828	[42]	1969	
Ru_2Si	—	9.749	[183]	1962	
$RuSi$	—	8.246	[183]	1962	
$OsSi$	—	13.704	[183]	1962	
$OsSi_{2.4}$	—	10.024	[183]	1962	
Rh_2Si	—	9.900	[183]	1962	
Rh_5Si_3	—	9.562	[42]	1969	
$Rh_{20}Si_{13}$	—	9.128	[42]	1969	
$RhSi$	8.3	8.513	[42]	1969	
Ir_3Si	—	18.511	[42]	1969	
Ir_2Si	—	17.066	[42]	1969	
$IrSi$	—	13.065	[42]	1969	
Pd_3Si	—	10.118	[187]	1960	
Pd_2Si	—	9.589	[261]	1958	
$PdSi$	—	7.693	[261]	1958	
Pt_3Si	—	17.549	[42]	1969	
Pt_2Si	—	16.268	[42]	1969	
Pt_6Si_5	—	13.373	[188]	1964	
$PtSi$	—	12.394	[189]	1950	
ScP	3.33	3.364	[42]	1969	
YP	—	4.388	[42]	1969	
LaP	—	5.157	[125]	1962	
CeP	—	5.506	[42]	1969	
PrP	—	5.636	[42]	1969	
NdP	—	5.847	[42]	1969	
SmP	6.28	6.34	[191]	1965	
GdP	—	6.668	[42]	1969	

DENSITY (continued)

Phase	Density, g/cm^3		Source	Year	Notes
	pycnometric	x-ray			
1	2	3	4	5	6
TbP	—	5,859	[42]	1969	
DyP	—	7,107	[42]	1969	
HoP	—	7,304	[42]	1969	
ErP	—	7,472	[42]	1969	
TmP	—	7,669	[42]	1969	
YbP	—	7,907	[42]	1969	
ThP	—	8,813	[42]	1969	
Th$_3$P$_4$	8.44	8,506	[125]	1962	
UP	—	10,232	[42]	1969	
UP	9.69	—	[125]	1962	
U$_3$P$_4$	—	10,039	[42]	1969	
UP$_2$	—	8,886	[42]	1969	
PuP	—	9,974	[42]	1969	
Ti$_3$P	4,68	4,624	[193]	1965	
Ti$_2$P	4,66	4,67	[193]	1965	
Ti$_5$P$_3$	4,85	4,82	[42]	1969	
Ti$_4$P$_3$	4,60	4,615	[193]	1965	
Ti$_3$P$_2$	4,66	4,646	[193]	1965	
TiP	4,08	4,269	[193]	1965	
TiP$_2$	—	4,272	[42]	1969	
Zr$_3$P	—	3,902	[195]	1966	
ZrP	5,35	5,498	[42]	1969	
ZrP$_2$	—	5,098	[195]	1966	
Hf$_3$P$_2$	—	11,1	[198]	1968	
HfP	—	9,742	[197]	1962	
HfP$_2$	—	8,165	[42]	1969	
V$_{12}$P$_7$	—	5,600	[200]	1970	
VP	4,72—4,98	4,993	[42]	1969	
VP$_2$	—	4,557	[42]	1969	
Nb$_3$P	—	7,875	[201]	1966	
NbP	6.48	6,504	[203]	1963	
NbP$_2$	—	5,392	[42]	1969	
Ta$_3$P	—	14,746	[201]	1966	
TaP	10,9	11,143	[42]	1969	
TaP$_2$	—	8,521	[42]	1969	
Cr$_3$P	6.25	6,501	[42]	1969	
CrP	5,25	5,484	[42]	1969	
Mo$_3$P	8.60	9,087	[42]	1969	
MoP	6,58—7,33	7,271	[42]	1969	
WP	11.7—12.3	12,365	[124]	1962	
WP$_2$	9,17	9,308	[42]	1969	
Mn$_3$P	—	6,753	[209]	1962	
Mn$_2$P	6,02	6,333	[125]	1962	
MnP	5,60	5,776	[42]	1969	
Re$_2$P	15,50	16,387	[42]	1969	
Fe$_3$P	6.92	7,132	[124]	1962	
Fe$_2$P	6,89	6,90	[124]	1962	
FeP	6,07	—	[124]	1962	
FeP	—	6,187	[204]	1962	
FeP$_2$	—	5,070	[42]	1969	

Phase	Density, g/cm³		Source	Year	Notes
	pycnometric	x-ray			
1	2	3	4	5	6
Co_2P	7.4	7.546	[42]	1969	
CoP	—	6.402	[125]	1962	
CoP_3	4.26	4,407	[42]	1969	
Ni_3P	7,9	7,833	[125]	1962	
Ni_2P	7.2	7.331	[42]	1969	
NiP_2	—	4,576	[215]	1961	
NiP_3	4.16	4.212	[42]	1969	
Ru_2P	—	9.855	[42]	1969	
RuP	—	8,194	[42]	1969	
RuP_2	5,88	6,261	[42]	1969	
Rh_2P	—	9.367	[125]	1962	
RhP_2	—	6.119	[215]	1961	
RhP_3	—	5.087	[42]	1969	
Pd_3P	—	10.275	[216]	1960	
PdP_2	—	5.638	[218]	1963	
PdP_3	—	5.787	[42]	1969	
OsP_2	—	9.541	[42]	1969	
Ir_2P	—	16,163	[125]	1962	
IrP_2	—	9,325	[215]	1961	
IrP_3	—	7.354	[42]	1969	
PtP_2	—	9,245	[42]	1969	
BeS	2,36	2,47	[33]	1972	
MgS	2.68	2.86	[33]	1972	
CaS	—	2,80	[33]	1972	
SrS	3.64	3,67	[33]	1972	
BaS	4.25	4.33	[33]	1972	
ScS	3.59	3.66	[33]	1972	
Sc_2S_3	2.897	2.917	[33]	1972	
YS	4,51	4,92	[33]	1972	
Y_5S_7	—	4,09	[220]	1964	
Y_2S_3	3.87	3.87	[33]	1972	
YS_2	4,25	4.35	[33]	1972	
LaS	—	5,61	[222]	1964	
La_3S_4	5.34	5,44	[33]	1972	
α-La_2S_3	4.90	5.00	[33]	1972	
	—	4.81	[224]	1968	
β-La_2S_3	4.93	4.98	[33]	1972	
γ-La_2S_3	4,997	—	[33]	1972	
La_5S_7	—	4,87	[32]	1968	
LaS_2	4.77	—	[33]	1972	
CeS	—	5,85	[222]	1964	
Ce_3S_4	5,51	5.67	[33]	1972	
α-Ce_2S_3	4,95	4,99	[224]	1968	
γ-Ce_2S_3	5.25	5,19	[33]	1972	
Ce_5S_7	—	5,27	[32]	1968	
CeS_2	4,96	5,07	[33]	1972	
PrS	—	5.96	[222]	1964	
Pr_3S_4	5,57	5.77	[33]	1972	
α-Pr_2S_3	5,09	5.14	[224]	1968	
γ-Pr_2S_3	5.27	5,27	[33]	1972	

Phase	Density, g/cm^3		Source	Year	Notes
	pycnometric	x-ray			
1	2	3	4	5	6
Pr_5S_7		5,39	[32]	1968	
NdS	—	6,23	[222]	1964	
Nd_3S_4	5,91	6,02	[33]	1972	
α-Nd_2S_3	5.26	5,45	[224]	1968	
β-Nd_2S_3	5,16	5,20	[33]	1972	
γ-Nd_2S_3	5,49	5,50	[33]	1972	
Nd_5S_7	—	5,68	[32]	1968	
Nd_4S_7	5,29	5,32	[33]	1972	
NdS_2	5.32	5,36	[33]	1972	
SmS	5.64	6,01	[33]	1972	
Sm_3S_4	6,11	6,14	[33]	1972	
α-Sm_2S_3	5,69	5,80	[224]	1968	
γ-Sm_2S_3	5,87	5,83	[262]	1965	
Sm_5S_7	—	5,93	[32]	1968	
EuS	5,71	5,75	[263]	1959	
Eu_3S_4	6.26	6,27	[263]	1959	
EuS_2	5.70	5.70	[263]	1959	$Eu_2S_{3,81}$
α-Gd_2S_3	6,14	6.187	[33]	1972	See [224]
γ-Gd_2S_3	6,06	6,15	[33]	1972	
GdS_2	5,90	5,98	[33]	1972	
Tb_2S_3	6,28	6,34	[229]	1965	Cub.
Dy_5S_7	6.14	6,35	[33]	1972	
α-Dy_2S_3	6,08	—	[33]	1972	
γ-Dy_2S_3	6,48	6,54	[33]	1972	
δ-Dy_2S_3	5,75	5,91	[33]	1972	
DyS_2	6,11	6,48	[33]	1972	
Ho_2S_3	5.92	6,06	[229]	1965	
ErS	6.75	7,10	[33]	1972	$ErS_{1,18}$
Er_5S_7	6.39	6.71	[33]	1972	
Er_2S_3	6.07	6.21	[33]	1972	
Tm_2S_3	6,27	—	[229]	1965	
Yb_2S_3	—	6,02	[33]	1972	
Lu_2S_3	7,30	—	[229]	1965	
ThS	—	9,56	[33]	1972	
Th_2S_3	7.57	7,87	[33]	1972	
Th_4S_7	6.91	7.65	[33]	1972	
ThS_2	7.3	7,36	[33]	1972	
US	10,51	10,86	[33]	1972	
U_2S_3	8.94	9,01	[33]	1972	
U_3S_5	8,30	8,34	[33]	1972	
α-US_2	7,57	7,60	[33]	1972	
β-US_2	8,07	8.09	[33]	1972	
γ-US_2	8,12	8.18	[33]	1972	
Np_2S_3	—	8,14	[33]	1972	
PuS	—	10,61	[33]	1972	
γ-Pu_2S_3	—	8.41	[33]	1972	
PuS_{2-x}	8,04	—	[228]	1966	$x = 0,1$
PuS_2	7,97	—	[228]	1966	
TiS	4.09	4,57	[264]	1964	
Ti_2S_3	3,56	3.87	[264]	1964	

DENSITY (continued)

Phase	Density, g/cm³		Source	Year	Notes
	pycnometric	x-ray			
1	2	3	4	5	6
TiS_2	3.27	3.22	[264]	1964	
TiS_3	3.22	3,25	[264]	1964	
ZrS_2	3.82	—	[238]	1973	
Hf_2S	—	11,0	[33]	1972	
HfS_2	6,18	—	[238]	1973	
VS	—	4.51	[33]	1972	
V_2S_3	—	4.7	[33]	1972	
Nb_2S_3	5,9	6.0	[243]	1966	
Ta_6S	15.18	15.21	[33]	1972	
$1s-TaS_2$	7.10	7.16	[33]	1972	
$2s-TaS_2$	6;65	6,93	[33]	1972	
$3s-TaS_2$	—	7.0	[33]	1972	$Ta_{1,15}S_2$
$6s-TaS_2$	6.85	7:02	[33]	1972	
TaS_3	—	5.959	[33]	1972	
CrS	—	4.091	[33]	1972	
Cr_7S_8	—	4,303	[33]	1972	$Cr_{0.87}S$
Cr_5S_6	—	4,261	[33]	1972	
Cr_3S_4	—	4.158	[33]	1972	$Cr_{0.76}S$
Cr_2S_3	—	3,922	[33]	1972	Rhombohed.
Cr_2S_3	—	3.77	[33]	1972	Trigon.
Mo_2S_3	5,75	5.806	[33]	1972	
WS_2	—	7.55	[33]	1972	Hex.
α-MnS	3.99	—	[33]	1972	
TcS_2	—	5,066	[33]	1972	
ReS	7,11	—	[265]	1963	
ReS_2	7,506	—	[33]	1972	Rhombohed.
Re_2S_7	4.866	—	[33]	1972	
FeS_2	4.7—5.2	—	[33]	1972	Pyrites
FeS_2	4,87	—	[33]	1972	Marcasite
Co_9S_8	5.321	—	[33]	1972	
β-CoS	5.45	—	[33]	1972	
CoS_2	4,269	4.8206	[33]	1972	$CoS_{1.97}$
β-NiS	5.6	—	[33]	1972	
γ-NiS	5,36	—	[33]	1972	
Ni_3S_4	4.5—4.8	—	[33]	1972	
NiS_2	4.45	—	[33]	1972	
PdS_2	4,92	—	[33]	1972	63 kbar; 1450° C
B_4C	2.51	2,517	[42]	1969	
β-SiC	—	3.208	[42]	1969	
α-BN	2,20	2.29	[248]	1969	
β-BN	—	3,45	[248]	1969	
γ-BN	—	~1.80	[248]	1969	
AlN	3,12	3.27	[249]	1964	
α-Si_3N_4	3,19	3.184	[248]	1969	
β-Si_3N_4	3.21	3.187	[248]	1969	

TEMPERATURE STABILITY RANGES

Phase	Maximum resistance temp., °C	Nature of conversion*	Source	Year	Notes
1	2	3	4	5	6
$MgBe_{13}$	950	D	[2]	1966	
$ScBe_{13}$	>1300	P	[2]	1966	
YBe_{13}	~1920	C	[266]	1967	
$LaBe_{13}$	1720	P	[2]	1966	
$CeBe_{13}$	1720	P	[2]	1966	
$ThBe_{13}$	~1930	C	[2]	1966	
UBe_{13}	~2000	C	[2]	1966	
$NpBe_{13}$	>1400	P	[2]	1966	
$PuBe_{13}$	1600	C	[2]	1966	
$AmBe_{13}$	1375	P	[2]	1966	
$TiBe_3$	1450	C	[267]	1970	
$TiBe_2$	1350	Pr	[267]	1970	
Ti_2Be_{17}	1632	P	[268]	1973	
$TiBe_{12}$	1593	Pr	[268]	1973	
$ZrBe_2$	1235	Pr	[267]	1970	
$ZrBe_5$	1475	Pr	[267]	1970	
Zr_2Be_{17}	1555	Pr	[267]	1970	
$ZrBe_{13}$	1645	C	[267]	1970	
$HfBe_2$	1330	Pr	[267]	1970	
$HfBe_5$	1585	Pr	[267]	1970	
Hf_2Be_{17}	1860	P	[267]	1970	
$HfBe_{13}$	1620	Pr	[267]	1970	
VBe_{12}	~1700	P	[267]	1970	
Nb_3Be_2	>1700	C	[268]	1973	
$NbBe_2$	1630	Pr	[268]	1973	
$NbBe_3$	>1650	C	[268]	1973	
Nb_2Be_{17}	1800	C	[268]	1973	
$NbBe_{12}$	1672	Pr	[268]	1973	
$TaBe_2$	1845	P	[2]	1966	
Ta_2Be_{17}	1980	P	[2]	1966	
$TaBe_{12}$	1850	P	[2]	1966	
$CrBe_2$	1840	C	[2]	1966	
$MoBe_2$	1840	P	[2]	1966	
$MoBe_{12}$	~1700	P	[2]	1966	
WBe_2	~2250	C	[2]	1966	
$FeBe_2$	1480	C	[2]	1966	
$FeBe_5$	1375	Pr	[2]	1966	
$FeBe_{12}$	1075	Pd	[2]	1966	
$CoBe$	1505	C	[2]	1966	
Co_5Be_{21}	1270	C	[2]	1966	
$NiBe$	1472	C	[2]	1966	
Ni_5Be_{21}	1262	C	[2]	1966	
Pd_3Be	960	Pr	[125]	1962	
Pd_2Be	1090	Pr	[125]	1962	
Pd_3Be_2	1170	Pr	[125]	1962	
Pd_4Be_3	1200	Pr	[125]	1962	
$Pd_{13}Be_{12}$	1055	Pd	[125]	1962	

TEMPERATURE STABILITY RANGES (continued)

Phase	Maximum resistance temp., °C	Nature of con-version	Source	Year	Notes
1	2	3	4	5	6
PdBe	1465	C	[125]	1962	
Be$_4$B	1140	Pr	[51]	1973	
Be$_2$B	985—1500	Pr	[51]	1973	
BeB$_2$	>2000	P	[51]	1973	
BeB$_4$	>2000	P	[51]	1973	
BeB$_6$	2020—2120	P	[51]	1973	
BeB$_{12}$	2300	P	[51]	1973	
MgB$_2$	800—1150	D	[268]	1973	
MgB$_6$	1150	—	[267]	1970	
MgB$_{12}$	1700	—	[267]	1970	
CaB$_6$	2235	P	[268]	1973	
SrB$_6$	2230	P	[269]	1961	
BaB$_6$	2270	P	[268]	1973	
ScB$_2$	1800	S	[56]	1970	
	2250	P	[270]	1960	
ScB$_{12}$	2040	C	[56]	1970	
YB$_2$	2100	C	[266]	1967	
YB$_4$	2800	C	[266]	1967	See [276]
YB$_6$	2600	Pr	[266]	1967	See [271]
YB$_{12}$	2200	Pr	[266]	1967	
YB$_{\sim 70}$	2000	Pr	[266]	1967	
LaB$_4$	1850	D	[276]	1972	
LaB$_6$	2530	C	[272]	1963	
CeB$_4$	2200	D	[276]	1972	
CeB$_6$	2290	C	[274]	1956	See [273]
PrB$_4$	1950	D	[276]	1972	
PrB$_6$	>2250	C	[268]	1973	
NdB$_4$	1850	D	[276]	1972	
NdB$_6$	2450	C	[267]	1970	
SmB$_4$	1650	D	[276]	1972	
SmB$_6$	2540	C	[275]	1959	
GdB$_2$	1300	—	[6]	1966	
GdB$_4$	>2500	C	[6]	1966	See [276]
GdB$_6$	~2500	Pr	[6]	1966	
TbB$_4$ ⎱ TbB$_6$ ⎰	>2100	P	[268]	1973	See [276]
DyB$_4$ ⎱ DyB$_6$ ⎰	>2090	P	[268]	1973	See [276]
YbB$_6$	>2000	P	[267]	1970	
ThB$_4$	>2200	C	[267]	1970	See [273]
ThB$_6$	2150	C	[277]	1956	See [273]
UB$_2$	2385	C	[64]	1959	
UB$_4$	2495	C	[64]	1959	
UB$_{12}$	2235	C	[64]	1959	
PuB	2050	Pr	[268]	1973	
PuB$_2$	1825	P	[268]	1973	
PuB$_4$	2050	Pr	[268]	1973	
PuB$_6$	2100	Pr	[268]	1973	

Phase	Maximum resistance temp., °C	Nature of conversion	Source	Year	Notes
1	2	3	4	5	6
Ti_2B	1800—2200	Pr	[125]	1962	
TiB	~2060	Pr	[125]	1962	
TiB_2	2790	C	[125]	1962	See [267]
ZrB_2	3200	C	[65]	1970	
ZrB_{12}	~2030	Pr	[65]	1970	
HfB_2	3250	C	[125]	1962	
V_3B_2	2070	C	[253]	1959	
VB	2250	C	[253]	1959	
V_3B_4	2350	Pr	[253]	1959	
VB_2	2400	C	[253]	1959	
Nb_3B_2	1850	Pr	[255]	1959	
NbB	2260	C	[255]	1959	
Nb_3B_4	2700	Pr	[255]	1959	
NbB_2	3000	C	[255]	1959	
Ta_2B	2040—2380	Pr	[278]	1971	
Ta_3B_2	1970	PD	[278]	1971	
TaB	3090	C	[278]	1971	
Ta_3B_4	3030	Pr	[278]	1971	
TaB_2	3037	C	[278]	1971	
Cr_2B	1870	Pr	[281]	1969	
Cr_5B_3	1900	Pr	[281]	1969	
CrB	2100	C	[281]	1969	
Cr_3B_4	2070	Pr	[281]	1969	
CrB_2	2200	C	[281]	1969	
CrB_4	1400—1600	Pr	[109]	1969	
Mo_2B	2270	Pr	[279]	1967	
Mo_3B_2	1850—2070	Pr	[125]	1962	
α-MoB	1900—2000	PI	[279]	1967	
β-MoB	2000—2550	C	[279]	1967	
MoB_2	1500—2350	Pr	[279]	1967	
Mo_2B_5	2200	Pr	[279]	1967	
MoB_{12}	2020	C	[279]	1967	
W_2B	2740	C	[282]	1967	
α-WB	2400	PI	[282]	1967	
β-WB	2400—2800	C	[282]	1967	
W_2B_5	2370	Pr	[282]	1967	
WB_{12}	2440	C	[282]	1967	
Mn_4B	1285	Pr	[9]	1967	
Mn_2B	1580	C	[9]	1967	
MnB	1890	C	[9]	1967	
Mn_3B_4	1750	Pr	[9]	1967	
MnB_2	1988	C	[9]	1967	
MnB_4	2160	C	[9]	1967	See [72]
Re_3B	~2150	Pr	[280]	1968	
Re_7B_3	~2000	Pr	[280]	1968	
ReB_2	~2400	C	[280]	1968	
Fe_2B	1410	Pr	[283]	1969	
FeB	1650	C	[283]	1969	

Phase	Maximum resistance temp., °C	Nature of conversion	Source	Year	Notes
1	2	3	4	5	6
$FeB_{\sim19}$	1980	Pr	[283]	1969	
Co_3B	1110	Pr	[284]	1960	
Co_2B	1265	C	[284]	1960	
Ni_3B	1175	C	[285]	1967	
Ni_2B	1240	C	[285]	1967	
Ni_4B_3	1580	C	[285]	1967	
NiB	1590	Pr	[285]	1967	
$NiB_{\sim12}$	2320	Pr	[285]	1967	
Ru_7B_3	~1660	Pr	[152]	1963	
RuB	1500	C	[152]	1963	
Ru_2B_3	1550	C	[152]	1963	
RuB_2	1600	Pr	[152]	1963	
Pd_3B_2	~1020	Pr	[267]	1970	
$\alpha\text{-}IrB_{0.9}$	~1200	PI	[286]	1971	
$\beta\text{-}IrB_{0.9}$	>1200	—	[286]	1971	
Be_2C	2400	P	[11]	1968	See [287]
Mg_2C_3	~660	D	[125]	1962	
MgC_2	~600	D	[125]	1962	
CaC_2I	450	PI	[125]	1962	
CaC_2IV	450—2300	P	[11]	1968	
CaC_2II	25	PI	[125]	1962	Metastable to 200°C
SrC_2	—30	PI	[125]	1962	Structure with low symmetry
	—30—370	PI	[125]	1962	Tetr.
	370—1900	D	[11]	1968	Cub.
BaC_2	150	PI	[125]	1962	Tetr.
	150—1770	D	[125]	1962	Cub.
ScC	1800	P	[11]	1968	
YC	1950	Pr	[267]	1970	
Y_2C_3	~1500	PD	[268]	1973	See [11]
YC_2	1320	PI	[288]	1967	Tetr.
	1320—2300	C	[288]	1967	Cub.
La_2C_3	1415	Pr	[267]	1970	
$\delta\text{-}LaC_2$	1800	PI	[267]	1970	See [288]
$\varepsilon\text{-}LaC_2$	1800—2356	C	[267]	1970	
Ce_2C_3	>1700	Pr	[268]	1973	
CeC_2	1090	PI	[288]	1967	Tetr.
	1090—2290	C	[288]	1967	Cub.
Pr_2C_3	1560	P	[11]	1968	
PrC_2	1135	PI	[288]	1967	Tetr.
	1135—2120	P	[288]	1967	Cub.
NdC_2	1150	PI	[288]	1967	Tetr.
	1150—2207	P	[288]	1967	Cub.
SmC_2	1170	PI	[288]	1967	Tetr.
	1170—2200	P	[288]	1967	Cub.
GdC_2	1270	PI	[288]	1967	Tetr.
	>1270	—	[288]	1967	Cub.

Phase	Maximum resistance temp., °C	Nature of conversion	Source	Year	Notes
1	2	3	4	5	6
TbC$_2$	1285	PI	[288]	1967	Tetr.
	>1285	—	[288]	1967	Cub.
DyC$_2$	1295	PI	[288]	1967	Tetr.
	>1295	—	[288]	1967	Cub.
HoC$_2$	1305	PI	[288]	1967	Tetr.
	>1305	—	[288]	1967	Cub.
ErC$_2$	1325	PI	[288]	1967	Tetr.
	>1325	—	[288]	1967	Cub.
TmC$_2$	1355	PI	[288]	1967	Tetr.
	>1355	—	[288]	1967	Cub.
YbC$_2$	800	DS	[94]	1958	
LuC$_2$	1500	PI	[288]	1967	Tetr.
	>1500	—	[288]	1967	Cub.
ThC	2625	P	[11]	1968	
ThC$_2$	2655	P	[11]	1968	At 1425 and 1485°C polymorphic inversions; at 1300°C ordered and disordered conversions [268]
UC	2520	P	[267]	1970	
U$_2$C$_3$	1880	D	[268]	1973	
α-UC$_2$	1820	PI	[11]	1968	
β-UC$_2$	>1820	—	[11]	1968	
Pu$_3$C$_2$	574	PD	[11]	1968	
PuC	1660	Pr	[292]	1969	
Pu$_2$C$_3$	2050	Pr	[11]	1968	
PuC$_2$	~1750—~2250	Pr	[11]	1968	
TiC	3257	C	[11]	1968	
ZrC	3530	C	[11]	1968	
HfC	3890	C	[11]	1968	
V$_2$C	2187	Pr	[289]	1973	
VC	2648	C	[289]	1973	43% (at.) C
V$_8$C$_7$	~1123	—	[290]	1972	Ordered and disordered conversions
V$_6$C$_5$	~1271				
β-Nb$_2$C	~2500	PI	[289]	1973	
γ-Nb$_2$C	~2500—3035	Pr	[289]	1973	
NbC	3613	C	[289]	1973	44% (at.) C
α-Ta$_2$C	2100	PI	[289]	1973	See [293]
β-Ta$_2$C	~2100—3330	Pr	[289]	1973	
TaC	3985	C	[289]	1973	47% (at.) C
Cr$_{23}$C$_6$	1518	Pr	[125]	1962	
Cr$_7$C$_3$	1782	Pr	[125]	1962	
Cr$_3$C$_2$	1895	Pr	[125]	1962	
α-Mo$_2$C	2480	C	[268]	1973	
β-Mo$_2$C	1475—2522	C	[268]	1973	

Phase	Maximum resistance temp., °C	Nature of con- version	Source	Year	Notes
1	2	3	4	5	6
η-MoC	1655—2550	C	[268]	1973	
α-MoC	1960—2600	C	[268]	1973	
W_2C	1300—2795	C	[268]	1973	30% (at.) C
α-WC	2755	PD	[268]	1973	
β-WC	2525—2785	Pr	[268]	1973	
$Mn_{23}C_6$	1010	PD	[125]	1962	
$Mn_{15}C_4$	850—1000	—	[76]	1967	
α-Mn_3C	1037	PI	[287]	1973	
β-Mn_3C	1037—1520	—	[287]	1973	
Mn_7C_3	>1340	—	[125]	1962	
Fe_3C	1650	P	[11]	1968	
ε-Fe_2C	420	P	[11]	1968	
χ-Fe_2C	550	D	[11]	1968	
Co_3C	477	D	[267]	1970	
Co_2C	450	D	[267]	1970	
Ni_3C	370	D	[11]	1968	
Be_3N_2	1367—1677	DS	[268]	1973	
α-Be_3N_2	1400	PI	[88]	1972	
β-Be_3N_2	1400—2200	PI	[88]	1972	
γ-Mg_3N_2	550	P	[267]	1970	
β-Mg_3N_2	550—788	PI	[267]	1970	
γ-Mg_3N_2	788—1520	DS	[267]	1970	
β-Ca_3N_2	700	PI	[14]	1969	
α-Ca_3N_2	700—1195	P	[14]	1969	
γ-Ca_3N_2	>1050	—	[14]	1969	
Sr_3N_2	1027	P	[14]	1969	⎫ Decompose at
Ba_3N_2	1000	P	[14]	1969	⎬ 400- 500°C in ⎭ vacuum
ScN	2650	P	[14]	1969	
YN	~1600	DS	[268]	1973	
YN	>2670	DS	[14]	1969	
CeN	2575	DS	[796]	1972	
NdN	~1600	DS	[268]	1973	
GdN	>2500	P	[268]	1973	
ThN	2790	C	[268]	1973	
Th_3N_4	1500	D	[14]	1969	
Th_3N_4	2100	P	[14]	1969	
UN	2800	D	[268]	1973	$p = 1$ atm.
UN	2850	C	[268]	1973	$p = 2.5$ atm.
U_2N_3	1345	D	[268]	1973	$p = 1$ atm.
NpN	2675	D	[268]	1973	$p = 1$ atm.
NpN	2830	C	[268]	1973	$p = \sim10$ atm.
PuN	1600	—	[14]	1969	Rapid volatiliza- tion
PuN	2750	P	[14]	1969	
α-Ti_2N	900	PI	[116]	1969	$\alpha \rightarrow \beta$
β-Ti_2N	900—1400	PI	[116]	1969	$\beta \rightarrow TiN_{0.5}$ (cub.)
TiN	2950	P	[14]	1969	

Phase	Maximum resistance temp., °C	Nature of conversion	Source	Year	Notes
1	2	3	4	5	6
ZrN	2980	P	[14]	1969	
	1800	D	[268]	1973	
HfN	~3000	P	[14]	1969	
VN	2050	D	[14]	1969	
Nb_2N	2400	—	[14]	1969	
ε-NbN	1300	PI	[268]	1973	
δ-NbN	1300—2300	D	[291]	1968	
γ-Nb_4N_3	1500	—	[14]	1969	
TaN	3087	P	[14]	1969	
Ta_3N_5	900	D	[118]	1968	
CrN	1500	D	[14]	1969	
Mo_3N	>600	D	[14]	1969	
Mo_2N, MoN	~700	D	[291]	1968	
W_2N, WN	~700	D	[291]	1968	
Mn_4N	>1000	—	[14]	1969	
Mn_5N_2	~750	D	[14]	1969	
Mn_3N_2	710	D	[14]	1969	
Mn_6N_5	580	D	[14]	1969	
Re_3N, Re_2N	280	D	[14]	1969	
Fe_4N	680	D	[14]	1969	
Fe_3N	~550	D	[111]	1967	
Fe_2N	~400	D	[111]	1967	
Co_2N	276	D	[14]	1969	
Ni_3N	360	D	[14]	1969	
$CaAl_2$	1079	C	[18]	1965	
$CaAl_4$	700	Pr	[18]	1965	
$BaAl_4$	1097	C	[267]	1970	
Y_2Al	1020	Pr	[266]	1967	
Y_2Al_3	1085	Pr	[266]	1967	
YAl	1165	Pr	[266]	1967	
YAl_2	1455	C	[266]	1967	
YAl_3	1355	Pr	[266]	1967	
LaAl	859	Pr	[18]	1965	
$LaAl_2$	1422	C	[18]	1965	
α-$LaAl_4$	813	PI	[18]	1965	
β-$LaAl_4$	813—1222	Pr	[18]	1965	
α-Ce_3Al	250	PI	[18]	1965	
β-Ce_3Al	250—655	C	[18]	1965	
CeAl	845	Pr	[18]	1965	
$CeAl_2$	1465	C	[18]	1965	
α-$CeAl_4$	1004	PI	[18]	1965	
β-$CeAl_4$	1004—1243	Pr	[18]	1965	
α-Pr_3Al	400—600	PI	[268]	1973	
β-Pr_3Al	>400—600	—	[268]	1973	
PrAl	892	Pr	[18]	1965	
$PrAl_2$	1442	C	[18]	1965	
α-$PrAl_4$	1018	PI	[18]	1965	
β-$PrAl_4$	1018—1244	Pr	[18]	1965	

Phase	Maximum resistance temp., °C	Nature of conversion	Source	Year	Notes
1	2	3	4	5	6
NdAl	935	Pr	[18]	1965	
$NdAl_2$	1450	C	[18]	1965	
α-$NdAl_4$	810	PI	[18]	1965	
β-$NdAl_4$	810—1250	Pr	[18]	1965	
α-Sm_3Al	460	PI	[268]	1973	
β-Sm_3Al	>460	Pr	[268]	1973	
$SmAl_4$	>1000	—	[268]	1973	
α-Gd_3Al	~645	PI	[268]	1973	
β-Gd_3Al	>645—>850	Pr	[268]	1973	
$GdAl_4$	~400	PD	[268]	1973	
$TbAl_4$	~400	PD	[268]	1973	
α-Dy_3Al	700	PI	[268]	1973	
β-Dy_3Al	700—975	C	[268]	1973	
Yb_2Al	1020	P	[18]	1965	
Yb_3Al_2	1085	P	[18]	1965	
YbAl	1165	P	[18]	1965	
$YbAl_2$	1455	P	[18]	1965	
$YbAl_3$	1355	P	[18]	1965	
Th_2Al	1307	C	[267]	1970	
Th_3Al_2	1075—1301	C	[267]	1970	
ThAl	1318	Pr	[267]	1970	
Th_2Al_3	1100—1394	Pr	[267]	1970	
$ThAl_2$	1600	C	[267]	1970	
$ThAl_3$	1120	Pr	[267]	1970	
UAl_2	1590	C	[18]	1965	
UAl_3	1350	Pr	[18]	1965	
UAl_4	730	Pr	[18]	1965	
$NpAl_3$	1375	D	[125]	1962	
$NpAl_4$	1150	D	[125]	1962	
Pu_3Al	560	PD	[268]	1973	
PuAl	193—590	PD	[268]	1973	
$PuAl_2$	1540	C	[268]	1973	
$PuAl_3$	1220	Pr	[268]	1973	
$PuAl_4$	926	Pr	[268]	1973	
Ti_6Al	~1020	—	[268]	1973	Forms congruently in solid state
Ti_3Al	~1100	—	[268]	1973	The same
Ti_2Al	1000	Pr	[18]	1965	
TiAl	1460	Pr	[18]	1965	
$TiAl_3$	1340	Pr	[18]	1965	
Zr_3Al	975	PD	[18]	1965	
Zr_2Al	1250	PD	[18]	1965	
Zr_5Al_3	1000—1395	Pr	[18]	1965	
Zr_3Al_2	1480	Pr	[18]	1965	
Zr_4Al_3	1530	C	[18]	1965	
ZrAl	1250	PD	[18]	1965	
Zr_2Al_3	1595	Pr	[18]	1965	

TEMPERATURE STABILITY RANGES (continued)

Phase	Maximum resistance temp., °C	Nature of conversion	Source	Year	Notes
1	2	3	4	5	6
$ZrAl_2$	1645	C	[18]	1965	
$ZrAl_3$	1580	C	[18]	1965	
V_5Al_8	1670	Pr	[18]	1965	
VAl_3	1360	Pr	[18]	1965	
V_4Al_{23}	736	Pr	[267]	1970	
V_7Al_{45}	688	Pr	[267]	1970	
VAl_{10}	670	Pr	[267]	1970	
Nb_3Al	1960	Pr	[268]	1973	
Nb_2Al	1870	Pr	[268]	1973	
$NbAl_3$	1605	C	[268]	1973	
Ta_2Al	2100	Pr	[268]	1973	
$TaAl_3$	>1700	—	[294]	1965	See [268]
Cr_2Al	910	—	[268]	1973	Forms congruently in solid state
Cr_5Al_8	1350	Pr	[268]	1973	
Cr_4Al_9	1170	Pr	[268]	1973	
$CrAl_4$	1030	Pr	[268]	1973	
Cr_2Al_{11}	940	Pr	[268]	1973	
$CrAl_7$	790	Pr	[268]	1973	
Mo_3Al	~2150	Pr	[295]	1971	
Mo_3Al_8	~1575	C	[295]	1971	
$MoAl_4$	715—1130	Pr	[268]	1973	
$MoAl_5$	700—900	Pr	[268]	1973	
$MoAl_6$	720	Pr	[268]	1973	
$MoAl_{12}$	706	Pr	[268]	1973	
WAl_4	1326	Pr	[125]	1962	
WAl_5	870	Pr	[125]	1962	
WAl_{12}	697	Pr	[125]	1962	
Mn_4Al_{11}	916	PI	[296]	1972	
	895—1002	Pr	[296]	1972	
$MnAl_4$	923	Pr	[296]	1972	
$MnAl_6$	708	Pr	[296]	1972	
$ReAl_2$	1485	Pr	[18]	1965	
$ReAl_{12}$	~600	PD	[18]	1965	
Fe_3Al	555	PD	[268]	1973	
$FeAl_2$	1158	PD	[125]	1962	
Fe_2Al_5	1165	C	[267]	1970	
$FeAl_3$	1150	C	[267]	1970	
Fe_2Al_7	560	Pr	[18]	1965	
$CoAl$	1628	C	[18]	1965	At 740-800°C disorderedness takes place [267]
Co_2Al_5	1180	Pr	[296]	1972	
$CoAl_3$	1135	Pr	[296]	1972	
Co_4Al_{13}	1093	Pr	[296]	1972	
Co_2Al_9	970	Pr	[296]	1972	

119

Phase	Maximum resistance temp., °C	Nature of conversion	Source	Year	Notes
1	2	3	4	5	6
Ni_3Al	1395	Pr	[18]	1965	
$NiAl$	1638	C	[18]	1965	
Ni_2Al_3	1132	Pr	[18]	1965	
$NiAl_3$	854	Pr	[18]	1965	
$RuAl$	~2000	C	[152]	1963	
Ru_2Al_3	~1000—1600	Pr	[152]	1963	
$RuAl_2$	~1100	PD	[152]	1963	
$RuAl_3$	~1200	PD	[152]	1963	
$RuAl_6$	~1350	C	[152]	1963	
$RuAl_{12}$	~750	Pr	[152]	1963	
Pd_2Al	~1430	C	[268]	1973	
α-$PdAl$	855	PI	[268]	1973	
β-$PdAl$	540—1645	C	[268]	1973	
Pd_2Al_3	940	Pr	[268]	1973	
$PdAl_3$	765	Pr	[268]	1973	
Pt_4Al	~1300	—	[268]	1973	
Pt_3Al	1556	C	[268]	1973	
Pt_5Al_3	1465	Pr	[268]	1973	
Pt_3Al_2	1397	Pr	[268]	1973	
$PtAl$	1554	C	[268]	1973	
Pt_2Al_3	1521	C	[268]	1973	
$PtAl_2$	1406	Pr	[268]	1973	
$PtAl_3$	1121	Pr	[268]	1973	
Pt_5Al_{21}	806	Pr	[268]	1973	
Mg_2Si	1100	C	[268]	1973	
Ca_2Si	910	Pr	[125]	1962	
$CaSi$	1245	C	[125]	1962	
$CaSi_2$	1020	Pr	[125]	1962	
Sr_2Si	1010	P	[268]	1973	
$SrSi$	1140	C	[268]	1973	
$SrSi_2$	1150	C	[268]	1973	
$BaSi$	~840	Pr	[268]	1973	
$BaSi_2$	1180	C	[268]	1973	
Y_5Si_3	1850	C	[266]	1967	
Y_5Si_4	1840	C	[266]	1967	
YSi	1835	Pr	[266]	1967	
α-Y_3Si_5	450	PI	[266]	1967	
β-Y_3Si_5	450—1635	Pr	[266]	1967	
α-$LaSi_2$	—160	PI	[297]	1970	
β-$LaSi_2$	—160—1520	P	[297]	1970	
$CeSi_2$	1620	C	[273]	1967	
α-$PrSi_2$	—120	PI	[298]	1959	
β-$PrSi_2$	>—120	—	[298]	1959	See [300]
$NdSi_2$	1525	P	[267]	1970	Polymorphic inversion occurs in 20-150°C range [298]
α-$SmSi_2$	380	PI	[298]	1959	
β-$SmSi_2$	>380	—	[298]	1959	

Phase	Maximum resistance temp., °C	Nature of conversion	Source	Year	Notes
1	2	3	4	5	6
α-EuSi$_2$	-150	PI	[298]	1959	
β-EuSi$_2$	$-150-1500$	P	[267]	1970	
α-GdSi$_2$	400	PI	[298]	1959	
β-GdSi$_2$	$400-1540$	P	[267]	1970	See [299]
α-DySi$_2$	540	PI	[298]	1959	
β-DySi$_2$	$540-1525$	P	[267]	1970	
Th$_3$Si$_2$	>1900	C	[268]	1973	
ThSi	~1780	Pr	[268]	1973	
α-ThSi$_2$	1200	PI	[267]	1970	
β-ThSi$_2$	~1900	C	[268]	1973	
δ-U$_3$Si	765	PI	[161]	1965	
δ'-U$_3$Si	$765-930$	PD	[161]	1965	
U$_3$Si$_2$	1665	C	[260]	1959	
USi	~1600	Pr	[260]	1959	
U$_3$Si$_5$	~1610	Pr	[267]	1970	β-USi$_2$
USi$_2$	450	PD	[267]	1970	USi$_3$ and USi$_{1.86}$ (α-USi$_2$) are formed
Pu$_5$Si$_3$	1377	Pr	[268]	1973	
Pu$_3$Si$_2$	1441	Pr	[268]	1973	
PuSi	1576	C	[268]	1973	
Pu$_3$Si$_5$	1646	P	[268]	1973	
PuSi$_2$	1638	P	[268]	1973	
Ti$_3$Si	1170	PD	[163]	1970	
Ti$_5$Si$_3$	2130	C	[163]	1970	
Ti$_5$Si$_4$	1920	Pr	[163]	1970	
TiSi	1570	Pr	[163]	1970	
TiSi$_2$	1500	C	[163]	1970	
Zr$_4$Si	1630	Pr	[125]	1962	
Zr$_2$Si	2110	Pr	[125]	1962	
Zr$_5$Si$_3$	2210	Pr	[125]	1962	
Zr$_4$Si$_3$	2225	Pr	[125]	1962	
Zr$_6$Si$_5$	2250	C	[125]	1962	
ZrSi	2095	Pr	[125]	1962	
ZrSi$_2$	1520	Pr	[125]	1962	
Hf$_2$Si	~2430	Pr	[267]	1970	
Hf$_5$Si$_3$	~2600	Pr	[267]	1970	
Hf$_3$Si$_2$	>2600	C	[267]	1970	
HfSi	2200	C	[267]	1970	
HfSi$_2$	1900	Pr	[267]	1970	
V$_3$Si	1935	C	[166]	1974	
V$_5$Si$_3$	2010	C	[166]	1974	
V$_5$Si$_4$	1670	Pr	[166]	1974	
VSi$_2$	1680	C	[166]	1974	
Nb$_4$Si	1950	Pr	[125]	1962	See [302]
α-Nb$_5$Si$_3$	2000	PI	[267]	1970	
β-Nb$_5$Si$_3$	$2000-2480$	C	[267]	1970	
NbSi$_2$	1930	C	[125]	1962	
Ta$_{4.5}$Si	2510	C	[125]	1962	

Phase	Maximum resistance temp., °C	Nature of conversion	Source	Year	Notes
1	2	3	4	5	6
Ta_2Si	2450	Pr	[125]	1962	
Ta_5Si_3	2500	C	[125]	1962	Polymorphism is noted
$TaSi_2$	2200	C	[125]	1962	
Cr_3Si	1770	C	[304]	1964	
Cr_5Si_3	1720	C	[304]	1964	
CrSi	1475	Pr	[304]	1964	
$CrSi_2$	1475	C	[304]	1964	
Mo_3Si	2025	Pr	[306]	1971	
Mo_5Si_3	2180	C	[306]	1971	
α-$MoSi_2$	1850	PI	[306]	1971	
β-$MoSi_2$	1850—2020	C	[306]	1971	
W_5Si_3	2370	Pr	[267]	1970	
WSi_2	2160	C	[267]	1970	
Mn_6Si	880	PD	[268]	1973	
Mn_9Si_2	1060	Pr	[268]	1973	
Mn_3Si	1075	Pr	[268]	1973	
Mn_5Si_2	~850	PD	[268]	1973	
Mn_5Si_3	1285	C	[268]	1973	
MnSi	1275	C	[268]	1973	
Mn_nSi_{2-n}	1152	Pr	[309]	1970	
$Mn_{11}Si_{19}$	1145	C	[268]	1970	
Re_5Si_3	1960	C	[267]	1970	
ReSi	1880	Pr	[267]	1970	
$ReSi_2$	1980	C	[267]	1970	
Fe_2Si	1040—1215	C	[308]	1968	
Fe_5Si_3	825—1090	PD	[308]	1968	
FeSi	1410	C	[22]	1971	
α-$FeSi_2$	955—1220	C	[175]	1968	
β-$FeSi_2$	986	PD	[175]	1968	
Co_3Si	1170—1212	Pr	[125]	1962	See [179]
α-Co_2Si	1320	Pr	[125]	1962	
β-Co_2Si	1238—1332	C	[125]	1962	
CoSi	1460	C	[125]	1962	
$CoSi_2$	1326	C	[125]	1962	
Ni_3Si	1035	PD	[22]	1971	
Ni_5Si_2	1282	C	[22]	1971	
α-Ni_2Si	1214	PD	[22]	1971	
β-Ni_2Si	806—1318	C	[22]	1971	
Ni_3Si_2	845	PD	[22]	1971	
NiSi	992	C	[22]	1971	
α-$NiSi_2$	981	PI	[22]	1971	
β-$NiSi_2$	981—993	Pr	[22]	1971	
Pd_3Si	960	Pr	[267]	1970	
Pd_2Si	1330	C	[267]	1970	
PdSi	1100	C	[267]	1970	
α-Pt_3Si	360	PI	[268]	1973	
β-Pt_3Si	360—~775	PI	[268]	1973	

Phase	Maximum resistance temp., °C	Nature of conversion	Source	Year	Notes
1	2	3	4	5	6
γ-Pt_3Si	~775—~870	Pr	[268]	1973	
Pt_7Si_3	986	Pr	[268]	1973	
α-Pt_2Si	695	PI	[268]	1973	
β-Pt_2Si	695—1100	C	[268]	1973	
Pt_6Si_5	~975	PD	[268]	1973	
PtSi	1229	C	[268]	1973	
Ba_3P_2	3080	P	[311]	1965	
ScP	800	DS	[27]	1973	
ScP	>2000	P	[27]	1973	
LaP	>750	DS	[27]	1973	
CeP	800	DS	[27]	1973	
CeP	>1800	P	[27]	1973	
CeP_2	800	D	[403]	1966	
PrP	750	DS	[27]	1973	
PrP	2850	P	[310]	1968	
NdP	380	DS	[27]	1973	
NdP	2220	P	[27]	1973	
SmP	350	DS	[27]	1973	
SmP	>2100	P	[27]	1973	
Eu_3P_2	>1300	P	[27]	1973	
EuP	730	DS	[27]	1973	
EuP	2200	P	[27]	1973	
EuP_2	840	P	[27]	1973	
GdP	>750	DS	[27]	1973	
GdP	>2500	P	[268]	1973	
TbP	260	DS	[27]	1973	
DyP	250	DS	[27]	1973	
YbP	820	DS	[27]	1973	
ThP	~3000	P	[268]	1973	
PuP	2000	D	[125]	1962	
Ti_3P	1760	Pr	[28]	1968	
Ti_2P	1920	Pr	[28]	1968	
TiP	1580	D	[312]	1962	
β-ZrP	1425	—	[268]	1973	In vacuum is changed into α-ZrP with loss of P
ZrP_2	850	D	[268]	1973	
VP	~1230	D	[312]	1962	
NbP	1730	D	[312]	1962	
TaP	1660	D	[312]	1962	
Cr_3P	1510	Pr	[125]	1962	
CrP	1360	D	[312]	1962	
CrP_2	700	D	[125]	1962	
$Mo_{\sim1.7}P$	>1720	—	[207]	1972	
Mo_8P_5	1580—1680	—	[207]	1972	
MoP	~1480	D	[312]	1962	
W_3P	900	—	[201]	1966	

Phase	Maximum resistance temp., °C	Nature of conversion	Source	Year	Notes
1	2	3	4	5	6
WP	$\sim$1450	D	[312]	1962	
Mn_3P	1105	Pr	[125]	1962	
Mn_2P	1327	C	[125]	1962	
Mn_3P_2	1002—1090	Pr	[125]	1962	See [208]
MnP	1147	C	[125]	1962	See [312]
Fe_3P	1166	Pr	[125]	1962	
Fe_2P	1365	C	[125]	1962	
Co_2P	1386	C	[125]	1962	
Ni_3P	970	Pr	[214]	1965	
α-Ni_5P_2	1025	PI	[214]	1965	
β-Ni_5P_2	1025—1175	C	[214]	1965	
$Ni_{12}P_5$	$\sim$1150	Pr	[214]	1965	
Ni_2P	1110	C	[214]	1965	
Ni_5P_4	$\sim$850	Pr	[214]	1965	
$Ni_{\sim1.22}P$	$\sim$770—$\sim$825	PD	[214]	1965	
NiP	>$\sim$850	—	[214]	1965	
Ru_2P	—	—	[267]	1970	Melting point higher than that of CO_2P
Rh_2P	>1500	C	[125]	1962	
Pd_3P	1047	C	[125]	1962	
PdP_2	$\sim$1150	C	[125]	1962	
Ir_2P	1350	C	[125]	1962	
IrP_2	$\sim$1230	—	[125]	1962	
$Pt_{20}P_7$	590	Pr	[125]	1962	
PtP_2	>1500	P	[125]	1962	
BeS	1400	—	[33]	1972	Volatilization in vacuum
MgS	1300	—	[33]	1972	
MgS	2000	P	[33]	1972	
CaS	2525	P	[267]	1970	
SrS	>2000	P	[33]	1972	
BaS	1725	—	[35]	1972	Volatilization in vacuum
BaS	>2200	P	[35]	1972	
BaS_2	664	PI	[125]	1962	
	664—925	Pr	[125]	1962	
BaS_3	554	Pr	[125]	1962	
Sc_2S_3	1100	D	[314]	1964	
	1775	P	[314]	1964	$p = 10^{-2}$ mm Hg
YS	2060	P	[267]	1970	
Y_5S_7	1630	P	[267]	1970	
Y_2S_3	850—1900	P	[267]	1970	Decomposes
YS_2	1660	P	[267]	1970	Dissociates
LaS	$\sim$2200	P	[267]	1970	
La_3S_4	2100	P	[267]	1970	
α-La_2S_3	900	PI	[225]	1971	
β-La_2S_3	1300	PI	[225]	1971	
γ-La_2S_3	1300—2095	P	[267]	1970	
LaS_2	1650	P	[267]	1970	Dissociates

TEMPERATURE STABILITY RANGES (continued)

Phase	Maximum resistance temp., °C	Nature of conversion	Source	Year	Notes
1	2	3	4	5	6
CeS	2450	P	[33]	1972	
Ce_3S_4	2100	P	[267]	1970	
α-Ce_2S_3	1100	PI	[225]	1971	
β-Ce_2S_3	1150	PI	[225]	1971	See [227], [316]
γ-Ce_2S_3	1150—1500	PI	[316]	1968	See [225]
β-Ce_2S_3	1500—2060	P	[267]	1970	
CeS_2	1700	P	[267]	1970	Decomposes; See [227]
PrS	2230	P	[267]	1970	
Pr_3S_4	2100	P	[267]	1970	
α-Pr_2S_3	925	PI	[267]	1970	
β-Pr_2S_3	925—1300	PI	[267]	1970	See [225]
γ-Pr_2S_3	1300—1795	P	[2 67]	1970	
PrS_2	1780	P	[267]	1970	Dissociates
NdS	2200	P	[267]	1970	
Nd_3S_4	2040	P	[267]	1970	
α-Nd_2S_3	1050	PI	[267]	1970	
β-Nd_2S_3	1050—1300	PI	[267]	1970	See [225]
γ-Nd_2S_3	1300—2200	P	[267]	1970	Dissociates
NdS_2	1760	P	[267]	1970	Decomposes
SmS	1940	P	[33]	1972	
Sm_3S_4	1800	P	[267]	1970	
α-Sm_2S_3	1110	PI	[225]	1971	
γ-Sm_2S_3	1110—1900	P	[33]	1972	Dissociates
EuS	1667	P	—**	1973	
GdS	2020	P	—	1973	
α-Gd_2S_3	1060	PI	[225]	1971	
γ-Gd_2S_3	1060—1885	P	[33]	1972	
GdS_2	800	DS	[33]	1972	
TbS	1970	P	—**	1973	
DyS	(1940)	P	—**	1973	
Dy_5S_7	1540	P	[33]	1972	
α-Dy_2S_3	950	PI	[267]	1970	
δ-Dy_2S_3	1470	P	[267]	1970	Forms γ-Dy_2S_3
γ-Dy_2S_3	1490	P	[267]	1970	
HoS	(1890)	P	—**	1973	
ErS	(1900)	P	—**	1973	
Er_5S_7	1620	P	[33]	1972	
Er_2S_3	1630	P	[33]	1972	
TmS	(1840)	P	—**	1973	
Yb_3S_4	800	D	[267]	1970	
Yb_2S_3	950—980	PI	[33]	1972	Cub.
Yb_2S_3	1300	DS	[267]	1970	Rhomb.
ThS	2335	P	[268]	1973	
Th_2S_3	1900—2000	P	[33]	1972	
Th_4S_7	2300	P	[33]	1972	
ThS_2	1904	P	[33]	1972	Decomposes

Phase	Maximum resistance temp., °C	Nature of con- version	Source	Year	Notes
1	2	3	4	5	6
US	2462	P	[267]	1970	
α-US$_2$	>1350	—	[33]	1972	
β-US$_2$	1350	PI	[33]	1972	
γ-US$_2$	425	PI	[33]	1972	
NpS	400	DS	[33]	1972	
Np$_3$S$_5$	900	D	[233]	1967	
α-Np$_2$S$_3$	1200	PI	[233]	1967	
β-Np$_2$S$_3$	1200—1500	PI	[233]	1967	
γ-Np$_2$S$_3$	>1500	—	[233]	1967	
NpS$_3$	500	DS	[233]	1967	
PuS	2350	P	[33]	1972	
α-Pu$_2$S$_3$	1300	PI	[227]	1966	
β-Pu$_2$S$_3$	1300—1400	PI	[33]	1972	
γ-Pu$_2$S$_3$	1400—1725	C	[268]	1973	
PuS$_2$	500—700	DS	[228]	1966	
Ti$_3$S	1305	Pr	[33]	1972	
Ti$_2$S	1410	Pr	[33]	1972	
δ-TiS	935	PI	[33]	1972	
δ'-TiS	935—1780	Pr	[33]	1972	
TiS$_3$	635	D	[267]	1970	
ZrS	1400	—	[33]	1972	
ZrS$_2$	1550	P	[33]	1972	
ZrS$_3$	700	D	[267]	1970	
α-V$_3$S	950—1400	Pr	[267]	1970	
β-V$_3$S	825	PI	[267]	1970	
V$_2$S$_3$	650	D	[33]	1972	
VS$_4$ }	300	D	[33]	1972	
VS$_5$ }					
NbS$_{1.6}$	800	DS	[33]	1972	
CrS	~1565	P	[125]	1962	Polymorphic in- version at 330°C [268]
Cr$_2$S$_5$	120	D	[317]	1967	
Mo$_2$S$_3$	1600	DS	[33]	1972	
MoS$_2$	>1800	P	[267]	1970	Polymorphism occurs
MoS$_2$	1000	D	[33]	1972	
MoS$_3$	350	D	[267]	1970	
WS$_2$	>1800	C	[267]	1970	Polymorphism occurs
WS$_2$	1100	D	[33]	1972	
WS$_3$	270—500	D	[268]	1973	
MnS	1610	C	[125]	1962	
MnS$_2$	304—345	D	[33]	1972	
ReS	680	D	[265]	1963	
ReS$_2$	1000	D	[33]	1972	
Re$_2$S$_7$	400	D	[267]	1970	

Phase	Maximum resistance temp., °C	Nature of conversion	Source	Year	Notes
1	2	3	4	5	6
FeS	∼1190	P	[268]	1973	Polymorphism occurs
FeS$_2$	450	PI	[33]	1972	Marcasite
FeS$_2$	520	D	[33]	1972	Pyrites
Co$_4$S$_3$	785—930	Pr	[33]	1972	
Co$_9$S$_8$	835	PD	[33]	1972	
β-CoS	>475	—	[33]	1972	
β′-CoS	360—1195	C	[33]	1972	
Co$_3$S$_4$	630	PD	[33]	1972	
CoS$_2$	950	Pr	[33]	1972	
Ni$_3$S$_2$	550	PI	[267]	1970	
	525—806	Pr	[267]	1970	
Ni$_7$S$_6$	∼400	PI	[267]	1970	
	400—573	PD	[267]	1970	
NiS	379	PI	[267]	1970	
	280—992	C	[267]	1970	
Ni$_3$S$_4$	303	PD	[267]	1970	
NiS$_2$	1010	C	[267]	1970	
RuS$_2$	1000	D	[33]	1972	
Pd$_4$S	761	Pr	[125]	1962	
Pd$_{2.8}$S	555—635	Pr	[125]	1962	
IrS	750	D	[33]	1972	
Ir$_2$S$_3$	700	DS	[33]	1972	
PtS$_2$	310	D	[33]	1972	
B$_4$C	2450	C	[267]	1970	18.5% (at.) C
SiC	2540	Pr	[267]	1970	Polymorphism occurs
β-BN	1650	PI	[267]	1970	
α-BN	∼3000	P	[14]	1969	
AlN	2450	K	[267]	1970	
α-Si$_3$N$_4$	1100	DS	[318]	1973	
α-Si$_3$N$_4$	1300—1450	PI	[267]	1970	
β-Si$_3$N$_4$	1800	D	[267]	1970	

*Key: (C) congruent melting; (Pr) peritectic-reaction melting; (P) melting (no indication of its type); (PD) peritectoid decomposition; (PI) polymorphic inversion; (D) decomposition; (S) sublimation; (DS) dissociation.

* B. D. Fenochka, Mass Spectrometric Study of High-Temperature Behaviour and Thermodynamic Properties of Rare-Earth Monosulfides [in Russian], Cand. Dissertation, Kiev (1973).

Chapter II

THERMAL AND THERMODYNAMIC PROPERTIES

HEAT OF FORMATION FROM THE ELEMENTS

Phase	Thermal effect $-\Delta H^{\circ}_{298}$, kcal/mole	Source	Year	Phase	Thermal effect $-\Delta H^{\circ}_{298}$, kcal/mole	Source	Year
1	2	3	4	1	2	3	4
$NbBe_2$	14.6 ± 1.9	[321]	1964	LuB_6	134,0	[3]	1975
$NbBe_5$	46.4 ± 3.8	[321]	1964	ThB_4	>52	[311]	1965
Nb_2Be_{17}	20.5 ± 2	[321]	1964	ThB_6	>66	[311]	1965
$NbBe_{12}$	29.8 ± 9.6	[321]	1964	TiB_2	70,0	[331]	1959
UBe_{13}	39	[324]	1971	Ti_2B_5	>105	[311]	1965
$PuBe_{13}$	$35.7 \pm 3,3$	[2]	1966	ZrB	>39	[311]	1965
Be_4B	$18.8 \pm 0,7$	—*	1972	ZrB_2	76,7	[330]	1965
Be_2B	$16,7 \pm 0,7$	—*	1972	ZrB_{12}	>120	[311]	1965
BeB_2	$15.5 \pm 0,7$	—*	1972	HfB	47	[323]	1971
BeB_4	20,8	—*	1972	HfB_2	85,6	[326]	1967
BeB_6	$26,0 \pm 0,7$	—*	1972	VB	31,0	[332]	1968
BeB_9	$38,6 \pm 0.7$	—*	1972	VB_2	62	[311]	1965
Mg_4B	16.8	—*	1972	NbB_2	59,0	[311]	1965
Mg_2B	14.1	—*	1972	TaB_2	46	[323]	1971
MgB_2	$13,3 \pm 0.8$	—*	1972	CrB_2	30	[333]	1960
MgB_4	$17,6 \pm 0.6$	—*	1972	Mo_2B	25,5	[311]	1965
MgB_6	22.4 ± 0.5	—*	1972	Mo_3B_2	42,0	[311]	1965
MgB_{12}	34,4	[3]	1975	MoB	16,3	[311]	1965
CaB_6	28,6	[326]	1967	MoB_2	23,0	[311]	1965
SrB_6	50,4	[327]	1960	Mo_2B_5	50,0	[311]	1965
BaB_6	79,6	[326]	1967	W_2B	24	[311]	1965
ScB_2	63.3	[326]	1967	WB	17	[311]	1965
YB_6	24	[271]	1958	W_2B_5	35	[311]	1965
LaB_6	112.3 ± 10	[329]	1961	MnB_2	19.0	[124]	1962
CeB_4	<84	[311]	1965	FeB	9,2	[3]	1975
CeB_6	81 ± 16	[274]	1956	Be_2C	$27,06 \pm 0.21$	[88]	1972
PrB_6	99,6	[3]	1975	BeC_2	57,4	[311]	1965
NdB_6	102,8	[3]	1975	Mg_2C_3	18 ± 8	[287]	1973
PmB_6	105,7	[3]	1975	MgC_2	21 ± 5	[287]	1973
SmB_6	108,6	[3]	1975	CaC_2	$14,2 \pm 0,4$	[287]	1973
EuB_6	112,2	[3]	1975	SrC_2	$20,2 \pm 3$	[334]	1973
GdB_6	114,7	[3]	1975	BaC_2	18	[324]	1971
TbB_6	117,8	[3]	1975	YC_2	27 ± 6	[334]	1973
DyB_6	120,0	[3]	1975	LaC_2	17	[325]	1973
HoB_6	123,0	[3]	1975	La_2C_3	19 ± 5	[257]	1974
ErB_6	126,0	[3]	1975	Ce_2C_3	$23,2 \pm 1.3$	[257]	1974
TmB_6	129,0	[3]	1975	CeC_2	15	[325]	1973
YbB_6	131.5	[3]	1975	PrC_2	$\sim15 \pm 5$	[257]	1974

HEAT OF FORMATION FROM THE ELEMENTS (continued)

Phase	Thermal effect $-\Delta H^{\circ}_{298}$, kcal/mole	Source	Year	Phase	Thermal effect $-\Delta H^{\circ}_{298}$, kcal/mole	Source	Year
1	2	3	4	1	2	3	4
NdC_2	12,5 ± 2,5	[257]	1974	Ba_3N_2	86.9 ± 8,0	[14]	1969
SmC_2	17	[325]	1973	ScN	68,0 ± 5,0	[14]	1969
EuC_2	15	[325]	1973	YN	71,5 ± 5,0	[14]	1969
GdC_2	30 ± 9	[334]	1973	LaN	71,5 ± 4,0	[14]	1969
DyC_2	11,1 ± 1,0	[257]	1974	CeN	78,0 ± 6,0	[14]	1969
Ho_2C_3	56	[325]	1973	SmN	75.0	[14]	1969
HoC_2	26	[325]	1973	GdN	75,0	[14]	1969
ErC_2	18,5 ± 0,4	[257]	1974	DyN	75,0	[14]	1969
TmC_2	20,0 ± 2,0	[334]	1973	ErN	75,0	[14]	1969
YbC_2	18,0 ± 1,0	[257]	1974	YbN	75,0	[14]	1969
LuC_2	~18,0 ± 5	[257]	1974	LuN	75,0	[14]	1969
ThC	29,6 ± 1,1	[287]	1973	Th_3N_4	310,0 ± 4.0	[14]	1969
$ThC_{1.91}$	29,9 ± 1,1	[287]	1973	UN	70.6	[338]	1969
UC	23,2 ± 1,0	[287]	1973	U_2N_3	213,0	[14]	1969
U_2C_3	49,0 ± 4,0	[287]	1973	PuN	7,8	[14]	1969
UC_2	21,9 ± 2,1	[287]	1973	TiN	80,4	[14]	1969
PuC	12,0 ± 2,0	[12]	1971	ZrN	87,3	[14]	1969
$PuC_{1.50}$	24,4 ± 2,7	[12]	1971	HfN	88,24 ± 0,34	[14]	1969
PuC_2	8	[334]	1973	VN	60.0 ± 5,0	[14]	1969
TiC	55,3 ± 0,3	[334]	1973	Nb_2N	61,1 ± 1,0	[14]	1969
ZrC	47,7 ± 5,0	[334]	1973	NbN	56,8 ± 1.5	[14]	1969
HfC	54,2 ± 0,5	[334]	1973	Ta_2N	64.7 ± 3.0	[14]	1969
V_2C	35,4 ± 5	[334]	1973	TaN	59,0	[14]	1969
$VC_{0.88}$	24,35 ± 0,4	[334]	1973	Cr_2N	25.2 ± 3.0	[14]	1969
Nb_2C	45,4	[323]	1971	CrN	28,2	[14]	1969
NbC	33,6 ± 0,6	[334]	1973	Mo_2N	16,6 ± 0.5	[14]	1969
Ta_2C	47.2 ± 3,4	[334]	1973	W_2N	17,2 ± 3.0	[14]	1969
TaC	34,6 ± 0,9	[334]	1973	Mn_4N	30.75	[322]	1969
$Cr_{23}C_6$	141,2 ± 20	[335]	1966	Mn_3N	31,2	[14]	1969
Cr_7C_3	48,8 ± 7	[335]	1966	Mn_5N_2	48.2 ± 0,6	[14]	1969
Cr_3C_2	23,4 ± 3	[335]	1966	Re_2N	−1,0	[14]	1969
Mo_2C	10,9	[322]	1969	Fe_8N	2.7	[14]	1969
MoC	2,4	[322]	1969	Fe_4N	2.6	[14]	1969
W_2C	11 ± 4	[257]	1973	Fe_2N	0,9 ± 2,0	[14]	1969
WC	8,4 ± 0,2	[257]	1973	Co_3N	−2,0 ± 5,0	[14]	1969
$Mn_{23}C_6$	3,3	[11]	1968	Ni_3N	−0,2 ± 0,1	[14]	1969
Mn_3C	3,6	[334]	1973	Mg_2Al_3	1.11	[340]	1968
Mn_7C_3	19,8 ± 1.2	[337]	1970	$CaAl_2$	52	[324]	1971
Fe_3C	−5,98 ± 0,5	[334]	1973	$CaAl_4$	51	[324]	1971
Fe_2C	−4,93 ± 1	[334]	1973	$BaAl_4$	28	[324]	1971
Co_3C	−3,95 ± 3	[334]	1973	$LaAl_2$	24,0	[340]	1968
Co_2C	−4,0	[311]	1965	$LaAl_4$	33,6	[340]	1968
Ni_3C	−9,2 ± 1.5	[334]	1973	Ce_3Al	22	[325]	1973
$\alpha\text{-}Be_3N_2$	140,6	[324]	1971	$CeAl_4$	39	[325]	1973
$\beta\text{-}Be_3N_2$	136,5	[324]	1971	$PrAl_4$	52,1	[325]	1973
Mg_3N_2	110,3 ± 3,0	[14]	1969	$PuAl_2$	22,6	[340]	1968
Ca_3N_2	105,0 ± 3,0	[14]	1969	$PuAl_3$	35.4	[340]	1968
Sr_3N_2	93,4 ± 5,0	[14]	1969	$PuAl_4$	34.4	[340]	1968

Phase	Thermal effect $-\Delta H^{\circ}_{298}$, kcal/mole	Source	Year	Phase	Thermal effect $-\Delta H^{\circ}_{298}$, kcal/mole	Source	Year
1	2	3	4	1	2	3	4
Ti_3Al	23,5	[323]	1971	U_3Si_2	40,8	[324]	1971
$TiAl$	18,0	[323]	1971	USi	20,2	[324]	1971
Ti_2Al_3	27,9	[340]	1968	USi_2	30,6	[324]	1971
$TiAl_3$	35,0	[323]	1971	$PuSi_2$	211	[311]	1965
V_5Al_8	70	[323]	1971	Ti_5Si_3	139,0 ± 7,0	[341]	1968
VAl_3	26	[323]	1971	$TiSi$	39,2 ± 3,0	[341]	1968
$NbAl_3$	28,4 ± 1,2	[340]	1968	$TiSi_2$	32,3 ± 2,0	[341]	1968
$TaAl$	19,2	[10]	1973	Zr_4Si	52	[323]	1971
$Ta_{12}Al_{17}$	109,0 ± 6,8	[340]	1968	Zr_2Si	50	[323]	1971
$TaAl_3$	26,1 ± 0,9	[340]	1968	Zr_5Si_3	138	[323]	1971
Cr_4Al_9	32,4	[340]	1968	Zr_3Si_2	92	[323]	1971
$CrAl_3$	12,0	[340]	1968	Zr_5Si_4	195,3	[344]	1968
$CrAl_4$	16,4	[340]	1968	Zr_6Si_5	205	[323]	1971
Cr_2Al_{11}	39,6	[340]	1968	$ZrSi$	37	[323]	1971
$CrAl_7$	22,4	[340]	1968	$ZrSi_2$	38	[323]	1971
$MnAl$	5,1	[340]	1968	Hf_2Si	45,0	[323]	1971
Mn_2Al_3	15,3	[340]	1968	Hf_5Si_2	134,4	[323]	1971
Mn_5Al_8	40,8	[340]	1968	$HfSi$	34,0	[323]	1971
Mn_4Al_{11}	56,1	[340]	1968	$HfSi_2$	54,0	[323]	1971
$MnAl_3$	15,3	[340]	1968	V_3Si	27,9	[345]	1960
$MnAl_4$	20,6	[340]	1968	V_2Si	37	[323]	1971
$MnAl_6$	21,6	[340]	1968	V_5Si_3	96 ± 4,5	[345]	1960
$FeAl$	12,0	[322]	1969	VSi_2	75	[345]	1960
$FeAl_2$	18,9	[322]	1969	Nb_4Si	21	[311]	1965
Fe_2Al_5	34,3	[340]	1968	Nb_5Si_3	116,0 ± 24,8	[346]	1973
$FeAl_3$	18,9	[322]	1969	$NbSi_2$	33 ± 11,4	[346]	1973
$CoAl$	26,4	[322]	1969	$Ta_{4,5}Si$	34,4	[311]	1965
$CoAl_2$	31,8	[322]	1969	Ta_3Si	37,0 ± 7,64	[346]	1973
Co_2Al_5	70,0	[322]	1969	Ta_2Si	30,0 ± 5,73	[346]	1973
$CoAl_4$	38,5	[322]	1969	Ta_5Si_3	80,0 ± 7,6	[346]	1973
$NiAl$	28,1	[322]	1969	$TaSi_2$	28,5 ± 2,85	[346]	1973
Ni_3Al_3	40,8	[340]	1968	Cr_3Si	33,0 ± 6	[341]	1968
$NiAl_3$	27,3	[340]	1968	Cr_5Si_3	78,0 ± 11,0	[341]	1968
Mg_2Si	18,6	[324]	1971	$CrSi$	19,0 ± 2,0	[341]	1968
Ca_2Si	50,0 ± 3,0	[341]	1968	$CrSi_2$	29,4	[341]	1968
$CaSi$	36,0 ± 2,0	[341]	1968	Mo_3Si	24,0	[341]	1968
$CaSi_2$	36,0 ± 3,0	[341]	1968	Mo_5Si_3	67,0	[341]	1968
Sr_2Si	116	[324]	1971	$MoSi_2$	26,0	[341]	1968
$SrSi$	133	[324]	1971	W_5Si_3	46,5	[311]	1965
$SrSi_2$	188	[324]	1971	WSi_2	22,2	[311]	1965
$BaSi$	200	[324]	1971	Mn_3Si	32,8	[341]	1968
$BaSi_2$	36.6	[324]	1971	Mn_5Si_3	55,2	[341]	1968
$BaSi_3$	458	[324]	1971	$MnSi$	23,2	[341]	1968
YSi	32.2	[311]	1965	$Mn_{0,37}Si_{0,63}$	6.7 ± 1,43	[346]	1973
$LaSi$	30,0	[343]	1960	$MnSi_{1,77}$	9,3	[325]	1973
$LaSi_2$	44.4	[343]	1960	Re_3Si	12,6	[311]	1965
$CeSi_2$	50.0	[311]	1965	Re_5Si_3	37,6 ± 1,53	[346]	1973
$ThSi_2$	41,7	[311]	1965	$ReSi$	12.6 ± 4,78	[346]	1973

Phase	Thermal effect $-\Delta H^\circ_{298}$, kcal/mole	Source	Year	Phase	Thermal effect $-\Delta H^\circ_{298}$, kcal/mole	Source	Year
1	2	3	4	1	2	3	4
$ReSi_2$	$21,6 \pm 7,17$	[346]	1973	OsP_2	33,0	[311]	1965
Fe_3Si	22,4	[322]	1969	BeS	56,0	[324]	1971
Fe_5Si_3	58,4	[311]	1965	MgS	82,7	[324]	1971
$FeSi$	17,6	[322]	1969	CaS	115,3	[324]	1971
$FeSi_2$	19,4	[322]	1969	SrS	108,3	[324]	1971
$FeSi_{2.33}$	14	[322]	1969	BaS	110	[324]	1971
Co_2Si	$27,6 \pm 2,0$	[341]	1968	LaS	109	[325]	1973
$CoSi$	$24,0 \pm 2,0$	[341]	1968	La_3S_4	401	[347]	1972
$CoSi_2$	$24,6 \pm 2,0$	[341]	1968	La_2S_3	289	[325]	1973
Ni_3Si	$33,5 \pm 3,0$	[341]	1968	LaS_2	145,0	[311]	1965
Ni_5Si_2	$72,0$	[311]	1965	CeS	109,8	[325]	1973
Ni_2Si	$35,5 \pm 3,0$	[341]	1968	Ce_3S_4	397	[325]	1973
Ni_3Si_2	53,5	[311]	1965	Ce_2S_3	284	[325]	1973
$NiSi$	$20,5 \pm 2,0$	[341]	1968	CeS_2	146,3	[325]	1973
$NiSi_2$	20,85	[322]	1969	PrS	111	[347]	1972
$RuSi$	(20,1)	[346]	1973	Pr_3S_4	391	[347]	1972
$RhSi$	(29,2)	[346]	1973	Pr_2S_3	273	[347]	1972
Pd_3Si	(60,2)	[346]	1973	PrS_2	141,0	—**	1969
Pd_2Si	(57,3)	[346]	1973	NdS	111	[347]	1972
$PdSi$	(34,0)	[346]	1973	Nd_3S_4	388	[347]	1972
$OsSi$	(22,0)	[346]	1973	Nd_2S_3	269	[347]	1972
$IrSi$	(32,0)	[346]	1973	NdS_2	138,8	—**	1969
Pt_3Si	(50.6)	[346]	1973	SmS	103,2	—***	1973
Pt_7Si_3	(153)	[346]	1973	EuS	97,5	—***	1973
Pt_2Si	(51.0)	[346]	1973	GdS	105	[347]	1972
Pt_6Si_5	(200)	[346]	1973	TbS	(104,8)	—***	1973
$PtSi$	(40,2)	[346]	1973	DyS	(102,4)	—***	1973
Mg_3P_2	128	[311]	1965	HoS	(101,8)	—***	1973
Ca_3P_2	121	[324]	1971	ErS	(100,4)	—***	1973
Sr_3P_2	152	[324]	1971	TmS	(100,7)	—***	1973
Ba_3P_2	118	[311]	1965	YbS	97,9	—***	1973
PrP	>176	[310]	1968	LuS	(127)	—***	1973
TiP	67,6	[323]	1971	ThS	100,0	[311]	1965
MnP	27	[322]	1969	Th_2S_3	258,6	[311]	1965
MnP_3	51	[322]	1969	Th_4S_7	665,0	[311]	1965
Fe_3P	39	[322]	1969	ThS_2	110,0	[311]	1965
Fe_2P	39	[322]	1969	US	90,0	[311]	1965
FeP	30	[322]	1969	TiS	57	[323]	1971
FeP_2	46	[322]	1969	TiS_2	80,0	[311]	1965
Co_2P	45	[322]	1969	ZrS_2	135,3	[323]	1971
CoP	38,0	[311]	1965	VS	22,5	[39]	1967
CoP_3	69,0	[311]	1965	V_2S_3	227	[323]	1971
Ni_3P	50,2	[322]	1969	$Nb_{1.136}S_2$	92,8	[323]	1971
Ni_5P_2	97,7	[322]	1969	TaS_2	111	[323]	1971
Ni_2P	44,0	[311]	1965	Mo_2S_3	87	[322]	1969
Ni_5P_6	159,4	[311]	1965	MoS_2	56,2	[322]	1969
NiP_2	40,0	[311]	1965	MoS_3	61,2	[311]	1965
NiP_3	48.0	[311]	1965	WS_2	50	[349]	1969

Phase	Thermal effect $-\Delta H_{298}^{\circ}$, kcal/mole	Source	Year	Phase	Thermal effect $-\Delta H_{298}^{\circ}$, kcal/mole	Source	Year
1	2	3	4	1	2	3	4
MnS	47.6—49,0	[311]	1965	NiS_2	34,0	[311]	1965
MnS_2	49.5	[311]	1965	RuS_2	47	[322]	1969
ReS_2	43	[322]	1969	Pd_4S	16	[322]	1969
Re_2S_7	107	[322]	1969	PdS	18	[322]	1969
α-FeS	23.9	[322]	1969	PdS_2	19,4	[322]	1969
Fe_7S_8	176,0	[322]	1969	OsS_2	34,9	[322]	1969
FeS_2	42.6 ****	[322]	1969	Ir_2S_3	56	[322]	1969
	37 0*****	[322]	1969	IrS_2	33	[322]	1969
Co_9S_8	197,0	[311]	1965	IrS_3	26,3	[311]	1965
CoS	19.8	[322]	1969	PtS	19,5	[322]	1969
Co_3S_4	75,0	[311]	1965	PtS_2	26.0	[322]	1969
Co_2S_3	51,0	[311]	1965	B_4C	17,1 $\pm$ 2,7	[287]	1973
CoS_2	33,5	[311]	1965	α-SiC	17,23 $\pm$ 0,46	[287]	1973
Ni_3S_2	48,5	[322]	1969	β-SiC	17,49 $\pm$ 0,49	[287]	1973
Ni_6S_5	105,6	[322]	1969	BN	60.7	[311]	1965
Ni_7S_6	133,5	[322]	1969	AlN	76,47 $\pm$ 0.20	[249]	1964
NiS	19,6	[322]	1969	Si_3N_4	179,0	[311]	1965

*G. A. Rybakova, Thermodynamic Study of Magnesium and Beryllium Borides [in Russian], Author's abstract, Cand. Dissertation, Leningrad (1972).

**A. D. Finogenov, Investigation of the Thermodynamic Properties of the Sulfides of Some Rare-Earth Elements in the Cerium Subgroup [in Russian], Author's abstract, Cand. Dissertation, Moscow (1969).

***B. V. Fenochka, see footnote on p. 127.

****Pyrites.

*****Marcasite.

ENTROPY OF COMPOUNDS

Phase	Standard entropy S_{298}°, cal/(mole·°C)	Entropy of formation from elements ΔS_{298}°, cal/(mole·°C)	Source	Year
1	2	3	4	5
MgB_2	8,60$\pm$0.04	—2,17	[3]	1975
MgB_4	12,41$\pm$0.06	—1,36	[3]	1975
MgB_{12}	21,32	—	[3]	1975
CaB_6	18,47	—	[326]	1967
SrB_6	21,52	—	[326]	1967
BaB_6	24,52	—	[326]	1967
ScB_2	11,8	—	[326]	1967

Phase	Standard entropy S°_{298}, cal/(mole·°C)	Entropy of formation from elements ΔS_{298}, cal/(mole·°C)	Source	Year
1	2	3	4	5
TiB	5,8	—3,1	[360]	1953
TiB$_2$	7,52	—3,12	[331]	1959
ZrB$_2$	10,7	—1,8	[360]	1953
HfB$_2$	14,2	—2,1	[360]	1953
VB	6,7	—	[332]	1968
VB$_2$	7,9	—2,3	[360]	1953
NbB$_2$	10,4	—1,3	[360]	1953
TaB$_2$	13,9	+0,8	[360]	1953
CrB	5,8	—1,5	[360]	1953
CrB$_2$	9,32	+0,24	[333]	1960
Mo$_2$B$_5$	12,21	+2,2	[360]	1953
W$_2$B$_5$	28,3	+8,3	[360]	1953
Be$_2$C	3,9±1,1	—	[88]	1972
Mg$_2$C$_3$	22,0	—	[11]	1968
MgC$_2$	14,0±2,5	+3,51	[361]	1958
CaC$_2$	16,8±0,5	+4,13	[361]	1968
BaC$_2$	3,4	—	[11]	1968
YC$_2$	13	—	[323]	1971
LaC$_2$	17	—	[325]	1973
CeC$_2$	20	—	[325]	1973
SmC$_2$	23	—	[325]	1973
EuC$_2$	24	—	[325]	1973
HoC$_2$	23	—	[325]	1973
YbC$_2$	19,2±1,4	—	[287]	1973
ThC	12,0	—	[11]	1968
ThC$_2$	19,2	+1,88	[361]	1958
UC	15,9	+3,4	[361]	1958
U$_2$C$_3$	29,3	+3,0	[361]	1958
UC$_2$	24,3	+2,0	[361]	1958
PuC$_{0,869}$	17,0	—	[334]	1973
TiC	5,8±0,1	—2,92	[361]	1958
ZrC	8,5±1,5	—2,16	[361]	1958
HfC	9,85	—	[339]	1970
V$_2$C	15,9	+0,49	[361]	1958
VC	6,77±0,1	—1,59	[361]	1958
Nb$_2$C	15,3	—	[323]	1971
NbC	8,9±0,7	—1,19	[363]	1960
Ta$_2$C	19,9	—	[11]	1968
TaC	10,1±0,2	—1,27	[361]	1958
Cr$_{23}$C$_6$	145,8	—	[322]	1969
Cr$_7$C$_3$	48,0±0,3	+4,16	[361]	1958
Cr$_3$C$_2$	20,4±0,2	+0,64	[361]	1958
Mo$_2$C	19,8±3,0	+4,78	[361]	1958
WC	8,5±1,5	—0,31	[361]	1958
Mn$_3$C	23,6±0,3	—0,56	[361]	1958
Mn$_7$C$_3$	57,1	—	[287]	1973
Fe$_3$C	24,2±1,2	+3,39	[361]	1958
Co$_3$C	23,5±1,5	+0,60	[361]	1958
Co$_2$C	17,8	—	[311]	1965
Ni$_3$C	25,4±1,5	+2,68	[361]	1958
Be$_3$N$_2$	12,0±2,0	—40,61	[361]	1958
Mg$_3$N$_2$	21,0±2,0	—48,08	[361]	1958
Ca$_3$N$_2$	25,4±1,5	—50,22	[361]	1958
Sr$_3$N$_2$	29,5±2,5	—53,77	[361]	1958

Phase	Standard entropy S°_{298}, cal/(mole·°C)	Entropy of formation from elements ΔS°_{298}, cal/(mole·°C)	Source	Year
1	2	3	4	5
Ba_3N_2	36,4±2,0	—57,97	[361]	1958
ScN	9,0±2,0	—22,88	[361]	1958
YN	11,0±2,5	—22,38	[361]	1958
CeN	11,7±1,5	—27,78	[361]	1958
ThN	13,4±0,239	—	[364]	1972
Th_3N_4	42,7±2,5	—86,10	[361]	1958
UN	18,0	—	[14]	1969
U_2N_3	29,0	—	[14]	1969
TiN	7,24±0,1	—22,98	[361]	1958
ZrN	9,3±0,1	—22,88	[361]	1958
HfN	13,1±1,5	—22,88	[361]	1958
VN	8,9±0,1	—20,98	[361]	1958
NbN	10,5±0,2	—21,11	[361]	1958
TaN	12,2±1,0	—20,18	[361]	1958
CrN	8,0±1,3	—19,66	[361]	1958
Cr_2N	18,0±2,0	—12.24	[361]	1958
Mo_2N	21,0	—	[14]	1969
Mn_5N_2	45,9±3,0	—37.86	[361]	1958
Fe_4N	37,3	—	[14]	1969
Fe_2N	24,2	—	[14]	1969
$LaAl_2$	23,6	—	[325]	1973
$CeAl_2$	25,87	—	[325]	1973
$PrAl_2$	27,42	—	[325]	1973
$NdAl_2$	27,67	—	[325]	1973
$GdAl_2$	26.62	—	[325]	1973
Mg_2Si	18	—	[324]	1971
CaSi	15,0±2,0	+0,60	[361]	1958
$CaSi_2$	22,0±2,5	+3,05	[361]	1958
VSi_2	—	—1,86±1,075	[346]	1973
Ta_3Si	—	0,0±3,82	[346]	1973
Ta_2Si	—	+1,5±2.86	[346]	1973
Ta_5Si_3	—	+4,01±6.32	[346]	1973
$TaSi_2$	—	—0,93±1,435	[346]	1973
Cr_3Si	—	—0,956±1,43	[346]	1973
Cr_5Si_3	—	+1,53±2,86	[346]	1973
CrSi	—	+0,286±0,72	[346]	1973
$CrSi_2$	—	—0,717±1.08	[346]	1973
Mo_3Si	25,4	—	[322]	1969
Mo_3Si	—	+0,382±1,15	[346]	1973
Mo_5Si_3	—	+1,91±2,3	[346]	1973
$MoSi_2$	24.5±0,2	+8.67	[361]	1958
$MoSi_2$	—	—0,286±0,86	[346]	1973
W_5Si_3	—	+2,49±2,86	[346]	1973
WSi_2	—	—1.5±1,075	[346]	1973
Mn_3Si	—	—2,68±1,43	[346]	1973
Mn_5Si_3	—	+5,36±2,86	[346]	1973
MnSi	14,1±2,5	+1,90	[361]	1958
MnSi	—	—1,05±0,72	[346]	1973
$Mn_{0,37}Si_{0.63}$	—	—0,694±0,358	[346]	1973
Re_5Si_3	—	+4,01±3,82	[346]	1973
ReSi	—	0,0±0,956	[346]	1973

ENTROPY OF COMPOUNDS (continued)

Phase	Standard entropy S°_{298}, cal/(mole·°C)	Entropy of formation from elements ΔS°_{298}, cal/(mole·°C)	Source	Year
1	2	3	4	5
$ReSi_2$	15,5	—	[311]	1965
$ReSi_2$	—	$0,0\pm1,435$	[346]	1973
Fe_3Si	24,8	—	[322]	1969
Fe_5Si_3	50,1	—	[322]	1969
$FeSi$	$12,0\pm1,5$	$+1,5$	[361]	1958
$FeSi_2$	13,3	—	[322]	1969
$FeSi_{2.33}$	16,6	—	[322]	1969
$CoSi$	$11,5\pm2,0$	$-0,18$	[361]	1958
$CoSi$	—	$-1.48\pm0,716$	[346]	1973
$CoSi_2$	—	$-0.861\pm1,075$	[346]	1973
$NiSi$	—	$-1,0\pm0,716$	[346]	1973
$Ni_{0.35}Si_{0.65}$	—	$-0,167\pm0.358$	[346]	1973
PrP	$20,09\pm0,05$	—	[27]	1973
OsP_2	20,0	—	[311]	1965
BeS	8,4	—	[311]	1965
MgS	12,03	—	[324]	1971
CaS	13,5	—	[324]	1971
SrS	16.3	—	[324]	1971
BaS	18,7	—	[324]	1971
LaS	16,00	—	[797]	1973
La_2S_3	31,5	—	[311]	1965
LaS_2	18,8	—	[311]	1965
CeS	17,61	—	[797]	1973
Ce_2S_3	31,5	—	[311]	1965
CeS_2	18,8	—	[311]	1965
PrS	18,62	—	[797]	1973
Pr_2S_3	44,8	—	—*	1969
NdS	17,61	—	[797]	1973
Nd_2S_3	44,9	—	—*	1969
SmS	24,15	—	[797]	1973
EuS	21,89	—	[797]	1973
GdS	20.69	—	[797]	1973

*A. D. Finogenov; see footnote on p. 132.

ENTROPY OF COMPOUNDS (continued)

Phase	Standard entropy S_{298}°, cal/(mole·°C)	Entropy of formation from elements ΔS_{298}°, cal/(mole·°C)	Source	Year
1	2	3	4	5
Th_2S_3	35.7	—	[311]	1965
US	45.0 (2300K)	—	[311]	1965
TiS	12.0	—	[311]	1965
TiS_2	18.73	—	[323]	1971
Mo_2S_3	28.0	—	[311]	1965
MoS_2	14.96	—	[322]	1969
MoS_3	18.0	—	[311]	1965
WS_2	20,0	—	[311]	1965
MnS	18,7	—	[322]	1969
ReS_2	20,0	—	[311]	1965
FeS	14,41	—	[322]	1969
Fe_7S_8	116,1	—	[322]	1969
FeS_2	12,65	—	[322]	1969
Co_9S_8	110,0	—	[311]	1965
$CoS_{0,89}$	13,6	—	[311]	1965
Ni_3S_2	32,0	—	[322]	1969
NiS	12,66	—	[322]	1969
RuS_2	17,5	—	[311]	1965
Pd_4S	43	—	[322]	1969
PdS	11	—	[322]	1969
PdS_2	19	—	[322]	1969
OsS_2	20,1	—	[311]	1965
Ir_2S_3	34.2	—	[311]	1965
IrS_2	25,2	—	[311]	1965
IrS_3	20,2	—	[311]	1965
PtS	13,16	—	[322]	1969
PtS_2	17,85	—	[322]	1969
B_4C	6,47+0,1	—0,98	[361]	1958
SiC	3.95+0.05	—1,91	[361]	1958
α-SiC	3.94+0,113	—	[287]	1973
β-SiC	3,97±0,113	—	[287]	1973
α-BN	3,67±0,05	—20,77	[361]	1958
AlN	5,0±1,0	—21,65	[361]	1958
Si_3N_4	23,0±2,5	—81,02	[361]	1958

FREE ENERGY OF FORMATION OF REFRACTORY COMPOUNDS

Phase	Reaction	Free energy, cal/mole	Temp. range, °K	Source	Year
1	2	3	4	5	6
MgB_{12}	—	−52 000	298	[324]	1971
CaB_6	$CaO + 3B_2O_3 + 10C = CaB_6 + 10CO$	−775 140 + 429.29 T	—	[326]	1967
	$CaO + B_4C + 2B = CaB_6 + CO$	−110 484 + 46.69 T	—	[326]	1967
	$CaO + 6B + C = CaB_6 + CO$	−96 684 + 45.55 T	—	[326]	1967
	$CaO + 7B = CaB_6 + BO$	−128 400 + 46.65 T	—	[326]	1967
SrB_6	$SrO + 3B_2O_3 + 10C = SrB_6 + 10CO$	−732 840 + 428.84 T	1275—2273	[326]	1967
	$SrO + B_4C + 2B = SrB_6 + CO$	−78 000 + 31.65 T	—	[269]	1961
	$SrO + 6B + C = SrB_6 + CO$	−54 384 + 45.10 T	—	[326]	1967
	$SrO + 7B = SrB_6 + BO$	−86 100 + 46.30 T	—	[326]	1967
BaB_6	$BaO + 3B_2O_3 + 10C = BaB_6 + 10CO$	−705 840 + 428.04 T	—	[326]	1967
	$BaO + B_4C + 2B = BaB_6 + CO$	−41 184 + 45.44 T	—	[326]	1967
	$BaO + 6B + C = BaB_6 + CO$	−27 384 + 44.30 T	—	[326]	1967
	$BaO + 7B = BaB_6 + BO$	−59 100 + 45.40 T	—	[326]	1967
ScB_2	$0.5Sc_2O_3 + B_2O_3 + 4.5C = ScB_2 + 4.5CO$	−328 726 + 206.54 T	—	[326]	1967
	$0.5Sc_2O_3 + 0.5B_4C + C = ScB_2 + 1.5CO$	−109 476 + 69.14 T	—	[326]	1967
	$0.5Sc_2O_3 + 2B + 1.5C = ScB_2 + 1.5CO$	−102 576 + 68.57 T	—	[326]	1967
	$0.5Sc_2O_3 + 3.5B = ScB_2 + 1.5BO$	−150 150 + 70.24 T	—	[326]	1967
YB_6	$0.5Y_2O_3 + 3B_2O_3 + 10.5C = YB_6 + 10.5CO$	−842 557 + 449.14 T	—	[326]	1967
	$0.5Y_2O_3 + 1.5B_4C = YB_6 + 1.5CO$	−184 701 + 67.06 T	—	[326]	1967
	$0.5Y_2O_3 + 6B + 1.5C = YB_6 + 1.5CO$	−164 101 + 66.36 T	—	[326]	1967
	$0.5Y_2O_3 + 7.5B = YB_6 + 1.5BO$	−211 675 + 67.83 T	—	[326]	1967

FREE ENERGY OF FORMATION OF REFRACTORY COMPOUNDS (continued)

Phase	Reaction	Free energy, cal/mole	Temp. range, °K	Source	Year
1	2	3	4	5	6
LaB_6	$0.5La_2O_3 + 3B_2O_3 + 10.5C = LaB_6 + 10.5CO$	$-740\ 767 + 458.54\ T$	—	[326]	1967
	$0.5La_2O_3 + 1.5B_4C = LaB_6 + 1.5CO$	$-83\ 011 + 76.51\ T$	—	[326]	1967
	$0.5La_2O_3 + 6B + 1.5C = LaB_6 + 1,5CO$	$-62\ 311 + 75.81\ T$	—	[326]	1967
	$0.5La_2O_3 + 7.5B = LaB_6 + 1.5BO$	$-109\ 885 + 76.50\ T$	—	[326]	1967
TiB_2	$TiO_2 + B_2O_3 + 5C = TiB_2 + 5CO$	$-328\ 880 + 212.47\ T$	—	[326]	1967
	$TiO_2 + 0.5B_4C + 1.5C = TiB_2 + 2CO$	$-109628 + 85.14\ T$	—	[326]	1967
	$TiO_2 + 2B + 2C = TiB_2 + 2BO$	$-102728 + 84.55\ T$	—	[326]	1967
	$TiO_2 + 4B = TiB_2 + B_2O_3$	$-166\ 160 + 86.75\ T$	—	[326]	1967
	$TiO + 4B = TiB_2 + B_2O_2$	$-48\ 935 + 44.48\ T$	1273—2273	[365]	1971
	$TiO + 0.5B_4C + 0.5C = TiB_2 + CO$	$-58\ 600 + 38.9\ T$	1273—2273	[320]	1969
ZrB_2	$ZrO + 0.5B_4C + 0.5C = ZrB_2 + CO$	$-57\ 600 + 40.7\ T$		[320]	1969
	$ZrO_2 + B_2O_3 + 5C = ZrB_2 + 5CO$	$-356\ 820 + 215.26\ T$		[326]	1967
	$ZrO_2 + 0.5B_4C + 1.5C = ZrB_2 + 2CO$	$-137\ 568 + 87.93\ T$		[326]	1967
	$ZrO_2 + 2B + 2C = ZrB_2 + 2BO$	$-130\ 668 + 87.35\ T$		[326]	1967
	$ZrO_2 + 4B = ZrB_2 + 2BO$	$-194\ 100 + 89.54\ T$		[326]	1967
	$ZrO_2 + 4B = ZrB_2 + B_2O_2$	$-74\ 906 + 44.85\ T$		[365]	1971
HfB_2	$HfO_2 + B_2O_3 + 5C = HfO_2 + 5CO$	$-353\ 770 + 216.70\ T$		[326]	1967
	$HfO_2 + 0.5B_4C + 1.5C = HfB_2 + 2CO$	$-134\ 518 + 89.37\ T$		[326]	1967
	$HfO_2 + 2B + 2C = HfB_2 + 2CO$	$-127\ 618 + 88.73\ T$		[326]	1967
	$HfO_2 + 4B = HfB_2 + 2BO$	$-191\ 050 + 90.98\ T$		[326]	1967
	$HfO_2 + 4B = HfB_2 + B_2O_2$	$-86\ 096 + 45.31\ T$		[365]	1971
VB_2	$0.5V_2O_3 + B_2O_3 + 4.5C = VB_2 + 4.5CO$	$-314\ 528 + 189.80\ T$		[326]	1967
	$0.5V_2O_3 + 0.5B_4C + C = VB_2 + 1.5CO$	$-95\ 276 + 62.47\ T$		[326]	1967
	$0.5V_2O_3 + 2B + 1.5C = VB_2 + 1.5CO$	$-88\ 376 + 60.91\ T$		[326]	1967
	$0.5V_2O_3 + 3.5B = VB_2 + 1.5BO$	$-135\ 950 + 63.55\ T$		[326]	1967
NbB_2	$0.5Nb_2O_5 + B_2O_3 + 5.5C = NbB_2 + 5.5CO$	$-351\ 612 + 233.65\ T$		[326]	1967
	$0.5Nb_2O_5 + 0.5B_4C + 2C = NbB_2 + 2.5CO$	$-132\ 360 + 106.31\ T$		[326]	1967
	$0.5Nb_2O_5 + 2B + 2.5C = NbB_2 + 2.5CO$	$-125\ 460 + 105.74\ T$		[326]	1967
	$0.5Nb_2O_5 + 4.5B = NbB_2 + 2.5BO$	$-204\ 750 + 108.48\ T$		[326]	1967

Compound	Reaction	Δ equation	Temp. range	Ref.	Year
TaB_2	$0{,}5Ta_2O_5 + B_2O_3 + 5{,}5C = TaB_2 + 5{,}5CO$	$-352\,512 + 236.35\,T$	—	[326]	1967
	$0{,}5Ta_2O_5 + 0{,}5B_4C + 2C = TaB_2 + 2{,}5CO$	$-133\,262 + 109.11\,T$	—	[326]	1967
	$0{,}5Ta_2O_5 + 2B + 2{,}5C = TaB_2 + 2{,}5CO$	$-126\,362 + 109.54\,T$	—	[326]	1967
	$0{,}5Ta_2O_5 + 4{,}5B = TaB_2 + 2{,}5BO$	$-205\,650 + 111.28\,T$	—	[326]	1967
CrB_2	$0{,}5Cr_2O_3 + B_2O_3 + 4{,}5C = CrB_2 + 4{,}5CO$	$-292\,853 + 193.41\,T$	—	[326]	1967
	$0{,}5Cr_2O_3 + 0{,}5B_4C + C = CrB_2 + 1{,}5CO$	$-73\,601 + 65.06\,T$	—	[326]	1967
	$0{,}5Cr_2O_3 + 2B + 1{,}5C = CrB_2 + 1{,}5CO$	$-66\,701 + 65.49\,T$	—	[326]	1967
	$0{,}5Cr_2O_3 + 3{,}5B = CrB_2 + 1{,}5BO$	$-114\,275 + 67.24\,T$	—	[326]	1967
	$Cr + 2B = CrB_2$	$-30\,000 - 0{,}24\,T$	298—2173	[333]	1960
Mo_2B_5	$2MoO_2 + 2{,}5B_2O_3 + 11{,}5C = Mo_2B_5 + 11{,}5CO$	$-673\,716 + 478.96\,T$	—	[326]	1967
	$2MoO_2 + 1{,}25B_4C + 2{,}75C = Mo_2B_5 + 4CO$	$-123\,586 + 160.59\,T$	—	[326]	1967
	$2MoO_2 + 5B + 4C = Mo_2B_5 + 4CO$	$-106\,336 + 159.19\,T$	—	[326]	1967
	$2MoO_2 + 9B = Mo_2B_5 + 4BO$	$-233\,200 + 163.57\,T$	—	[326]	1967
W_2B_5	$2WO_2 + 2{,}5B_2O_3 + 11{,}5C = W_2B_5 + 11{,}5CO$	$-697\,827 + 468.51\,T$	—	[326]	1967
	$\frac{1}{2}\,WO_2 + \frac{5}{16}\,B_4C + \frac{11}{16}\,C = \frac{1}{4}\,W_2B_5 + CO$	$-55\,000 + 40{,}3\,T$	1273—2273	[320]	1969
	$2WO_2 + 5B + 4C = W_2B_5 + 4CO$	$-131\,936 + 175.28\,T$	—	[326]	1967
	$2WO_2 + 9B = W_2B_5 + 4BO$	$-258\,800 + 178.66\,B$	—	[326]	1967
Be_2C	$2Be + C = Be_2C$	$-7\,830$	2400	[311]	1965
	$2BeO + 3C = Be_2C + 2CO$	$-205\,542$	298	[287]	1973
Mg_2C_3	$2Mg + 3C = Mg_2C_3$	$+18\,000$	298—923	[311]	1965
MgC_2	$Mg + 2C = MgC_2$	$+21\,000$	298—923	[311]	1965
CaC_2	$\alpha\text{-}Ca + 2C = CaC_2$	$-13\,600 - 5.9\,T$	298—720	[311]	1965
	$\beta\text{-}Ca + 2C = CaC_2$	$-11\,620 - 8.64\,T$	720—1123	[311]	1965
YC_2	—	$-26\,000$	298	[323]	1971
LaC_2	—	$-17\,300$	298	[325]	1973
CeC_2	—	$-15\,200$	298	[325]	1973
SmC_2	—	$-18\,100$	298	[325]	1973
EuC_2	—	$-16\,000$	298	[325]	1973
HoC_2	—	$-26\,700$	298	[325]	1973
YbC_2	—	$-18\,500$	298	[325]	1973
ThC	$Th + C = ThC$	$-7\,000 + 2{,}0\,T$	298—2000	[311]	1965

FREE ENERGY OF FORMATION OF REFRACTORY COMPOUNDS (continued)

Phase	Reaction	Free energy, cal/mole	Temp. range, °K	Source	Year
1	2	3	4	5	6
ThC_2	$Th + 2C = ThC_2$	$-43\,800 + 4.0\,T$	1500—2100	[361]	1958
	$ThO_2 + 4C = ThC_2 + 2CO$	$-213\,736 + 87.49\,T$	2100—2500	[366]	1971
UC	$U + C = UC$	$-21\,530 - 1.57\,T$	298—1406	[311]	1965
U_2C_3	$2U + 3C = U_2C_3$	$-48\,872 - 0.428\,T$	298—1406	[311]	1965
UC_2	$U + 2C = UC_2$	$-23\,160 - 1,14\,T$	298—1406	[311]	1965
$PuC_{0.77}$	$Pu + 0.77C = PuC_{0.77}$	$+3\,700 - 4.0\,T$	298—2000	[311]	1965
Pu_2C_3	$2Pu + 3C = Pu_2C_3$	$-1700 - 3.0\,T$	298—2000	[311]	1965
PuC_2	$Pu + 2C = PuC_2$	$+7\,500$	298	[311]	1965
TiC	$Ti + C = TiC$	$-43\,750 + 2.41\,T$	298—1155	[361]	1958
		$-44\,600 + 3.16T$	1155—2000	[361]	1958
ZrC	$Zr + C = ZrC$	$-44\,100 + 2.2\,T$	298—2200	[311]	1965
HfC	$Hf + C = HfC$	$-48\,500 + 3.262\,T$	298—2600	[311]	1965
V_2C	$2V + C = V_2C$	$-21\,550 + 26.16\,T$	973—1273	[311]	1965
VC	$V + C = VC$	$-12\,500 + 1.6\,T$	298—2000	[311]	1965
Nb_2C	$Nb + \dfrac{1}{2}\,C = NbC_{0,5}$	$-21\,400 - 2.187\,T \log T + 7.75T + 0.385 \cdot 10^{-3}\,T^2 + 0.22 \cdot 10^{1}\,T^{-1}$	298—1800	[311]	1965
NbC	$Nb + C = NbC$	$-33\,980 - 5.64\,T \log T + 17.79\,T + 0.927 \cdot 10^{-3}\,T^2 + 0.49 \cdot 10^{-1}\,T^{-1}$	298—1800	[311]	1965
Ta_2C	$2Ta + C = Ta_2C$	$-33\,997 + 1.332\,T$	298—3000	[311]	1965
TaC	$Ta + C = TaC$	$-38\,497 + 1.332\,T$	298—3000	[311]	1965
$Cr_{23}C_6$	$\dfrac{23}{6}\,Cr + C = \dfrac{1}{6}\,Cr_{23}C_6$	$-16\,380 - 1.54\,T$	973—1273	[311]	1965
Cr_7C_3	$7Cr + 3C = Cr_7C_3$	$-41\,655 - 6.192\,T$	298—2171	[311]	1965
	$7Cr + 3C = Cr_7C_3$	$-86\,200 + 14.35\,T$	2171—2938	[311]	1965

Compound	Reaction	Expression	Temperature range	Ref.	Year
Cr_3C_2	$\frac{7}{27}Cr_{23}C_6 + C = \frac{23}{27}Cr_7C_3$	$-10\,050 - 2.85\,T$	298—1673	[361]	1958
	$3Cr + 2C = Cr_3C_2$	$-8\,550 - 5.03\,T$	298—2171	[311]	1965
	$\frac{3}{5}Cr_7C_3 + C = \frac{7}{5}Cr_3C_2$	$-3\,200 - 0.20\,T$	298—1673	[361]	1958
Mo_2C	$2Mo + C = Mo_2C$	$-6\,700$	298—1273	[361]	1958
	$2Mo + C = Mo_2C$	$-11\,700 - 1.83\,T$	1200—1340	[311]	1965
MoC	$Mo + C = MoC$	$+2\,000$	298	[311]	1965
W_2C	$2W + C = W_2C$	$+11\,700$	298	[311]	1965
WC	$W + C = WC$	$-9\,100 + 0.4\,T$	298—2000	[361]	1958
WC	$W + C = WC$	$-5\,385 - 3,82\,T$	298—3000	[311]	1965
$Mn_{23}C_6$	$\frac{23}{6}Mn + C = \frac{1}{6}Mn_{23}C_6$	$-3\,300 - 3.35\,T$	973—1173	[376]	1961
Mn_3C	$3Mn + C = Mn_3C$	$-3\,300 - 0.26\,T$	298—1010	[256]	1957
	$3Mn + C = Mn_3C$	$-3\,870 + 3.63\,T$	1010—1516	[311]	1965
Mn_7C_3	$7\,(\alpha\text{-Mn}) + 3C = Mn_7C_3$	$19185 - 38.855\,T$	298—990	[311]	1965
	$7\,(\beta\text{-Mn}) + 3C = Mn_7C_3$	$19\,550 - 34.9\,T$	990—1360	[311]	1965
	$7\,(\gamma\text{-Mn}) + 3C = Mn_7C_3$	$11\,400 - 32,0\,T$	1360—1410	[311]	1965
	$7\,(\delta\text{-Mn}) + 3C = Mn_7C_3$	$6\,200 - 28.3\,T$	1410—1516	[311]	1965
Fe_3C	$3Fe + C = Fe_3C$	$+6\,200 - 5.56\,T$	298—463	[361]	1958
		$+6\,380 - 5.92\,T$	463—1115	[361]	1958
		$+2\,475 - 2.43\,T$	1115—1808	[361]	1958
Fe_2C	$2Fe + C = Fe_2C$	$4\,751 - 2.52\,T$	298—1115	[311]	1965
		$4\,406 - 2.17\,T$	1115—1808	[311]	1965
Co_3C	$3Co + C = Co_3C$	$-395 + 1.006\,T\log T - 3.43\,T$	298—1273	[361]	1958
Co_2C	$2Co + C = Co_2C$	$+3\,950 - 2.08\,T$	298—1200	[361]	1958
Ni_3C	$3Ni + C = Ni_3C$	$+8\,110 - 1.70\,T$	298—1726	[361]	1958
Be_3N_2	$3Be + N_2 = Be_3N_2$	$-134\,700 + 40.6\,T$	298—1000	[361]	1958
Mg_3N_2	$3Mg + N_2 = Mg_3N_2$	$-115\,500 + 48.3\,T$	298—923	[361]	1958
Ca_3N_2	$3Ca + N_2 = Ca_3N_2$	$-103\,200 + 50.2\,T$	298—923	[361]	1958
Sr_3N_2	$3Sr + N_2 = Sr_3N_2$	$-91\,400 + 50,9\,T$	—	[311]	1965

FREE ENERGY OF FORMATION OF REFRACTORY COMPOUNDS (continued)

Phase	Reaction	Free energy, cal/mole	Temp. range, °K	Source	Year
1	2	3	4	5	6
Ba_3N_2	$3Ba + N_2 = Ba_3N_2$	$-87\,000 + 57.4\,T$	298—1000	[361]	1958
LaN	$La + \frac{1}{2}\,N_2 = LaN$	$-72\,100 + 25.0\,T$	298—1000	[361]	1958
CeN	$Ce + \frac{1}{2}\,N_2 = CeN$	$-78\,000 + 25.0\,T$	298—1000	[361]	1958
Th_3N_4	$3Th + 2N_2 = Th_3N_4$	$-310\,400 + 89.7\,T$	298—2000	[361]	1958
UN	$U + \frac{1}{2}\,N_2 = UN$	$-68\,500 + 21,5\,T$	298—2000	[361]	1958
U_3N_4	$3U + 2N_2 = U_3N_4$	$-275\,000 + 81.9\,T$	—	[311]	1965
TiN	$\alpha\text{-Ti} + \frac{1}{2}\,N_2 = TiN$	$-80\,250 + 22.2\,T$	298—1155	[361]	1958
TiN	$\beta\text{-Ti} + \frac{1}{2}\,N_2 = TiN$	$-80\,850 + 22.78\,T$	1155—1500	[361]	1958
ZrN	$\alpha\text{-Zr} + \frac{1}{2}\,N_2 = ZrN$	$-87\,000 + 22.3\,T$	298—1135	[361]	1958
ZrN	$\beta\text{-Zr} + \frac{1}{2}\,N_2 = ZrN$	$-87\,925 + 46,22\,T$	1135—1500	[361]	1958
HfN	$Hf + \frac{1}{2}\,N_2 = HfN$	$-81\,400$	298	[311]	1965
VN	$V + \frac{1}{2}\,N_2 = VN$	$-60\,000 - 1.75\,T\log T + 26,3\,T$	298—2000	[361]	1958
NbN	—	$-49\,200$	298	[323]	1971
TaN	$2Ta + N_2 = 2TaN$	$-117\,800 + 13,8\,T\log T + 79.7\,T$	298—2240	[361]	1958

Compound	Reaction	ΔG	Temperature range	Ref.	Year
Cr_2N	$4Cr + N_2 = 2Cr_2N$	$-51\,900 - 11.5\,T\log T + 66.0\,T$	298−1400	[361]	1958
CrN	$2Cr_2N + 2N_2 = 4CrN$	$-64\,000 - 11{,}5\,T\log T + 83.2\,T$	298−1400	[361]	1958
Mo_2N	$4Mo + N_2 = 2Mo_2N$	$-34\,400 - 9.2\,T\log T + 57.9\,T$	298−1300	[361]	1958
Mn_5N_2	$5Mn + N_2 = Mn_5N_2$	$-33\,200 + 42{,}0\,T$	1500−2000	[361]	1958
		$-57\,770 + 36.4\,T$	—	[311]	1965
Fe_4N	$4Fe + \tfrac{1}{2}\,N_2 = Fe_4N$	$-200 + 11.62\,T\log T - 24.85\,T$	298−950	[361]	1958
Ni_3Al	—	$-32\,120$	1045	—*	1965
Mg_2Si	—	$-19\,150 + 7.065\,T$	400−600	—**	1964
$CaSi$	$Ca + Si = CaSi$	$-36\,000 - 0.5\,T$	298−1123	[311]	1965
Ti_5Si_3	$5TiO_2 + 13Si = Ti_5Si_3 + 10SiO$	$-726\,000 + 432\,T$	1700−2000	[377]	1965
	$5TiCl_4 + 3SiCl_4 + 16H_2 = Ti_5Si_3 + 32HCl$	$-540\,000 + 348\,T$		[375]	1969
$TiSi$	$TiO_2 + 3Si = TiSi + 2SiO$	$-135\,000 + 87\,T$	1600−2000	[377]	1965
	$TiCl_4 + SiCl_4 + 4H_2 = TiSi + 8HCl$	$-120\,000 + 78\,T$	—	[375]	1969
$TiSi_2$	$TiO_2 + 4Si = TiSi_2 + 2SiO$	$-131\,400 + 87\,T$	1550−2000	[377]	1965
	$TiCl_4 + 2SiCl_4 + 6H_2 = TiSi_2 + 12HCl$	$-180\,000 + 708\,T$	—	[375]	1969
Nb_5Si_3	$10NbCl_5 + 6SiCl_4 + 37H_2 = 2Nb_5Si_3 + 74HCl$	$-860\,000 + 870\,T$	—	[375]	1969
$NbSi_2$	$2NbCl_5 + 4SiCl_4 + 13H_2 = 2NbSi_2 + 26HCl$	$-340\,000 + 494\,T$	—	[375]	1969
Ta_2Si	$2TaCl_5 + SiCl_4 + 7H_2 = Ta_2Si + 14HCl$	$-170\,000 + 146\,T$	—	[375]	1969
Ta_5Si_3	$10TaCl_5 + 6SiCl_4 + 37H_2 = 2Ta_5Si_3 + 74HCl$	$-931\,600 + 818\,T$	—	[375]	1969
$TaSi_2$	$2TaCl_5 + 4SiCl_4 + 13H_2 = 2TaSi_2 + 26HCl$	$-360\,000 + 480\,T$	—	[375]	1969
Mo_3Si	—	$-23\,000$	298	[322]	1969
Re_3Si	$3Re + Si = Re_3Si$	$-24\,600 - 5.0\,T$	1750−1970	[361]	1958
$ReSi$	$Re + Si = ReSi$	$-30\,000 + 0{,}5\,T$	1750−1970	[361]	1958
$ReSi_2$	$Re + 2Si = ReSi_2$	$-62\,100 + 1.7\,T$	1750−1970	[361]	1958
Fe_3Si	—	$-22\,600$	298	[322]	1969
$FeSi$	—	$-17\,600$	298	[322]	1969

* V. V. Pokidyshev, Study of the Thermodynamic Properties of the Systems Ni−Al, Ni−Al−Cr, Ni−Al−W, Ni−Al−Co, and Ni−Al−Fe, by the emf Method [in Russian], Author's abstract, Cand. Dissertation, Moscow (1965).

** G. M. Lukashenko, Study of the Thermodynamic Properties of Binary Metallic Alloys Based on Magnesium [in Russian], Author's abstract, Cand. Dissertation, Kiev (1964).

FREE ENERGY OF FORMATION OF REFRACTORY COMPOUNDS (continued)

Phase	Reaction	Free energy, cal/mole	Temp. range, °K	Source	Year
1	2	3	4	5	6
$FeSi_2$	—	$-18\ 700$	298	[322]	1969
$FeSi_{2.33}$	—	$-14\ 000$	298	[322]	1969
CoSi	—	$-23\ 600$	298	[322]	1969
Fe_3P	$3Fe + \frac{1}{2} P_2 = Fe_3P$	$-51\ 000 + 11,3\ T$	298—1439	[311]	1965
MgS	$Mg + \frac{1}{2} S_2 = MgS$	$-99\ 650 + 22,8\ T$	298—923	[311]	1965
CaS	$\alpha\text{-}Ca + \frac{1}{2} S_2 = CaS$	$-129\ 435 + 22,81\ T$	298—673	[311]	1965
CaS	$\beta\text{-}Ca + \frac{1}{2} S_2 = CaS$	$-129\ 550 + 22,96\ T$	673—1124	[311]	1965
SrS	$Sr + S = SrS$	$-110\ 380 - 3.02\ T \log T + 11.16\ T + 2.88 \cdot 10^{-3} T^2$	—	[311]	1965
BaS	$Ba + S = BaS$	$-102\ 640 - 3.02\ T \log T + 11,18\ T + 2.88 \cdot 10^{-3} T^2$	—	[311]	1965
LaS	—	$-107\ 900$	298	[325]	1973
CeS	$Ce + \frac{1}{2} S_2 = CeS$	$-133\ 500 + 20,0\ T$	298—2200	[311]	1965
Nd_2S_3	—	$-280\ 200$	298	[325]	1973
TiS_3	$2TiS_2 + S_2 = 2TiS_3$	$-31\ 400 - 34.5\ T$	465—600	[378]	1969
Mo_2S_3	$2Mo + \frac{3}{2} S_2 = Mo_2S_3$	$-145\ 200 + 62.7\ T$	1300—1425	[311]	1965
Mo_2S_3	$2Mo + 3S = Mo_2S_3$	$-105\ 250 + 31.45\ T$	371—1187	[311]	1965
MoS_2	$Mo + 2S = MoS_2$	$-57\ 640 - 15,78\ T \log T + 49.24\ T + 5.60 \cdot 10^{-3} T^2$	—	[311]	1965

	Reaction		T (K)	Ref.	Year
MoS_3	$Mo + 3S = MoS_3$	$-60\,440 + 8.1\,T\log T - 9.78\,T$	—	[311]	1965
WS_2	$\frac{1}{2}W + \frac{1}{2}S_2 = \frac{1}{2}WS_2$	$-40\,400 + 17{,}8\,T$	1023—1273	[349]	1970
	$W + S_2 = WS_2$	$-62\,360 + 23{,}0\,T$	298—1400	[311]	1965
	$W + 2S = WS_2$	$-45\,840 + 3{,}08\,T\log T - 10.14\,T$	—	[311]	1965
MnS	$\alpha\text{-}Mn + \frac{1}{2}S_2 = MnS$	$-64\,000 + 15.32\,T$	298—1000	[311]	1965
	$\beta\text{-}Mn + \frac{1}{2}S_2 = MnS$	$-64\,540 + 15.86\,T$	1000—1374	[311]	1965
	$\gamma\text{-}Mn + \frac{1}{2}S_2 = MnS$	$-65\,510 + 16.56\,T$	1410—1517	[311]	1965
ReS_2	$Re + S_2 = ReS_2$	$-64\,200 - 9.2\,T\log T + 68{,}6\,T$	298—1500	[311]	1965
$\alpha\text{-}FeS$	$\alpha\text{-}Fe + \frac{1}{2}S_2 = FeS$	$-37\,160 + 15.59\,T$	298—412	[311]	1965
	$\alpha\text{-}Fe + S = FeS$	$-22\,270 + 13{,}08\,T\log T - 32{,}19\,T - 13{,}19\cdot10^{-3}\,T^2$		[311]	1965
$\beta\text{-}FeS$	$\alpha\text{-}Fe + \frac{1}{2}S_2 = FeS$	$-35\,910 + 12{,}56\,T$	412—1179	[311]	1965
Fe_7S_8	$\gamma\text{-}Fe + \frac{1}{2}S_2 = FeS$	$-36\,070 + 12.74\,T$	1179—1261	[311]	1695
		$-178\,900$	298	[322]	1969
FeS_2	$FeS + \frac{1}{2}S_2 = FeS_2$	$-43\,350 + 45.0\,T$	600—1100	[311]	1965
	$\alpha\text{-}Fe + 2S = FeS_2$	$-38\,350 + 1.36\,T\log T + 3.48\,T + 2{,}75\cdot10^{5}\,T^2$	—	[311]	1965
Co_9S_8	$9Co + 4S_2 = Co_9S_8$	$-316\,960 + 159.24\,T$	298—1048	[311]	1965
CoS	$Co + S = CoS$	$-20\,820 - 4.38\,T\log T + 8.97\,T + 3.53\cdot10^{-3}\,T^2$	—	[311]	1965

FREE ENERGY OF FORMATION OF REFRACTORY COMPOUNDS (continued)

Phase	Reaction	Free energy, cal/mole	Temp. range, °K	Source	Year
1	2	3	4	5	6
Co_3S_4	$\frac{1}{3} Co_9S_8 + \frac{2}{3} S_2 = Co_3S_4$	$-34\,427 + 22.687\,T$	600—750	[311]	1965
CoS_2	$\frac{1}{3} Co_3S_4 + \frac{1}{3} S_2 = CoS_2$	$-16\,720 + 16.803\,T$	600—900	[311]	1965
Ni_3S_2	$3Ni + S_2 = Ni_3S_2$	$-79\,240 + 39.01\,T$	650—800	[311]	1965
NiS	$Ni + \frac{1}{2} S_2 = NiS$	$-34\,980 + 17.205\,T$	670—850	[311]	1965
RuS_2	$Ru + 2S = RuS_2$	$-48\,140 - 12.87\,T\log T + 43.6\,T + 5.80 \cdot 10^{-3} T^2$	—	[311]	1965
Pd_4S	—	$-16\,000$	298	[322]	1969
PdS	—	$-16\,000$	298	[322]	1969
PdS_2	—	$-17\,800$	298	[322]	1969
Ir_2S_3	$2Ir + \frac{3}{2} S_2 = Ir_2S_3$	$-96\,750 + 66.99\,T$	1000—1600	[311]	1965
PtS	$Pt + \frac{1}{2} S_2 = PtS$	$-33\,050 + 21.8\,T$	1000—1700	[311]	1965
PtS_2	$PtS + \frac{1}{2} S_2 = PtS_2$	$-21\,875 + 21.9\,T$	700—1100	[311]	1965
B_4C	$4B + C = B_4C$	$-12\,845 + 2.5\,T$	298—2300	[311]	1965
SiC	$Si + C = SiC$	$-12\,770 + 1.66\,T$	298—1683	[311]	1965
BN	$B + \frac{1}{2} N_2 = BN$	$-26000 + 9.7\,T$	1200—2300	[311]	1965
AlN	$Al + \frac{1}{2} N_2 = AlN$	$-77\,000 + 22.3\,T$	298—923	[361]	1958
Si_3N_4	$3Si + 2N_2 = Si_3N_4$	$-180\,000 + 80.4\,T$	298—1680	[311]	1965

SPECIFIC HEAT

Phase	$c_p = a + bT + cT^2$, cal/(mole·°C)	Temp., K	Source	Year	Specific heat at 293°K, c_p cal/(mole·°C)	Source	Year	Notes
1	2	3	4	5	6	7	8	9
ZrBe13	64.6	373	[10]	1973	—	—	—	—
	79	811	[2]	1966				
	98	1755	[2]	1966				
NbBe12	60.3	373	[10]	1973	—	—	—	
	78.5	811	[2]	1966				
	88.5	1755	[2]	1966				
Ta2Be17	113.5	811	[2]	1966	—	—	—	
	134	1755	[2]	1966				
TaBe12	62.7	373	[10]	1973	—	—	—	
	78	811	[2]	1966				
	86.5	1755	[2]	1966				
WBe13	63.3	373	[10]	1973	—	—	—	
MgB2	—	—	—	—	11.59	[324]	1971	
MgB4	—	—	—	—	16.78	[324]	1971	
MgB12	—	—	—	—	36.2	[324]	1971	
LaB6	$21.73 + 20.4\cdot10^{-3}T$	293—1483	[329]	1961	27.85	[329]	1961	
NdB6	$24.1 + 21.8\cdot10^{-3}T - 5.675\cdot10^5 T^{-2}$	298—1300	[379]	1970	24.2	[379]	1970	
SmB6	$30.0 + 17.4\cdot10^{-3}T - 11.75\cdot10^5 T^{-2}$	298—1300	[379]	1970	22.0	[379]	1970	
EuB6	$24.3 + 24.0\cdot10^{-3}T - 9.33\cdot10^5 T^{-2}$	298—1300	[379]	1970	21.45	[379]	1970	
ThB6	$1.15\cdot10^{-3}T + 0.717\cdot10^{-4}T^3$	2—12	[57]	1968	—	—	—	
UB2	$10.9 + 3.6\cdot10^{-3}T$	298—2000	[380]	1964	11.97	[380]	1964	

SPECIFIC HEAT (continued)

Phase	$c_p = a + bT + cT^2$, cal/(mole·°C)	Temp., K	Source	Year	Specific heat at 293°K, c_p cal/(mole·°C)	Source	Year	Notes
1	2	3	4	5	6	7	8	9
UB$_4$	$15.3 + 7.0\cdot10^{-3}\,T$	298—2000	[380]	1964	17.38	[380]	1964	
TiB$_2$	$7.219 + 1.147\cdot10^{-2}\,T$	291—1073	[331]	1959	10.57	[331]	1959	
ZrB$_2$	$11.78 + 9.986\cdot10^{-3}\,T - 4.028\cdot10^{-6}\,T^2$	291—1073	[381]	1958	12.00	[381]	1958	
HfB$_2$	$17.632 + 1.867\cdot10^{-3}\,T - 5.501\cdot10^{5}\,T^{-2}$	298—2813	[400]	1966	11.89	[323]	1971	
NbB$_{1.963}$	—	—	—	—	11.42	[323]	1971	
TaB$_2$	—	—	—	—	13.98	[10]	1973	
CrB$_2$	$7.808 + 1.517\cdot10^{-2}\,T$	291—1073	[333]	1960	12.24	[333]	1960	
MoB	—	—	—	—	16.7	[10]	1973	
WB	—	—	—	—	15.1	[10]	1973	
Be$_2$C	$10.2 + 5.1\cdot10^{-3}\,T$	293—1373	[311]	1965	11.69	[311]	1965	
Mg$_2$C$_3$	$28.38 + 2.56\cdot10^{-3}\,T - 0.598\cdot10^{6}\,T^{-2}$	298—2500	[287]	1973	22.42	[287]	1973	
MgC$_2$	$17.02 + 1.54\cdot10^{-3}\,T - 0.359\cdot10^{6}\,T^{-2}$	298—2500	[287]	1973	13.44	[287]	1973	
α-CaC$_2$	$16.40 + 2.84\cdot10^{-3}\,T - 2.07\cdot10^{5}\,T^{-2}$	298—720	[361]	1958	14.66	[361]	1958	
β-CaC$_2$	$15.40 + 2.00\cdot10^{-3}\,T$	720—1273	[361]	1958				
CeC$_2$	$10.13 + 14.75\cdot10^{-3}\,T$	298—1500	[287]	1973	14.53	[287]	1973	
ThC$_2$	$12.6 + 2.0\cdot10^{-3}\,T - 2.62\cdot10^{5}\,T^{-2}$		[311]	1965	13.55	[334]	1973	
UC	$14.315 - 3.026\cdot10^{-4}\,T + 1.05\cdot10^{-6}\,T^2 - 2.0828\cdot10^{5}\,T^{-2}$	298—2823	[334]	1973	11.98	[334]	1973	
U$_2$C$_3$	—	—	—	—	25.69	[334]	1973	
α-UC$_{1.93}$	$22.352 - 8.946\cdot10^{-3}\,T + 6.92\cdot10^{-6}\,T^2 - 5.1379\cdot10^{5}\,T^{-2}$	298—2038	[334]	1973	14.52	[334]	1973	See [382, 383]
β-UC$_{1.93}$	29.436	2038—2800	[334]	1973				
PuC$_{0.87}$	$13.08 + 11.44\cdot10^{-4}\,T - 3.232\cdot10^{5}\,T^{-2}$	400—1300	[334]	1973				
TiC	$11.83 + 0.8\cdot10^{-3}\,T - 3.58\cdot10^{5}\,T^{-2}$	400—1800	[334]	1973	8.04	[323]	1971	See [362]
TiC$_{0.99}$	$8.178 + 3.16\cdot10^{-3}\,T$	1300—2500	[384]	1968				
TiC$_{0.82}$	$8.144 + 2.784\cdot10^{-3}\,T$	1300—2500	[384]	1968				

Formula	Equation	Temp. range	Ref.	Year	Value	Ref.	Year
TiC$_{0.71}$	$8{,}110 + 2{,}384\cdot10^{-3}\,T + 2519\cdot10^{8}\,T^{-2}\cdot\exp(-24\,513/T)$	1300—2200	[384]	1968	—	—	—
TiC$_{0.64}$	$8{,}076 + 2{,}182\cdot10^{-3}\,T + 3\,591\cdot10^{9}\,T^{-2}\cdot\exp(-29\,860/T)$	1300—2200	[384]	1968	—	—	—
ZrC	$12{,}6 + 0{,}83\cdot10^{-3}\,T - 0{,}32\cdot10^{6}\,T^{-2}$	500—2400	[385]	1973	9,06	[323]	1971
ZrC$_{0.99}$	$7{,}329 + 3{,}326\cdot10^{-3}\,T$	1300—2500	[386]	1968	—	—	—
ZrC$_{0.76}$	$7{,}762 + 2{,}348\cdot10^{-3}\,T + 3\,311\cdot10^{7}\,T^{-2}\cdot\exp(-19\,957/T)$	1300—2500	[386]	1968	—	—	—
ZrC$_{0.69}$	$8{,}057 + 1{,}928\cdot10^{-3}\,T + 7\,623\cdot10^{9}\,T^{-2}\cdot\exp(-31\,397/T)$	1300—2500	[386]	1968	—	—	—
HfC$_{0.99}$	$10{,}322 + 1{,}594\cdot10^{-3}\,T$	1300—2500	[387]	1967	13,05	[11]	1968
HfC$_{0.85}$	$10{,}267 + 1{,}452\cdot10^{-3}\,T$	1300—2500	[388]	1971	—	—	—
HfC$_{0.71}$	$10{,}216 + 1{,}322\cdot10^{-3}\,T$	1300—2500	[388]	1971	—	—	—
VC$_{0.46}$	$7{,}027 + 0{,}926\cdot10^{-3}\,T - 915\,T^{-1}$	300—1300	[334]	1973	13,3	[323]	1971
V$_2$C	—	—	—	—	7,66	[323]	1971
VC$_{0.88}$	$9{,}82 + 2{,}430\cdot10^{-3}\,T$	1300—2500	[334]	1973	—	—	—
VC$_{0.86}$	$9{,}80 + 2{,}320\cdot10^{-3}\,T$	1300—2500	[334]	1973	—	—	—
VC$_{0.75}$	$9{,}78 + 2{,}230\cdot10^{-3}\,T$	1300—2500	[334]	1973	—	—	—
VC$_{0.72}$	$7{,}94 + 1{,}50\cdot10^{-3}\,T - 1{,}025\cdot10^{5}\,T^{-2}$	298—1800	[363]	1960	7,25	[363]	1960
NbC$_{0.5}$	$11{,}58 + 1{,}044\cdot10^{-3}\,T$	1300—2500	[389]	1967	8,81	[323]	1971
NbC$_{0.99}$	$11{,}51 + 0{,}920\cdot10^{-3}\,T + 1\,529\cdot10^{6}\,T^{-2}\exp(-14\,261/T)$	1300—2500	[389]	1967	—	—	—
NbC$_{0.91}$	$11{,}30 + 0{,}832\cdot10^{-3}\,T + 7\,628\cdot10^{6}\,T^{-2}\exp(-16\,300/T)$	1300—2500	[389]	1967	—	—	—
NbC$_{0.86}$	$10{,}10 + 1{,}138\cdot10^{-3}\,T + 1\,989\cdot10^{8}\,T^{-2}\exp(-22\,907/T)$	1300—2500	[389]	1967	—	—	—
NbC$_{0.75}$	—	1300—2500	[389]	1967	—	—	—
TaC	$7{,}28 + 1{,}65\cdot10^{-3}\,T$	298—2073	[361]	1958	8,79	[323]	1971
TaC$_{1-x}$	$7{,}3 - 1{,}5\cdot10^{-3}\,T + (0{,}773 + 1{,}476\cdot10^{-3}\,T)(1-x)$	1200—2200	[390]	1968	—	—	—
Cr$_{23}$C$_6$	$169{,}12 + 42{,}64\cdot10^{-3}\,T - 28{,}915\cdot10^{5}\,T^{-2}$	470—1700	[334]	1973	149,2	[322]	1969
Cr$_7$C$_3$	$57{,}0 + 14{,}38\cdot10^{-3}\,T - 10{,}104\cdot10^{5}\,T^{-2}$	298—1473	[361]	1958	49,92	[322]	1969
Cr$_3$C$_2$	$30{,}03 + 5{,}58\cdot10^{-3}\,T - 7{,}396\cdot10^{5}\,T^{-2}$	298—1473	[361]	1958	23,53	[322]	1969
Mo$_2$C	—	—	—	—	14,55	[334]	1973
WC	$7{,}98 + 2{,}17\cdot10^{-3}\,T$	298—1973	[361]	1958	8,5?	[361]	1958
WC	$10{,}00 + 1{,}786\cdot10^{-3}\,T$	1275—2640	[334]	1973	—	—	—
α-Mn$_3$C	$25{,}26 + 5{,}60\cdot10^{-3}\,T - 0{,}407\cdot10^{6}\,T^{-2}$	298—1310	[287]	1973	22,33	[322]	1969
β-Mn$_3$C	$38{,}00$	1310—1800	[287]	1973	—	—	—

SPECIFIC HEAT (continued)

Phase	$c_p = a + bT + cT^2$, cal/(mole·°C)	Temp., K	Source	Year	Specific heat at 293°K, c_p cal/(mole·°C)	Source	Year	Notes
1	2	3	4	5	6	7	8	9
Mn_7C_3	$52.5 + 24 \cdot 10^{-3}\,T$	298—1517	[287]	1973	59.65	[287]	1973	
α-Fe_3C	$19.64 + 20.00 \cdot 10^{-3}\,T$	273—463	[361]	1958	25.50	[361]	1958	
β-Fe_3C	$25.62 + 3.00 \cdot 10^{-3}\,T$	463—1026	[361]	1958	—	—	—	
Be_3N_2	$7.32 + 30.8 \cdot 10^{-3}\,T$	273—800	[361]	1958	16.6	[361]	1958	
α-Mg_3N_2	$20.77 + 11.20 \cdot 10^{-3}\,T$	298—823	[361]	1958	24.05	[361]	1958	
β-Mg_3N_2	$20.07 + 10.66 \cdot 10^{-3}\,T$	823—1061	[361]	1958	—	—	—	
γ-Mg_3N_2	28.50	1061—1300	[361]	1958	—	—	—	
Ca_3N_2	$20.44 + 22.0 \cdot 10^{-3}\,T$	293—800	[361]	1958	26.89	[361]	1958	
LaN	—	—	—	—	11.487	—	1975	[807]
CeN	—	—	—	—	11.589	—	1975	
PrN	—	—	—	—	11.522	—	1975	
NdN	—	—	—	—	11.530	—	1975	
SmN	—	—	—	—	10.866	—	1975	
EuN	—	—	—	—	12.162	—	1975	
GdN	—	—	—	—	11.584	—	1975	
ThN	—	—	—	—	10.75	[364]	1972	
Th_3N_4	$27.78 + 31.8 \cdot 10^{-3}\,T$	273—773	[14]	1969	37.1	[14]	1969	
UN	$10.6 + 7.0 \cdot 10^{-3}\,T$	273—2973	[14]	1969	12.65	[14]	1969	See [382]
TiN	$11.91 + 0.94 \cdot 10^{-3}\,T - 2.96 \cdot 10^5\,T^{-2}$	298—1823	[14]	1969	8.86	[323]	1971	
ZrN	$11.10 + 1.58 \cdot 10^{-3}\,T - 1.72 \cdot 10^5\,T^{-2}$	298—1823	[14]	1969	10.88	[14]	1969	
HfN	$9.84 + 2.22 \cdot 10^{-3}\,T$	273—1973	[14]	1969	10.49	[14]	1969	
VN	$10.94 + 2.10 \cdot 10^{-3}\,T - 2.21 \cdot 10^5\,T^{-2}$	298—1623	[14]	1966	9.08	[14]	1966	
Nb_2N	$14.9109 + 4.0891 \cdot 10^{-3}\,T$	298—1000	[400]	1966	16.13	[400]	1966	
NbN	$16.908 + 2.092 \cdot 10^{-3}\,T$	1000—2673	[400]	1966	10.41	[14]	1969	See [402]
NbN	$8.69 + 5.40 \cdot 10^{-3}\,T$	273—900	[14]	1969				
Ta_2N	$16.845 + 4.2193 \cdot 10^{-3}\,T - 1.6868 \cdot 10^5\,T^{-2}$	298—3000	[400]	1966	16.205	[400]	1966	

150

Compound	c_p equation	ΔT range	Ref.	Year	value	Ref.	Year	Notes
TaN	$7.73 + 7.80 \cdot 10^{-3}T$	273—773	[14]	1969	9.7	[323]	1971	
Cr₂N	$15.24 + 6.8 \cdot 10^{-3}T$	273—800	[361]	1958	17.23	[361]	1958	
CrN	$9.84 + 3,9 \cdot 10^{-3}T$	273—800	[361]	1958	10.98	[361]	1958	
Mo₂N	$8,20 + 26.25 \cdot 10^{-3}T - 12.85 \cdot 10^{-6}T^2$	273—800	[361]	1958	16.99	[361]	1968	
Mn₄N	$22.3 + 27,2 \cdot 10^{-3}T$	273—800	[361]	1958	30.26	[361]	1958	
Mn₅N₂	$32.5 + 35,0 \cdot 10^{-3}T$	273—800	[361]	1958	43.0	[361]	1958	
Mn₃N₂	$22.5 + 22.5 \cdot 10^{-3}T$	273—800	[361]	1958	29.10	[361]	1958	
Fe₄N	$26.84 + 8.16 \cdot 10^{-3}T$	273—800	[361]	1958	31.55	[361]	1958	
Fe₂N	$14.91 + 6.09 \cdot 10^{-3}T$	273—800	[361]	1958	16.8	[361]	1958	
LaAl₂	—	—	—	—	17.62	[325]	1973	
CeAl₂	—	—	—	—	17.85	[325]	1973	
PrAl₂	—	—	—	—	17.64	[325]	1973	
NdAl₂	—	—	—	—	17.45	[325]	1973	
GdAl₂	—	—	—	—	17.71	[325]	1973	
Mg₂Si	$17.52 + 3.58 \cdot 10^{-3}T - 2.11 \cdot 10^{5}T^{-2}$	298—873	[311]	1965	17.6	[324]	1971	
U₃Si	$33,06 + 1.02 \cdot 10^{-3}T + 1.06 \cdot 10^{5}T^{-2}$	298—900	[311]	1965	34.55	[311]	1965	
USi₂	$17.24 + 2.04 \cdot 10^{-3}T + 2.12 \cdot 10^{5}T^{-2}$	1100—1873	[311]	1965		—		
USi₃	$26.84 + 3.06 \cdot 10^{-3}T + 3.18 \cdot 10^{5}T^{-2}$	298—1500	[311]	1965	31,32	[311]	1965	
Ti₅Si₃	$58.22 + 5.742 \cdot 10^{-3}T - 2.646 \cdot 10^{6}T^{-2}$	298—1173	[392]	1959	33.44	[392]	1959	
TiSi	$15.43 - 0,8832 \cdot 10^{6}T^{-2}$	298—1400	[392]	1959	6,63	[392]	1959	
TiSi₂	$14.94 + 8.32 \cdot 10^{-3}T + 0,455 \cdot 10^{6}T^{-2}$	298—1173	[392]	1959	12.88	[392]	1959	
Zr₂Si	$-5,77 + 18,1 \cdot 10^{-3}T + 4 \cdot 10^{6}T^{-2}$	1200—1700	[391]	1973	~18	[344]	1968	
Zr₅Si₃	—	—	—	—	~45	[344]	1968	
Zr₃Si₂	—	—	—	—	~27	[344]	1968	
Zr₅Si₄	—	—	—	—	~43	[344]	1968	
ZrSi	—	—	—	—	~11.5	[344]	1968	
ZrSi₂	$6.02 + 11,45 \cdot 10^{-3}T + 1,925 \cdot 10^{6}T^{-2}$	1100—1750	[385]	1973	~16.5	[344]	1968	
HfSi₂	$-3,71 + 16,8 \cdot 10^{-3}T + 3,65 \cdot 10^{6}T^{-2}$	1200—1600	[385]	1973		—		
V₃Si	$26.536 + 2.788 \cdot 10^{-3}T - 5,0322 \cdot 10^{5}T^{-2}$	298—1300	[393]	1962	21,7	[393]	1962	See [346]
V₅Si₃	$57.624 + 0,2144 \cdot 10^{-3}T - 19,685 \cdot 10^{5}T^{-2}$	298—1300	[393]	1962	35,6	[393]	1962	See [346]
VSi₂	$23.271 - 2,169 \cdot 10^{-3}T - 14,924 \cdot 10^{5}T^{-2}$	298—1300	[393]	1962	5,88	[393]	1962	
Nb₄Si	$9.77 + 7,37 \cdot 10^{-3}T + 1,19 \cdot 10^{6}T^{-2}$; $-8.8 + 30,6 \cdot 10^{-3}T + 6,81 \cdot 10^{6}T^{-2}$	1200—1950	[391]	1973		—		See [385]
Nb₅Si₃	$45.2 + 7,35 \cdot 10^{-3}T - 3,6 \cdot 10^{5}T^{-2}$	1200—2100	[385]	1973	43,35	[346]	1973	
NbSi₂	$15.09 + 3,675 \cdot 10^{-3}T - 0,666 \cdot 10^{5}T^{-2}$	298—2000	[346]	1973	15.44	[346]	1973	
TaSi₂	$17.6 + 1.84 \cdot 10^{-3}T - 2,16 \cdot 10^{5}T^{-2}$	298—2000	[346]	1973	20.65	[346]	1973	

SPECIFIC HEAT (continued)

Phase	$c_p = a + bT + cT^2$, cal/(mole·°C)	Temp, K	Source	Year	Specific heat at 293°K, c_p cal/(mole·°C)	Source	Year	Notes
1	2	3	4	5	6	7	8	9
Cr₃Si	$22{,}62 + 8{,}80 \cdot 10^{-3}\,T - 5{,}31 \cdot 10^5\,T^{-2}$	298—873	[394]	1961	39,36	[394]	1961	
Cr₅Si₃	$59{,}144 + 6{,}42 \cdot 10^{-3}\,T - 23{,}2536 \cdot 10^5\,T^{-2}$	298—873	[394]	1961	35,26	[394]	1961	See [346]
CrSi	$12{,}51 + 3{,}420 \cdot 10^{-3}\,T - 3{,}8466 \cdot 10^5\,T^{-2}$	298—873	[394]	1961	9,35	[394]	1961	
CrSi₂	$14.30 + 10.53 \cdot 10^{-3}\,T - 4{,}1763 \cdot 10^5\,T^{-2}$	298—873	[394]	1961	12,64	[394]	1961	
Mo₃Si	$20.52 + 5{,}42 \cdot 10^{-3}\,T + 0{,}0765 \cdot 10^5\,T^{-2}$	298—2000	[346]	1973	22,33	[322]	1969	
Mo₅Si₃	$43{,}8 + 8{,}4 \cdot 10^{-3}\,T - 2{,}86 \cdot 10^5\,T^{-2}$	298—2200	[346]	1973	43,09	[346]	1973	See [395]
MoSi₂	$16{,}2 + 2{,}86 \cdot 10^{-3}\,T - 1{,}57 \cdot 10^5\,T^{-2}$	298—2200	[346]	1973	15,29	[346]	1973	
W₅Si₃	$42{,}9 + 9{,}36 \cdot 10^{-3}\,T - 2{,}12 \cdot 10^5\,T^{-2}$	298—2200	[346]	1973	43,31	[346]	1973	
WSi₂	$16{,}2 + 2{,}64 \cdot 10^{-3}\,T - 1{,}455 \cdot 10^5\,T^{-2}$	298—2200	[346]	1973	15,35	[346]	1973	See [396]
Mn₃Si	$24{,}12 + 12{,}44 \cdot 10^{-3}\,T - 3{,}52 \cdot 10^5\,T^{-2}$	298—950	[346]	1973	23,87	[346]	1973	
Mn₅Si₃	$48{,}1 + 12{,}95 \cdot 10^{-3}\,T - 4{,}68 \cdot 10^5\,T^{-2}$	298—1573	[346]	1973	46,7	[346]	1973	
MnSi	$11{,}8 + 3{,}05 \cdot 10^{-3}\,T - 1{,}53 \cdot 10^5\,T^{-2}$	298—1550	[346]	1973	10,99	[346]	1973	
Mn₃Si₅	$22{,}726 + 24{,}762 \cdot 10^{-3}\,T - 31{,}2272 \cdot 10^5\,T^{-2}$	610—1400	[396]	1963	—	—	—	
Mn₀.₃₇Si₀.₆₃	$6{,}37 + 0{,}409 \cdot 10^{-3}\,T - 1{,}155 \cdot 10^5\,T^{-2}$	298—1425	[346]	1973	5,19	[346]	1973	
ReSi₂	$7{,}98 + 7{,}92 \cdot 10^{-3}\,T + 1{,}25 \cdot 10^6\,T^{-2}$	1473—2173	[385]	1973	—	—	—	
Fe₃Si	$17{,}04 + 20{,}87 \cdot 10^{-3}\,T + 35{,}57 \cdot 10^2\,T + 93.09 \cdot 10^4\,T^{-2}$	273—800	*	1962	23,5	*	1962	
	34,6	850—1525	*	1962	—	—	—	
Fe₅Si₃	$42{,}13 + 21{,}24 \cdot 10^{-3}\,T - 96{,}91 \cdot 10^3\,T^{-2}$	1535—1800	*	1962	47,7	*	1962	
FeSi	$10{,}66 + 36{,}60 \cdot 10^{-4}\,T - 94{,}94 \cdot 10^2\,T^{-2}$	273—1360	*	1962	11,6	*	1962	
	19,5	1675—1925	*	1962	—	—	—	
FeSi₂	$14.56 + 4{,}108 \cdot 10^{-3}\,T$	273—1238	*	1962	15,8	*	1962	
FeSi₂.₃₃	$8{,}932 + 15{,}88 \cdot 10^{-3}\,T + 17{,}24 \cdot 10^4\,T^{-2}$	273—1480	*	1962	17,6	*	1962	
	30,08	1485—1723	*	1962	—	—	—	
CoSi	$11{,}74 + 2{,}89 \cdot 10^{-3}\,T - 1{,}8 \cdot 10^5\,T^{-2}$	298—1733	[346]	1973	10,6	[322]	1969	

Compound	c_p equation	Temp. range	Ref.	Year	Value	Ref.	Year	Notes
CoSi$_2$	$16{,}95 + 4{,}45 \cdot 10^{-3}T - 2{,}37 \cdot 10^{5}T^{-2}$	298—1600	[346]	1973	15,62	[346]	1973	
Ni$_2$Si	$11{,}64 + 1{,}46 \cdot 10^{-3}T - 1{,}56 \cdot 10^{5}T^{-2}$				16,8	[322]	1969	
NiSi	$5{,}98 + 0{,}88 \cdot 10^{-3}T - 0{,}862 \cdot 10^{5}T^{-2}$	298—1265	[346]	1973	10,9	[322]	1969	
Ni$_{0{,}85}$Si$_{0{,}65}$		298—1200	[346]	1973	5,27	[346]	1973	
PrP					12,42	[404]	1972	
MgS					10,89	[324]	1971	
CaS	$10{,}20 + 3{,}80 \cdot 10^{-3}T$	273—1000	[311]	1965	11,33	[324]	1971	
SrS					11,64	[324]	1971	
BaS					11,80	[324]	1971	
LaS					14	[325]	1973	
La$_2$S$_3$					~28,5	[397]	1972	See [401]
CeS					11,94	[325]	1973	
Ce$_2$S$_3$					~30,5	[397]	1973	See [397]
PrS					11,485	[401]	1973	
NdS					11,454	[401]	1973	
Nd$_2$S$_3$					29,28	[325]	1973	See [397]
SmS					13,633	[401]	1973	
EuS					11,352	[401]	1973	
GdS					11,547	[401]	1973	
US	$13{,}0 + 1{,}5 \cdot 10^{-3}T$	298—2000	[311]	1965	17,47	[311]	1965	
α-TiS$_2$	$8{,}08 + 27{,}34 \cdot 10^{-3}T$	298—420	[311]	1965	16,23	[323]	1971	
β-TiS$_2$	$14{,}99 + 5{,}14 \cdot 10^{-3}T$	420—1010	[311]	1965				
Nb$_{1{,}136}$S$_2$					20	[323]	1971	
MoS$_2$	$11{,}20 + 13{,}50 \cdot 10^{-3}T$	298—729	[311]	1965	15,19	[322]	1969	
MnS	$11{,}40 + 1{,}8 \cdot 10^{-3}T$	298—1803	[311]	1965	11,94	[322]	1969	
α-FeS	$5{,}19 + 26{,}40 \cdot 10^{-3}T$	298—411	[311]	1965	12,08	[322]	1969	
β-FeS	17,40	411—598	[311]	1965				
γ-FeS	$12{,}20 + 2{,}38 \cdot 10^{-3}T$	598—1468	[311]	1965				
Fe$_7$S$_8$					95,26	[322]	1969	
FeS$_2$	$17{,}88 + 1{,}32 \cdot 10^{-3}T - 3{,}05 \cdot 10^{5}T^{-2}$	298—1000	[311]	1965	14,86	[322]	1969	
CoS$_{0{,}89}$	$10{,}6 + 2{,}51 \cdot 10^{-3}T$	273—1373	[311]	1965	11,35	[311]	1965	
Ni$_3$S$_2$					28,12	[322]	1969	

*R. P. Krentsis, Specific Heat, Enthalpy, and Entropy of Iron Silicides and Some Steels [in Russian], Cand. Dissertation, Sverdlovsk (1962).

SPECIFIC HEAT (continued)

Phase	$c_p = a + bT + cT^2$, cal/(mole·°C)	Temp., K	Source	Year	Specific heat at 298°K, c_p cal/(mole·°C)	Source	Year	Notes
1	2	3	4	5	6	7	8	9
NiS	$9.25 + 12.80·10^{-3} T$	298—600	[311]	1965	11.26	[322]	1969	
PtS	$11.4 + 2.86·10^{-3} T$	298—1000	[311]	1965	10.37	[322]	1969	
PtS$_2$	$13.86 + 7.14·10^{-3} T$	298—1000	[311]	1965	15.75	[322]	1969	
B$_4$C	$-6.0491 + 0.06224 T + 0.9445·10^4 T^{-2}$	55—295	[287]	1973	12.545	[287]	1973	See [334]
B$_4$C	$23.077 + 5.24·10^{-3} T - 1.075·10^6 T^{-2}$	298—1800	[287]	1973	—	—	—	
α-SiC	$9.93 + 1.92·10^{-3} T - 0.366·10^6 T^{-2}$	298—1800	[287]	1973	6.39	[287]	1973	
β-SiC	$9.97 + 1.82·10^{-3} T - 0.364·10^6 T^{-2}$	298—1700	[287]	1973	6.42	[287]	1973	
α-BN	$1,82 + 3.62·10^{-3} T$	273—1173	[361]	1958	2.88	[361]	1958	
AlN	$5.47 + 7.80·10^{-3} T$	293—900	[14]	1969	7.75	[14]	1969	
Si$_3$N$_4$	$32.074 + 4.7867·10^{-3} T - 0.23122·10^6 T^{-2}$	298—4000	[399]	1974	30.9	[399]	1974	

Phase	Heat of reaction kcal/mole	Reaction***	Temp., K	Source	Year
1	2	3	4	5	6
SrB_6	97.2	—	298	[327]	1960
LaB_6	133.7	LaB_6 (sl.) = La (gas) + 6B (sl.)	298	[405]	1971
CeB_6	123.0	CeB_6 (sl.) = Ce (gas) + 6B (sl.)	2033	[405]	1971
PrB_6	113.0	PrB_6 (sl.) = Pr (gas) + 6B (sl.)	2190	[405]	1971
NdB_6	107.0	NdB_6 (sl.) = Nd (gas) + 6B (sl.)	2136	[405]	1971
SmB_4	93.2	$3SmB_4$ (sl.) = $2SmB_6$ (gas) + Sm (gas)	298	[405]	1971
SmB_6	102.8	SmB_6 (sl.) = Sm (gas) + 6B (sl.)	2210	[405]	1971
GdB_6	127	GdB_6 (sl.) = Gd (gas) + 6B (sl.)	2350	[405]	1971
TbB_6	128.7	TbB_6 (sl.) = Tb (gas) + 6B (sl.)	2162	[405]	1971
TiB	279	TiB (sl.) = Ti (gas) + B (gas)	298	[406]	1964
TiB_2	435	TiB_2 (sl.) = Ti (gas) + 2B (gas)	298	[406]	1964
ZrB_2	467	ZrB_2 (sl.) = Zr (gas) + 2B (gas)	2280	[407]	1966
Fe_2B	20.7	—	298	[408]	1971
FeB	51.1	—	298	[408]	1971
Be_2C	92.2	$\frac{1}{2}$ Be_2C (sl.) = Be (gas) + $\frac{1}{2}$ C (gas)	1500	[334]	1973
CaC_2	54.3	CaC_2 (sl.) = Ca (gas) + 2C (sl.)	298	[334]	1973
SrC_2	59.3	SrC_2 (sl.) = Sr (gas) + 2C (sl.)	298	[334]	1973
BaC_2	61.2	BaC_2 (sl.) = Ba (gas) + 2C (sl.)	298	[334]	1973
YC_2	125	YC_2 (sl.) = Y (gas) + 2C (sl.)	298	[334]	1973
YC_2	169.1	YC_2 (sl.) = YC_2 (gas)	298	[334]	1973
LaC_2	118	LaC_2 (sl.) = La (gas) + 2C (sl.)	0	[405]	1971
LaC_2	157	LaC_2 (sl.) = LaC_2 (gas)	0	[405]	1971
CeC_2	116.2	CeC_2 (sl.) = Ce (gas) + 2C (sl.)	298	[334]	1973
CeC_2	131.9	CeC_2 (sl.) = CeC_2 (gas)	0	[405]	1971
NdC_2	95.8	NdC_2 (sl.) = Nd (gas) + 2C (sl.)	0	[405]	1971
NdC_2	140.7	NdC_2 (sl.) = NdC_2 (gas)	0	[405]	1971
SmC_2	67	SmC_2 (sl.) = Sm (gas) + 2C (sl.)	0	[405]	1971
EuC_2	58.4	EuC_2 (sl.) = Eu (gas) + 2C (sl.)	0	[405]	1971
GdC_2	100.7	GdC_2 (sl.) = Gd (gas) + 2C (sl.)	2150	[406]	1964
GdC_2	142.6	GdC_2 (sl.) = GdC_2 (gas)	2150	[406]	1964
DyC_2	82.9	DyC_2 (sl.) = Dy (gas) + 2C (sl.)	0	[405]	1971
HoC_2	93.0	HoC_2 (sl.) = Ho (gas) + 2C (sl.)	0	[405]	1971
ErC_2	101.1	ErC_2 (sl.) = Er (gas) + 2C (sl.)	0	[405]	1971
YbC_2	54	YbC_2 (sl.) = Yb (gas) + 2C (sl.)	0	[405]	1971
ThC_2	188.1 * 212.8**	ThC_2 (sl.) = ThC_2 (gas)	298	[406]	1964
	160.3 * 184.7**	ThC_2 (sl.) = Th (gas) + 2C (sl.)	298	[434]	1973
UC	199	UC (sl.) = U (gas) + C (gas)	298	[383]	1966

*According to 2nd law of thermodynamics.
** According to 3rd law of thermodynamics.
*** sl. = solid.

Phase	Heat of reaction, kcal/mole	Reaction***	Temp., K	Source	Year
1	2	3	4	5	6
UC	235,1	$3UC$ (sl.) $= 3U$ (gas) $+ C_1$ (gas) $+$ $+ \frac{1}{2} C_2$ (gas) $+ \frac{1}{3} C_3$ (gas)	298	[334]	1973
UC_2	127.3	UC_2 (sl.) $= U$ (gas) $+ 2C$ (sl.)	2425	[334]	1973
UC_2	185	UC_2 (sl.) $= UC_2$ (gas)	2425	[334]	1973
PuC_2	95,1	PuC_2 (sl.) $= Pu$ (gas) $+ 2C$ (sl.)	298	[334]	1973
TiC	155.9	TiC (sl.) $= Ti$ (gas) $+ C$ (sl.)	2500	[334]	1973
VC	44,64	VC (sl.) $= V$ (gas) $+ C$ (sl.)	298	[334]	1973
NbC	150 * 188 **	NbC (sl.) $= NbC_x$ (sl.) $+ (1-x) C$ (gas)	2800	[409]	1962
$Cr_{23}C$	98,7— 100,7	$\frac{1}{9} Cr_{23}C_6$ (sl.) $= \frac{2}{9} Cr_7C_3$ (sl.) $+$ $+ Cr$ (gas)	298	[334]	1973
Cr_7C_3	97,2— 103,0	$\frac{2}{5} Cr_7C_3$ (sl.) $= \frac{3}{5} Cr_3C_2$ (sl.) $+$ $+ Cr$ (gas)	298	[410]	1966
Cr_3C_2	98,2— 104,8	$\frac{1}{3} Cr_3C_2$ (sl.) $= -\frac{2}{3} C$ (sl.) $+ Cr$ (gas)	298	[410]	1966
Mo_2C	239	$MoC_{0.49}$ (sl.) $= Mo$ (gas) $+ 0,49C$ (gas)	298	[334]	1973
Mn_7C_3	493.9	Mn_7C_3 (sl.) $= 7Mn$ (gas) $+ 3C$ (sl.)	298	[287]	1973
Be_3N_2	136	Be_3N_2 (sl.) $= 3Be$ (gas) $+ N_2$	—	[14]	1969
CeN	123	—	298	[798]	1971
U_2N_3	57—58	$\frac{1}{2} U_2N_3$ (sl.) $= UN$ (sl.) $+ \frac{1}{2} N$	—	[14]	1969
TiN	191.2	TiN (sl.) $= Ti$ (gas) $+ \frac{1}{2} N_2$	—	[411]	1954
ZrN	79,5	ZrN (sl.) $= Zr$ (sl.) $+ \frac{1}{2} N_2$	—	[411]	1954
Mo_3Si	130,4	Mo_3Si (sl.) $= 3Mo$ (sl.) $+ Si$ (gas)	298	[406]	1964
Mo_5Si_3	131	$\frac{3}{4} Mo_5Si_3$ (sl.) $= \frac{5}{4} Mo_3Si$ (sl.) $+$ $+ Si$ (gas)	298	[412]	1966
$MoSi_2$	118,0	$\frac{5}{7} MoSi_2$ (sl.) $= \frac{1}{7} Mo_5Si_3$ (sl.) $+$ $+ Si$ (gas)	0	[406]	1964

*According to 2nd law of thermodynamics.
** According to 3rd law of thermodynamics.
*** sl. = solid.

Phase	Heat of reaction, kcal/mole	Reaction	Temp., K	Source	Year
1	2	3	4	5	6
PrP	79.4 [*1]	PrP (sl.) $= PrP_{1-x}$ (sl.) $+ \dfrac{x}{2} P_2$ (gas)	1270	[405]	1971
	91.2 [*2]		1265		
	110.4 [*3]		1424		
	120.5 [*4]		1555		
UP	251.6	UP (sl.) $=$ U (gas) $+$ P (gas)	0	[406]	1964
UP	184.4	UP (sl.) $=$ U (gas) $+ \dfrac{1}{2} P_2$ (gas)	0	[406]	1964
SrS	212	SrS (sl.) $=$ Sr (gas) $+$ S (gas)	0	[33]	1972
SrS	123	SrS (sl.) $=$ SrS (gas)	0	[33]	1972
ScS	113.4	ScS (gas) $=$ Sc (gas) $+$ S (gas)	0	[405]	1971
YS	125.7	YS (gas) $=$ Y (gas) $+$ S (gas)	0	[405]	1971
LaS	139.7	LaS (sl.) $=$ LaS (gas)	298	[405]	1971
LaS	281.0	LaS (sl.) $=$ La (gas) $+$ S (gas)	298	[405]	1971
CeS	136.0	CeS (gas) $=$ Ce (gas) $+$ S (gas)	0	[405]	1971
PrS	143.2	PrS (sl.) $=$ PrS (gas)	298	[405]	1971
PrS	262.0	PrS (sl.) $=$ Pr (gas) $+$ S (gas)	298	[405]	1971
NdS	138.7	NdS (sl.) $=$ NdS (gas)	298	[405]	1971
NdS	251.7	NdS (sl.) $=$ Nd (gas) $+$ S (gas)	298	[405]	1971
SmS	84	SmS (gas) $=$ Sm (gas) $+$ S (gas)	0	[405]	1971
EuS	76	EuS (gas) $=$ Eu (gas) $+$ S (gas)	0	[405]	1971
EuS	204.8	EuS (sl.) $=$ Eu (gas) $+$ S (gas)	298	[405]	1971
GdS	136	GdS (gas) $=$ Gd (gas) $+$ S (gas)	0	[405]	1971
TbS	122.4	TbS (gas) $=$ Tb (gas) $+$ S (gas)	0	[405]	1971
DyS	98.1	DyS (gas) $=$ Dy (gas) $+$ S (gas)	0	[405]	1971
HoS	101.4	HoS (gas) $=$ Ho (gas) $+$ S (gas)	0	[405]	1971
ErS	94	ErS (gas) $=$ Er (gas) $+$ S (gas)	0	[405]	1971
TmS	86.9	TmS (gas) $=$ Tm (gas) $+$ S (gas)	0	[405]	1971
YbS	63	YbS (gas) $=$ Yb (gas) $+$ S (gas)	0	[405]	1971
LuS	120.2	LuS (gas) $=$ Lu (gas) $+$ S (gas)	0	[405]	1971

[*1] For $x = 0.969$.
[*2] For $x = 0.891$.
[*3] For $x = 0.863$.
[*4] For $x = 0.855$.

HEATS OF SUBLIMATION AND DISSOCIATION (continued)

Phase	Heat of reaction, kcal/mole	Reaction	Temp., K	Source	Year
1	2	3	4	5	6
US	150.3	$US (sl.) = US (gas)$	2300	[406]	1964
TiS	127.8	$TiS (sl.) = TiS (gas)$	298	[33]	1972
TiS$_3$	31.4	$2TiS_3 (sl.) = 2TiS_2 (sl.) + S_2$	—	[33]	1972
B$_4$C	138.0	$\frac{1}{4} B_4C (sl.) = B (gas) + \frac{1}{4} C (sl.)$	298	[334]	1973
SiC	125	$SiC (sl.) = Si (gas) + C (sl.)$	298	[334]	1973
SiC	172	$2SiC (sl.) = Si_2 (gas) + 2C (sl.)$	298	[334]	1973
SiC	197	$SiC (sl.) = SiC (gas)$	0	[334]	1973
SiC	212	$2SiC (sl.) = Si_2C_2 (gas)$	0	[334]	1973
SiC	168	$2SiC (sl.) = Si_2C (gas) + C (sl.)$	298	[334]	1973
SiC	182	$2SiC (sl.) = SiC_2 (gas) + Si (sl, liq.)$	298	[334]	1973
BN	120.5	$BN (sl.) = B (sl.) + \frac{1}{2} N_2$	298	[412]	1966
AlN	63	$AlN (sl.) = Al (sl.) + \frac{1}{2} N_2$	298	[412]	1966
AlN	153.65	$AlN (sl.) = Al (gas) + \frac{1}{2} N_2$	298	[412]	1966

HEAT OF FUSION

Phase	Heat of fusion, kcal/mole	Source	Year	Phase	Heat of fusion, kcal/mole	Source	Year
Be$_2$C	(18)	[287]	1973	FeSi$_{2.33}$	25.1	—*	1962
CaC$_2$	7.7	[350]	1974	Co$_2$Si	15.3	[346]	1973
Mg$_2$Si	20.5	[311]	1965	CoSi	16.52	[346]	1973
VSi$_2$	37.8	[346]	1973	CoSi$_2$	23.94	[346]	1973
Cr$_5$Si$_3$	64.24	[346]	1973	Ni$_2$Si	12.96	[346]	1973
CrSi$_2$	30.6	[346]	1973	NiSi	10.28	[346]	1973
Mn$_5$Si$_3$	41.2	[346]	1973	MnS	6.3	[311]	1965
MnSi	14.2	[346]	1973	FeS	7.73	[311]	1965
Fe$_3$Si	13.7	—*	1962	Ni$_3$S$_2$	5.8	[311]	1965
FeSi	16.0	[311]	1965	B$_4$C	(25)	[287]	1973

*R. P. Krentsis; see footnote on p. 153.

Phase	Melting point		Nature of fusion	Source	Year	Notes
	°C	K				
1	2	3	4	5	6	7
$ScBe_{13}$	>1300	>1573	—	[2]	1966	
YBe_{13}	~1920	~2193	C	[266]	1967	
$LaBe_{13}$	1720	1993	—	[2]	1966	
$CeBe_{13}$	1720	1993	—	[2]	1966	
$ThBe_{13}$	~1930	~2203	C	[2]	1966	
UBe_{13}	~2000	~2273	C	[2]	1966	
$NpBe_{13}$	>1400	>1673	—	[2]	1966	
$PuBe_{13}$	1600	1873	C	[2]	1966	
$AmBe_{13}$	1375	1648	—	[2]	1966	
$TiBe_3$	1450	1723	C	[267]	1970	
$TiBe_2$	1350	1623	Pr	[267]	1970	
Ti_2Be_{17}	1632	1905	—	[268]	1973	
$TiBe_{12}$	1593	1866	Pr	[268]	1973	
$ZrBe_2$	1235	1508	Pr	[267]	1970	
$ZrBe_5$	1475	1748	Pr	[267]	1970	
Zr_2Be_{17}	1555	1828	Pr	[267]	1970	
$ZrBe_{13}$	1645	1918	C	[267]	1970	
$HfBe_2$	1330	1603	Pr	[267]	1970	
$HfBe_5$	1585	1858	Pr	[267]	1970	
Hf_2Be_{17}	1860	2113	—	[267]	1970	
$HfBe_{13}$	1620	1893	Pr	[267]	1970	
VBe_{12}	~1700	~1973	—	[267]	1970	
Nb_3Be_2	>1700	>1973	C	[268]	1973	
$NbBe_2$	1630	1903	Pr	[268]	1973	
$NbBe_3$	>1650	>1923	C	[268]	1973	
Nb_2Be_{17}	1800	2073	C	[268]	1973	
$NbBe_{12}$	1672	1945	Pr	[268]	1973	
$TaBe_2$	1845	2118	—	[2]	1966	
Ta_2Be_{17}	1980	2253	—	[2]	1966	
$TaBe_{12}$	1850	2123	—	[2]	1966	
$CrBe_2$	1840	2113	C	[2]	1966	
$MoBe_2$	1840	2113	—	[2]	1966	
$MoBe_{12}$	~1700	~1973	—	[2]	1966	
WBe_2	~2250	~2523	C	[2]	1966	
$FeBe_2$	1480	1753	C	[2]	1966	
$FeBe_5$	1375	1648	Pr	[2]	1966	
$CoBe$	1505	1778	C	[2]	1966	
Co_5Be_{21}	1270	1543	C	[2]	1966	
$NiBe$	1472	1745	C	[2]	1966	
Ni_5Be_{21}	1262	1535	C	[2]	1966	
Pd_3Be	960	1233	Pr	[125]	1962	
Pd_2Be	1090	1363	Pr	[125]	1962	
Pd_3Be_2	1170	1443	Pr	[125]	1962	
Pd_4Be_3	1200	1473	Pr	[125]	1962	
$PdBe$	1465	1738	C	[125]	1962	

Phase	Melting point		Nature of fusion	Source	Year	Notes
	°C	K				
1	2	3	4	5	6	7
Be$_4$B	1140	1413	Pr	[51]	1973	
Be$_2$B	1500	1773	Pr	[51]	1973	
BeB$_2$	>2000	>2273	—	[51]	1973	
BeB$_4$	2020—2120	2293—2393	—	[51]	1973	
BeB$_6$						
BeB$_{12}$	2300	2573	—	[51]	1973	
CaB$_6$	2235	2508	—	[268]	1973	
SrB$_6$	2230	2503	—	[269]	1961	
BaB$_6$	2270	2543	—	[268]	1973	
ScB$_2$	2250	2523	—	[270]	1960	
ScB$_{12}$	2040	2313	C	[56]	1970	
YB$_2$	2100	2373	C	[266]	1967	
YB$_4$	2800	3073	C	[266]	1967	See [276]
YB$_6$	2600	2873	Pr	[266]	1967	See [271]
YB$_{12}$	2200	2473	Pr	[266]	1967	
YB$_{\sim 70}$	2000	2273	Pr	[266]	1967	
LaB$_6$	2530	2803	C	[272]	1963	
CeB$_6$	2290	2563	C	[274]	1956	See [273]
PrB$_6$	>2250	>2523	C	[268]	1973	
NdB$_6$	2450	2723	C	[267]	1970	
SmB$_6$	2540	2813	C	[275]	1959	
GdB$_4$	>2500	>2773	C	[6]	1966	See [276]
GdB$_6$	~2500	~2773	Pr	[6]	1966	
TbB$_4$	>2100	>2373	—	[268]	1973	See [276]
TbB$_6$						
DyB$_4$	>2090	>2363	—	[268]	1973	See [276]
DyB$_6$						
YbB$_6$	>2000	>2273	—	[267]	1970	
ThB$_4$	>2200	>2473	C	[267]	1970	See [273]
ThB$_6$	2150	2423	C	[277]	1956	See [273]
UB$_2$	2385	2658	C	[64]	1959	
UB$_4$	2495	2768	C	[64]	1959	
UB$_{12}$	2235	2508	C	[64]	1959	
PuB	2050	2323	Pr	[268]	1973	
PuB$_2$	1825	2098	—	[268]	1973	
PuB$_4$	2050	2323	Pr	[268]	1973	
PuB$_6$	2100	2373	Pr	[268]	1973	
Ti$_2$B	2200	2473	Pr	[125]	1962	
TiB$_2$	2790	3063	C	[125]	1962	See [267]
ZrB$_2$	3200	3473	C	[65]	1970	
ZrB$_{12}$	~2030	~2303	Pr	[65]	1970	
HfB$_2$	3250	3523	C	[125]	1962	
V$_3$B$_2$	2070	2343	C	[253]	1959	
VB	2250	2523	C	[253]	1959	
V$_3$B$_4$	2350	2623	Pr	[253]	1959	
VB$_2$	2400	2673	C	[253]	1959	
Nb$_3$B$_2$	1850	2123	Pr	[255]	1959	
NbB	2260	2533	C	[255]	1959	
Nb$_3$B$_4$	2700	2973	Pr	[255]	1959	
NbB$_2$	3000	3273	C	[255]	1959	

Phase	Melting point		Nature of fusion	Source	Year	Notes
	°C	K				
1	2	3	4	5	6	7
Ta_2B	2380	2653	Pr	[278]	1971	
TaB	3090	3363	C	[278]	1971	
Ta_3B_4	3030	3303	Pr	[278]	1971	
TaB_2	3037	3310	C	[278]	1971	
Cr_2B	1870	2143	Pr	[281]	1969	
Cr_5B_3	1900	2173	Pr	[281]	1969	
CrB	2100	2373	C	[281]	1969	
Cr_3B_4	2070	2343	Pr	[281]	1969	
CrB_2	2200	2473	C	[281]	1969	
CrB_4	1400—1600	1673—1873	Pr	[109]	1969	
Mo_2B	2270	2543	Pr	[279]	1967	
Mo_3B_2	2070	2343	Pr	[125]	1962	
MoB	2550	2823	C	[279]	1967	
MoB_2	2350	2623	Pr	[279]	1967	
Mo_2B_5	2200	2473	Pr	[279]	1967	
MoB_{12}	2020	2293	C	[279]	1967	
W_2B	2740	3013	C	[282]	1967	
WB	2800	3073	C	[282]	1967	
W_2B_5	2370	2643	Pr	[282]	1967	
WB_{12}	2440	2713	C	[282]	1967	
Mn_4B	1285	1558	Pr	[9]	1967	
Mn_2B	1580	1853	C	[9]	1967	
MnB	1890	2163	C	[9]	1967	
Mn_3B_4	1750	2023	Pr	[9]	1967	
MnB_2	1988	2261	C	[9]	1967	
MnB_4	2160	2433	C	[9]	1967	See [72]
Re_3B	~2150	~2423	Pr	[280]	1968	
Re_7B_3	~2000	~2273	Pr	[280]	1968	
ReB_2	~2400	~2673	C	[280]	1968	
Fe_2B	1410	1683	Pr	[283]	1969	
FeB	1650	1923	C	[283]	1969	
$FeB_{\sim19}$	1980	2253	Pr	[283]	1969	
Co_3B	1110	1383	Pr	[284]	1960	
Co_3B	1265	1538	C	[284]	1960	
Ni_3B	1175	1448	C	[285]	1967	
Ni_2B	1240	1513	C	[285]	1967	
Ni_4B_3	1580	1853	C	[285]	1967	
NiB	1590	1863	Pr	[285]	1967	
$NiB_{\sim12}$	2320	2593	Pr	[285]	1967	
Ru_7B_3	~1660	~1933	Pr	[152]	1963	
RuB	1500	1773	C	[152]	1963	
Ru_2B_3	1550	1823	C	[152]	1963	
RuB_2	1600	1873	Pr	[152]	1963	
Pd_3B_2	~1020	~1293	Pr	[267]	1970	
Be_2C	2400	2673	—	[11]	1968	See [287]
CaC_2	2300	2573	—	[11]	1968	
ScC	1800	2073	—	[11]	1968	
YC	1950	2223	Pr	[267]	1970	
YC_2	2300	2573	C	[288]	1967	

Phase	Melting point		Nature of fusion	Source	Year	Notes
	°C	K				
1	2	3	4	5	6	7
La_2C_3	1415	1688	Pr	[267]	1970	
LaC_2	2356	2629	C	[267]	1970	
Ce_2C_3	>1700	>1973	Pr	[268]	1973	
CeC_2	2290	2563	C	[288]	1967	
Pr_2C_3	1560	1833	—	[11]	1968	
PrC_2	2120	2393	—	[288]	1967	
NdC_2	2207	2480	—	[288]	1967	
SmC_2	2200	2473	—	[288]	1967	
ThC	2625	2898	—	[11]	1968	
ThC_2	2655	2928	—	[11]	1968	
UC	2520	2793	—	[267]	1970	
PuC	1660	1933	Pr	[292]	1969	
Pu_2C_3	2050	2323	Pr	[11]	1968	
PuC_2	~2250	~2523	Pr	[11]	1968	
TiC	3257	3530	C	[11]	1968	
ZrC	3530	3803	C	[11]	1968	
HfC	3890	4163	C	[11]	1968	
V_2C	2187	2460	Pr	[289]	1973	
VC	2648	2921	C	[289]	1973	43% (at.) C
Nb_2C	3035	3308	Pr	[289]	1973	
NbC	3613	3886	C	[289]	1973	44% (at.) C
Ta_2C	3330	3603	Pr	[289]	1973	
TaC	3985	4258	C	[289]	1973	47% (at.) C
$Cr_{23}C_6$	1518	1791	Pr	[125]	1962	
Cr_7C_3	1782	2055	Pr	[125]	1962	
Cr_3C_2	1895	2168	Pr	[125]	1962	
$\alpha\text{-}Mo_2C$	2480	2753	C	[268]	1973	
$\beta\text{-}Mo_2C$	2522	2795	C	[268]	1973	
$\eta\text{-}MoC$	2550	2823	C	[268]	1973	
$\alpha\text{-}MoC$	2600	2873	C	[268]	1973	
W_2C	2795	3068	C	[268]	1973	30% (at.) C
$\beta\text{-}WC$	2785	3058	Pr	[268]	1973	
Fe_3C	1650	1923	—	[11]	1968	
Be_3N_2	2200	2473	—	[88]	1972	
Ca_3N_2	1195	1468	—	[14]	1969	
Sr_3N_2	1027	1300	—	[14]	1969	Decomposes
Ba_3N_2	1000	1273	—	[14]	1969	The same
ScN	2650	2923	—	[14]	1969	
YN	>2670	>2943	—	[14]	1969	
CeN	2575	2848	—	[796]	1972	
GdN	>2500	>2773	—	[268]	1973	
ThN	2790	3063	C	[268]	1973	
Th_3N_4	2100	2373	—	[14]	1969	
UN	2850	3123	C	[268]	1973	$p = 2.5$ atm.

Phase	Melting point		Nature of fusion	Source	Year	Notes
	°C	K				
1	2	3	4	5	6	7
NpN	2830	3103	C	[268]	1973	$p = \sim 10$ atm.
PuN	2750	3023	—	[14]	1969	
TiN	2950	3223	—	[14]	1969	
ZrN	2980	3253	—	[14]	1969	
HfN	~ 3000	~ 3273	—	[14]	1969	
VN	2050	2323	—	[14]	1969	Decomposes
NbN	2300	2573	—	[14]	1969	The same
TaN	3087	3360	—	[14]	1969	
$CaAl_2$	1079	1352	C	[18]	1965	
$CaAl_4$	700	973	Pr	[18]	1965	
$BaAl_4$	1097	1370	C	[267]	1970	
Y_2Al	1020	1293	Pr	[266]	1967	
Y_2Al_3	1085	1358	Pr	[266]	1967	
YAl	1165	1438	Pr	[266]	1967	
YAl_2	1455	1728	C	[266]	1967	
YAl_3	1355	1628	Pr	[266]	1967	
$LaAl$	859	1132	Pr	[18]	1965	
$LaAl_2$	1422	1695	C	[18]	1965	
$LaAl_4$	1222	1495	Pr	[18]	1965	
Ce_3Al	655	928	C	[18]	1965	
$CeAl$	845	1118	Pr	[18]	1965	
$CeAl_2$	1465	1738	C	[18]	1965	
$CeAl_4$	1243	1516	Pr	[18]	1965	
$PrAl$	892	1165	Pr	[18]	1965	
$PrAl_2$	1442	1715	C	[18]	1965	
$PrAl_4$	1244	1517	Pr	[18]	1965	
$NdAl$	935	1208	Pr	[18]	1965	
$NdAl_2$	1450	1723	C	[18]	1965	
$NdAl_4$	1250	1523	Pr	[18]	1965	
Dy_3Al	975	1248	C	[268]	1973	
Yb_2Al	1020	1293	—	[18]	1965	
Yb_3Al_2	1085	1358	—	[18]	1965	
$YbAl$	1165	1438	—	[18]	1965	
$YbAl_2$	1455	1728	—	[18]	1965	
$YbAl_3$	1355	1628	—	[18]	1965	
Th_2Al	1307	1580	C	[267]	1970	
Th_3Al_2	1301	1574	C	[267]	1970	
$ThAl$	1318	1591	Pr	[267]	1970	
Th_2Al_3	1394	1667	Pr	[267]	1970	
$ThAl_2$	1600	1873	C	[267]	1970	
$ThAl_3$	1120	1393	Pr	[267]	1970	
UAl_2	~ 1590	~ 1863	C	[18]	1965	
UAl_3	1350	1623	Pr	[18]	1965	
UAl_4	730	1003	Pr	[18]	1965	
$PuAl_2$	1540	1813	C	[268]	1973	
$PuAl_3$	1220	1493	Pr	[268]	1973	
$PuAl_4$	926	1199	Pr	[268]	1973	
Ti_2Al	1000	1273	Pr	[18]	1965	

Phase	Melting point		Nature of fusion	Source	Year	Notes
	°C	K				
1	2	3	4	5	6	7
TiAl	1460	1733	Pr	[18]	1965	
TiAl$_3$	1340	1613	Pr	[18]	1965	
Zr$_5$Al$_3$	1395	1668	Pr	[18]	1965	
Zr$_3$Al$_2$	1480	1753	Pr	[18]	1965	
Zr$_4$Al$_3$	1530	1803	C	[18]	1965	
Zr$_2$Al$_3$	1595	1868	Pr	[18]	1965	
ZrAl$_2$	1645	1918	C	[18]	1965	
ZrAl$_3$	1580	1853	C	[18]	1965	
V$_5$Al$_8$	1670	1943	Pr	[18]	1965	
VAl$_3$	1360	1633	Pr	[18]	1965	
V$_4$Al$_{23}$	736	1009	Pr	[267]	1970	
V$_7$Al$_{45}$	688	961	Pr	[267]	1970	
VAl$_{10}$	670	943	Pr	[267]	1970	
Nb$_3$Al	1960	2233	Pr	[268]	1973	
Nb$_2$Al	1870	2143	Pr	[268]	1973	
NbAl$_3$	1605	1878	C	[268]	1973	
Ta$_2$Al	2100	2373	Pr	[268]	1973	
Cr$_5$Al$_8$	1350	1623	Pr	[268]	1973	
Cr$_4$Al$_9$	1170	1443	Pr	[268]	1973	
CrAl$_4$	1030	1303	Pr	[268]	1973	
Cr$_2$Al$_{11}$	940	1213	Pr	[268]	1973	
CrAl$_7$	790	1063	Pr	[268]	1973	
Mo$_3$Al	~2150	~2423	Pr	[295]	1971	
Mo$_3$Al$_8$	~1575	~1848	C	[295]	1971	
MoAl$_4$	1130	1403	Pr	[268]	1973	
MoAl$_5$	900	1173	Pr	[268]	1973	
MoAl$_6$	720	993	Pr	[268]	1973	
MoAl$_{12}$	706	979	Pr	[268]	1973	
WAl$_4$	1326	1599	Pr	[125]	1962	
WAl$_5$	870	1143	Pr	[125]	1962	
WAl$_{12}$	697	970	Pr	[125]	1962	
Mn$_4$Al$_{11}$	1002	1275	Pr	[296]	1972	
MnAl$_4$	923	1196	Pr	[296]	1972	
MnAl$_6$	708	981	Pr	[296]	1972	
ReAl$_2$	1485	1738	Pr	[18]	1965	
Fe$_2$Al$_5$	1165	1438	C	[267]	1970	
FeAl$_3$	1150	1423	C	[267]	1970	
Fe$_2$Al$_7$	560	833	Pr	[18]	1965	
CoAl	1628	1901	C	[18]	1965	
Co$_2$Al$_5$	1180	1453	Pr	[296]	1972	
CoAl$_3$	1135	1408	Pr	[296]	1972	
Co$_4$Al$_{13}$	1093	1366	Pr	[296]	1972	
Co$_2$Al$_9$	970	1243	Pr	[296]	1972	
Ni$_3$Al	1395	1668	Pr	[18]	1965	
NiAl	1638	1911	C	[18]	1965	
Ni$_2$Al$_3$	1132	1405	Pr	[18]	1965	
NiAl$_3$	854	1127	Pr	[18]	1965	
RuAl	~2000	~2273	C	[152]	1963	
Ru$_2$Al$_3$	1600	1873	Pr	[152]	1963	
RuAl$_6$	~1350	~1623	C	[152]	1963	

MELTING POINTS (continued)

Phase	Melting point		Nature of fusion	Source	Year	Notes
	°C	K				
1	2	3	4	5	6	7
$RuAl_{12}$	~750	~1023	Pr	[152]	1963	
Pd_2Al	~1430	~1703	C	[268]	1973	
$PdAl$	1645	1918	C	[268]	1973	
Pd_2Al_3	940	1213	Pr	[268]	1973	
$PdAl_3$	765	1038	Pr	[268]	1973	
Pt_3Al	1556	1829	C	[268]	1973	
Pt_5Al_3	1465	1738	Pr	[268]	1973	
Pt_3Al_2	1397	1670	Pr	[268]	1973	
$PtAl$	1554	1827	C	[268]	1973	
Pt_2Al_3	1521	1794	C	[268]	1973	
$PtAl_2$	1406	1679	Pr	[268]	1973	
$PtAl_3$	1121	1394	Pr	[268]	1973	
Pt_5Al_{21}	806	1079	Pr	[268]	1973	
Mg_2Si	1100	1373	C	[268]	1973	
Ca_2Si	910	1183	Pr	[125]	1962	
$CaSi$	1245	1518	C	[125]	1962	
$CaSi_2$	1020	1293	Pr	[125]	1962	
Sr_2Si	1010	1283	—	[268]	1973	
$SrSi$	1140	1413	C	[268]	1973	
$SrSi_2$	1150	1423	C	[268]	1973	
$BaSi$	~840	~1113	Pr	[268]	1973	
$BaSi_2$	1180	1453	C	[268]	1973	
Y_5Si_3	1850	2123	C	[266]	1967	
Y_5Si_4	1840	2113	C	[266]	1967	
YSi	1835	2108	Pr	[266]	1967	
Y_3Si_5	1635	1908	Pr	[266]	1967	
$LaSi_2$	1520	1793	—	[297]	1970	
$CeSi_2$	1620	1893	C	[273]	1967	
$NdSi_2$	1525	1798	—	[267]	1970	
$EuSi_2$	1500	1773	—	[267]	1970	
$GdSi_2$	1540	1813	—	[267]	1970	See [299]
$DySi_2$	1525	1798	—	[267]	1970	
Th_3Si_2	>1900	>2173	C	[268]	1973	
$ThSi$	~1780	~2053	Pr	[268]	1973	
$ThSi_2$	~1900	~2173	C	[268]	1973	
U_3Si_2	1665	1938	C	[260]	1959	
USi	~1600	~1873	Pr	[260]	1959	
U_3Si_5	~1610	~1883	Pr	[267]	1970	
Pu_5Si_3	1377	1650	Pr	[268]	1973	
Pu_3Si_2	1441	1714	Pr	[268]	1973	
$PuSi$	1576	1849	C	[268]	1973	
Pu_3Si_5	1646	1919	—	[268]	1973	
$PuSi_2$	1638	1911	—	[268]	1973	
Ti_5Si_3	2130	2403	C	[163]	1970	
Ti_5Si_4	1920	2193	Pr	[163]	1970	
$TiSi$	1570	1843	Pr	[163]	1970	
$TiSi_2$	1500	1773	C	[163]	1970	
Zr_4Si	1630	1903	Pr	[125]	1962	
Zr_2Si	2110	2383	Pr	[125]	1962	
Zr_5Si_3	2210	2483	Pr	[125]	1962	

MELTING POINTS (continued)

Phase	Melting point		Nature of fusion	Source	Year	Notes
	°C	K				
1	2	3	4	5	6	7
Zr_4Si_3	2225	2498	Pr	[125]	1962	
Zr_6Si_5	2250	2523	C	[125]	1962	
ZrSi	2095	2368	Pr	[125]	1962	
$ZrSi_2$	1520	1793	Pr	[125]	1962	
Hf_2Si	~2430	~2703	Pr	[267]	1970	
Hf_5Si_3	~2600	~2873	Pr	[267]	1970	
Hf_3Si_2	>2600	>2873	C	[267]	1970	
HfSi	2200	2473	C	[267]	1970	
$HfSi_2$	1900	2173	Pr	[267]	1970	
V_3Si	1935	2208	C	[166]	1974	
V_5Si_3	2010	2283	C	[166]	1974	
V_5Si_4	1670	1943	Pr	[166]	1974	
VSi_2	1680	1953	C	[166]	1974	
Nb_4Si	1950	2223	Pr	[125]	1962	
Nb_5Si_3	2480	2753	C	[267]	1970	
$NbSi_2$	1930	2203	C	[125]	1962	
$Ta_{4.5}Si$	2510	2783	C	[125]	1962	
Ta_2Si_3	2450	2723	Pr	[125]	1962	
Ta_5Si	2500	2773	C	[125]	1962	
$TaSi_2$	2200	2473	C	[125]	1962	
Cr_3Si	1770	2043	C	[304]	1964	
Cr_5Si	1720	1993	C	[304]	1964	
CrSi	1475	1748	Pr	[304]	1964	
$CrSi_2$	1475	1748	C	[304]	1964	
Mo_3Si	2025	2298	Pr	[306]	1971	
Mo_5Si_3	2180	2453	C	[306]	1971	
$MoSi_2$	2020	2293	C	[306]	1971	
W_5Si_3	2370	2643	Pr	[267]	1970	
WSi_2	2160	2433	C	[267]	1970	
Mn_9Si_2	1060	1333	Pr	[268]	1973	
Mn_3Si	1075	1348	Pr	[268]	1973	
Mn_5Si_3	1285	1558	C	[268]	1973	
MnSi	1275	1548	C	[268]	1973	
$Mn_{11}Si_{2-n}$	1152	1425	Pr	[309]	1970	
$Mn_{11}Si_{19}$	1145	1418	C	[268]	1970	
Re_5Si_3	1960	2233	C	[267]	1970	
ReSi	1880	2153	Pr	[267]	1970	
$ReSi_2$	1980	2253	C	[267]	1970	
Fe_2Si	1215	1488	C	[308]	1968	
FeSi	1410	1683	C	[22]	1971	
$\alpha\text{-}FeSi_2$	1220	1493	C	[175]	1968	
Co_3Si	1212	1485	Pr	[125]	1962	
$\alpha\text{-}Co_2Si$	1320	1593	Pr	[125]	1962	
$\beta\text{-}Co_2Si$	1332	1605	C	[125]	1962	
CoSi	1460	1733	C	[125]	1962	
$CoSi_2$	1326	1599	C	[125]	1962	
Ni_5Si_2	1282	1555	C	[22]	1971	
$\beta\text{-}Ni_2Si$	1318	1591	C	[22]	1971	
NiSi	992	1265	C	[22]	1971	
$\beta\text{-}NiSi_2$	993	1266	Pr	[22]	1971	

Phase	Melting point		Nature of fusion	Source	Year	Notes
	°C	K				
1	2	3	4	5	6	7
Pd_3Si	960	1233	Pr	[267]	1970	
Pd_2Si	1330	1603	C	[267]	1970	
$PdSi$	1100	1373	C	[267]	1970	
Pt_3Si	~870	~1143	Pr	[268]	1973	
Pt_7Si_3	986	1259	Pr	[268]	1973	
Pt_2Si	1100	1373	C	[268]	1973	
$PtSi$	1229	1502	C	[268]	1973	
Ba_3P_2	3080	3353	—	[311]	1965	
ScP	>2000	>2273	—	[27]	1973	
CeP	>1800	>2073	—	[27]	1973	
PrP	2850	3123	—	[310]	1968	
NdP	2220	2493	—	[27]	1973	
SmP	>2100	>2373	—	[27]	1973	
Eu_3P_2	>1300	>1573	—	[27]	1973	
EuP	2200	2473	—	[27]	1973	
EuP_2	840	1113	—	[27]	1973	
GdP	>2500	>2773	—	[268]	1973	
ThP	~3000	~3273	—	[268]	1973	
Ti_3P	1760	2033	Pr	[28]	1968	
Ti_2P	1920	2193	Pr	[28]	1968	
Cr_3P	1510	1783	Pr	[125]	1962	
Mn_3P	1105	1378	Pr	[125]	1962	
Mn_2P	1327	1600	C	[125]	1962	
Mn_3P_2	1090	1363	Pr	[125]	1962	See [208]
MnP	1147	1420	C	[125]	1962	See [312]
Fe_3P	1166	1439	Pr	[125]	1962	
Fe_2P	1365	1638	C	[125]	1962	
Co_2P	1386	1659	C	[125]	1962	
Ni_3P	970	1243	Pr	[214]	1965	
Ni_5P_2	1175	1448	C	[214]	1965	
$Ni_{12}P_5$	~1150	~1423	Pr	[214]	1965	
Ni_2P	1110	1383	C	[214]	1965	
Ni_5P_4	~850	~1123	Pr	[214]	1965	
Ru_2P	—	—	—	[267]	1970	The fusion point is higher than that of Co_2P
Rh_2P	>1500	>1773	C	[125]	1962	
Pd_3P	1047	1320	C	[125]	1962	
PdP_2	1150	1423	C	[125]	1962	
Ir_2P	1350	1623	C	[125]	1962	
$Pt_{20}P_7$	590	863	Pr	[125]	1962	
PtP_2	>1500	>1773	—	[125]	1962	
MgS	2000	2273	—	[33]	1972	
CaS	2525	2798	—	[267]	1970	
SrS	>2000	>2273	—	[33]	1972	
BaS	>2200	>2473	—	[35]	1972	
BaS_2	925	1198	Pr	[125]	1962	

Phase	Melting point		Nature of fusion	Source	Year	Notes
	°C	K				
1	2	3	4	5	6	7
BaS_3	554	827	Pr	[125]	1962	
Sc_2S_3	1775	2048	—	[314]	1964	$p =$ $= 10^{-2}$ mm Hg
YS	2060	2333	—	[267]	1970	
Y_5S_7	1630	1903	—	[267]	1970	
Y_2S_3	1900	2173	—	[267]	1970	Decomposes
YS_2	1660	1933	—	[267]	1970	Dissociates
LaS	~2200	~2473	—	[267]	1970	
La_3S_4	2100	2373	—	[267]	1970	
La_2S_3	2095	2368	—	[267]	1970	
LaS_2	1650	1923	—	[267]	1970	Dissociates
CeS	2450	2723	—	[33]	1972	
Ce_3S_4	2100	2373	—	[267]	1970	
Ce_2S_3	2060	2333	—	[267]	1970	
CeS_2	1700	1973	—	[267]	1970	Decomposes; see [227]
PrS	2230	2503	—	[267]	1970	
Pr_3S_4	2100	2373	—	[267]	1970	
Pr_2S_3	1795	2268	—	[267]	1970	
PrS_2	1780	2253	—	[267]	1970	Dissociates
NdS	2200	2473	—	[267]	1970	
Nd_3S_4	2040	2313	—	[267]	1970	
Nd_2S_3	2200	2473	—	[267]	1970	Dissociates
NdS_2	1760	2033	—	[267]	1970	Decomposes
SmS	1940	2213	—	[33]	1972	
Sm_3S_4	1800	2073	—	[267]	1970	
Sm_2S_3	1900	2173	—	[33]	1972	Dissociates
EuS	1667	1940	—	—**	1973	
GdS	2020	2293	—	—**	1973	
Gd_2S_3	1885	2158	—	[33]	1972	
TbS	1970	2243	—	—**	1973	
DyS	(1940)	(2213)	—	—**	1973	
Dy_5S_7	1540	1813	—	[33]	1972	
Dy_2S_3	1490	1763	—	[267]	1970	
HoS	(1890)	(2163)	—	—**	1973	
ErS	(1900)	(2173)	—	—**	1973	
Er_5S_7	1620	1893	—	[33]	1972	
Er_2S_3	1630	1903	—	[33]	1972	
TmS	(1840)	(2113)	—	—**	1973	

** B. V. Fenochka; see footnote on p. 127.

Phase	Melting point		Nature of fusion	Source	Year	Notes
	°C	K				
1	2	3	4	5	6	7
ThS	2335	2608	—	[268]	1973	
Th_2S_3	1900—2000	2173—2273	—	[33]	1972	
Th_4S_7	2300	2573	—	[33]	1972	
ThS_2	1904	2177	—	[33]	1972	Decomposes
US	2462	2735	—	[267]	1970	
PuS	2350	2623	—	[33]	1972	
Pu_2S_3	1725	1998	C	[268]	1973	
Ti_3S	1305	1578	Pr	[33]	1972	
Ti_2S	1410	1683	Pr	[33]	1972	
TiS	1780	2053	Pr	[33]	1972	
ZrS_2	1550	1823	—	[33]	1972	
V_3S	1400	1673	Pr	[267]	1970	
CrS	~1565	~1838	—	[125]	1962	
MoS_2	>1800	>2073	—	[267]	1970	
WS_2	>1800	>2073	C	[267]	1970	
MnS	1610	1883	C	[125]	1962	
FeS	~1190	~1463	—	[268]	1973	
Co_4S_3	930	1203	Pr	[33]	1972	
CoS	1195	1468	C	[33]	1972	
CoS_2	950	1223	Pr	[33]	1972	
Ni_3S_2	806	1079	Pr	[267]	1970	
NiS	992	1265	C	[267]	1970	
NiS_2	1010	1283	C	[267]	1970	
Pd_4S	761	1034	Pr	[125]	1962	
$Pd_{2.8}S$	635	908	Pr	[125]	1962	
B_4C	2450	2723	C	[267]	1970	18,5% (at.) C
SiC	2540	2813	Pr	[267]	1970	
α-BN	~3000	~3273	—	[14]	1969	

*Key: (C) congruent melting; (Pr) peritectic-reaction melting.

BOILING POINTS

Phase	Boiling point		Source	Year
	°C	K		
SrB_6	5100	5373	[413]	1961
Be_2C	2537	2810	[311]	1965
ThC_2	5000	5273	[311]	1965
U_2C_3	4100	4373	[311]	1965
UC_2	4370	4643	[311]	1965
TiC	4300	4573	[311]	1965
ZrC	5100	5373	[311]	1965
HfC	5400	5673	[413]	1961
VC	3900	4173	[311]	1965
NbC	4500	4773	[413]	1961
TaC	5500	5773	[311]	1965
Cr_3C_2	3800	4073	[311]	1965
W_2C	6000	6273	[311]	1965
WC	6000	6273	[311]	1965
$Ta_{4.5}Si$	4000	4273	[311]	1965
Ta_2Si	3727	4000	[311]	1965
$TaSi_2$	5347	5620	[311]	1965
BaS	~3000	~3273	[33]	1972
SiC	2607	2880	[261]	1958
BN	5067	5340	[254]	1960

VAPOR PRESSURE AND EVAPORATION RATES

Phase	Temp., °C	Evap. rate g/(cm²·sec)	Vapor pressure, mm Hg	Vapor pressure equation	Source	Year	Notes
1	2	3	4	5	6	7	8
SrB_6	1500 1600 1700 1800 2000	$0.336 \cdot 10^{-7}$ $2.128 \cdot 10^{-7}$ $3.25 \cdot 10^{-7}$ $17.25 \cdot 10^{-7}$ $188.9 \cdot 10^{-7}$	$0.197 \cdot 10^{-5}$ $1\,266 \cdot 10^{-5}$ $1.98 \cdot 10^{-5}$ $10.79 \cdot 10^{-5}$ $125.0 \cdot 10^{-5}$	$\log p_{SrB_6} =$ $= 6.43 - \dfrac{21\,423}{T}$ (mm Hg)	[269]	1961	
LaB_6	1720 1810 1910 2027	—	$6.92 \cdot 10^{-4}$ $23.3 \cdot 10^{-4}$ $111 \cdot 10^{-4}$ $526 \cdot 10^{-4}$	$\log p_{La} = 8.75 - 29\,300/T$ (at.)	[407]	1966	See [405]
SmB_4	1027—2027	—	—	$\log p_{Sm} = 5.38 - 20000/T$ (at.)	[405]	1971	
GdB_6	1908 1950 2000 2032	—	$18.9 \cdot 10^{-4}$ $49.3 \cdot 10^{-4}$ $139 \cdot 10^{-4}$ $242.5 \cdot 10^{-4}$	$\log p_{Gd} = 7.40 - 27\,750/T$ (at.)	[405]	1971	
TiB_2	1700 1800 1900 2100	$1{,}78 \cdot 10^{-7}$ $5.26 \cdot 10^{-7}$ $9.78 \cdot 10^{-7}$ $16{,}73 \cdot 10^{-7}$	$\left.\begin{array}{l} 1.45 \cdot 10^{-5} \\ 4.41 \cdot 10^{-5} \\ 8.36 \cdot 10^{-5} \\ 14.6 \cdot 10^{-5} \end{array}\right\} p_{Ti}$	—	[412]	1966	

VAPOR PRESSURE AND EVAPORATION RATES (continued)

Phase	Temp., °C	Evap. rate g/(cm²·sec)	Vapor pressure, mm Hg	Vapor pressure equation	Source	Year	Notes
1	2	3	4	5	6	7	8
TiB_2	1700	$1.78 \cdot 10^{-7}$	$1.37 \cdot 10^{-5}$	—	[412]	1966	
	1800	$5.26 \cdot 10^{-7}$	$4.18 \cdot 10^{-5}$ $\left.\right\} p_B$				
	1900	$9.78 \cdot 10^{-7}$	$7.9 \cdot 10^{-5}$				
	2100	$16.73 \cdot 10^{-7}$	$13.9 \cdot 10^{-5}$				
ZrB_2	1400	$0.15 \cdot 10^{-7}$	$0.1 \cdot 10^{-5}$	—	[413]	1961	See [407, 412]
	1500	$1.248 \cdot 10^{-7}$	$0.834 \cdot 10^{-5}$				
	1600	$1.03 \cdot 10^{-7}$	$0.72 \cdot 10^{-5}$				
	1700	$1.969 \cdot 10^{-7}$	$1.417 \cdot 10^{-5}$				
	1800	$3.843 \cdot 10^{-7}$	$3.227 \cdot 10^{-5}$				
TaB_2	1247	10^{-11}	—	—	[320]	1969	P_{Ta} substantially less
	1277	—	$750 \cdot 10^{-7}$				
	1427	—	$750 \cdot 10^{-5}$				
	1447	10^{-9}	— $\left.\right\} p_B$				
	1657	10^{-7}	—				
	1677	—	$750 \cdot 10^{-3}$				
	1927	10^{-5}	$750 \cdot 10^{-1}$				
	2327	—	7500				
	2677	10^{-3}	$750 \cdot 10^{3}$				
CrB_2	1200	$2.89 \cdot 10^{-7}$	$1.71 \cdot 10^{-5}$	—	[413]	1961	
	1300	$4.52 \cdot 10^{-7}$	$3.56 \cdot 10^{-5}$				
	1400	$4.72 \cdot 10^{-7}$	$3.82 \cdot 10^{-5}$				
	1500	$1.45 \cdot 10^{-7}$	$1.20 \cdot 10^{-5}$				
	1600	$7.68 \cdot 10^{-7}$	$6.64 \cdot 10^{-5}$				
	1800	$4.98 \cdot 10^{-7}$	$4.62 \cdot 10^{-5}$				

Compound	T		p	Equation	Ref.	Year	Note
Be_2C	1157	—	$14.4 \cdot 10^{-5}$				
	1173	—	$13.07 \cdot 10^{-5}$				
	1263	—	$15.8 \cdot 10^{-4}$				
	1305	—	$16.56 \cdot 10^{-4}$	$\log p_{Be} = 7.026 - \dfrac{19\,720}{T}$ (at.)	[334]	1973	
	1307	—	$25.08 \cdot 10^{-4}$				
	1317	—	$28.88 \cdot 10^{-4}$				
	1370	—	$6.46 \cdot 10^{-3}$				
	1486	—	$13.9 \cdot 10^{-3}$				
	1496	—	$13.6 \cdot 10^{-3}$				
CaC_2	979	—	$3.12 \cdot 10^{-3}$				
	1197		$27.4 \cdot 10^{-3}$				
	1355		$127 \cdot 10^{-3}$	$\log p_{Ca} = 2.9 - 10\,710/T$ (at.)	[334]	1973	
	1519		$601 \cdot 10^{-3}$				
	1687		$2\,620 \cdot 10^{-3}$				
SrC_2	929	—	$0.532 \cdot 10^{-3}$				
	1088		$7.67 \cdot 10^{-3}$	$\log p_{Sr} = 3.7 - 11\,840/T$ (at.)	[414]	1966	See [334]
	1192		$31.2 \cdot 10^{-3}$				
	1234		$79.5 \cdot 10^{-3}$				
BaC_2	1208	—	$5.62 \cdot 10^{-3}$				
	1312		$21.0 \cdot 10^{-3}$	$\log p_{Ba} = 3.15 - 12\,260/T$ (at.)	[414]	1966	See [334]
	1408		$54.1 \cdot 10^{-3}$				
	1802	—	$3.04 \cdot 10^{-5}$				
	1911		$12.9 \cdot 10^{-5}$	$\log p_{Y} = 4.44 - 24\,230/T$ (at.)	[334]	1973	
	1995		$33.5 \cdot 10^{-5}$				
	2067		$56.3 \cdot 10^{-5}$				
YC_2	1802	—	$1.75 \cdot 10^{-6}$				
	1911		$11.4 \cdot 10^{-6}$	$\log p_{YC_2} = 6.94 - 32\,860/T$ (at.)	[334]	1973	
	1995		$41.8 \cdot 10^{-6}$				
	2067		$98.9 \cdot 10^{-6}$				

VAPOR PRESSURE AND EVAPORATION RATES (continued)

Phase	Temp., °C	Evap. rate g/(cm²·sec)	Vapor pressure, mm Hg	Vapor pressure equation	Source	Year	Notes
1	2	3	4	5	6	7	8
YC_2	1802—2067	—	—	$\log p_{YC_4} = 9.71 - 4\,456/T$ (at.)	[334]	1973	
LaC_2	1994—2327	—	—	$\log p_{La} = 5{,}152 - 26\,060/T$ (at.) $\log p_{LaC_2} = 7{,}803 - 33\,270/T$ (at.) $\log p_{LaC_3} = 9{,}547 - 44\,140/T$ (at.) $\log p_{LaC_4} = 10{,}057 - 43\,010/T$ (at.)	[415]	1971	
LaC_2	1650—2000	—	—	$\log p_{\Sigma} = 5.8 - 27\,300/T$ (at.)	[416]	1965	
CeC_2	1717—2027	—	—	$\log p_{Ce} = 5{,}49 - 25\,100/T$ (at.) $\log p_{CeC_2} = 8.21 - 31\,800/T$ (at.) $\log p_{CeC_4} = 9.6 - 38\,800/T$ (at.)	[417]	1965	See [416]
NdC_2	1397—2057	—	—	$\log p_{Nd} = 3.57 - 19\,700/T$ (at.)	[416]	1965	
SmC_2	1201—1802	—	—	$\log p_{Sm} = 3.61 - 13\,800/T$ (at.)	[416]	1965	
EuC_2	1024—1454	—	—	$\log p_{Eu} = 3.27 - 11\,567/T$ (at.)	[416]	1965	
GdC_2	1868—2365	—	—	$\log p_{Gd} = 5.08 - 25\,666/T$ (at.)	[418]	1967	

Compound	Temperature range			Equation	Reference	Year	Note
HoC₂	1327—1897 2077—2227 2077—2227	—		$\log p_{\mathrm{Ho}} = 3.84 - 19\,480/T$ (at.) $\log p_{\mathrm{Ho}} = 3.43 - 18\,470/T$ (at.) $\log p_{\mathrm{HoC_2}} = 6.63 - 32\,120/T$ (at.)	[405] [419] [419]	1971 1969 1969	See [417]
ErC₂	1477—2227	—		$\log p_{\mathrm{Er}} = 4{,}567 - 20\,441/T$ (at.)	[420]	1969	
TmC₂	—	—		$\log p_{\mathrm{Tm}} = 3.89 - 15\,445/T$ (at.)	[334]	1973	
YbC₂	827—1277	—		$\log p_{\mathrm{Yb}} = 4.15 - 11\,120/T$ (at.)	[405]	1973	See [334]
ThC₂	2155 2267 2368	—	$0{,}791 \cdot 10^{-6}$ $3.70 \cdot 10^{-6}$ $15.15 \cdot 10^{-6}$	$\log p_{\mathrm{ThC_2}} = 7.20 - 39\,364/T$ (at.)	[334]	1973	
	2155 2267 2368	—	$0{,}625 \cdot 10^{-6}$ $2.80 \cdot 10^{-6}$ $9.54 \cdot 10^{-6}$	$\log p_{\mathrm{Th}} = 5{,}74 - 36\,025/T$ (at.)	[334]	1973	
UC	1605 1720 1845 2010 2085	$1.623 \cdot 10^{-7}$ $2.905 \cdot 10^{-7}$ $7.830 \cdot 10^{-7}$ $1.834 \cdot 10^{-6}$ $4.156 \cdot 10^{-6}$	$3.482 \cdot 10^{-5}$ $6.240 \cdot 10^{-5}$ $1.781 \cdot 10^{-4}$ $4.335 \cdot 10^{-4}$ $1{,}0 \cdot 10^{-3}$	$\log p_{\mathrm{C}} = 2.515 - 13\,240/T$ (mm Hg)	[383]	1966	See [334]
UC	1605—2085	—		$\log p_{\mathrm{U}} = 10{,}054 - 28\,090/T$ (mm Hg)	[383]	1966	See [334]

VAPOR PRESSURE AND EVAPORATION RATES (continued)

Phase	Temp., °C	Evap. rate g/(cm²·sec)	Vapor pressure, mm Hg	Vapor pressure equation	Source	Year	Notes
1	2	3	4	5	6	7	8
UC_2	1657 1790 1890 2006 2088	—	$0.532 \cdot 10^{-7}$ $2,96 \cdot 10^{-7}$ $17,4 \cdot 10^{-7}$ $10,3 \cdot 10^{-6}$ $23,3 \cdot 10^{-6}$	$\log p_U = 12,14 - 67\,164/T - 0,111\,(\ln T + 2\,000/T)$ (at.)	[334]	1973	
UC_2	1877—2427	—	—	$\log p_{UC_2} = 8,38 - 40\,400/T$ (at.)	[334]	1973	
$PuC_{0,9}$	1150—1650	—	—	$\log p_{Pu} = 4.3 - 18\,800/T$ (at.)	[367]	1970	
Pu_2C_3	1500—2000	—	—	$\log p_{Pu} = 4,23 - 20\,200/T$ (at.)	[367]	1970	
Pu_2C_3	2000—2300	—	—	$\log p_{Pu} = 6,8 - 25\,200/T$ (at.)	[367]	1970	
PuC_2	2000—2500	—	—	$\log p_{Pu} = 3,15 - 18\,150/T$ (at.)	[367]	1970	
TiC	2500 2245—2517	$99.0 \cdot 10^{-6}$	$8.36 \cdot 10^{-3} p_C$ $8.14 \cdot 10^{-3} p_{Ti}$ —	— $\log p_{Ti} = 6,27 - 28\,950/T$ (at.) $\log p_{TiC_2} = 8.76 - 41\,620/T$ (at.) $\log v_{TiC_4} = 10,04 - 50\,380/T$ (at.)†	[334]	1973	See [407]

Compound	T, °K	$p_{Me/C}$		log equation	Ref.	Year	
ZrC	2500	$2.22 \cdot 10^{-6}$	$0.1595 \cdot 10^{-3}$ ⎫				
	2600	$7.06 \cdot 10^{-6}$	$0.502 \cdot 10^{-3}$ ⎪				
	2700	$35.8 \cdot 10^{-6}$	$2.61 \cdot 10^{-3}$ ⎬ p_{Zr}	—	[407]	1966	See [334]
	2800	$64.4 \cdot 10^{-6}$	$4.77 \cdot 10^{-3}$ ⎪				
	2900	$19.0 \cdot 10^{-5}$	$1.43 \cdot 10^{-2}$ ⎭				
	2500	$2.22 \cdot 10^{-6}$	$0.1445 \cdot 10^{-3}$ ⎫				
	2600	$7.06 \cdot 10^{-6}$	$0.479 \cdot 10^{-3}$ ⎪				
	2700	$35.8 \cdot 10^{-6}$	$2.47 \cdot 10^{-3}$ ⎬ p_{C}	—	[407]	1966	See [334]
	2800	$64.4 \cdot 10^{-6}$	$4.52 \cdot 10^{-3}$ ⎪				
	2900	$19.0 \cdot 10^{-5}$	$1.36 \cdot 10^{-2}$ ⎭				
HfC	2500	$0.98 \cdot 10^{-6}$	$0.532 \cdot 10^{-4}$ ⎫				
	2600	$2.71 \cdot 10^{-6}$	$1.5 \cdot 10^{-4}$ ⎪				
	2700	$8.31 \cdot 10^{-6}$	$4.69 \cdot 10^{-4}$ ⎬ p_{Hf}	—	[407]	1966	See [334]
	2800	$25.9 \cdot 10^{-6}$	$14.8 \cdot 10^{-4}$ ⎪				
	2900	$64.1 \cdot 10^{-6}$	$37.4 \cdot 10^{-4}$ ⎭				
	2500	$0.98 \cdot 10^{-6}$	$0.517 \cdot 10^{-4}$ ⎫				
	2600	$2.71 \cdot 10^{-6}$	$1.45 \cdot 10^{-4}$ ⎪				
	2700	$8.31 \cdot 10^{-6}$	$4.55 \cdot 10^{-4}$ ⎬ p_{C}	—	[407]	1966	See [334]
	2800	$25.9 \cdot 10^{-6}$	$14.35 \cdot 10^{-4}$ ⎪				
	2900	$64.1 \cdot 10^{-6}$	$36.3 \cdot 10^{-4}$ ⎭				
VC	2073	—	$2.73 \cdot 10^{-3}$	$\log p_V = 7{,}5 - 30\,400/T$ (at.)	[334]	1973	
	2188	—	$13.8 \cdot 10^{-3}$				
	2272	—	$22.3 \cdot 10^{-3}$				
NbC$_{0.77}$	2500	$0.72 \cdot 10^{-6}$	$0.418 \cdot 10^{-4}$ ⎫				
	2600	$2.18 \cdot 10^{-6}$	$1.63 \cdot 10^{-4}$ ⎪				
	2700	$6.57 \cdot 10^{-6}$	$3.96 \cdot 10^{-4}$ ⎬ p_{Nb}	—	[407]	1966	See [334]
	2800	$20.9 \cdot 10^{-6}$	$16.1 \cdot 10^{-4}$ ⎪				
	2900	$56.8 \cdot 10^{-6}$	$44.5 \cdot 10^{-4}$ ⎭				

VAPOR PRESSURE AND EVAPORATION RATES (continued)

Phase	Temp., °C	Evap. rate g/(cm²·sec)	Vapor pressure, mm Hg	Vapor pressure equation	Source	Year	Notes
1	2	3	4	5	6	7	8
$NbC_{0.77}$	2500 2600 2700 2800 2900	$0.72 \cdot 10^{-6}$ $2.18 \cdot 10^{-6}$ $6.57 \cdot 10^{-6}$ $20.9 \cdot 10^{-6}$ $56.8 \cdot 10^{-6}$	$0.41 \cdot 10^{-4}$ $1.255 \cdot 10^{-4}$ $3.85 \cdot 10^{-4}$ $12.7 \cdot 10^{-4}$ $34.4 \cdot 10^{-4}$ } p_C	—	[407]	1966	See [334]
Ta_2C	2742 2877 2977 3080	$0.33 \cdot 10^{-5}$ $1.4 \cdot 10^{-5}$ $3.5 \cdot 10^{-5}$ $9.0 \cdot 10^{-5}$	—	—	—*	1970	
	2700 2800 2900 3000 3100	$1.25 \cdot 10^{-6}$ $4.0 \cdot 10^{-6}$ $6.2 \cdot 10^{-6}$ $22.4 \cdot 10^{-6}$ $12.2 \cdot 10^{-5}$	$0.737 \cdot 10^{-4}$ $2.39 \cdot 10^{-4}$ $3.76 \cdot 10^{-4}$ $1.38 \cdot 10^{-3}$ $7.6 \cdot 10^{-3}$ } p_{Ta}	—	[407]	1966	See [334]
$TaC_{0.71}$	2700 2800 2900 3000 3100	$1.25 \cdot 10^{-6}$ $4.0 \cdot 10^{-6}$ $6.2 \cdot 10^{-6}$ $22.4 \cdot 10^{-6}$ $12.2 \cdot 10^{-5}$	$0.525 \cdot 10^{-4}$ $0.17 \cdot 10^{-3}$ $0.268 \cdot 10^{-3}$ $0.98 \cdot 10^{-3}$ $5.42 \cdot 10^{-3}$ } p_C	—	[407]	1966	See [334]
$Cr_{23}C_6$	1327 1377 1427 1527	—	$7.6 \cdot 10^{-4}$ $1.58 \cdot 10^{-3}$ $5.41 \cdot 10^{-3}$ $19.4 \cdot 10^{-3}$ } p_{Cr}	—	[334]	1973	

Cr_7C_3	1327 1427 1527 1627	— — — —	$\left.\begin{array}{l}0{,}988\cdot10^{-3}\\3{,}91\cdot10^{-3}\\1{,}55\cdot10^{-2}\\3{,}62\cdot10^{-2}\end{array}\right\}\,p_{Cr}$	—	[410]	1966
Cr_3C_2	1635 1728 1822 1964	$1.10\cdot10^{-6}$ $3.25\cdot10^{-6}$ $9.30\cdot10^{-6}$ $31.6\cdot10^{-6}$	$2.28\cdot10^{-2}$ $6.91\cdot10^{-2}$ $20.2\cdot10^{-2}$ $71\cdot10^{-2}$	$\log p_{Cr} = 6.525 - 21\,194/T$ (at.)	[334]	1973
Mo_2C	1977—2227	—	—	$\log p_{Mo} = 7.112 - 34\,200/T$ (at.) $\log p_{C} = 6.351 - 34\,200/T$ (at.)	[334]	1973
Mn_7C_3	800—950	—	—	$\log p_{Mn} = 6.07 - 14\,290/T$ (at.)	[334]	1973
Be_3N_2	1704 1848 2015 2208 2400	— — — — —	10^{-3} 10^{-2} 10^{-1} 1 760	—	[311]	1965
Mg_3N_2	1260 1359 1472 1602	— — — —	10^{-3} 10^{-2} 10^{-1} 1	—	[311]	1965

*T. A. Nikol'skaya, Investigation of High-Temperature Volatilization from an Open Surface in Vacuum of Carbide Phases of Transition Metals of Groups IV–V [in Russian], Author's abstract, Cand. Dissertation, Leningrad (1970).

VAPOR PRESSURE AND EVAPORATION RATES (continued)

Phase	Temp., °C	Evap. rate g/(cm² · sec)	Vapor pressure, mm Hg	Vapor pressure equation	Source	Year	Notes
1	2	3	4	5	6	7	8
Ca_3N_2	1000 1069 1147 1235	— — — —	10^{-3} 10^{-2} 10^{-1} 1	—	[311]	1965	
Sr_3N_2	302 975 1058 1154	— — — —	10^{-3} 10^{-2} 10^{-1} 1	—	[311]	1965	
Ba_3N_2	794 855 923 1002	—	10^{-3} 10^{-2} 10^{-1} 1	—	[311]	1965	
ScN	1230 1495 1607 1734 1880	—	$7,6 \cdot 10^{-5}$ 10^{-3} 10^{-2} 10^{-1} 1	—	[311]	1965	
YN	1587 1704 1838 1990	—	10^{-3} 10^{-2} 10^{-1} 1	—	[311]	1965	

Compound	Temperature		p_{N_2}		Ref.	Year
LaN	1602 1721 1855 2010	—	10^{-3} 10^{-2} 10^{-1} 1	—	[311]	1965
CeN	1756 1884 2029 2196	—	10^{-3} 10^{-2} 10^{-1} 1	—	[311]	1965
ThN	1480 1730 2030	10^{-7} 10^{-5} 10^{-3}	0.75 75 7500	—	[320]	1969
Th$_3$N$_4$	1890 2037 2206 2401	—	10^{-3} 10^{-2} 10^{-1} 1	—	[311]	1965
UN	1177 1327 1547 1747 2057 2107 2407 2827	— — 10^{-5} 10^{-3} — —	$0.75\cdot10^{-4}$ $0.75\cdot10^{-2}$ 0.75 75 7500 — $75\cdot10^{4}$ $75\cdot10^{6}$	—	[320]	1969
UN	1600—1900	—	—	$\log p_U = 10{,}15 - 27\,500/T$ (mm Hg)	[338]	1969

VAPOR PRESSURE AND EVAPORATION RATES (continued)

Phase	Temp., °C	Evap. rate g/(cm²·sec)	Vapor pressure, mm Hg	Vapor pressure equation	Source	Year	Notes
1	2	3	4	5	6	7	8
PuN	2017—2497	—	—	$\log p_{N_2} = 8.193 - 29\,570/T + 11.28 \cdot 10^{-18} T^5$ (at.)	[412]	1966	
TiN	1714 1744 1777 1785 1882 1884 1939 1968	$1{,}510 \cdot 10^{-5}$ $1{,}972 \cdot 10^{-5}$ $3{,}129 \cdot 10^{-5}$ $3{,}360 \cdot 10^{-5}$ $15{,}963 \cdot 10^{-5}$ $20{,}510 \cdot 10^{-5}$ $33{,}254 \cdot 10^{-5}$ $70{,}555 \cdot 10^{-5}$	—	—	[411]	1954	See [412]
	1714 1744 1777 1882 1939 1968	$1.168 \cdot 10^{-5}$ $1.526 \cdot 10^{-5}$ $2.421 \cdot 10^{-5}$ $12.351 \cdot 10^{-5}$ $25.728 \cdot 10^{-5}$ $54.589 \cdot 10^{-5}$	$1.29 \cdot 10^{-3}$ $1.75 \cdot 10^{-3}$ $2.70 \cdot 10^{-3}$ $1.42 \cdot 10^{-2}$ $3.00 \cdot 10^{-2}$ $6.38 \cdot 10^{-2}$	$\log p_{Ti} = -\dfrac{27\,859}{T} - 0.40 \times 10^{-4} T + 8.263$ (at.)	[411]	1954	
TiN	1714 1744 1777 1882 1939 1968	$0{,}342 \cdot 10^{-5}$ $0{,}446 \cdot 10^{-5}$ $0{,}708 \cdot 10^{-5}$ $3.612 \cdot 10^{-5}$ $7.525 \cdot 10^{-5}$ $15{;}767 \cdot 10^{-5}$	$5.31 \cdot 10^{-4}$ $6.82 \cdot 10^{-4}$ $1.06 \cdot 10^{-3}$ $5.40 \cdot 10^{-3}$ $1.15 \cdot 10^{-2}$ $2.44 \cdot 10^{-2}$	$\log p_{N_2} = -\dfrac{27\,859}{T} - 0.40 \times 10^{-4} T + 7.963$ (at.)	[411]	1954	

	T						Ref.	Year	
ZrN	1963	$0,53\cdot10^{-6}$	$0,072\cdot10^{-5}$	$6,95\cdot10^{-4}$		—	[411]	1954	See [412]
	1986	$1,30\cdot10^{-6}$	$0,094\cdot10^{-5}$	$1,81\cdot10^{-3}$					
	2045	$2,98\cdot10^{-6}$	$0,208\cdot10^{-5}$	$4,94\cdot10^{-3}$					
	2060	$2,71\cdot10^{-6}$	$0,251\cdot10^{-5}$	$3,78\cdot10^{-3}$	p_{Zr}				
	2071	$3,02\cdot10^{-6}$	$0,287\cdot10^{-5}$	$4,20\cdot10^{-3}$		—	[411]	1954	
	2178	$14,98\cdot10^{-6}$	$1,001\cdot10^{-5}$	$2,2\cdot10^{-2}$					
	2193	$16,13\cdot10^{-6}$	$1,184\cdot10^{-5}$	$2,36\cdot10^{-2}$					
						$\log p_{N_2} = -\dfrac{34\,816}{T} +$ $+ 2{,}96\cdot10^{-4}\,T + 8{,}934$ (at.)	[411]	1954	
HfN	1427	—	—	$0,234\cdot10^{-10}$		—	[412]	1966	
	1527			$3,92\cdot10^{-10}$					
	1627			$4,79\cdot10^{-9}$	p_{N_2}				
	1727			$4,53\cdot10^{-8}$					
	1827			$3,450\cdot10^{-7}$					
	1927			$2,180\cdot10^{-6}$					
VN	1300—1650	—	—	—		$\log p_{N_2} = -\dfrac{9690}{T} +$ $+ 3{,}13$ (mm Hg)	—*	1973	

*V. M. Zhikharev, Investigation of the Thermodynamic Properties of Vanadium Mononitride [in Russian], Cand. Dissertation, Leningrad (1972).

VAPOR PRESSURE AND EVAPORATION RATES (continued)

Phase	Temp., °C	Evap. rate g/(cm². sec)	Vapor pressure, mm Hg	Vapor pressure equation	Source	Year	Notes
1	2	3	4	5	6	7	8
NbN	1527 1627 1727 1827 1927 2027 2127 2227	—	0.0545 0.278 1.20 4.58 15.05 44.9 121 305 } p_{N_2}	—	[412]	1966	
Ta₂N	1 200 1 400 1 600 1 800	$0.189 \cdot 10^{-7}$ $1.80 \cdot 10^{-7}$ $4.926 \cdot 10^{-7}$ $11.39 \cdot 10^{-7}$	$0.061 \cdot 10^{-5}$ $0.648 \cdot 10^{-5}$ $1.845 \cdot 10^{-5}$ $4.63 \cdot 10^{-5}$	—	[413]	1961	See [412]
Ta₂N	1600—2380	—	—	$\log p_{N_2} = -\dfrac{21\,700}{T} + 8.65$ (mm Hg)	[412]	1966	
TaN	1527 1627 1727 1827 1927 2027 2127 2227 2327 2427	—	$0.219 \cdot 10^{-3}$ $0.113 \cdot 10^{-2}$ $0.494 \cdot 10^{-2}$ $1.87 \cdot 10^{-2}$ $6.22 \cdot 10^{-2}$ $18.45 \cdot 10^{-2}$ 0.503 1.26 2.91 6.32 } p_{N_2}	—	[412]	1966	

	Temperature		p_{N_2}			
Cr₂N	1000 1100 1200 1300 1400	—	$\left.\begin{matrix}0.5\\1.96\\6.25\\17.6\\42\end{matrix}\right\}$	—	[412]	1966
CrN	533 582 646 719	—	10^{-3} 10^{-2} 10^{-1} 1	—	[311]	1965
Mo₂N	203 243 283 329	—	10^{-3} 10^{-2} 10^{-1} 1	—	[311]	1965
W₂N	221 256 296 344	—	10^{-3} 10^{-2} 10^{-1} 1	—	[311]	1965
Ta₄.₅Si	1960 1986 2010 2036 2118 2128	—	$1.98\cdot10^{-4}$ $2.207\cdot10^{-4}$ $3.18\cdot10^{-4}$ $5.43\cdot10^{-4}$ $26.4\cdot10^{-4}$ $28.27\cdot10^{-4}$	$\log p_{Si} = 9.5 -$ $-\dfrac{28\,600}{T}$ (mm Hg)	[422]	1957

VAPOR PRESSURE AND EVAPORATION RATES (continued)

Phase	Temp., °C	Evap. rate g/(cm²·sec)	Vapor pressure, mm Hg	Vapor pressure equation	Source	Year	Notes
1	2	3	4	5	6	7	8
TaSi	1943	—	$5.75 \cdot 10^{-4}$	$\log p_{Si} = 9.8 - \dfrac{28\,000}{T}$ (mm Hg)	[422]	1957	
	1968		$8.52 \cdot 10^{-4}$				
	1983		$13.3 \cdot 10^{-4}$				
	2027		$29.0 \cdot 10^{-4}$				
	2043		$32.55 \cdot 10^{-4}$				
	2051		$28.7 \cdot 10^{-4}$				
	2073		$38.67 \cdot 10^{-4}$				
	2100		$39 \cdot 10^{-4}$				
	2114		$70.1 \cdot 10^{-4}$				
	2143		$74.5 \cdot 10^{-4}$				
TaSi$_{0.6}$	1760	—	$3.57 \cdot 10^{-4}$	$\log p_{Si} = 9{,}1 - \dfrac{25\,000}{T}$ (mm Hg)	[422]	1957	
	1832		$9.35 \cdot 10^{-4}$				
	1874		$13.6 \cdot 10^{-4}$				
Mo$_3$Si	1891	—	$0.86 \cdot 10^{-3}$	$\log p_{Si} = -\dfrac{33\,690}{T} - 5.75 \log T + 28.84$ (at.)	[412]	1966	
	1914		$2.06 \cdot 10^{-3}$				
	1955		$2.06 \cdot 10^{-3}$				
	1968		$2.36 \cdot 10^{-3}$				
Mo$_5$Si$_3$	1777		$3.10 \cdot 10^{-4}$	$\log p_{Si} = -\dfrac{32\,940}{T} - 5.75 \log T + 28.67$ (at.)	[412]	1966	
	1796		$4.55 \cdot 10^{-4}$				
	1811		$4.02 \cdot 10^{-4}$				
	1818		$3.90 \cdot 10^{-4}$				
	1853		$7.50 \cdot 10^{-4}$				
	1863		$1.08 \cdot 10^{-3}$				

Compound	T	p		Equation	Ref.	Year
	1877		$1.24 \cdot 10^{-3}$			
	1911		$1.71 \cdot 10^{-3}$			
	1933		$2.35 \cdot 10^{-3}$			
	1955	—	$3.28 \cdot 10^{-3}$			
	1966		$3.19 \cdot 10^{-3}$			
	1988		$5.24 \cdot 10^{-3}$			
$MoSi_2$	1653		$1.56 \cdot 10^{-4}$			
	1686		$1.73 \cdot 10^{-4}$			
	1697		$2.82 \cdot 10^{-4}$			
	1731		$4.23 \cdot 10^{-4}$	$\log p_{Si} = -\dfrac{29\,800}{T} -$		
	1749	—	$6.03 \cdot 10^{-4}$	$-\,5.75 \log T + 28.62 \ (\text{at.})$	[412]	1966
	1800		$0.94 \cdot 10^{-3}$			
	1809		$1.45 \cdot 10^{-3}$			
	1856		$1.93 \cdot 10^{-3}$			
	1866		$2.94 \cdot 10^{-3}$			
	1883		$3.95 \cdot 10^{-3}$			
Re_3Si	—	—	—	$\log p_{Si} = 5.953 -$ $-\,\dfrac{24\,040}{T}$ (mm Hg)	[412]	1966
$ReSi$	—	—	—	$\log p_{Si} = 7{,}444 -$ $-\,\dfrac{25\,800}{T}$ (mm Hg)	[412]	1966
$ReSi_2$	—	—	—	$\log p_{Si} = 7{,}512 -$ $-\,\dfrac{25\,610}{T}$ (mm Hg)	[412]	1966

VAPOR PRESSURE AND EVAPORATION RATES (continued)

Phase	Temp., °C	Evap. rate g/(cm²·sec)	Vapor pressure, mm Hg	Vapor pressure equation	Source	Year	Notes
1	2	3	4	5	6	7	8
SmP	527	—	42,99 $\left.\begin{array}{l} \end{array}\right\} p_{P_4}$	—	[423]	1971	
	627	—	71,97					
	727	—	104,89					
	827	—	142,56					
	927	—	179,62					
	1027	—	221,44					
	527	—	$2,4 \cdot 10^{-3}$ $\left.\begin{array}{l} \end{array}\right\} p_{P_2}$	—	[423]	1971	
	627	—	$2,1 \cdot 10^{-2}$					
	727	—	0,11					
	827	—	0,44					
	927	—	1,38					
	1027	—	3,56					
GdP	527	—	55,49 $\left.\begin{array}{l} \end{array}\right\} p_{P_4}$	—	[423]	1971	
	627	—	96,97					
	727	—	149,87					
	827	—	205,47					
	927	—	273,30					
	1027	—	345,53					
	527	—	$2,8 \cdot 10^{-3}$ $\left.\begin{array}{l} \end{array}\right\} p_{P_2}$	—	[423]	1971	
	627	—	$2,4 \cdot 10^{-2}$					
	727	—	$1,3 \cdot 10^{-1}$					
	827	—	$5,3 \cdot 10^{-1}$					
	927	—	1,70					
	1027	—	4,47					

Compound	T, °C		p_{P_4} / p_{P_2}		[ref]	Year
TbP	527	—	77.99		[423]	1971
	627	—	124.97			
	727	—	180.86	p_{P_4}		
	827	—	249.42			
	927	—	316.67			
	1027	—	395.22			
	527	—	$3.2 \cdot 10^{-3}$		[423]	1971
	627		$9.7 \cdot 10^{-2}$			
	727		0.14	p_{P_2}		
	827		0.58			
	927		1.83			
	1027		4.78			
DyP	527	—	92.99		[423]	1971
	627		144.97			
	727		202.85	p_{P_4}		
	827		274.39			
	927		358.05			
	1027		444.93			
	527	—	$3.6 \cdot 10^{-3}$		[423]	1971
	627		$2.9 \cdot 10^{-2}$			
	727		0.15	p_{P_2}		
	827		0,61			
	927		1,95			
	1027		5,07			
BaS	1620	—	$8.36 \cdot 10^{-4}$	—	[311]	1965
LaS	1793—2075	—	—	$\log p_{LaS} = 7{,}719 - 28\,730/T$	[405]	1971
Ce$_3$S$_4$	1700	—	$7.6 \cdot 10^{-3}$	—	[311]	1965

VAPOR PRESSURE AND EVAPORATION RATES (continued)

Phase	Temp., °C	Evap. rate g/(cm²·sec)	Vapor pressure, mm Hg	Vapor pressure equation	Source	Year	Notes
1	2	3	4	5	6	7	8
Ce_2S_3	1840	—	10^{-3}	—	[311]	1965	
Nd_2S_3	1900	—	10^{-3}	—	[311]	1965	
ThS	1700	—	$7.6 \cdot 10^{-7}$	—	[311]	1965	
US	1700	—	$7.6 \cdot 10^{-7}$	—	[311]	1965	
TiS	1530—1913		—	$\log p_\Sigma = 7{,}21 - 25\,750/T$ (at.)	[33]	1972	
TiS_3	465 505 550 600	— —	13.5 41.5 130 366	$\log p_{S_2} = 10{,}42 - 6\,850/T$ (mm Hg)	[378]	1969	
ZrS_3	783 814 847 874	—	$\left.\begin{array}{l}52\\97\\185\\307\end{array}\right\} p_{S_2}$	—	[33]	1972	
$TaS_{1.96}$	1430 1287—1607	$1.37 \cdot 10^{-4}$ —	— —	$\log p = 10{,}72 - 26\,600/T$ (at.)	[37]	1971	

Compound				Reference	Year	Notes
ReS₂	—	—	$\log p = 3.214 - 4976/T$	[33]	1972	
Re₂S₇	—	—	$\log p = 8.86 - 4800/T$	[33]	1972	
FeS	804—1006	—	$\log p = 4.162 - 10850/T$ (mm Hg)	[33]	1972	
CoS	804—1006	—	$\log p = 5.84 - 12630/T$ (mm Hg)	[33]	1972	
B₄C	1911 2010 2103 2148 2249	$0,684 \cdot 10^{-3}$ $2,2 \cdot 10^{-3}$ $9,63 \cdot 10^{-3}$ $16,75 \cdot 10^{-3}$ $61,3 \cdot 10^{-3}$	$\log p_\mathrm{B} = 7.506 - 29630/T$ (at.)	[334]	1973	
SiC	—	—	$\log p_\Sigma = 7.78 - 27400/T$ (at.)	[334]	1973	
BN	2227 2327 2427 2527 2627 2727	$2,13 \cdot 10$ $4,75 \cdot 10$ $1,02 \cdot 10^2$ $2,03 \cdot 10^2$ $3,96 \cdot 10^2$ 760	—	[424]	1961	See [412]
AlN	1177 1317 1489 1597	$0,606 \cdot 10^{-6}$ $1,18 \cdot 10^{-6}$ $1,04 \cdot 10^{-4}$ $6,0 \cdot 10^{-4}$	—	[412]	1966	
Si₃N₄	927—1427 1487—1727	—	$\log p_{N_2} = 16.5 - 37806/T$ $\log p_{N_2} = 21,2 - 45673/T$	[412]	1966	

Phase	Thermal conductivity, cal/(cm·sec·°C)	Temp., °C	Source	Year	Notes
1	2	3	4	5	6
UBe_{13}	0,0885	650	[320]	1969	
$ZrBe_{13}$	0,067	760	[2]	1966	
$ZrBe_{13}$	0.085	1370	[2]	1966	
$ZrBe_{13}$	0,082	1385	[2]	1966	
$ZrBe_{13}$	0,0869	1482	[2]	1966	
Nb_2Be_{17}	0,0778	760	[2]	1966	
Nb_2Be_{17}	0,0819	1482	[2]	1966	
$NbBe_{12}$	0,0723	760	[2]	1966	
$NbBe_{12}$	0,0785	1482	[2]	1966	
Ta_2Be_{17}	0,0715	760	[2]	1966	
Ta_2Be_{17}	0,0781	1482	[2]	1966	
$TaBe_{12}$	0,0732	760	[2]	1966	
$TaBe_{12}$	0.0897	1482	[2]	1966	
CaB_6	0,055±0,004	20	[429]	1963	
SrB_6	0,063±0,005	20	[429]	1963	
BaB_6	0,087±0,004	20	[429]	1963	
YB_4	0,0697	27	—*	1974	
YB_6	0,070±0,010	20	[429]	1963	
LaB_6	0,114±0,010	20	[429]	1963	
CeB_6	0,081±0,002	20	[429]	1963	
PrB_6	0,098±0,005	20	[429]	1963	
NdB_6	0,113±0,008	20	[429]	1963	
SmB_6	0,033±0,004	20	[429]	1963	
EuB_6	0,055±0,002	20	[429]	1963	
GdB_4	0,355	27	—*	1974	
GdB_6	0,049±0,003	20	[429]	1963	
TbB_4	0.302	27	—*	1974	
TbB_6	0,048±0,003	20	[429]	1963	
DyB_4	0,282	27	—*	1974	
HoB_4	0,305	27	—*	1974	
TmB_4	0,378	27	—*	1974	
YbB_6	0,060±0,004	20	[429]	1963	
ThB_4	~0,060	20	[320]	1969	
ThB_4	~0,067	730	[320]	1969	
ThB_4	~0,074	1230	[320]	1969	
ThB_4	~0,098	1730	[320]	1969	
ThB_6	0,107±0,012	20	[429]	1963	
UB_2	0.124	20	[320]	1969	
UB_4	0.00955	20	[320]	1969	
TiB_2	0,154	27	[430]	1973	See [432]
TiB_2	0,167	1027	[430]	1973	
TiB_2	0,292	2027	[430]	1973	
ZrB_2	0,1385	27	[430]	1973	See [432]
ZrB_2	0,154	1027	[430]	1973	
ZrB_2	0,320	2027	[430]	1973	
ZrB_{12}	0,029	—	[291]	1968	

*E. N. Severyanina, Producing and Studying the Physical Properties of Rare-Earth Metal Tetraborides [in Russian], Author's abstract, Cand. Dissertation, Kiev (1974).

Phase	Thermal conductivity, cal/(cm · sec · °C)	Temp., °C	Source	Year	Notes
1	2	3	4	5	6
HfB_2	0.122	27	[430]	1973	
HfB_2	0.1435	1027	[430]	1973	
HfB_2	0.342	2027	[430]	1973	
VB_2	0.101	27	[430]	1973	See [432]
VB_2	0.1005	1027	[430]	1973	
VB_2	0.161	2027	[430]	1973	
Nb_3B_2	0.0287	27	[3]	1975	
NbB	0.0373	27	[3]	1975	
Nb_3B_4	0.049	27	[3]	1975	
NbB_2	0.0574	27	[430]	1973	
NbB_2	0.0561	1027	[430]	1973	
NbB_2	0.0963	2027	[430]	1973	
TaB_2	0.0382	27	[430]	1973	
TaB_2	0.0384	1027	[430]	1973	
TaB_2	0.0866	2027	[430]	1973	
Cr_4B	0.0262 ± 0.0009	20	[431]	1962	
Cr_2B	0.026	20	[3]	1975	
Cr_5B_3	0.0377	20	[3]	1975	
CrB	0.048	20	[3]	1975	
Cr_3B_4	0.049	20	[3]	1975	
CrB_2	0.076	20	[3]	1975	See [432]
CrB_2	0.0812	1027	[430]	1973	
CrB_2	0.1315	2027	[430]	1973	
Cr_2B_5	0.043 ± 0.002	20	[431]	1962	
Mo_2B_5	~0.12	20	[432]	1972	
Mo_2B_5	~0.065	900	[432]	1972	
W_2B_5	~0.125	20	[432]	1972	
W_2B_5	~0.062	800	[432]	1972	
Mn_4B	0.012 ± 0.001	—	[433]	1968	
Mn_2B	0.0157 ± 0.001	—	[433]	1968	
MnB	0.0185 ± 0.0005	—	[433]	1968	
Mn_3B_4	0.0205 ± 0.0008	—	[433]	1968	
MnB_2	0.0244 ± 0.0005	—	[433]	1968	
Fe_2B	0.072	20	[434]	1971	See [408]
FeB	0.0287	20	[434]	1971	See [408]
Co_3B	0.0406	—	[435]	1972	
Co_2B	0.0334	—	[435]	1972	
CoB	0.0406	—	[435]	1972	
Ni_3B	0.1	—	[435]	1972	
Ni_2B	0.131	—	[435]	1972	
NiB	0.0525	—	[435]	1972	
Be_2C	0.102	25	[11]	1968	
ThC	0.069	25	[11]	1968	
ThC_2	0.057	25	[11]	1968	
UC	>0.06	20	[320]	1969	
UC	0.05	730	[320]	1969	
UC	~0.045	2230	[320]	1969	
UC_2	<0.0285	20	[320]	1969	
UC_2	~0.04	730	[320]	1969	
UC_2	~0.0525	2230	[320]	1969	

Phase	Thermal conductivity, cal/(cm·sec·°C)	Temp., °C	Source	Year	Notes
1	2	3	4	5	6
PuC	0,0232	200	[320]	1969	
$TiC_{0.96}$	~0,0162	20	[436]	1965	
$TiC_{0.88}$	~0,0170	20	[436]	1965	
$TiC_{0.76}$	~0,0167	20	[436]	1965	
$TiC_{0.61}$	~0,0167	20	[436]	1965	
$TiC_{0.54}$	~0,0158	20	[436]	1965	
$TiC_{0.48}$	~0,0153	20	[436]	1965	
TiC	~0,0717	1500	[427]	1973	See [432, 437]
TiC	~0,0956	2400	[427]	1973	
$ZrC_{0.903}$	0,0277	20	[438]	1964	See [439]
$ZrC_{0.820}$	0,0246	20	[438]	1964	
$ZrC_{0.756}$	0,0263	20	[438]	1964	
$ZrC_{0.718}$	0,0225	20	[438]	1964	
$ZrC_{0.628}$	0,0196	20	[438]	1964	
ZrC	~0,065	1500	[440]	1970	See [432, 437, 441]
ZrC	~0,095	2500	[440]	1970	
HfC	0,015	25	[11]	1968	
HfC	0,0717	1600	[442]	1973	
HfC	0,0885	2600	[442]	1973	
$VC_{0.90}$	0,0234±0,0046	20	[443]	1970	
$VC_{0.85}$	0,0203±0,0022	20	[443]	1970	
$VC_{0.79}$	0,0210±0,0017	20	[443]	1970	
$VC_{0.76}$	0,0196±0,0029	20	[443]	1970	
$NbC_{0.908}$	0,0268	20	[438]	1968	
$NbC_{0.855}$	0,0256	20	[438]	1968	
$NbC_{0.808}$	0,0244	20	[438]	1968	
$NbC_{0.759}$	0,0232	20	[438]	1968	
$NbC_{0.710}$	0,0215	20	[438]	1968	
NbC	~0,0765	1500	[427]	1973	See [432, 441]
NbC	~0,112	2400	[427]	1973	
$TaC_{1.0}$	~0,0525	20	[257]	1974	
$TaC_{0.97}$	~0,0475	20	[257]	1974	
$TaC_{0.86}$	~0,033	20	[257]	1974	
$TaC_{0.7}$	~0,027	20	[257]	1974	
TaC	0,0382	1230	[320]	1969	See [432]
TaC	0,0525	1730	[320]	1969	
TaC	0,0695	2230	[320]	1969	
TaC	0,086	2730	[320]	1969	
$Cr_{23}C_6$	0,0470±0,0060	20	[431]	1962	
Cr_7C_3	0,0364±0,0015	20	[431]	1962	
Cr_3C_2	0,0458±0,0009	20	[431]	1962	See [432]
Mo_2C	0,076	25	[11]	1968	
W_2C	0,07	25	[11]	1968	
WC	0,07	25	[11]	1968	See [320]
Be_3N_2	0.00172	20	[320]	1969	
UN	0,0405	20	[320]	1969	
UN	0,0502	730	[320]	1969	

Phase	Thermal conductivity, cal/(cm·sec·°C)	Temp., °C	Source	Year	Notes
1	2	3	4	5	6
PuN	0,0358	470	[320]	1969	
TiN	~0,162	1500	[427]	1973	See [445]
TiN	~0,167	1600	[427]	1973	
TiN	~0,165	1700	[427]	1973	
TiN	~0,136	2300	[427]	1973	
$TiN_{0.83}$	0,018	−173	—*	1974	
$TiN_{0.83}$	0,030	27	—*	1974	
$TiN_{0.83}$	0,060	1000	—*	1974	
$ZrN_{0.92}$	0,0612	−173	—*	1974	
$ZrN_{0.92}$	0,0674	27	—*	1974	
$ZrN_{0.92}$	0,065	1000	—*	1974	
HfN	0,0442	−173	—*	1974	
HfN	0,0454	27	—*	1974	
HfN	0,0492	1000	—*	1974	
HfN	~0,15	1600	[442]	1973	
HfN	~0,105	2200	[442]	1973	
$VN_{0.75}$	0,0275	−173	—*	1974	
$VN_{0.75}$	0,0325	27	—*	1974	
$VN_{0.75}$	0,0373	1000	—*	1974	
Nb_2N	0,0200±0,008	20	[444]	1962	
$NbN_{0.75}$	0,033	−173	—*	1974	
$NbN_{0.75}$	0,0382	27	—*	1974	
$NbN_{0.75}$	0,0254	1000	—*	1974	
NbN	0,009±0,002	20	[444]	1962	
Ta_2N	0,0240±0,005	20	[444]	1962	
TaN	0,00525	−173	—*	1974	
TaN	0,0131	27	—*	1974	
TaN	0,029	1000	—*	1974	
Cr_2N	0,0519±0,004	20	[446]	1961	See [431]
CrN	0,0284±0,0023	20	[446]	1961	See [431]
Mo_2N	0,0427±0,007	20	[444]	1962	
TiAl	0,063	—	[18]	1965	
$TiAl_3$	0,080	—	[18]	1965	
Zr_3Al	0,036	—	[18]	1965	
Zr_2Al	0,025	—	[18]	1965	
Zr_3Al_2	0,0375	—	[18]	1965	
Zr_4Al_3	0,0333	—	[18]	1965	
$ZrAl_3$	0,100	—	[18]	1965	
$HfAl_3$	0,070	—	[18]	1965	
V_5Al_8	0,0204	—	[18]	1965	
VAl_3	0,0562	—	[18]	1965	
VAl_6	0,0578	—	[18]	1965	
VAl_{11}	0,0522	—	[18]	1965	
Nb_3Al	0,0301	—	[18]	1965	
Nb_2Al	0,0282	—	[18]	1965	
$NbAl_3$	0,070	—	[18]	1965	
Ta_3Al	0,0211	—	[18]	1965	

*L. K. Shvedova, Investigation of the Properties of Nitrides in the Groups IV-V Transition Metals [in Russian], Author's abstract, Cand. Dissertation, Kiev (1974).

Phase	Thermal conductivity, cal/(cm·sec·°C)	Temp., °C	Source	Year	Notes
1	2	3	4	5	6
Ta$_2$Al	0.0265	—	[18]	1965	
TaAl$_3$	0.0272	—	[18]	1965	
Mo$_3$Al	0,0767	—	[18]	1965	
MoAl$_2$	0,0165	—	[18]	1965	
BaSi$_2$	0,0037	20	[449]	1963	
U$_3$Si	0,04	—	[447]	1960	
VSi$_2$	0,12	—203	[451]	1970	
VSi$_2$	0,06	20	[451]	1970	
VSi$_2$	0,0656	827	[451]	1970	
VSi$_2$	0,06	1327	[451]	1970	
CrSi	0,0363	—193	—*	1970	
CrSi	0,0311	—123	—*	1970	
CrSi	0,0318	20	—*	1970	
CrSi	0,044	727	—*	1970	
CrSi	0,0552	1427	—*	1970	
CrSi$_2$	0,015	—	[448]	1958	
Mo$_3$Si	0,095	20	[450]	1963	
Mo$_5$Si$_3$	0,052	20	[450]	1963	
MoSi$_2$	0.116	20	[450]	1963	
MnSi	0,0203	—193	—*	1970	
MnSi	0,0239	20	—*	1970	
MnSi	0.0294	527	—*	1970	
MnSi	0,0363	1127	—*	1970	
MnSi$_{1.73}$	0.018	—203	—*	1970	
MnSi$_{1.73}$	0.00955	20	—*	1970	
MnSi$_{1.73}$	0,01	427	—*	1970	
MnSi$_{1.73}$	0,0143	827	—*	1970	
MnSi$_{1.72}$	0,00885	20	—**	1972	Direction [100]
MnSi$_{1.72}$	0,0043	20	—**	1972	Direction [001]
FeSi	0,0593	—208	[452]	1970	Direction [210]
FeSi	0.0234	20	[452]	1970	
FeSi	0,026	627	[452]	1970	
FeSi	0,041	1327	[452]	1970	
FeSi$_2$	0,1575	—203	—*	1970	Polycrystal
FeSi$_2$	0,0292	20	—*	1970	
FeSi$_2$	0,00885	727	—*	1970	
FeSi$_{2.43}$	0,0382	—203	—*	1970	Direction [110]
FeSi$_{2.43}$	0,0232	27	—*	1970	
CoSi	0,0968	—208	—	1970	Direction [210]

*F. N. Ostrovskii, Thermal Conductivity and Temperature Conductivity of the Monosilicides and Higher Silicides of 3d-Transition Metals [in Russian], Cand. Dissertation, Sverdlovsk (1970).

**L. D. Ivanova, Investigation of Higher Manganese Silicide-Based Semiconductor Materials [in Russian], Author's abstract, Cand. Dissertation, Moscow (1972).

Phase	Thermal conductivity, cal/(cm · sec · °C)	Temp., °C	Source	Year	Notes
1	2	3	4	5	6
CoSi	0,0342	20	—*	1970	Direction [210]
CoSi	0,0296	427	—*	1970	Direction [210]
CoSi	0,0463	1327	—*	1970	Direction [210]
CoSi	0,031	20	[249]	1964	Polycrystal
$CoSi_2$	~0,079	—213	[453]	1970	Direction [111]
$CoSi_2$	~0,091	87	[453]	1970	
$CoSi_2$	~0,172	—193	[453]	1970	Polycrystal
$CoSi_2$	~0 122	77	[453]	1970	
LaS	~0,05	—200	[397]	1972	
LaS	~0 0665	27	[397]	1972	
LaS	~0.0715	177	[397]	1972	
La_3S_4	~0.006	27	[397]	1972	
La_3S_4	~0,005	1000	[397]	1972	
La_2S_3	0.0055	20	[320]	1969	
CeS	0,0225	20	[320]	1969	
Ce_3S_4	0,00835	20	[320]	1969	
Ce_2S_3	0,00835	20	[320]	1969	
Pr_3S_4	0,00583	20	[320]	1969	
Pr_2S_3	0,00583	20	[320]	1969	
NdS	0,00263	20	[320]	1969	
Nd_3S_4	~0,003	—173	[397]	1972	
Nd_3S_4	~0,0054	27	[397]	1972	
Nd_3S_4	~0,0087	427	[397]	1972	
$\alpha\text{-}Nd_2S_3$	~0,007	—173	[397]	1972	
$\alpha\text{-}Nd_2S_3$	~0,0054	27	[397]	1972	
$\alpha\text{-}Nd_2S_3$	~0,007	100	[397]	1972	
$\beta\text{-}Nd_2S_3$	~0,0027	—173	[397]	1972	
$\beta\text{-}Nd_2S_3$	~0,003	27	[397]	1972	
$\beta\text{-}Nd_2S_3$	~0,004	100	[397]	1972	
SmS	0.005—0,015	27	[397]	1972	
GdS	~0,03	—173	[397]	1972	
GdS	~0,05	27—427	[397]	1972	
$\alpha\text{-}TiS$	0.0115	—	—**	1966	
Ti_2S_3	0,0042	—	[264]	1964	
ZrS_2	0,0053	—	—**	1966	
Nb_2S_3	0,0189	—	[243]	1966	
$\alpha\text{-}TaS_2$	0,011	—	—**	1966	
Cr_2S_3	0,0058	—	—**	1966	
$\alpha\text{-}MoS_2$	0,0085—0,0205	15	—***	1968	
$\beta\text{-}MoS_2$	0,0045—0,0115	15	—***	1968	
$\gamma\text{-}MoS_2$	0,0085—0,017	15	—***	1968	
FeS	0,0025	—	—**	1966	

*See footnote on page 196.

** V. Kh. Oganesyan, Study of the Physical Properties of Some Transition Metal Sulfides in Groups IV-V [in Russian], Author's abstract, Cand. Dissertation, Kiev (1966).

*** M. V. Teslitskaya, Synthesis, Structure, and Some Properties of Molybdenum Disulfide [in Russian], Author's abstract, Cand. Dissertation, Moscow (1968).

THERMAL CONDUCTIVITY (continued)

Phase	Thermal conductivity, cal/(cm·sec·°C)	Temp., °C	Source	Year	Notes
1	2	3	4	5	6
B_4C	0,11	20	[320]	1969	
β-SiC	~0,1	20	[320]	1969	
β-SiC	~0,0465	730	[320]	1969	
β-SiC	~0,031	1230	[320]	1969	
BN	0,0222	300	[454]	1972	
BN	0,0258	500	[454]	1972	
BN	0,0217	1000	[454]	1972	
BN	0,0203	1300	[454]	1972	
BN	0,0191	1500	[454]	1972	
BN	0,0193	1800	[454]	1972	
AlN	0,0382	20	[249]	1964	
AlN	0,0251	300	[454]	1972	
AlN	0,0164	700	[454]	1972	
AlN	0,0144	1100	[454]	1972	
AlN	0,0097	1600	[454]	1972	
AlN	0,01	1800	[454]	1972	
Si_3N_4	0,0478	20	[320]	1969	

THERMAL EXPANSION

Phase	Coef. therm. exp.·10^6, deg^{-1}	Temp. range, °C	Source	Year	Notes
1	2	3	4	5	6
UBe_{13}	16.7	20—1060	[320]	1969	
Zr_2Be_{17}	8,39	27—1510	[2]	1966	
$ZrBe_{13}$	9,86	27—1510	[2]	1966	
Hf_2Be_{21}	5,1	20—1500	[320]	1969	
Nb_2Be_{17}	8,83	27—1510	[2]	1966	
$NbBe_{12}$	9,36	27—1510	[2]	1966	
Ta_2Be_{17}	8,72	27—1510	[2]	1966	
$TaBe_{12}$	8,42	27—1510	[2]	1966	
$MoBe_{12}$	12	—	[10]	1973	
$MoBe_{22}$	21,5±0,6	До 200	[2]	1966	
CaB_6	6,5±0,5	20—800	[54]	1961	
SrB_6	6,7±0,5	20—800	[54]	1961	
BaB_6	6,8±0,5	20—800	[54]	1961	
ScB_2	7,6±0,5	20—600	[455]	1964	Along axis c
ScB_2	6,8±0,5	20—600	[455]	1964	Along axis a
ScB_4	4,1	—	[320]	1969	
YB_2	9,4±1,0	—	[3]	1975	Along axis a
YB_2	8,5±0,9	—	[3]	1975	Along axis c
YB_4	7,68	20—1000	[456]	1973	

THERMAL EXPANSION (continued)

Phase	Coef. therm. exp.·10^6, deg^{-1}	Temp. range, °C	Source	Year	Notes
1	2	3	4	5	6
YB_4	7,6±0,5	—	[3]	1975	Along axis a
YB_4	6,4±0,6	—	[3]	1975	Along axis c
YB_6	6,02±0,6	—	[3]	1975	
YB_{12}	6,6±0,6	—	[3]	1975	
LaB_4	7,17±1,16	—	[3]	1975	Along axis a
LaB_4	8,36±1,03	—	[3]	1975	Along axis c
LaB_6	6,4±0,5	20—800	[54]	1961	
CeB_6	7,3±0,5	20—800	[54]	1961	
PrB_4	5,0	27—1027	—*	1974	
PrB_6	7,50±0,5	20—800	[54]	1961	
NdB_4	5,84	27—1027	—*	1974	
NdB_6	7.30±1,0	20—800	[54]	1961	
SmB_6	6,8±0.5	20—800	[54]	1961	
EuB_6	6,9±0,5	20—800	[54]	1961	
GdB_4	7,0	20—1000	[456]	1973	
GdB_6	8,7±0,5	20—800	[54]	1961	
TbB_4	6,55	27—1027	—*	1974	
TbB_6	7,8±1,0	20—800	[54]	1961	
TbB_{12}	3,2	—	[3]	1975	
DyB_4	5,93	27—1027	—*	1974	
DyB_{12}	4,2	—	[3]	1975	
HoB_4	7,85	20—1000	[456]	1973	
HoB_6	3.0	—	[320]	1969	
HoB_{12}	3,6	—	[3]	1975	
ErB_4	7,6	20—1000	[456]	1973	
ErB_{12}	3,7	—	[3]	1975	
TmB_4	6,5	27—1027	—*	1974	
TmB_{12}	3,85	—	[3]	1975	
YbB_6	5,8 ± 0,5	20—800	[54]	1961	
YbB_{12}	2,0	—196—20	[59]	1971	
YbB_{12}	3,9	20—300	[59]	1971	
YbB_{12}	5,8	300—1000	[59]	1971	
LuB_{12}	3,4	—	[3]	1975	
ThB_4	7,9	20—1770	[320]	1969	
ThB_6	7,8 ± 0,5	20—800	[54]	1961	
UB_2	9	20—205	[457]	1956	Along axis a
UB_2	8	20—205	[457]	1956	Along axis c
UB_4	7,0	20—1000	[320]	1969	
UB_{12}	4,6	—	[458]	1971	
TiB_2	4,6	27—1027	[430]	1973	
TiB_2	5,2	1027—2027	[430]	1973	
ZrB_2	5,9	27—1027	[430]	1973	
ZrB_2	6,5	1027—2027	[430]	1973	See [459, 460]
HfB_2	6,3	27—1027	[430]	1973	
HfB_2	6,8	1027—2027	[430]	1973	
VB_2	7,6	27—1027	[430]	1973	
VB_2	8,3	1027—2027	[430]	1973	

*E. N. Severyanina, Producing and Studying the Physical Properties of Rare-Earth Metal Tetraborides [in Russian], Author's abstract, Cand. Dissertation, Kiev (1974).

Phase	Coef. therm. exp.$\cdot 10^6$, deg^{-1}	Temp. range, °C	Source	Year	Notes
1	2	3	4	5	6
Nb_3B_2	13,8	27—1027	[3]	1975	
Nb_3B_2	13,9	1027—2027	[3]	1975	
NbB	12,9	27—1027	[3]	1975	
NbB	13,4	1027—2027	[3]	1975	
Nb_3B_4	9,9	27—1027	[3]	1975	
Nb_3B_4	10,3	1027—2027	[3]	1975	
NbB_2	8,0	27—2027	[430]	1973	
NbB_2	8,5	1027—2027	[430]	1973	See [459, 460]
TaB_2	8,2	27—1027	[430]	1973	
TaB_2	8,8	1027—2027	[430]	1973	
Cr_2B	14,2	27—1027	[3]	1975	
Cr_2B	15,0	1027—2027	[3]	1975	
Cr_5B_3	13,7	27—1027	[3]	1975	
Cr_5B_3	14,2	1027—2027	[3]	1975	
CrB	12,3	27—1027	[3]	1975	
CrB	12,6	1027—2027	[3]	1975	
Cr_3B_4	11,8	27—1027	[3]	1975	
Cr_3B_4	12,1	1027—2027	[3]	1975	
CrB_2	10,5	27—1027	[430]	1973	
CrB_2	11,8	1027—2027	[430]	1973	
MoB_2	7,7	—	[320]	1969	
Mo_2B_5	8,6	27—1027	[3]	1975	
Mo_2B_5	9,9	1027—2027	[3]	1975	
MoB_4	6,5	—	[3]	1975	
W_2B	6,7	—	[3]	1975	
W_2B_5	7,8	27—1027	[3]	1975	
W_2B_5	8,8	1027—2027	[3]	1975	
WB_4	5,8	—	[3]	1975	
Fe_2B	11,8±0,3	20—800	[79]	1968	Along [100]
Fe_2B	8,9±0,6	20—800	[79]	1968	Along [001]
Fe_2B	~8	To ~700	[434]	1971	Close to the
FeB	~9,5	To ~100	[434]	1971	Curie point the
FeB	~12	400—1000	[434]	1971	coef. of therm. exp. increases
Be_2C	10,5	20—600	[320]	1969	by about 40 %
ScC	11,4	—	[461]	1971	
YC	11—13.6	20—1100	[320]	1969	
Y_2C_3	10,3±0,3	100—900	—*	1968	
YC_2	7,0	20—1000	[462]	1965	
La_2C_3	14,6±0,3	100—900	—*	1968	
LaC_2	8,4	20—1000	[462]	1965	
Ce_2C_3	14,7±0,3	100—900	—*	1968	
CeC_2	9,9	20—1000	[462]	1965	
PrC_2	6.2	20—1000	[462]	1965	
Nd_2C_3	13,5±0,3	100—900	—*	1968	

*V. L. Yupko, Physical Properties of Rare-Earth Metal Carbides [in Russian], Author's abstract, Cand. Dissertation, Kiev (1968).

Phase	Coef. therm. exp.·10^6, deg^{-1}	Temp. range, °C	Source	Year	Notes
1	2	3	4	5	6
NdC_2	8.8	20—1000	[462]	1965	
SmC_2	11,4	100—900	—*	1968	
GdC_2	13.5	100—900	—*	1968	
TbC_2	10,2	100—900	—*	1968	
DyC_2	11,4	100—900	—*	1968	
ErC_2	12,4	100—900	—*	1968	
TmC_2	10.7	100—900	—*	1968	
$ThC_{0.76}$	5.8	20—750	[99]	1969	At 750-800°C a superstructure forms
$ThC_{0.76}$	18,4	800	[99]	1969	
ThC	6,53	20	[11]	1968	
ThC_2	8,5	20	[11]	1968	
UC	10,4	20—1000	[463]	1961	
U_2C_3	11,3	20—1800	[320]	1969	
PuC	10,7	20—780	[320]	1969	
Pu_2C_3	14,7	20—780	[320]	1969	
$TiC_{0.96}$	7.95	25—1000	[464]	1970	See [430]
$TiC_{0.89}$	8,04	25—1000	[464]	1970	
$TiC_{0.79}$	8.34	25—1000	[464]	1970	
$TiC_{0.72}$	8,45	25—1000	[464]	1970	
$TiC_{0.62}$	8,58	25—1000	[464]	1970	
$ZrC_{0.97}$	7,01	25—1000	[464]	1970	
$ZrC_{0.90}$	7,11	25—1000	[464]	1970	
$ZrC_{0.80}$	7,22	25—1000	[464]	1970	
$ZrC_{0.75}$	7,36	25—1000	[464]	1970	
$ZrC_{0.70}$	7,42	25—1000	[464]	1970	
$ZrC_{0.65}$	7,42	25—1000	[464]	1970	
$HfC_{0.98}$	6.80	25—1000	[464]	1970	
$HfC_{0.91}$	6.89	25—1000	[464]	1970	
$HfC_{0.80}$	7,02	25—1000	[464]	1970	
$HfC_{0.71}$	7,13	25—1000	[464]	1970	
$VC_{0.41}$	6.94	25—1000	[466]	1975	Hex.
$VC_{0.42}$	6,72	25—1000	[466]	1975	
$VC_{0.44}$	6,18	25—1000	[466]	1975	
$VC_{0.46}$	5,92	25—1000	[466]	1975	
$VC_{0.49}$	5,44	25—1000	[466]	1975	
$VC_{0.87}$	7,25	25—1000	[465]	1974	Cub.
$VC_{0.82}$	7,31	25—1000	[465]	1974	
$VC_{0.77}$	7,38	25—1000	[465]	1974	
$VC_{0.72}$	7,45	25—1000	[465]	1974	
$NbC_{0.42}$	8,42	25—1000	[466]	1975	Hex.
$NbC_{0.45}$	8.16	25—1000	[466]	1975	
$NbC_{0.47}$	7,81	25—1000	[466]	1975	
$NbC_{0.49}$	7,72	25—1000	[466]	1975	
$NbC_{0.50}$	7,62	25—1000	[466]	1975	

*V. L. Yupko, see footnote on page 200.

Phase	Coef. therm. exp. $\cdot 10^6$, deg^{-1}	Temp. range, °C	Source	Year	Notes
1	2	3	4	5	6
Nb_2C	7,0	—	[257]	1974	Along axis a
Nb_2C	8,7	—	[257]	1974	Along axis c
$NbC_{0.99}$	7,21	25—1000	[464]	1970	
$NbC_{0.92}$	6,99	25—1000	[464]	1970	
$NbC_{0.82}$	6,88	25—1000	[464]	1970	Cub., see [430]
$NbC_{0.76}$	6,95	25—1000	[464]	1970	
$NbC_{0.72}$	7,25	25—1000	[464]	1970	
Ta_2C	~7,8	—	[257]	1974	
$TaC_{0.99}$	7,09	25—1000	[464]	1970	
$TaC_{0.91}$	6.77	25—1000	[464]	1970	
$TaC_{0.85}$	6,50	25—1000	[464]	1970	
$TaC_{0.82}$	6,52	25—1000	[464]	1970	
$TaC_{0.75}$	7,12	25—1000	[464]	1970	
$Cr_{23}C_6$	10,1	—	[291]	1968	
Cr_7C_3	9,4	20—1100	[320]	1969	
Cr_3C_2	11,7	20—1100	[320]	1969	
Mo_2C	7,8	—	[257]	1974	Along axis a
Mo_2C	9,3	—	[257]	1974	Along axis c
W_2C	6,4	—	[257]	1974	Along axis a
W_2C	8,1	—	[257]	1974	Along axis c
WC	3,84	—	[257]	1974	Along axis a
WC	3,90	—	[257]	1974	Along axis c
Fe_3C	$4,11 + 24,2 \cdot 10^{-3}t$	—200—1000	[467]	1971	[100]
Fe_3C	$0,98 + 30,6 \cdot 10^{-3}t$	—200—1000	[467]	1971	[010]
Fe_3C	$13,5 + 4,2 \cdot 10^{-3}t$	—200—1000	[467]	1971	[001]
Fe_3C	$6,2 + 20,0 \times \times 10^{-3}t$	—200—1000	[467]	1971	α_{mean}
Fe_3C	13,2	20—700	[467]	1971	See [111]
Fe_3C	17,5	20—1130	[467]	1971	
Fe_5C_2	4,9	—177—28	[111]	1967	Along axis a
Fe_5C_2	5,9	—177—28	[111]	1967	Along axis b
Fe_5C_2	5,6	—177—28	[111]	1967	Along axis c
LaN	9	—	[320]	1969	
CeN	30	—	[320]	1969	
PrN	13	—	[320]	1969	
ThN	(7,39)	—	[320]	1969	
UN	8,61	45—1000	[320]	1969	
TiN	9,35±0,04	25—1100	[468]	1955	
$TiN_{0.98}$	4,76	—180—27	—*	1974	
$TiN_{0.92}$	4,48	—180—27	—*	1974	
$TiN_{0.85}$	3,58	—180—27	—*	1974	
$TiN_{0.78}$	3,02	—180—27	—*	1974	
ZrN	7,24	20—1100	[14]	1969	

*L. K. Shvedova, Investigation of the Physical Properties of Transition Metal Nitrides in Groups IV-V [in Russian], Kiev (1974).

Phase	Coef. therm. exp. $\cdot 10^6$, deg^{-1}	Temp. range, °C	Source	Year	Notes
1	2	3	4	5	6
$ZrN_{0.99}$	4.09	—180—27	—*	1974	
$ZrN_{0.86}$	4.19	—180—27	—*	1974	
$ZrN_{0.77}$	4.62	— 180—27	—*	1974	
HfN	6.9	20—1100	[14]	1969	
V_3N	8.1	20—1100	[14]	1969	
$VN_{0.93}$	9.2	20—1100	[14]	1969	
$VN_{0.98}$	5.01	—180—27	—*	1974	
$VN_{0.85}$	5.73	—180—27	—*	1974	
$VN_{0.71}$	6.02	—180—27	—*	1974	
Nb_2N	3.26	20—1100	[469]	1961	
NbN	10.1	20—1000	[469]	1961	
$NbN_{0.95}$	4.27	—180—27	—*	1974	
$NbN_{0.89}$	4.38	—180—27	—*	1974	
Ta_2N	5.2	20—1000	[14]	1969	
TaN	3.6	20—700	[14]	1969	
TaN	2.95	—180—27	—*	1974	Along axis a
TaN	3.05	—180—27	—*	1974	Along axis c
Cr_2N	9.41	20—1100	[14]	1969	
CrN	2.3	20—800	[14]	1969	
CrN	7.5	850—1040	[14]	1969	
Mo_2N	6.2	20—1100	[14]	1969	
TiAl	11.43	20—1070	[470]	1962	
$TiAl_3$	10.65	20—870	[470]	1962	
Zr_3Al	8.46	—	[18]	1965	
$ZrAl_3$	12.1	—	[18]	1965	
V_2Al_8	13.2	—	[18]	1965	
VAl_3	13.5	—	[18]	1965	
VAl_6	15.9	—	[18]	1965	
VAl_{11}	18	—	[18]	1965	
Nb_3Al	9.15	—	[18]	1965	
Nb_2Al	9.5	—	[18]	1965	
$NbAl_3$	11.2	—	[18]	1965	
Ta_3Al	9.7	—	[18]	1965	
TaAl	11.43	—	[10]	1973	
$TaAl_3$	11.7	—	[18]	1965	
Mo_3Al	~8	—	[10]	1973	
$MoAl_2$	10.1	—	[18]	1965	
NiAl	15.1	—	[10]	1973	
Mg_2Si	14.8	—	[471]	1963	
$BaSi_2$	8.2	90—690	[471]	1963	
$BaSi_2$	8.6	690—1090	[471]	1963	
Sc_5Si_3	9.8±0.20	20—450	—**	1975	
Sc_5Si_3	10.7±0.21	450—900	—**	1975	
ScSi	9.4±0.19	20—580	—**	1975	
ScSi	9.5±0.19	580—900	—**	1975	

*L. K. Shvedova; see footnote on page 202.

**V. I. Lazorenko, Investigation of the Physical Properties of Rare-Earth Metal Silicides in the Cerium Subgroup [in Russian], Author's abstract, Cand. Dissertation, Kiev (1975).

Phase	Coef. therm. exp.$\cdot 10^6$, deg^{-1}	Temp. range, °C	Source	Year	Notes
1	2	3	4	5	6
ScSi$_{1.66}$	8,6±0,17	20—600	—*	1975	
ScSi$_{1.66}$	9,0±0,18	600—900	—*	1975	
Y$_5$Si$_3$	~8,1	200	[472]	1974	Break on the
Y$_5$Si$_3$	~9	900	[472]	1974	α (T) curve at 500-600°C
YSi	~7,5	150	[472]	1974	
YSi	~8	300—430	[472]	1974	
YSi	~8,2	450	[472]	1974	
YSi	~8,4	900	[472]	1974	
YSi$_2$	~9	200	[472]	1974	Break on the α (T) curve at 450 and 600°C
YSi$_2$	~9,8	900	[472]	1974	
La$_5$Si$_3$	8,8±0,17	20—320	—*	1975	
La$_5$Si$_3$	11,0±0,22	320—900	—*	1975	
LaSi	8,3±0,16	20—520	—*	1975	
LaSi	10,2±0,20	520—900	—*	1975	
LaSi$_2$	7,8±0,15	20—475	—*	1975	
LaSi$_2$	8,1±0,16	475—900	—*	1975	
Ce$_5$Si$_3$	14,8±0.29	20—450	—*	1975	
Ce$_5$Si$_3$	16,0±0,32	450—900	—*	1975	
CeSi	13,9±0,27	20—600	—*	1975	
CeSi	14,5±0,29	600—900	—*	1975	
CeSi$_2$	12,0±0,24	20—500	—*	1975	
CeSi$_2$	13,1±0,26	500—900	—*	1975	
Pr$_5$Si$_3$	12,3±0,24	20—500	—*	1975	
Pr$_5$Si$_3$	12,5±0,25	500—900	—*	1975	
PrSi	11,4±0,22	20—550	—*	1975	
PrSi	11,6±0,23	550—900	—*	1975	
PrSi$_2$	10,9±0,21	20—600	—*	1975	
PrSi$_2$	11,1±0,22	600—900	—*	1975	
Sm$_5$Si$_3$	11,8±0,23	20—250	—*	1975	
Sm$_5$Si$_3$	12,2±0,24	250—900	—*	1975	
SmSi	10,8±0,21	20—700	—*	1975	
SmSi	11,6±0,23	700—900	—*	1975	
SmSi$_2$	10,1±0,20	20—380	—*	1975	
SmSi$_2$	11,2±0,22	380—900	—*	1975	
δ-U$_3$Si	8,7+0,0171 t	0—760	[161]	1965	
δ-U$_3$Si	17,2	770—930	[161]	1965	
β-USi$_2$	57?	20—205	[457]	1956	Along axis a
β-USi$_2$	—26?	20—205	[457]	1956	Along axis c
Ti$_5$Si$_3$	11,0	170—1070	[471]	1963	
TiSi	8,8	20—370	[471]	1963	
TiSi	10,4	370—1070	[471]	1963	
TiSi$_2$	12,5	200—1200	[375]	1969	

*V. L. Lazorenko, see footnote on p. 203.

Phase	Coef. therm. exp.·10^6, deg^{-1}	Temp. range, °C	Source	Year	Notes
1	2	3	4	5	6
TiSi$_2$	~0,1	15—900	[473]	1966	Monocrystal; along axis c
Zr$_2$Si	~7	200—1000	[344]	1968	Recorded from
Zr$_5$Si$_3$	~7,2	200—1000	[344]	1968	graph for coef.
Zr$_3$Si$_2$	~9	200—1000	[344]	1968	therm. exp.
Zr$_5$Si$_4$	~10	200—1000	[344]	1968	and composi-
ZrSi	~8	200—1000	[344]	1968	tion relation-
ZrSi$_2$	~8,3	200—1000	[344]	1968	ship in Zr—Si
V$_3$Si	8.0	20—620	[471]	1963	system
V$_3$Si	12.0	620—820	[471]	1963	
V$_3$Si	14,1	820—1070	[471]	1963	
V$_5$Si$_3$	9.5	20—770	[471]	1963	
V$_5$Si$_3$	11,1	770—1070	[471]	1963	
VSi$_2$	11,2	20—770	[471]	1963	
VSi$_2$	14,65	770—1070	[471]	1963	
Nb$_5$Si$_3$	7,3	20—650	[463]	1961	Along axis a
Nb$_5$Si$_3$	4,6	20—650	[463]	1961	Along axis c
NbSi$_2$	8,4	20—370	[471]	1963	
NbSi$_2$	11,7	370—1070	[471]	1963	
Ta$_5$Si$_3$	5,5	20—1000	[463]	1961	Along axis a ⎫ T2
Ta$_5$Si$_3$	8,0	20—1000	[463]	1961	Along axis c ⎬
Ta$_5$Si$_3$	6,3	20—1000	[463]	1961	Along axis a ⎫ D8$_8$
Ta$_5$Si$_3$	6,6	20—1000	[463]	1961	Along axis c ⎬
TaSi$_2$	8,9	20—1000	[463]	1961	Along axis a
TaSi$_2$	8,8	20—1000	[463]	1961	Along axis c
TaSi$_2$	9,5	20—320	[471]	1963	
TaSi$_2$	10.7	320—1070	[471]	1963	
Cr$_3$Si	10,5	20—1070	[471]	1963	
Cr$_5$Si$_3$	6,0	20—170	[471]	1963	
Cr$_5$Si$_3$	10.6	170—720	[471]	1963	
Cr$_5$Si$_3$	14,2	720—1070	[471]	1963	
CrSi	11,3	20—770	[471]	1963	
Mo$_3$Si	3,4	20—170	[471]	1963	
Mo$_3$Si	6,5	170—1070	[471]	1963	
Mo$_5$Si$_3$	4,3	20—270	[471]	1963	
Mo$_5$Si$_3$	6,7	270—1070	[471]	1963	
MoSi$_2$	8,25	20—1070	[471]	1963	
WSi$_2$	6.25	20—420	[471]	1963	
WSi$_2$	7.90	420—1070	[471]	1963	
MnSi	16,3±1,0	20—800	[474]	1962	
ReSi	2.7	20—220	[471]	1963	
ReSi	4,9	220—1070	[471]	1963	
ReSi$_2$	6,6	20—1070	[471]	1963	
Fe$_3$Si	$12.0 + 1.5 \times 10^{-2}t$	20—820	[475]	1965	
Fe$_3$Si	14,4	20—870	[471]	1963	
Fe$_3$Si	11,0	870—1070	[471]	1963	
Fe$_5$Si$_3$	$13,0 + 0,8 \times 10^{-2}t$	20—600	[475]	1965	

Phase	Coef. therm. exp.·10^6, deg^{-1}	Temp. range, °C	Source	Year	Notes
1	2	3	4	5	6
FeSi	$15.7 + +0.25 \cdot 10^{-2}t$	20—1000	[475]	1965	
FeSi$_2$	$7.7 + 0.6 \times \times 10^{-2}t$	20—900	[475]	1965	β-leboite
FeSi$_{2.5}$	$8.3 + 0.67 \times \times 10^{-2}t$	20—500	[475]	1965	α-leboite
Co$_3$Si	13,4	20—520	[471]	1963	
Co$_3$Si	16,6	520—970	[471]	1963	
CoSi	11,1$\pm$1,0	20—800	[474]	1962	
CoSi$_2$	10,14	—	—*	1968	
Ni$_3$Si	9,0	20—370	[471]	1963	
Ni$_3$Si	11,5	370—770	[471]	1963	
Ni$_3$Si	14,85	770—1070	[471]	1963	
Ni$_2$Si	16,5	20—870	[471]	1963	
Ni$_2$Si	19,0	870—1070	[471]	1963	
Ni$_{1.04}$Si$_{1.93}$	12,06	—	—*	1968	
Th$_3$P$_4$	11	—	[320]	1969	
Ti$_3$P	8,8	20	—**	1966	
Ti$_2$P	8,2	20	—**	1966	
Ti$_5$P$_3$	6,5	20	—**	1966	
Ti$_4$P$_3$	10,1	20	—**	1966	
TiP	(6,5)	20	—**	1966	
Ni$_3$P	12,65	20—800	—***	1969	
Ni$_{12}$P$_5$	14,05	20—800	—***	1969	
Ni$_2$P	15,05	20—800	—***	1969	
LaS	11,5	—	[347]	1972	
La$_3$S$_4$	11,4	—	[347]	1972	
La$_2$S$_3$	10,4	—	[347]	1972	
CeS	12,3	—	[347]	1972	
Ce$_3$S$_4$	12,5	—	[347]	1972	
Ce$_2$S$_3$	12,7	—	[347]	1972	
PrS	13,9	–	[347]	1972	
Pr$_3$S$_4$	12,5	—	[347]	1972	
Pr$_2$S$_3$	12,0	—	[347]	1972	
NdS	14,4	—	[347]	1972	
Nd$_3$S$_4$	14,5	—	[347]	1972	
Nd$_2$S$_3$	14,6	—	[347]	1972	
SmS	11,7	—	[347]	1972	
Sm$_3$S$_4$	11,8	—	[347]	1972	

*L. A. Miroshnikov, Physical Properties of Solid Solutions of Iron and Cobalt Disilicides in Higher Nickel Silicide [in Russian], Author's abstract, Cand. Dissertation, Sverdlovsk (1968).

**V. E. Listovnichii, Physicochemical Investigation of Interactions between Titanium, Phosphorus, and Sulfur [in Russian], Author's abstract, Cand. Dissertation, Kiev (1966).

***P. A. Vityaz', Investigation of Nickel Phosphides Solid and Their Possible Use for Alloying Iron-Based Cermets [in Russian], Author's abstract, Cand. Dissertation, Minsk (1969).

Phase	Coef. therm. exp.·10^6, deg^{-1}	Temp. range, °C	Source	Year	Notes
1	2	3	4	5	6
Sm_2S_3	11,9	—	[347]	1972	
EuS	14,6	—	[347]	1972	
GdS	12,0	—	[347]	1972	
Gd_2S_3	12,2	—	[347]	1972	
Dy_2S_3	14,4	—	[347]	1972	
Er_2S_3	11,9	—	[347]	1972	
ThS	10,2	20—980	[320]	1969	
US	11,9	20—980	[320]	1969	
Ti_5S	10,5	20	—*	1966	
Ti_3S (ξ)	10,5	20	—*	1966	
Ti_2S (η)	10,5	20	—*	1966	
TiS_{1-x}(δ)	9,1	20	—*	1966	
$Ti_{1-x}S$ (γ)	—2,0	20	—*	1966	
Ti_2S_3	17±6	20—1000	[476]	1966	
ZrS_2	11,7	20—1000	[476]	1966	
Nb_2S_3	10,0	20—1000	[476]	1966	
α-TaS_2	13,9	20—1000	[476]	1966	
Cr_2S_3	12,3	20—1000	[476]	1966	
MoS_2	10,7	20—1000	[476]	1966	
MnS	~17,3	—173	[477]	1971	
MnS	~26,9	—123	[477]	1971	
MnS	~49,9	—116,5	[477]	1971	
MnS	~11,5	—87	[477]	1971	
MnS	~15,8	—23	[477]	1971	
MnS	~17.3	27	[477]	1971	
MnS_2	13,0	70—252	[33]	1972	
FeS	22,1	20—1000	[476]	1966	
B_4C	4,5	25	[478]	1965	
β-SiC	4,7	20—2127	[320]	1969	
α-BN	0,5—1,7	—	[248]	1969	
Si_3N_4	2,75	20—1000	[248]	1969	
AlN	4,8	20—300	[454]	1972	See [249]
AlN	5,0	300—1000	[454]	1972	

*V. E. Listovnichii; see footnote on p. 206.

CRYSTAL LATTICE ENERGY

Phase	E a, kcal/mole	Source	Year	Phase	E a, kcal/mole	Source	Year
1	2	3	4	1	2	3	4
CaB_6	1230	[353]	1962	Ta_2C	2390	[353]	1962
SrB_6	1220	[353]	1962	Mo_2C	2280	[353]	1962
BaB_6	1200	[353]	1962	WC	2760	[353]	1962
YB_6	1780	[353]	1962	W_2C	2000	[353]	1962
LaB_6	1770	[353]	1962	ScN	1062	[479]	1959
CeB_6	2410	[353]	1962	LaN	860	[479]	1959
PrB_6	1780	[353]	1962	TiN	3900	[479]	1959
NdB_6	1780	[353]	1962	ZrN	3540	[479]	1959
PmB_6	~1770	[353]	1962	HfN	2840	[353]	1962
SmB_6	1780	[353]	1962	VN	3820	[479]	1959
EuB_6	1220	[353]	1962	NbN	3560	[479]	1959
GdB_6	1790	[353]	1962	TaN	3320	[353]	1962
TbB_6	1790	[353]	1962	CrN	2640	[479]	1959
DyB_6	1790	[353]	1962	Mo_2N	2670	[353]	1962
HoB_6	1790	[353]	1962	W_2N	2390	[353]	1962
TmB_6	1790	[353]	1962	WN	3300	[353]	1962
ThB_6	2430	[353]	1962	$TiSi_2$	2230	[353]	1962
TiB_2	3260	[353]	1962	$ZrSi_2$	2490	[353]	1962
ZrB_2	2540	[352]	1957	VSi_2	2730	[353]	1962
VB_2	2880	[352]	1957	Nb_5Si_3	3735	[353]	1962
NbB_2	3060	[352]	1957	$NbSi_2$	2490	[353]	1962
TaB	1640	[353]	1962	$TaSi_2$	2318	[353]	1962
TaB_2	2940	[352]	1957	LaS	768	—*	1969
CrB	2140	[353]	1962	La_2S_3	2634	—*	1969
Mo_2B	2620	[352]	1957	CeS	778	—*	1969
MoB	1870	[353]	1962	Ce_2S_3	2655	—*	1969
W_2B	2470	[352]	1957	CeS_2	2326	[481]	1961
ThC_2	1960	[353]	1962	PrS	782	—*	1969
UC_2	1970	[353]	1962	Pr_2S_3	2672	—*	1969
TiC	3890	[353]	1962	NdS	789	—*	1969
ZrC	3470	[353]	1962	Nd_2S_3	2687	—*	1969
HfC	2800	[353]	1962	SmS	702,0	—**	1973
VC	3900	[353]	1962	EuS	691,2	—**	1973
NbC	3220	[353]	1962	GdS	804	—*	1969
Nb_2C	2570	[353]	1962	YbS	715,5	—**	1973
TaC	2770	[353]	1962				

*A. D. Finogenov, Investigation of the Thermodynamic Properties of Sulfides of Some Rare-Earth Metals in the Cerium Subgroup [in Russian], Author's abstract, Cand. Dissertation, Moscow (1969).

**B. V. Fenochka, Mass-Spectrometric Investigation of the High-Temperature Behavior and Thermodynamic Properties of Rare-Earth Metal Sulfides [in Russian], Author's abstract, Cand. Dissertation, Kiev (1973).

Phase	Atomization energy, kcal/mole	Source	Year	Phase	Atomization energy, kcal/mole	Source	Year
1	2	3	4	1	2	3	4
TiC	328	[480]	1969	PrS	265.32	—*	1969
ZrC	360	[480]	1969	Pr_3S_4	919.2	—*	1969
HfC	380	[480]	1969	Pr_2S_3	646.86	—*	1969
VC	338	[480]	1969	PrS_2	360.4	—*	1969
NbC	384	[480]	1969	NdS	254.56	—*	1969
TaC	391	[480]	1969	Nd_3S_4	884.0	—*	1969
WC	376	[480]	1969	Nd_2S_3	621.34	—*	1969
TiN	305	[480]	1969	NdS_2	347.6	—*	1969
ZrN	335	[480]	1969	SmS	217.9	—**	1973
HfN	371	[480]	1969	EuS	204.8	—**	1973
VN	293	[480]	1969	GdS	266	—**	1973
NbN	347	[480]	1969	TbS	(263)	—**	1973
LaS	281.16	—*	1969	DyS	(239)	—**	1973
La_3S_4	971.2	—*	1969	HoS	(237)	—**	1973
La_2S_3	683.60	—*	1969	ErS	(232)	—**	1973
LaS_2	377.81	—*	1969	TmS	(225)	—**	1973
CeS	288.83	—*	1969	YbS	199.4	—**	1973
Ce_3S_4	987.70	—*	1969	β-SiC	300	[480]	1969
Ce_2S_3	695.88	—*	1969	BN	314.6	[14]	1969
CeS_2	385.25	—*	1969	AlN	263.6	[14]	1969

*A. D. Finogenov, see footnote on p. 208.
**B. V. Fenochka, see footnote on p. 208.

CHARACTERISTIC TEMPERATURE

Phase	θ, K	Test method[1]	Source	Year	Notes
1	2	3	4	5	6
CaB_6	700	x-ray	—[2]	1972	
SrB_6	889	CTE	[482]	1961	
BaB_6	824	CTE	[482]	1961	
ScB_2	550	LT	[483]	1969	
YB_4	670	CTE	[456]	1973	
	695	MP	[456]	1973	
YB_6	570	x-ray	—[2]	1972	

[1]CTE is the coefficient of thermal expansion; LT is the low-temperature specific heat; MP means from melting point; OP means from other properties.
[2]Ya. I. Fedyshin, X-ray Investigation of the Thermal Oscillations of Atoms of Certain Cubic Structures [in Russian], Author's abstract, Cand. Dissertation, L'vov (1972).

Phase	θ, K	Test method [*1]	Source	Year	Notes
1	2	3	4	5	6
LaB_6	710	x-ray	—[*2]	1972	
	250	LT	[484]	1969	
CeB_6	520	x-ray	—[*2]	1972	
PrB_4	784	CTE	—[*3]	1974	
PrB_6	780	x-ray	—[*2]	1972	
NdB_4	733	CTE	—[*3]	1974	
NdB_6	650	x-ray	—[*2]	1972	
SmB_6	755	CTE	[482]	1961	
EuB_6	735	CTE	[54]	1961	
GdB_4	632	CTE	[456]	1973	
GdB_6	745	CTE	[482]	1961	
TbB_4	661	CTE	—[*3]	1974	
TbB_6	690	CTE	[54]	1961	
TbB_{12}	810	x-ray	—[*2]	1972	
DyB_4	698	CTE	—[*3]	1974	
DyB_{12}	920	x-ray	—[*2]	1972	
HoB_4	517	CTE	[456]	1973	
HoB_{12}	870	x-ray	—[*2]	1972	
ErB_4	543	CTE	[456]	1973	
ErB_{12}	1000	x-ray	—[*2]	1972	
TmB_4	657	CTE	—[*2]	1974	
TmB_{12}	790	x-ray	—[*2]	1972	
YbB_6	450	x-ray	—[*2]	1972	
	193	LT	[484]	1969	
LuB_{12}	830	x-ray	—[*2]	1972	
ThB_2	600	CTE	[54]	1961	
ThB_6	530	x-ray	—[*2]	1972	
	188	LT	[484]	1969	
UB_{12}	740	x-ray	—[*2]	1972	
TiB	662	LT	[368]	1969	Traces TiB_2
TiB_2	807	LT	[368]	1969	See [483, 485]
	1100	CTE	[430]	1973	
	970	MP	[430]	1973	
ZrB_2	585	LT	[368]	1969	
	765	CTE	[430]	1973	
	730	MP	[430]	1973	
HfB_2	499	LT	[368]	1969	
	550	CTE	[430]	1973	
	530	MP	[430]	1973	
VB_2	850	LT	[483]	1969	
	880	CTE	[430]	1973	
	880	MP	[430]	1973	
NbB	566	LT	[368]	1969	Traces Nb_3B_2
NbB_2	701	CTE	[430]	1973	
	720	MP	[430]	1973	
Ta_2B	328	LT	[368]	1969	Traces γ-TaB

*A. D. Finogenov, see footnote on p. 208.
**E. N. Severyanina, see footnote on p. 199.

Phase	θ, K	Test method[*1]	Source	Year	Notes
1	2	3	4	5	6
TaB_2	545	CTE	[430]	1973	
	570	MP	[430]	1973	
CrB_2	545	LT	[483]	1969	
CrB_2	726	CTE	[430]	1973	
	780	MP	[430]	1973	
Mo_2B	478	LT	[368]	1969	
MoB	553	LT	[368]	1969	
MoB_2	534	LT	[368]	1969	
W_2B	373	LT	[368]	1969	
WB	391	LT	[368]	1969	
Mn_2B	320	LT	[486]	1970	
MnB	356	LT	[486]	1970	
MnB_2	632	LT	[486]	1970	See [483]
Co_3B	345	LT	[487]	1968	
Co_2B	270	LT	[487]	1968	
CoB	696	LT	[487]	1968	
YC_2	381	MP	—*	1968	
La_2C_3	189	MP	—*	1968	
LaC_2	316	MP	—*	1968	
Ce_2C_3	199	MP	—*	1968	
CeC_2	310	MP	—*	1968	
PrC_2	298	MP	—*	1968	
Nd_2C_3	203	MP	—*	1968	
NdC_2	311	MP	—*	1968	
GdC_2	337	MP	—*	1968	
TbC_2	294	MP	—*	1968	
DyC_2	304	MP	—*	1968	
ErC_2	308	MP	—*	1968	
TmC_2	296	MP	—*	1968	
ThC	308	MP	[292]	1969	
UC	330	OP	[292]	1969	
	320	LT	[488]	1963	
PuC	271	MP	[292]	1969	
TiC	934	MP	[292]	1969	
	614	LT	[319]	1974	
$ZrC_{0.964}$	690	OP	[292]	1969	
	694	MP	[292]	1969	
ZrC	491	LT	[319]	1974	
$ZrC_{1.0}$	~700	OP	[491]	1973	
$ZrC_{0.9}$	~670	OP	[491]	1973	
$ZrC_{0.8}$	~640	OP	[491]	1973	
$ZrC_{0.7}$	~610	OP	[491]	1973	
$HfC_{0.967}$	552	OP	[292]	1969	
	549	MP	[292]	1969	
HfC	436	LT	[319]	1974	
V_2C	490	LT	[319]	1974	
$VC_{0.89}$	896	MP	[292]	1969	
$VC_{0.88}$	659	LT	[319]	1974	

*V. L. Yupko, see footnote on p. 200.

Phase	θ, K	Test method*1	Source	Year	Notes
1	2	3	4	5	6
$VC_{0.83}$	466	LT	[319]	1974	
$NbC_{0.48}$	464	LT	[489]	1968	Type ε-Fe_2N
$NbC_{0.5}$	662	LT	[319]	1974	Rhomb.
$NbC_{1.0}$	~780	OP	[491]	1973	
$NbC_{0.9}$	~770	OP	[491]	1973	
$NbC_{0.8}$	~720	OP	[491]	1973	
$NbC_{0.73}$	~650	OP	[491]	1973	
$NbC_{0.964}$	740	OP	[292]	1969	
	740	MP	[292]	1969	
$NbC_{0.98}$	604	LT	[489]	1968	See [319]
$NbC_{0.91}$	555	LT	[489]	1968	
$NbC_{0.86}$	542	LT	[489]	1968	
$NbC_{0.83}$	521	LT	[489]	1968	
$NbC_{0.77}$	500	LT	[489]	1968	
$TaC_{0.47}$	378	LT	[489]	1968	C6
$TaC_{0.994}$	572	OP	[292]	1969	
	572	MP	[292]	1969	
$TaC_{0.95}$	489	LT	[489]	1968	
$TaC_{0.93}$	483	LT	[489]	1968	
$TaC_{0.83}$	434	LT	[489]	1968	
$TaC_{0.78}$	418	LT	[489]	1968	
$MoC_{0.667}$	666	MP	[292]	1969	
α-$MoC_{0.69}$	620	LT	[319]	1974	B1
η-$MoC_{0.64}$	536	LT	[319]	1974	Hex.
α-$MoC_{0.54}$	473	LT	[319]	1974	Rhomb.
β-$MoC_{0.5}$	492	LT	[319]	1974	Hex.
$WC_{1.007}$	617	OP	[292]	1969	
WC	493	LT	[319]	1974	
Fe_3C	259.5	LT	[490]	1969	
LaN	300	LT	[319]	1974	
NdN	300	—	[492]	1965	
ThN	284	LT	[364]	1972	
UN	366	OP	[493]	1972	
TiN	636	LT	[319]	1974	
$TiN_{0.98}$	899	CTE	—*	1974	
$TiN_{0.92}$	920	CTE	—*	1974	
$TiN_{0.85}$	1010	CTE	—*	1974	
$TiN_{0.78}$	1075	CTE	—*	1974	
ZrN	515	LT	[319]	1974	
$ZrN_{0.99}$	753	CTE	—*	1974	
$ZrN_{0.86}$	746	CTE	—*	1974	
$ZrN_{0.77}$	657	CTE	—*	1974	
HfN	618	CTE	—*	1974	
	421	LT	[319]	1974	
VN	420	LT	[319]	1974	
$VN_{0.98}$	900	CTE	—*	1974	
$VN_{0.85}$	860	CTE	—*	1974	
$VN_{0.71}$	840	CTE	—*	1974	
$NbN_{0.95}$	750	CTE	—*	1974	

*L. K. Shvedorn, see footnote on p. 202.

Phase	θ, K	Test method[*1]	Source	Year	Notes
1	2	3	4	5	6
$NbN_{0.91}$	307	LT	[319]	1974	B1
$NbN_{0.84}$	331	LT	[319]	1974	
Ta_2N	231	—	[14]	1969	
TiAl *	234—243	—	[495]	1973	
$TiAl_3^*$	112—142	—	[495]	1973	
Nb_3Al	290	—	[494]	1969	
Mg_2Si	398	—	[260]	1959	
Sc_5Si_3	384	CTE	—**	1975	
ScSi	572	CTE	—**	1975	
$ScSi_{1.66}$	527	CTE	—**	1975	
Y_5Si_3	312	CTE	—**	1975	
YSi	493	CTE	—**	1975	
YSi_2	569	CTE	—**	1975	
La_5Si_3	316	CTE	—**	1975	
LaSi	420	CTE	—**	1975	
$LaSi_2$	430	CTE	—**	1975	
Ce_5Si_3	348	CTE	—**	1975	
CeSi	402	CTE	—**	1975	
$CeSi_2$	498	CTE	—**	1975	
Pr_5Si_3	307	CTE	—**	1975	
PrSi	342	CTE	—**	1975	
$PrSi_2$	408	CTE	—**	1975	
Sm_5Si_3	287	CTE	—**	1975	
SmSi	332	CTE	—**	1975	
$SmSi_2$	406	CTE	—**	1975	
$V_{3.15}Si$	316	LT	[374]	1969	
$V_{3.05}Si$	340	LT	[374]	1969	
$V_{2.97}Si$	340	LT	[374]	1969	
$V_{2.5}Si$	362	LT	[374]	1969	
VSi_2	560	LT	[22]	1971	
CoSi	510	LT	[24]	1971	
Fe_3Si	435	MP	—*	1965	

*The predominant phase in the melt.
**V. I. Lazorenko, see footnote on p. 203.

Phase	θ, K	Test method [1]	Source	Year	Notes
1	2	3	4	5	6
Fe_3Si	437	LT	—*	1965	
	477	OP	—*	1965	
Fe_5Si_3	423	LT	—*	1965	
	418	OP	—*	1965	
FeSi	520	MP	—*	1965	
	515	LT	—*	1965	
	638	OP	—*	1965	
$FeSi_2$	613	LT	—*	1965	
	587	OP	—*	1965	
$FeSi_{2.3}$	550	MP	—*	1965	
	533	LT	—*	1965	
LaS	740	—	[347]	1972	See [497]
La_3S_4	900	—	[347]	1972	
La_2S_3	1000	—	[347]	1972	See [497]
CeS	640	—	[347]	1972	
Ce_3S_4	870	—	[347]	1972	
Ce_2S_3	870	—	[347]	1972	
PrS	660	—	[347]	1972	
Pr_3S_4	880	—	[347]	1972	
Pr_2S_3	900	—	[347]	1972	
NdS	590	—	[347]	1972	
Nd_3S_4	740	—	[347]	1972	
Nd_2S_3	750	—	[347]	1972	
SmS	730	—	[347]	1972	
Sm_3S_4	900	—	[347]	1972	
Sm_2S_3	900	—	[347]	1972	See [262]
EuS	640	—	[347]	1972	
GdS	700	—	[347]	1972	
Gd_2S_3	880	—	[347]	1972	
Dy_2S_3	830	—	[347]	1972	
Er_2S_3	905	—	[347]	1972	
β-SiC	930—1180	—	[496]	1971	
AlN	850	—	[249]	1964	

*L. P. Andreeva, Mechanical, Thermal, and Electrical Properties of Iron Silicides [in Russian], Cand. Dissertation, Sverdlovsk (1965).

MEAN-SQUARE AMPLITUDE OF THERMAL VIBRATIONS OF ATOM COMPLEXES

Phase	$\sqrt{\overline{u^2_{291}}}$, Å	Source	Year	Phase	$\sqrt{\overline{u^2_{291}}}$, Å	Source	Year
1	2	3	4	1	2	3	4
CaB_6	0,050	[54]	1961	Nb_3B_4	0,100	—*3	1971
SrB_6	0,049	[54]	1961	NbB_2	0,088	[459]	1971
BaB_6	0,045	[54]	1961	TaB_2	0,090	[459]	1971
YB_4	0,028	—*1	1974	Cr_2B	0,117	[3]	1975
YB_6	0,047	[54]	1961	Cr_5B_3	0,106	[3]	1975
YB_{12}	0,043	—*2	1970	CrB	0,103	[3]	1975
LaB_6	0,042	[54]	1961	Cr_3B_4	0,100	[3]	1975
CeB_6	0,047	[54]	1961	CrB_2	0,088	[3]	1975
PrB_4	0,021	—*1	1974	Mo_2B_5	0,101	—*3	1971
PrB_6	0,049	[54]	1961	W_2B_5	0,103	—*3	1971
NdB_4	0,022	—*1	1974	YC_2	0,17	—*4	1968
NdB_6	0,047	[54]	1961	La_2C_3	0,20	—*4	1968
SmB_6	0,045	[54]	1961	LaC_2	0,16	—*4	1968
EuB_6	0,047	[54]	1961	Ce_2C_3	0,19	—*4	1968
GdB_4	0,024	—*1	1974	CeC_2	0,16	—*4	1968
GdB_6	0,045	[54]	1961	PrC_2	0,17	—*4	1968
TbB_4	0,023	—*1	1974	Nd_2C_3	0,19	—*4	1968
TbB_6	0,047	[54]	1961	NdC_2	0,16	—*4	1968
TbB_{12}	0,042	—*2	1970	GdC_2	0,15	—*4	1968
DyB_4	0,022	—*1	1974	TbC_2	0,16	—*4	1968
DyB_{12}	0,047	—*2	1970	DyC_2	0,15	—*4	1968
HoB_4	0,029	—*1	1974	ErC_2	0,15	—*4	1968
HoB_{12}	0,042	—*2	1970	TmC_2	0,15	—*4	1968
ErB_4	0,027	—*1	1974	TiC	0,067	[498]	1957
ErB_{12}	0,042	—*2	1970	$ZrC_{1.0}$	~0,076	[491]	1973
TmB_4	0,023	—*1	1974	$ZrC_{0.9}$	~0,0775	[491]	1973
TmB_{12}	0,043	—*2	1970	$ZrC_{0.8}$	~0,079	[491]	1973
YbB_6	0,043	[54]	1961	$ZrC_{0.7}$	~0,081	[491]	1973
YbB_{12}	0,043	—*2	1970	HfC	0.083	[499]	1962
LuB_{12}	0,041	—*2	1970	VC	0,088	[498]	1957
ThB_6	0,045	[54]	1961	$NbC_{1.0}$	~0,072	[491]	1973
UB_{12}	0,044	—*2	1970	$NbC_{0.9}$	~0,0735	[491]	1973
TiB_2	0.075	[459]	1971	$NbC_{0.8}$	~0,075	[491]	1973
ZrB_2	0,081	[459]	1971	$NbC_{0,73}$	~0,084	[491]	1973
ZrB_{12}	0,043	—*2	1970	TaC	0,082	[498]	1957
HfB_2	0,083	[459]	1971	Mo_2C	0,055	[498]	1957
VB_2	0,096	[459]	1971	W_2C	0,062	[498]	1957
Nb_3B_2	0,108	—*3	1971	WC	0,058	[498]	1957
NbB	0,109	—*3	1971	$TiN_{0.98}$	0,056	—*	1974

*1 E. N. Severyanina, see footnote on p. 199.

*2 V. V. Odintsov, see footnote on p. 225.

*3 B. A. Kovenskaya, Investigation of Some Physical Properties, and the Electron Structure of, Transition Metal Borides of Groups IV-VI [in Russian], Author's abstract, Cand. Dissertation, Kiev (1971).

*4 V. L. Yupko, see footnote on p. 200.

MEAN-SQUARE AMPLITIDE OF THERMAL VIBRATIONS OF
ATOM COMPLEXES (continued)

Phase	$\sqrt{\overline{u^2_{291}}}$, Å	Source	Year	Phase	$\sqrt{\overline{u^2_{291}}}$, Å	Source	Year
1	2	3	4	1	2	3	4
$TiN_{0.92}$	0,056	—*1	1974	Ce_5Si_3	0,104	—*3	1975
$TiN_{0.85}$	0,054	—*1	1974	$CeSi$	0,091	—*3	1975
$TiN_{0.78}$	0,053	—*1	1974	$CeSi_2$	0,089	—*3	1975
$ZrN_{0.99}$	0,050	—*1	1974	Pr_5Si_3	0,114	—*3	1975
$ZrN_{0.86}$	0,052	—*1	1974	$PrSi$	0,111	—*3	1975
$ZrN_{0.77}$	0,058	—*1	1974	$PrSi_2$	0,108	—*3	1975
$HfN_{1.0}$	0,045	—*1	1974	Sm_5Si_3	0,118	—*3	1975
$VN_{0.98}$	0,056	—*1	1974	$SmSi$	0,111	—*3	1975
$VN_{0.85}$	0,057	—*1	1974	$SmSi_2$	0,109	—*3	1975
$VN_{0.71}$	0,058	—*1	1974	$TiSi_2$	0,087	[561]	1959
$NbN_{0.95}$	0,051	—*1	1974	$ZrSi_2$	0,087	[561]	1959
$TiAl$*2	0.224—0,249	[495]	1973	$TaSi_2$	0,073	[561]	1959
$TiAl_3$ *2	0,360—0,450	[495]	1973	$CrSi_2$	0,082	[561]	1959
Sc_5Si_3	0,144	—*3	1975	$MoSi_2$	0,076	[561]	1959
$ScSi$	0,104	—*3	1975	WSi_2	0,073	[561]	1959
$ScSi_{1.66}$	0,114	—*3	1975	Ti_3P	0.144	—*4	1966
Y_5Si_3	0,164	—*3	1975	Ti_2P	0,141	—*4	1966
YSi	0,094	—*3	1975	Ti_5P_3	0,124	—*4	1966
YSi_2	0,092	—*3	1975	Ti_4P_3	0,155	—*4	1966
La_5Si_3	0,109	—*3	1975	TiP	0,130	—*4	1966
$LaSi$	0,098	—*3	1975	Ti_5S	0,162	—*4	1966
$LaSi_2$	0,101	—*3	1975	Ti_3S	0,158	—*4	1966
				Ti_2S	0,158	—*4	1966
				TiS_{1-x}	0,149	—*4	1966

*1 L. K. Shvedova, see footnote on p. 202.
*2 Predominant phase in melt.
*3 V. I. Lazorenko, see footnote on p. 203.
*4 V. E. Listovnichii, see footnote on p. 206.

HEAT OF PHASE INVERSION

Phase	Inversion heat, kcal/mole	Inversion temp., °C	Source	Year
CaC_2 (I) → CaC_2 (II)	1,33	447	[287]	1973
α-UC_2 → β-UC_2	2.57	1767	[334]	1973
$\frac{1}{6}$ V_6C_5 → $VC_{0.833}$ *	0,402±0,177	1271	[290]	1972
$\frac{1}{8}$ V_8C_7 → $VC_{0.875}$ *	0,43±0,19	1123	[290]	1972
α-Mn_3C → β-Mn_3C	3,57	1037	[287]	1973
α-Mg_3N_2 → β-Mg_3N_2	0,22	550	[14]	1969
β-Mg_3N_2 → γ-Mg_3N_2	0.26	788	[14]	1969
α-Mn_3Si → β-Mn_3Si	87,2	600	[260]	1959

*Ordered — disordered inversion.

PARAMETERS OF DIFFUSION OF NONMETALS INTO METALS

System	Temp., °C	Resulting phase	Pre-exponential factor D_0, cm²/sec	Activation energy Q, cal/mole	Source	Year	Notes
1	2	3	4	5	6	7	8
Be → Nb	900—1300	NbBe₁₂	$7.66 \cdot 10^{-4}$	3 200	[2]	1966	
Be → W	1000—1200	WBe₁₂	2.36	66 950	[2]	1966	
B → β-Ti	1100—1500	TiB₂	$8.9 \cdot 10^{-5}$	30 600	[804]	1963	
B → Ti	1100—1400	TiB₂	$2.36 \cdot 10^{-4}$	41 500	—*	1969	
B → Zr	1100—1500	ZrB₂	$1.26 \cdot 10^{-4}$	34 500	[806]	1964	
B → Zr	1100—1400	ZrB₂	$5.12 \cdot 10^{-4}$	42 400	—*	1969	
B → Hf	1100—1400	HfB₂	$5 \cdot 10^{-3}$	47 700	—*	1969	
B → V	1100—1400	VB₂	0.316	51 200	—*	1969	
B → Nb	1100—1500	NbB₂	2.94	59 000	[804]	1963	
B → Nb	1100—1400	NbB₂	1.36	57 200	—*	1969	
B → Ta	1100—1500	TaB	$9.44 \cdot 10^{-4}$	48 000	[804]	1963	
B → Ta	1100—1400	TaB₂	1.37	60 000	—*	1969	
B → Mo	1100—1800	Mo₂B	$2.53 \cdot 10^{-6}$	24 000	[501]	1962	
B → Mo	1100—1500	Mo₂B	$6.96 \cdot 10^{-2}$	44 000	[804]	1963	
B → Mo	1100—1400	α-MoB	9.7	67 000	—*	1969	
B → W	1100—1500	W₂B	$1.48 \cdot 10^{-2}$	64 250	[804]	1963	
B → W	1100—1400	α-WB	11	72 100	—**	1969	
B → α-Fe	—	Solid solution	$1.2 \cdot 10^{-2}$	20 300	—**	1968	
B → α-Fe	—	Fe₂B, FeB	—	21 200	[500]	1956	
B → γ-Fe	—	Solid solution	10^{-5}	15 000	[805]	1968	
B → γ-Fe	—	Fe₂B, FeB	$2 \cdot 10^{-3}$	21 000	[805]	1968	
C → Th	1000—1200	Solid solution	—	38 000	[805]	1968	
C → α-Ti	600—800	—	$7.9 \cdot 10^{-4}$	30 500	[257]	1974	
C → β-Ti	1000—1700	Solid solution	$3.18 \cdot 10^{-3}$	19 100	[504]	1963	
C → β-Ti	900—1300	TiC	$2.08 \cdot 10^{-3}$	33 000	[806]	1964	
C → Ti	1200—1400	TiC	77.8	81 000	—*	1969	
C → Ti	1100—1600	Solid solution	$3.02 \cdot 10^{-3}$	20 000	[808]	1968	
C → α-Zr	600—800	—	$3.51 \cdot 10^{-5}$	30 700	[257]	1974	
C → β-Zr	1100—1600	"	$3.57 \cdot 10^{-2}$	34 200	[505]	1966	
C → β-Zr	900—1300	"	$4.8 \cdot 10^{-3}$	26 700	[506]	1965	

System	Temp., °C	Resulting phase	Pre-exponential factor D_0, cm²/sec	Activation energy Q, cal/mole	Source	Year	Notes
1	2	3	4	5	6	7	8
C→Zr	1200—1400	ZrC	100	88 000	—*	1969	
C→Zr	900—1300	ZrC	$3.44 \cdot 10^{-2}$	41 000	[806]	1964	
C→Zr	1100—1700	—	$1.4 \cdot 10^{-6}$	30 000	[809]	1969	
C→α-Hf	—	—	74	74 600	[810]	1968	
C→β-Hf	—	—	$4.2 \cdot 10^{-2}$	40 000	[810]	1968	
C→Hf	1100—1300	HfC	$6.1 \cdot 10^{-5}$	24 000	—*	1969	
C→V	845—1130	—	0.0049	27 300	[507]	1968	
C→V	1200—1400	VC	5.78	71 800	—*	1969	
C→Nb	1200—2000	—	0.033	38 000	[507]	1968	See [503]
C→Nb	1200—1400	—	$9.32 \cdot 10^{-3}$	35 000	[808]	1966	
C→Nb	1920—2340	—	0.026	37 800	[811]	1972	Mean parameters from various authors
C→Nb	130—2340	—	0.01	33 920	[811]	1972	
C→Nb	1300—1600	Nb₂C	$9.6 \cdot 10^{-2}$	70 000	—*	1969	
C→Nb	1300—1600	NbC	0.183	71 400	—*	1969	
C→Ta	1450—2200	—	0.012	40 300	[507]	1968	
C→Ta	600—1400	Solid solution	$2.78 \cdot 10^{-3}$	24 600	[812]	1969	Mean parameters from various authors
C→Ta	2200—2680	—	0.038	48 200	[811]	1972	
C→Ta	130—2340	—	0.0067	38 600	[811]	1972	
C→Ta	1100—1600	—	$2.57 \cdot 10^{-2}$	43 000	[808]	1966	
C→Ta	1600—1900	—	$8.54 \cdot 10^{-4}$	52 500	[809]	1969	
C→Ta	1300—1600	TaC	84.6	98 800	—*	1969	
C→Cr	1200—1400	Cr₃C₂	—	26 100	[352]	1957	
C→Cr	800—1500	Solid solution	$8.3 \cdot 10^{-3}$	28 000	[805]	1968	
C→Cr	140—162, 1150—1600	"	$8.7 \cdot 10^{-3}$	26 500	[813]	1966	
C→Mo	1400—1500	Mo₂C	0.00164	83 000	[502]	1971	See [509, 510, 804]
C→Mo	1000—1780	Mo₂C	800	83 000	[839]	1971	

Diffusion	Temperature, °C	Phase	D_0	Ref.	Year	Notes
C → Mo	—	—	$3{,}4\cdot10^{-2}$	[814]	1967	
C → Mo	1200—2000	Mo$_2$C	$4{,}15\cdot10^{-5}$	[501]	1962	
C → Mo	1100—1600	—	$2{,}04\cdot10^{-5}$	[808]	1966	
C → W	1700—2050	W$_2$C	$3\cdot10^{5}$	[839]	1971	
C → W	1800—2800	Solid solution	$9{,}22\cdot10^{-3}$	[812]	1969	See [509, 510, 806]
C → W	1600—1700	W$_2$C, WC	$1{,}56\cdot10^{-3}$	[502]	1971	
C → W	1100—1600	—	$8{,}91\cdot10^{-2}$	[808]	1966	
C → W	1500—1800	—	$3{,}45\cdot10^{-3}$	[815]	1972	
C → Re	1500—2000	—	$0{,}1$	[511]	1968	
C → α-Fe	350—850	Solid solution	$6{,}2\cdot10^{-3}$	[512]	1955	
C → γ-Fe	900—1050	"	$0{,}051$	[512]	1955	See [513]
C → Ni	900—900	"	$2{,}2\cdot10^{4}$	[512]	1955	
C → Mg	>500	"	$2{,}1\cdot10^{-3}$	[805]	1968	
N → Th	845—1890	Solid solution	$1{,}2\cdot10^{-2}$	[14]	1969	
N → α-Ti	900—1400	Solid solution	$5{,}5\cdot10^{-2}$	[805]	1968	
N → β-Ti	900—1600	"	$4{,}5\cdot10^{-3}$	[805]	1968	
N → Ti	600—900	TiN	—	—***	1962	
N → α-Zr	400—825	Solid solution	$1{,}5\cdot10^{-2}$	[805]	1968	
N → β-Zr	920—1640	"	$7{,}47\cdot10^{-5}$	[805]	1968	
N → Zr	500—600	"	—	—***	1962	
N → Hf	876—1034	HfN	$0{,}20$	[805]	1968	
N → Hf	100—200	Solid solution	$9{,}2\cdot10^{-3}$	[805]	1968	
N → V	360—660	Solid solution	$9{,}8\cdot10^{-2}$	[514]	1971	
N → Nb	500—600	Solid solution	$8{,}05\cdot10^{-4}$	[805]	1971	
N → Nb	600—900	"	$6{,}16\cdot10^{-4}$	—***	1962	
N → Nb	900—1200	Nb$_2$N	$4{,}5\cdot10^{-3}$	—***	1962	
N → Nb	800—1600	NbN	$6{,}1\cdot10^{-2}$	[514]	1962	
N → Ta	800—1300	Solid solution	$0{,}24$	[805]	1971	
N → Ta	500—700	"	$22{,}19$	—***	1968	
N → Ta	800—900	Ta$_2$N	$4{,}914\cdot10^{3}$	—***	1962	
N → Ta	1000—1200	TaN	$1{,}22$	—***	1962	
N → Cr	500—900	CrN	14.8	[816]	1962	
N → Cr	140—180	~	$1{,}6\cdot10^{-2}$	[514]	1967	
N → Mo	1200—2100	—	$4{,}3\cdot10^{-3}$	[514]	1971	
N → W	1400—2200	—	$2{,}4\cdot10^{-3}$	[514]	1971	See [515]
N → Re	1300—1900	—	$0{,}14$	[514]	1971	

PARAMETERS OF DIFFUSION OF NONMETALS INTO METALS (continued)

System	Temp., °C	Resulting phase	Pre-exponential factor D_0, cm^2/sec	Activation energy Q, cal/mole	Source	Year	Notes
1	2	3	4	5	6	7	8
N → α-Fe	—	—	0.003	18 000	[513]	1967	See [805, 817]
N → γ-Fe	—	—	0.1	34 000	[513]	1967	
Al → Pu	350—517	—	$2,25 \cdot 10^{-4}$	25 500	[818]	1969	
Al → Ti	590	TiAl$_3$	$0,066 \cdot 10^{-10}$	62 180	[516]	1972	
Al → Ti	611	TiAl$_3$	$0,235 \cdot 10^{-10}$				
Al → Ti	648	TiAl$_3$	$0,9 \cdot 10^{-10}$				
Al → Ti	560	Solid solution	$0,17 \cdot 10^{-10}$	~30 000	[516]	1972	
Al → Ti	590	"	$0,20 \cdot 10^{-10}$				
Al → Ti	611	"	$0,35 \cdot 10^{-10}$				
Al → Ti	648	"	$0,57 \cdot 10^{-10}$				
Al → Nb	900—1300	NbAl$_3$	$7,18 \cdot 10^{-6}$	6 700	[805]	1968	
Al → Fe	715—880	Solid solution	$3,23 \cdot 10^{-2}$	13 000	[805]	1968	
Al → Fe	950—1100	"	0,72	45 000	[805]	1968	
Al → Ni	800—970	"	1,1	59 500	[805]	1968	
Al → Ni	1100—1305	"	1,87	136 000	[805]	1968	
Si → Ti	800—1000	TiSi, TiSi$_2$	2,99	5 216	[517]	1959	
Si → β-Ti	900—1200	TiSi$_2$	$8,1 \cdot 10^{-2}$	39 700	[518]	1959	
Si → Zr	1000—1200	ZrSi$_2$	$1,1 \cdot 10^{-5}$	55 750	[518]	1959	
Si → V	1000—1200	VSi$_2$	$6,2 \cdot 10^{-5}$	61 200	[518]	1959	
Si → Nb	700—1500	Nb$_5$Si$_3$	$0,51 \cdot 10^{-2}$	48 000	[520]	1971	
Si → Nb	900—1100	NbSi$_2$	56,1	36 840	[518]	1959	
Si → Ta	900—1200	TaSi$_2$	$36,23 \cdot 10^{-3}$	21 150	[517]	1959	

Process	Temperature range	Phase			Ref.	Year	Note
Si → Ta	800—1200	$TaSi_2$	93,1	34 600	[518]	1959	
Si → Cr	900—1100	$CrSi_2$	3,92	22 760	[518]	1959	
Si → Mo	1200—1500	Mo_5Si_3	447	78 000	[520]	1971	
Si → Mo	900—1100	$MoSi_2$	56,1	36 840	[518]	1959	See [519]
Si → W	900—1100	WSi_2	$4,4 \cdot 10^6$	63 000	[518]	1959	
Si → α-Fe	700—800	Solid solution	2,52	1 550	[517]	1959	
Si → γ-Fe	900—1100	$FeSi_2$	$14,55 \cdot 10^3$	22 030	[517]	1959	
Si → Co	—	$CoSi_2$	—	13 090	[519]	1957	
Si → Ni	—	Solid solution	10,6	64 800	[805]	1968	
Si → Ni	—	$NiSi_2$	—	24 950	[519]	1957	
P → α-Fe	850—1040	Solid solution	2,9	55 000	[819]	1963	
P → γ-Fe	1280—1350	"	28,3	70 000	[819]	1963	
S → α-Fe	750—900	—	2,7	49 000	[521]	1970	See [805]
S → γ-Fe	935—1025	—	0,5	50 000	[521]	1970	
S → Ni	800—1000	—	0,2	46 000	[521]	1970	
B → C	1940—2400	Solid solution	3,02	57 250	[522]	1970	
B → C	1700—2400	"	$6,32 \cdot 10^3$	157 000	[805]	1968	
B → C	1700—2400	"	7,1	153 000	[805]	1968	
N → B	600—1200	BN	$30,1 \cdot 10^{-3}$	61 300	[523]	1959	
N → B	1200—1500	BN	$20,3 \cdot 10^{-5}$	4 000	[523]	1959	
N → Al	530—625	Solid solution	$4,2 \cdot 10^{10}$	23 700	[805]	1968	$\perp c$
C → Si	1050—1400	"	1,9	3 100	[805]	1968	$\parallel c$

*G. I. Zhunkhovskii, Investigation of Diffusion Saturation Processes of Refractory Metals by Carbon and Boron in a Vacuum [in Russian], Cand. Dissertation, Kiev (1969).

**A. N. Svobodov, Relaxation Characteristics — The Structure and Properties of Iron with Boron in a Wide Concentration Range [in Russian], Author's abstract, Cand. Dissertation, Tula (1968).

***T. S. Verkhoglyadova, Technology for Producing, and Some Physical Properties of, Transition Metal Nitrides [in Russian], Author's abstract, Cand. Dissertation, Kiev (1962).

SELF-DIFFUSION PARAMETERS

System	Temp., °C	Pre-exponential factor, D_0, cm²/sec	Activation energy Q, cal/mole	Source	Year	Notes
1	2	3	4	5	6	7
UC ← C	1065—1500	0.1	62 500	[820]	1969	
UC ← C	1266—1684	1.75	63 000	[821]	1969	As C content rises D_0 rises and Q falls
UC ← U	1505—1863	8.47	104 000	[821]	1969	
UC ← U	1450—2000	$7.5 \cdot 10^{-5}$	81 000	[822]	1968	Fused
UC ← U	1500—1775	$3.18 \cdot 10^{-2}$	95 400	[823]	1971	Sintered
UC ← U	1500—1775	26.1	118 500	[823]	1971	
$TiC_{0.67}$ ← C	1745—2080	$2.85 \cdot 10^{-4}$	49 600	[824]	1968	At 2080°C anomalous change in self-diffusion parameters
$TiC_{0.67}$ ← C	2080—2720	$1.44 \cdot 10^{2}$	109 900	[824]	1968	
$TiC_{0.887}$ ← C	1450—2280	45.44	106 800	[825]	1968	See [829]
$TiC_{0.970}$ ← C	1450—2280	6.98	95 300	[825]	1968	
$TiC_{0.970}$ ← C	2000—2100	220	97 700	[826]	1970	
TiC_x ← Ti	1920—2215	$4.36 \cdot 10^{4}$	176 400	[827]	1969	$0.67 \leqslant x \leqslant 0.97$; parameters of self-diffusion of Ti into TiC_x are not dependent on x
TiC ← Ti	—	$1.43 \cdot 10^{4}$	148 300	[828]	1969	
ZrC ← C	1350—2150	$1.32 \cdot 10^{2}$	113 200	[830]	1967	
ZrC_x ← C	1827—2627	0.34	69 000	[831]	1967	$0.56 \leqslant x \leqslant 1.0$; See [829, 832]
$ZrC_{0.85}$ ← C	2200—2800	56.4	124 000	[833]	1969	
ZrC ← Zr	—	$6.67 \cdot 10^{5}$	161 800	[828]	1969	

Reaction	Temperature range	a	Q	Ref.	Year	Remarks
$HfC_{0.97} \leftarrow C$	2200—2800	63	130 300	[833]	1969	
$HfC \leftarrow Hf$	—	$1.48 \cdot 10^{8}$	182 800	[828]	1969	
$V_6C_5 \leftarrow C$	1700—2200	2,65	85 000	[834]	1972	Disordered V_6C_5
$VC_{0.87} \leftarrow C$	—	0,05	67 000	[843]	1967	
$VC_{0.87} \leftarrow V$	—	0,013	95 000	[844]	1966	
$Nb_2C \leftarrow C$	1700—2200	$2,3 \cdot 10^{-3}$	39 900	[508]	1970	
$NbC \leftarrow C$	1700—2200	0,35	74 200	[508]	1970	
$NbC \leftarrow C$	1647—2837	$1,49 \cdot 10^{-2}$	76 600	[835]	1971	
$NbC_{0.97} \leftarrow C$	2200—2800	0,5	102 000	[833]	1969	
$NbC_x \leftarrow C$	1827—2627	0,66	73 000	[831]	1967	$0.64 \leqslant x \leqslant 0.9$; see [829]
$NbC \leftarrow Nb$	—	1,25	107 500	[828]	1969	
$Ta_2C \leftarrow C$	2100—2650	10^{3}	115 000	[836]	1968	
$TaC \leftarrow C$	2100—2650	2,00	90 700	[836]	1968	
$TaC \leftarrow C$	1700—2700	0,18	85 000	[837]	1966	
$TaC_{0.98} \leftarrow C$	2200—2800	3,9	118 700	[833]	1969	
$TaC_x \leftarrow C$	1827—2627	0,089	64 000	[831]	1967	$0.66 \leqslant x \leqslant 0.9$
$TaC \leftarrow Ta$	—	24,6	121 600	[828]	1969	
$Mo_2C \leftarrow C$	1000—1775	810	83 000	[838]	1968	Study of contact interaction
$Mo_2C \leftarrow C$	1470—1727	80	91 500	[838]	1968	
$Mo_2C \leftarrow C$	1180—1730	300	95 000	[839]	1971	
$W_2C \leftarrow C$	1500—1850	2	85 000	[839]	1971	
$WC \leftarrow C$	1965—2370	$1,9 \cdot 10^{-6}$	88 000	[840]	1971	
$ZrN_{0.78} \leftarrow N$	1600—2200	0,75	78 300	[841]	1969	
$ZrN_{0.95} \leftarrow C$	2030—2690	$1,59 \cdot 10^{-3}$	92 100	[842]	1972	
$SiC \leftarrow C$	—	300	141 500	[845]	1966	p-type; along [0001]
$SiC \leftarrow C$	—	$2 \cdot 10^{17}$	302 400	[845]	1966	n-type; along [0001]

Chapter III

ELECTRICAL AND MAGNETIC PROPERTIES

ELECTRICAL CONDUCTIVITY

Phase	Resistivity, $\mu\Omega \cdot cm$	Temp., °C	Specific electric conductivity, $\Omega^{-1} \cdot cm^{-1}$	Source	Year	Notes
1	2	3	4	5	6	7
$ZrBe_{13}$	16,4	27	61 000	[2]	1966	
$NbBe_{12}$	55,6	27	18 000	[2]	1966	
$TaBe_{12}$	43,5	27	23 000	[2]	1966	
$MoBe_{12}$	20	—	50 000	[10]	1973	
$MoBe_{22}$	30	20	33 300	[2]	1966	
Be_5B	30	20	33 300	[562]	1971	
Be_2B	$12 \cdot 10^2$	20	834	[562]	1971	
BeB_2	$2 \cdot 10^4$	20	50	[562]	1971	
BeB_4	$11 \cdot 10^{11}$	20	$91 \cdot 10^{-8}$	[562]	1971	
BeB_6	$11 \cdot 10^{12}$	20	$91 \cdot 10^{-9}$	[562]	1971	
BeB_9	$60 \cdot 10^{12}$	20	$16,6 \cdot 10^{-9}$	[562]	1971	
CaB_6	222,0	20	4 500	[563]	1961	See [564]
SrB_6	111	20	9 000	[563]	1961	
BaB_6	77	20	13 000	[563]	1961	See [564]
ScB_2	7—15	20	143 000— 67 000	[270]	1960	
YB_2	39	—	25 600	[3]	1975	
YB_4	28,5	—	35 100	[3]	1975	
YB_6	40,0	20	25 000	[563]	1961	
YB_{12}	94,8	—	10 550	[3]	1975	
LaB_4	24 ± 12	20	41 800	[565]	1961	
LaB_4	~ 12	—190	$\sim83\ 200$	[565]	1961	
LaB_6	15,0	20	66 700	[563]	1961	See [564]
CeB_6	29,4	20	34 000	[563]	1961	
PrB_4	40,3	27	24 800	—*	1974	
PrB_6	19,5	20	51 300	[563]	1961	See [564]
NdB_4	39,2	27	25 500	—*	1974	
NdB_6	20,0	20	50 000	[563]	1961	See [564]
SmB_6	207	20	4 830	[563]	1961	See [564, 566, 567]
EuB_6	84,7	20	11 800	[563]	1961	See [252]
GdB_4	31,1	27	32 200	—*	1974	
GdB_6	44,7	20	22 400	[563]	1961	
TbB_4	32,0	27	31 200	—*	1974	
TbB_6	37,4	20	26 700	[563]	1961	

*E. N. Severyanina, Production and Study of the Physical Properties of Rare-Earth Metal Tetraborides [in Russian], Author's abstract, Cand. Dissertation, Kiev (1974).

Phase	Resistivity, $\mu\Omega \cdot cm$	Temp., °C	Specific electric conductivity, $\Omega^{-1} \cdot cm^{-1}$	Source	Year	Notes
1	2	3	4	5	6	7
TbB_{12}	12.0 ± 0.5	20	83 200	—**	1970	
DyB_4	35.2	27	28 400	—*	1974	
DyB_{12}	14.4 ± 0.6	20	69 400	—**	1970	
HoB_4	30.0	27	33 300	—*	1974	
HoB_{12}	14.7 ± 0.6	20	68 000	—**	1970	
ErB_4	49.5	27	20 200	—*	1974	
ErB_{12}	16.1 ± 0.7	20	62 000	—**	1970	
TmB_4	34.7	27	28 800	—*	1974	
TmB_{12}	17.0 ± 0.7	20	58 800	—**	1970	
YbB_6	46.6	20	21 500	[563]	1961	See [252, 564]
YbB_{12}	185	20	5 400	[59]	1971	
LuB_{12}	13.6 ± 0.6	20	73 500	—**	1970	
ThB_6	14.8	20	67 600	[563]	1961	
UB_4	98	—	10 200	[3]	1975	
UB_{12}	23.0 ± 1.0	—	43 500	[458]	1971	
TiB	40	20	25 000	[291]	1968	
TiB_2	9.0	25	111 000	[459]	1971	
	29—87	—	34 500—11 500	[485]	1973	Monocrystal; obtained by gas-phase method
	6.4—9.1	20	156 000—111 000	[485]	1973	Monocrystal; obtained by Verneuil method
	114	—	8 770	[485]	1973	Monocrystal; obtained by Czochralski method
ZrB_2	9,7	25	103 000	[459]	1971	
	32,2	1027	31 100	[459]	1971	
	13	—	77 000	[485]	1973	Monocrystal; obtained by Czochralski method
ZrB_{12}	22.0 ± 0.9	20	45 500	—**	1970	
HfB_2	10,6	25	94 400	[459]	1971	
	45,8	1027	21 800	[459]	1971	
VB	35	20	28 600	[291]	1968	

*E. N. Severyanina, see footnote on p. 224.

**V. V. Odintsov, Production and Physical Properties of Metal Dodecaborides with the UB_{12} Structure [in Russian], Author's abstract, Cand. Dissertation, Kiev (1970).

ELECTRICAL CONDUCTIVITY (continued)

Phase	Resistivity, $\mu\Omega \cdot cm$	Temp., °C	Specific electric conductivity, $\Omega^{-1} \cdot cm^{-1}$	Source	Year	Notes
1	2	3	4	5	6	7
VB_2	22.7	20	44 000	[459]	1971	
	70,8	1 027	14 100	[459]	1971	
Nb_3B_2	45,0	27	22 200	[3]	1975	
NbB	40	27	25 000	[3]	1975	
Nb_3B_4	33,5	27	29 850	[3]	1975	
NbB_2	25,7	27	38 900	[459]	1971	
	52,7	1027	19 000	[459]	1971	
TaB	100	20	10 000	[291]	1968	
TaB_2	32,5	27	30 800	[459]	1971	
	75,3	1027	13 300	[459]	1971	
Cr_2B	107	27	9 350	[3]	1975	See [431]
Cr_5B_3	49	27	20 400	[3]	1975	
CrB	45,5	27	22 000	[3]	1975	See [431]
Cr_3B_4	60	27	16 670	[3]	1975	
CrB_2	30	27	33 300	[3]	1975	See [431]
Mo_2B	40	20	25 000	[291]	1968	
α-MoB	45	20	22 250	[291]	1968	
β-MoB	25	20	40 000	[291]	1968	
MoB_2	45	20	22 250	[291]	1968	
Mo_2B_5	26,0	—	38 500	[3]	1975	
W_2B_5	22,0	—	45 500	[3]	1975	
Mn_4B	85±5	20	11 760	[433]	1968	
Mn_2B	40±5	20	25 000	[433]	1968	
MnB	57±8	20	17 300	[433]	1968	
Mn_3B_4	62±5	20	16 000	[433]	1968	
MnB_2	71±5	20	14 100	[433]	1968	
Fe_2B	38	20	26 300	[568]	1972	See [434]
FeB	80	20	12 500	[568]	1972	See [434, 569]
Co_3B	28	20	35 700	[568]	1972	
Co_2B	33	20	30 300	[568]	1972	
CoB	76	20	13 150	[568]	1972	See [569]
Ni_3B	21	—	47 600	[435]	1972	
Ni_2B	14	—	71 500	[435]	1972	
NiB	50	—	20 000	[435]	1972	
Rh_7B_3	88	20	11 350	[570]	1971	
$RhB_{\sim 1,1}$	880	20	1 135	[570]	1971	
Pd_3B	10	20	100 000	[571]	1972	
Pd_5B_2	26	20	38 400	[571]	1972	
Be_2C	$1.1 \cdot 10^6$	20	0,98	[11]	1968	
ScC	274	20	3 650	[11]	1968	
YC	$4,54 \cdot 10^2$	20	22,2	[11]	1968	
Y_2C	446±13	20	2 240	[573]	1970	
YC_2	30±2	20	33 300	[573]	1970	
La_2C_3	340±10	20	2 940	[573]	1970	
LaC_2	45±2	20	22 200	[573]	1970	
Ce_2C_3	398±12	20	25 100	[573]	1970	

Phase	Resistivity, $\mu\Omega \cdot cm$	Temp., °C	Specific electric conductivity, $\Omega^{-1} \cdot cm^{-1}$	Source	Year	Notes
1	2	3	4	5	6	7
CeC_2	60 ± 2	20	16 670	[573]	1970	
Pr_2C_3	396	20	25 200	[573]	1966	
PrC_2	36 ± 3	20	27 800	[573]	1970	
Nd_2C_3	322 ± 10	20	3 100	[573]	1970	
NdC_2	49 ± 5	20	20 400	[573]	1970	
SmC_2	34 ± 1	20	29 400	[573]	1970	
GdC_2	35 ± 5	20	28 600	[573]	1970	
TbC_2	35 ± 1	20	28 600	[573]	1970	
DyC_2	32 ± 2	20	31 200	[573]	1970	
ErC_2	57 ± 10	20	17 540	[573]	1970	
TmC_2	451 ± 15	20	2 220	[573]	1970	
ThC	25	20	40 000	[574]	1961	See [369]
ThC_2	30	20	33 300	[574]	1961	
UC	47	20	21 300	[369]	1964	See [575]
U_2C_3	~250	27	~4 000	[576]	1965	
PuC	~250	20	~4 000	[369]	1964	
$TiC_{0.96}$	61 ± 5	25	16 400	—*	1971	
$TiC_{0.89}$	78 ± 5	25	12 800	—*	1971	
$TiC_{0.79}$	101 ± 5	25	9 900	—*	1971	
$TiC_{0.72}$	120 ± 5	25	8 330	—*	1971	
$TiC_{0.62}$	147 ± 5	25	6 800	—*	1971	See [438, 440, 577, 578]
$ZrC_{0.97}$	49 ± 5	25	20 400	—*	1971	
$ZrC_{0.90}$	96 ± 5	25	10 400	—*	1971	
$ZrC_{0.80}$	131 ± 5	25	7 630	—*	1971	
$ZrC_{0.75}$	146 ± 5	25	6 850	—*	1971	
$ZrC_{0.70}$	155 ± 5	25	6 450	—*	1971	
ZrC	~170	1 000	~5 890	[579]	1966	
	~235	2 800	~4 250	[579]	1966	
$HfC_{0.98}$	39 ± 5	25	25 600	—*	1971	
$HfC_{0.91}$	81 ± 5	25	12 350	—*	1971	
$HfC_{0.80}$	124 ± 5	25	8 060	—*	1971	
$HfC_{0.71}$	148 ± 5	25	6 750	—*	1971	
$HfC_{0.62}$	159 ± 5	25	6 280	—*	1971	
$VC_{0.41}$	182	20	5 500	[466]	1975	
$VC_{0.42}$	174	20	5 750	[466]	1975	
$VC_{0.44}$	165	20	6 050	[466]	1975	Hex.
$VC_{0.46}$	152	20	6 580	[466]	1975	
$VC_{0.49}$	139	20	7 200	[466]	1975	
$VC_{0.72}$	146 ± 5	25	6 850	[465]	1974	Cub.; see [443, 578, 580, 581]
$VC_{0.77}$	136 ± 5	25	7 350	[465]	1974	
$VC_{0.82}$	102 ± 5	25	9 800	[465]	1974	
$VC_{0.87}$	78 ± 5	25	12 800	[465]	1974	

*V. Ya. Naumenko, Technology for Obtaining and Studying Some Properties of Transition Metal Carbides in Groups IV-V in Their Homogeneity Regions [in Russian], Author's abstract, Cand. Dissertation, Kiev (1971).

Phase	Resistivity, $\mu\Omega \cdot cm$	Temp., °C	Specific electric conductivity, $\Omega^{-1} \cdot cm^{-1}$	Source	Year	Notes
1	2	3	4	5	6	7
V_8C_7	~115	1100	~8 700	[290]	1972	At 1123°C the jump in ρ is connected with disorientation
	~135	1150	~7 400	[290]	1972	
V_6C_5	~146	1270	~6 850	[290]	1972	At 1271°C the jump in ρ is due to ordering
	~153	1280	~6 540	[290]	1972	
$NbC_{0.42}$	162	20	6 170	[466]	1975	Hex.
$NbC_{0.45}$	158	20	6 330	[466]	1975	
$NbC_{0.47}$	154	20	6 500	[466]	1975	
$NbC_{0.49}$	148	20	6 750	[466]	1975	
$NbC_{0.50}$	144	20	6 950	[466]	1975	
$NbC_{0.72}$	144±5	25	6 950	—*	1971	Cub.; see [438, 578, 580, 581]
$NbC_{0.76}$	140±5	25	7 150	—*	1971	
$NbC_{0.82}$	120±5	25	8 340	—*	1971	
$NbC_{0.92}$	77±5	25	13 000	—*	1971	
$NbC_{0.99}$	44±5	25	22 700	—*	1971	
NbC	~115	1200	~8 700	[579]	1966	
	~175	2800	~5 710	[579]	1966	
Ta_2C	49	20	20 400	[581]	1968	
$TaC_{0.73}$	155±5	25	6 450	—*	1971	See [578, 580, 581]
$TaC_{0.82}$	135±5	25	7 400	—*	1971	
$TaC_{0.85}$	128±5	25	7 810	—*	1971	
$TaC_{0.91}$	80±5	25	12 500	—*	1971	
$TaC_{0.99}$	22±5	25	45 500	—*	1971	
TaC	~100	1200	~10 000	[579]	1966	
	~145	3100	~6 900	[579]	1966	
$Cr_{23}C_6$	127±2	20	7 880	[431]	1962	
Cr_7C_3	109±4	20	9 180	[431]	1962	
Cr_3C_2	75±5	20	13 330	[431]	1962	
Mo_2C	71	20	14 100	[582]	1960	
Mo_2C	279—331	—	3580—3020	[485]	1973	Monocrystal
MoC	49	20	20 400	[11]	1968	
W_2C	75,7±0,1	20	13 200	[582]	1960	
WC	19,2±0,3	20	52 200	[582]	1960	
α-WC	17	27	58 800	[583]	1971	Monocrystal
Ca_2N	$5 \cdot 10^6$	—	0,2	[115]	1972	
Sr_2N	$2 \cdot 10^5$	—	5	[115]	1972	
ScN	25,4	25	39 400	[584]	1963	
YN	93	25	10 750	[585]	1963	
LaN	100	25	10 000	[585]	1963	

*V. Ya. Naumenko, see footnote on p. 227.

ELECTRICAL CONDUCTIVITY (continued)

Phase	Resistivity, $\mu\Omega \cdot cm$	Temp., °C	Specific electric conductivity, $\Omega^{-1} \cdot cm^{-1}$	Source	Year	Notes
1	2	3	4	5	6	7
CeN	17	25	58 800	[586]	1963	See [587]
PrN	110	25	9 100	[585]	1963	
NdN	75	25	13 330	[586]	1963	See [585, 492]
SmN	$\sim$120	25	$\sim$8 330	[586]	1963	
EuN	$\sim$120	25	$\sim$8 330	[586]	1963	
GdN	$\sim$200	25	$\sim$5 000	[586]	1963	
TbN	$\sim$200	25	$\sim$5 000	[586]	1963	
DyN	100	25	10 000	[585]	1963	
HoN	110	25	9 100	[585]	1963	
ErN	79	25	12 650	[585]	1963	
TmN	180	25	5 550	[586]	1963	
YbN	$9 \cdot 10^3$	25	111	[586]	1963	
LuN	360	25	2 780	[586]	1963	
ThN	—	—	—	[364]	1972	Metal-type conductivity
UN	183	22	5 460	[575]	1972	
NpN	85	4K	11 750	[588]	1968	
	$\sim$430	82K	$\sim$2 380	[588]	1968	Curie point
	380	647	2 630	[588]	1968	
$TiN_{0.79}$	85	27	11 750	—*	1974	
$TiN_{0.83}$	78	27	12 800	—*	1974	
$TiN_{0.87}$	70	27	14 300	—*	1974	
$TiN_{0.97}$	40	27	25 000	—*	1974	
$ZrN_{0.82}$	30	27	33 330	—*	1974	
$ZrN_{0.85}$	28	27	35 700	—*	1974	
$ZrN_{0.92}$	20	27	50 000	—*	1974	
$ZrN_{0.97}$	18	27	55 500	—*	1974	
HfN	32	27	31 300	—*	1974	
V_3N	123.0 ± 10	20	8 140	[444]	1962	
$VN_{0.71}$	127	27	7 880	—*	1974	
$VN_{0.80}$	96	27	10 400	—*	1974	
$VN_{0.85}$	81	27	12 350	—*	1974	
$VN_{0.87}$	77	27	13 000	—*	1974	
$VN_{0.93}$	66	27	15 150	—*	1974	
$VN_{0.96}$	60	27	16 660	—*	1974	
Nb_2N	142.0	20	7 042	[14]	1969	
$NbN_{0.75}$	109	27	9 180	—*	1974	
$NbN_{0.80}$	92	27	10 900	—*	1974	
$NbN_{0.89}$	72	27	13 900	—*	1974	
$NbN_{0.95}$	65	27	15 400	—*	1974	
NbN	54	27	18 500	—*	1974	
Ta_2N	263.0	20	3 800	[14]	1969	

*L. K. Shvedova, Investigation of the Physical Properties of Nitrides in Groups IV- V Transition Metals [in Russian], Author's abstract, Cand. Dissertation, Kiev (1974).

ELECTRICAL CONDUCTIVITY (continued)

Phase	Resistivity, $\mu\Omega \cdot cm$	Temp., °C	Specific electric conductivity, $\Omega^{-1} \cdot cm^{-1}$	Source	Year	Notes
1	2	3	4	5	6	7
TaN	198	27	5 050	—*	1974	
Ta_3N_5	$\sim 10^{10}$	—	$\sim 10^{-4}$	[16]	1969	
Cr_2N	84±5	20	11 900	[431]	1962	
CrN	640±40	20	1 562	[431]	1962	
Mo_2N	19,8	20	50 500	[14]	1969	
Ni_3N	$2,8 \cdot 10^3$	25	357.1	[589]	1956	
TiAl	54	—	18 500	[470]	1962	Sintered
	48,5	—	20 600	[470]	1962	Fused
$TiAl_3$	17—22	—	58 800—45 500	[470]	1962	Sintered
	16,5	—	60 500	[470]	1962	Fused
Zr_3Al	93	20	10 750	[18]	1965	
Zr_2Al	80	20	12 500	[18]	1965	
Zr_3Al_2	85	20	11 750	[18]	1965	
Zr_4Al_3	105,6	20	9 470	[18]	1965	
ZrAl	105	20	9 530	[18]	1965	
$ZrAl_3$	17	20	58 900	[18]	1965	
$HfAl_3$	16,8	20	59 500	[18]	1965	
V_5Al_8	254	—	3 940	[590]	1964	
VAl_3	71	—	14 100	[590]	1964	
VAl_6	40,3	—	24 800	[590]	1964	
VAl_{11}	45,3	—	22 100	[590]	1964	
Nb_3Al	103,1	—	9 700	[149]	1964	
Nb_2Al	137,9	—	7 250	[149]	1964	
$NbAl_3$	49	—	20 400	[149]	1964	
Ta_3Al	489	—	2 045	[149]	1964	
Ta_2Al	462	—	2 160	[149]	1964	
TaAl	27	—	37 000	[10]	1973	
$TaAl_3$	141,9	—	7 050	[149]	1964	
$CrAl_7$	40	20	25 000	[18]	1965	
Mo_3Al	45.9	20	21 800	[18]	1965	
$MoAl_2$	241	20	4 150	[18]	1965	
$MoAl_3$	640	20	1 560	[18]	1965	
$MnAl_6$	63	20	15 900	[18]	1965	
Fe_3Al	80	—	12 500	[18]	1965	
CoAl	16,6	—	60 300	[10]	1973	
β-NiAl	8,4	25	119 000	[591]	1969	
Ni_3Al	37,4	20	26 750	[18]	1965	
$NiAl_3$	20	20	50 000	[18]	1965	
Mg_2Si	—	—	—	[260]	1959	Semiconductor
$BaSi_2$	$38 \cdot 10^4$	20	2,635	[592]	1960	
Sc_5Si_3	143,0	20	7 000	—**	1975	
ScSi	30	20	33 330	[297]	1970	
$ScSi_{1.66}$	40,6	20	24 600	—**	1975	

*L. K. Shvedova, see footnote on p. 229.

**V. I. Lazorenko, Investigation of the Physical Properties of Rare-Earth Metal Silicides of the Cerium Subgroup [in Russian], Author's abstract, Cand. Dissertation, Kiev (1975).

Phase	Resistivity, $\mu\Omega \cdot cm$	Temp., °C	Specific electric conductivity, $\Omega^{-1} \cdot cm^{-1}$	Source	Year	Notes
1	2	3	4	5	6	7
Y_5Si_3	102,0	20	9 800	[472]	1974	
YSi	51,0	20	19 600	—*	1975	
YSi_2	69,0	20	14 500	[472]	1974	
La_5Si_3	365,0	20	2 740	—*	1975	
LaSi	340,0	20	2 940	[593]	1973	
$LaSi_2$	350,0	20	2 860	[297]	1970	
Ce_5Si_3	222,0	20	4 500	—*	1975	
CeSi	127,0	20	7 870	—*	1975	
$CeSi_2$	183,0	20	5 460	—*	1975	
Pr_5Si_3	220,0	20	4 550	—*	1975	
PrSi	102,0	20	9 800	—*	1975	
$PrSi_2$	126,0	20	7 930	—*	1975	
$NdSi_2$	349	20	2 870	[405]	1971	
Sm_5Si_3	225,0	20	4 450	—*	1975	
SmSi	205,0	20	4 880	—*	1975	
$SmSi_2$	210,0	20	4 760	—*	1975	
$GdSi_2$	263	20	3 800	[405]	1971	
U_3Si	55	20	18 200	[260]	1959	
Ti_5Si_3	55±4	20	18 200	[592]	1960	
TiSi	63±6	20	15 900	[592]	1960	
$TiSi_{1.93}$	12,8±0,4	20	78 000	[473]	1966	Monocrystal
$TiSi_2$	16,9±0,5	20	59 250	[592]	1960	
ZrSi	49,4	20	20 250	[594]	1958	
$ZrSi_2$	75,8±3,1	20	13 200	[592]	1960	
V_3Si	203,5± ±37,5	20	4 910	[592]	1960	
V_5Si_3	114,5± ±8,5	20	8 780	[592]	1960	
VSi_2	66,5±2,5	20	15 050	[592]	1960	See [451]
$NbSi_2$	50,4±2,3	20	19 900	[592]	1960	
$Ta_{1.5}Si$	174,5	20	5 740	[592]	1960	
Ta_2Si	124	20	8 070	[592]	1960	
Ta_5Si_3	108	20	9 270	[592]	1960	
$TaSi_2$	46,1±1,3	20	21 700	[592]	1960	
Cr_3Si	35±5	20	28 600	[592]	1960	
Cr_3Si_2	80±5	20	12 500	[592]	1960	
Cr_5Si_3	153	20	6 540	[592]	1960	
CrSi	129,5± ±7,5	20	7 730	[592]	1960	
$CrSi_2$	914± ±74,5	20	1 095	[592]	1960	
Mo_3Si	21,6±0,7	20	46 300	[592]	1960	} See [450]
Mo_5Si_3	45,9±1,2	20	21 800	[592]	1960	
$MoSi_2$	21,6±0,9	20	46 300	[592]	1960	

*V. I. Lazorenko, see footnote on p. 230.

Phase	Resistivity, $\mu\Omega \cdot cm$	Temp., °C	Specific electric conductivity, $\Omega^{-1} \cdot cm^{-1}$	Source	Year	Notes
1	2	3	4	5	6	7
W_3Si	93	20	10 760	[592]	1960	
WSi_2	12.5±0.2	20	80 000	[592]	1960	
Mn_3Si	160±3	20	6 250	[592]	1960	
Mn_5Si_3	257±14	20	3 900	[592]	1960	See [595]
$MnSi$	259±12	20	3 860	[592]	1960	
$MnSi_{1.72}$	1073	20	932	—*	1972	[100]
$MnSi_{1.72}$	4550	20	220	—*	1972	[001]
Re_3Si	129	20	7 730	[592]	1960	
$ReSi$	736±36	20	1 360	[592]	1960	
$ReSi_2$	7000± ±1000	20	143	[592]	1960	
Fe_3Si	130±19	20	7 700	[592]	1960	
Fe_5Si_3	199	20	5 020	[22]	1971	
$FeSi$	~260	25	~3 840	[452]	1970	Monocrystal, direction [210]
α-$FeSi_2$	455	20	2 200	—**	1973	
β-$FeSi_2$	$6.67 \cdot 10^5$	20	1.5	—**	1973	
Co_3Si	129±9	20	7 760	[592]	1960	
Co_2Si	66.2	—	15 100	[448]	1958	
$CoSi$	86±15.5	20	11 620	[592]	1960	
$CoSi_2$	~18	27	~55 500	[453]	1970	Monocrystal, direction [111]
	~14	27	~71 400	[453]	1970	Polycrystal
Ni_3Si	93±7.5	20	10 770	[592]	1960	
Ni_5Si_2	149.5	20	6 700	[448]	1958	
Ni_3Si_2	79±7	20	12 660	[592]	1960	
Ni_2Si	20	20	50 000	[596]	1963	
$NiSi$	20.2	20	49 500	[448]	1958	
$Ni_{1.04}Si_{1.93}$	35.8	20	28 000	[22]	1971	
PrP	$\sim15 \cdot 10^3$	20	~66.7	[597]	1969	
SmP	~220	20	~4 550	[598]	1970	
GdP	$5 \cdot 10^3$	20	200	[478]	1968	
TbP	$4 \cdot 10^3$	20	250	[478]	1968	
DyP	$8 \cdot 10^3$	20	125	[478]	1968	
UP_2	~205	20	~4 880	[599]	1969	Monocrystal, $\perp c$
	~730	20	1 370	[599]	1969	Monocrystal, $\parallel c$

*L. D. Ivanova, Investigation of Semiconductor Materials Based on Higher Manganese Silicide [in Russian], Author's abstract, Cand. Dissertation, Moscow (1972).

**T. V. Ivanova, On the Problem of the Physicochemical Nature of Iron Disilicide [in Russian], Author's abstract, Cand. Dissertation, Voronezhe (1973).

Phase	Resistivity, $\mu\Omega \cdot$ cm	Temp., °C	Specific electric conductivity, $\Omega^{-1} \cdot$ cm^{-1}	Source	Year	Notes
1	2	3	4	5	6	7
	$\cdot \sim 380$	20	$\sim 2\,030$	[599]	1969	Polycrystal
Ti_3P	160	20	6 250	—[*1]	1966	
Ti_2P	180	20	5 550	—[*1]	1966	
Ti_5P_3	145	20	6 900	—	1966	
Ti_4P_3	120	20	8 330	—[*1]	1966	
TiP	60	20	16 670	—[*1]	1966	
FeP_2	$\sim 0.04 \times \\ \times 10^6$	25	~ 25	[600]	1971	Monocrystal
Ni_3P	48.5	20	20 600	—[*2]	1969	
$Ni_{12}P_5$	61.5	20	16 250	—[*2]	1969	
Ni_2P	32.0	20	31 200	—[*2]	1969	
LaS	$38.4 \cdot 10^8$	20	$2,6 \cdot 10^{-4}$	[222]	1964	
La_3S_4	$24 \cdot 10^4$	20	4,17	[221]	1956	
La_2S_3	$2 \cdot 10^6$	20	0,5	—[*3]	1963	
CeS	$22,2 \cdot 10^8$	20	$4,5 \cdot 10^{-4}$	[222]	1964	
Ce_3S_4	$58 \cdot 10^4$	20	1,725	[221]	1956	
Ce_2S_3	$1,19 \cdot 10^6$	20	0,84	—[*3]	1963	
PrS	$60 \cdot 10^8$	20	$1,67 \cdot 10^{-4}$	—[*4]	1968	
Pr_2S_3	$1,1 \cdot 10^{12}$	20	$9,1 \cdot 10^{-7}$	—[*3]	1963	
NdS	$45 \cdot 10^8$	20	$2,22 \cdot 10^{-4}$	—[*4]	1968	
Nd_3S_4	$1,2 \cdot 10^6$	20	0,835	[221]	1956	
Nd_2S_3	$0,7 \cdot 10^{12}$	20	$14,3 \cdot 10^{-7}$	[601]	1970	
Sm_3S_4	$66,4 \cdot 10^6$	20	0,015	[221]	1956	
Sm_2S_3	$7 \cdot 10^{13}$	27	$14,3 \cdot 10^{-9}$	[262]	1965	
EuS	$\sim 10^{13}$	20	$\sim 10^{-7}$	[602]	1965	
GdS	—	—	—	[603]	1968	Metal-type conductivity
$\alpha\text{-}Gd_2S_3$	$5 \cdot 10^8$	20	$2 \cdot 10^{-3}$	[604]	1970	Monocrystal
$\gamma\text{-}Gd_2S_3$	—	—	—	[603]	1968	Polycrystal
YbS	$\sim 10^{10}$	20	$\sim 10^{-4}$	[605]	1970	
ThS	$20 \cdot 10^4$	20	5,0	[33]	1972	
Th_2S_3	10^7	20	10^{-1}	[33]	1972	
Th_4S_7	$25 \cdot 10^9$	20	$4 \cdot 10^{-5}$	[33]	1972	
ThS_2	10^{16}	20	10^{-10}	[33]	1972	
US	530	20	1 885	[606]	1974	Sintered
	200	20	5 000	[606]	1974	Fused

[*1] V. E. Listovnichii, see footnote on p. 206.

[*2] P. A. Vityaz, Investigation of Nickel Phosphides and the Possibilities of Using Them for Alloying Iron-Based Cermets [in Russian], Author's abstract, Cand. Dissertation, Minsk (1969).

[*3] V. I. Marchenko, Physical Properties of Rare-Earth Metal Sulfides of the Cerium Group [in Russian], Author's abstract, Cand. Dissertation, Kiev (1963).

[*4] E. V. Goncharova, Electrical and Other Properties of Monocompounds of Rare-Earth Metals with Elements in Groups V and VI [in Russian], Author's abstract, Cand. Dissertation, Leningrad (1968).

Phase	Resistivity, $\mu\Omega \cdot$ cm	Temp., °C	Specific electric conductivity, $\Omega^{-1} \cdot$ cm^{-1}	Source	Year	Notes
1	2	3	4	5	6	7
U_2S_3	$1.2 \cdot 10^{10}$	20	$8.33 \cdot 10^{-5}$	[606]	1974	
α-US_2	$1.5 \cdot 10^{10}$	20	$6.67 \cdot 10^{-5}$	[606]	1974	
$US_{2.33}$	$0.2 \cdot 10^6$	20	5	[606]	1974	
PuS	2600	25	385	[607]	1969	
Pu_3S_4	$0.3 \cdot 10^6$	25	3.33	[607]	1969	
Ti_5S	130	20	7 700	—*	1966	
Ti_3S	150	20	6 670	—*	1966	
Ti_2S	300	20	3 330	[264]	1964	
δ-TiS_{1-x}	250	20	4 000	—*	1966	
γ-$Ti_{1-x}S$	860	20	1 165	—*	1966	
Ti_2S_3	1600	20	625	[264]	1964	
TiS_2	8000	20	125	[264]	1964	
TiS_3	$6 \cdot 10^6$	20	0.166	[264]	1964	
ZrS_2	10^7	20	0.1	—**	1966	
HfS_2	—	—	—	[237]	1967	Semiconductor
V_3S_4	—	—	—	[608]	1967	Metal-type conductivity
$NbS_{1.6}$	$5 \cdot 10^3$	20	200	[243]	1966	
α-TaS_2	$8 \cdot 10^3$	20	125	—**	1966	
Cr_7S_8	$\sim 10^3$	(−150)— —(+600)	$\sim 10^3$	[244]	1967	
Cr_5S_6	$\sim 10^3$	(−150)— —(+600)	$\sim 10^3$	[244]	1967	
Cr_3S_4	$\sim 10^3$	(−150)— —(+600)	$\sim 10^3$	[244]	1967	
Cr_2S_3	$\sim 10^4$	(−150)— —(+500)	$\sim 10^2$	[244]	1967	Trigon.
Cr_2S_3	$\sim 2 \cdot 10^6$— 10^5	(−150)— —(+600)	~ 0.5—10	[244]	1967	Rhomohed.
MoS_2	$6.2 \cdot 10^4$	20	16.1	—**	1966	
FeS	$1.3 \cdot 10^3$	20	$7.7 \cdot 10^2$	—**	1966	
FeS_2	—	—	—	[609]	1967	Semiconductor (pyrites, marcasite)
Co_3S_4	—	—	—	[608]	1967	Metal-type conductivity
B_4C	10^6	20	1	[610]	1961	
B_4C	$38 \cdot 10^3$	600	26.4	[610]	1961	
B_4C	$30 \cdot 10^3$	1000	33.3	[610]	1961	
B_4C	$22 \cdot 10^3$	2000	44.5	[610]	1961	

*V. E. Listovnichii, see footnote on p. 206.

**V. Kh. Oganesyan, Investigation of the Physical Properties of Transition Metal Sulfides in Groups IV-V [in Russian], Cand. Dissertation, Kiev (1966).

Phase	Resistivity, $\mu\Omega \cdot cm$	Temp., °C	Specific electric conductivity, $\Omega^{-1} \cdot cm^{-1}$	Source	Year	Notes
1	2	3	4	5	6	7
SiC	$>0.13 \times \times 10^6$	25	<7.7	[611]	1961	
SiC	$>0,05 \times \times 10^6$	1100	<20	[611]	1961	
α-BN	$1,7 \cdot 10^{19}$	25	$5,9 \cdot 10^{-14}$	[14]	1969	
α-BN	$2,3 \cdot 10^{16}$	500	$4,35 \cdot 10^{-11}$	[14]	1969	
α-BN	$3,1 \cdot 10^{10}$	1000	$3,23 \cdot 10^{-5}$	[14]	1969	
α-BN	$6 \cdot 10^8$	1500	$1,67 \cdot 10^{-3}$	[14]	1969	
β-BN	$2 \cdot 10^8 - 10^9$	25	$5 \cdot 10^{-3} - 10^{-3}$	[14]	1969	Borazon p-type
β-BN	$10^{11} - 10^{15}$	25	$10^{-5} - 10^{-9}$	[14]	1969	Borazon n-type
Si_3N_4	$10^{19} - 10^{20}$	20	$10^{-13} - 10^{-14}$	[612]	1960	
Si_3N_4	10^{15}	350	10^{-9}	[612]	1960	
Si_3N_4	$5 \cdot 10^{12}$	600	$2 \cdot 10^{-7}$	[612]	1960	
Si_3N_4	$2 \cdot 10^9$	1000	$5 \cdot 10^{-4}$	[612]	1960	
AlN	$>10^{19}$	20	$<10^{-13}$	[249]	1964	
AlN	$2,25 \cdot 10^{17}$	400	$4,45 \cdot 10^{-12}$	[249]	1964	
AlN	10^{15}	500	10^{-9}	[249]	1964	
AlN	$8 \cdot 10^{13}$	600	$1,25 \cdot 10^{-8}$	[249]	1964	Specimens
AlN	$4 \cdot 10^{12}$	800	$2,5 \cdot 10^{-7}$	[249]	1964	contain
AlN	$7 \cdot 10^{11}$	900	$1,43 \cdot 10^{-6}$	[249]	1964	4% C_{free},
AlN	10^{11}	1000	10^{-5}	[249]	1964	$P = 10\text{-}20\%$
AlN	$4 \cdot 10^{10}$	1100	$2,5 \cdot 10^{-5}$	[249]	1964	
AlN	$9 \cdot 10^9$	1200	$1,11 \cdot 10^{-4}$	[249]	1964	

THERMAL COEFFICIENT OF ELECTRICAL RESISTANCE

Phase	Resistance coefficient, $deg^{-1} \cdot 10^3$	Temp. range, °C	Source	Year
1	2	3	4	5
$MoBe_{22}$	$-3,2$	200—1000	[2]	1966
BeB_2	$-0,9$	20—80	[3]	1975
BeB_6	$-0,23$	20—80	[3]	1975
CaB_6	$+1,16$	0—100	[563]	1961
SrB_6	$+0.83$	0—100	[563]	1961
BaB_6	$+1.08$	0—100	[563]	1961
YB_6	$+1,24$	0—100	[563]	1961

Phase	Resistance coefficient, $deg^{-1} \cdot 10^3$	Temp. range, °C	Source	Year
1	2	3	4	5
YB_{12}	$+2,7\pm0,3$	20—900	—*	1970
LaB_6	$+2,68$	0—100	[563]	1961
CeB_6	$+1,00$	0—100	[563]	1961
PrB_6	$+1,92$	0—100	[563]	1961
NdB_6	$+1,93$	0—100	[563]	1961
SmB_6	$-0,42$	0—100	[563]	1961
EuB_6	$+0,90$	0—100	[563]	1961
GdB_6	$+1,40$	0—100	[563]	1961
TbB_6	$+1,31$	0—100	[563]	1961
TbB_{12}	$+2,1\pm0,2$	20—900	—*	1970
DyB_{12}	$+1,8\pm0,2$	20—900	—*	1970
HoB_{12}	$+1,7\pm0,2$	20—900	—*	1970
ErB_{12}	$+2,1\pm0,2$	20—900	—*	1970
TmB_{12}	$+1,6\pm0,2$	20—900	—*	1970
YbB_6	$+2,34$	0—100	[563]	1961
LuB_{12}	$+2,3\pm0,2$	20—900	—*	1970
ThB_6	$+2,31$	0—100	[563]	1961
UB_{12}	$+2,3\pm0,2$	20—900	[458]	1971
TiB_2	$+2,0$	20—1250	[459]	1971
ZrB_2	$+2,3$	20—1250	[459]	1971
ZrB_{12}	$+1,40\pm0.1$	20—900	—*	1970
HfB_2	$+3,3$	20—1250	[459]	1971
VB_2	$+2,1$	20—1250	[459]	1971
Nb_3B_2	2,4	—	[3]	1975
NbB	2,2	—	[3]	1975
Nb_3B_4	1,6	—	[3]	1975
NbB_2	1,0	20—1250	[459]	1971
TaB_2	1,2	20—1250	[459]	1971
Cr_4B	$1,13\pm0,01$	0—100	[431]	1962
Cr_2B	0,6	—	[3]	1975
Cr_5B_3	1,9	—	[3]	1975
CrB	1,1	—	[3]	1975
Cr_3B_4	3,7	—	[3]	1975
CrB_2	1,2	—	[3]	1975
Cr_2B_5	$2,06\pm0,01$	0—100	[431]	1962
Mo_2B_5	$+3,3$	100—1100	[768]	1958
W_2B_5	$+4,26$	100—1100	[768]	1958
Mn_4B	$1,1\pm0,1$	—	[433]	1968
Mn_2B	$1,8\pm0,1$	—	[433]	1968
MnB	$2,2\pm0,2$	—	[433]	1968
Mn_3B_4	$2,3\pm0,3$	—	[433]	1968
MnB_2	$2,4\pm0,2$	—	[433]	1968
Y_2C_3	$0,45\pm0,04$	20—800	[573]	1970
YC_2	$+3,0$	20—800	[573]	1970
La_2C_3	$+0,56$	—	[573]	1970
LaC_2	$+3,9\pm0,4$	20—800	[573]	1970

*V. V. Odintsov, Preparation and Physical Properties of Metal Dodecaborides with the UB_{12} Structure [in Russian], Author's abstract, Cand. Dissertation, Kiev (1970).

THERMAL COEFFICIENT OF ELECTRICAL RESISTANCE (continued)

Phase	Resistance coefficient, $deg^{-1} \cdot 10^3$	Temp. range, °C	Source	Year
1	2	3	4	5
CeC_2	$+2,5\pm0,3$	20—800	[573]	1970
Pr_2C_3	$+0,24$	—	[769]	1966
PrC_2	$+3,9$	20—800	[573]	1970
Nd_2C_3	$+0,26$	—	[769]	1966
NdC_2	$+2,4\pm0,3$	20—800	[573]	1970
SmC_2	$+3,1\pm0,3$	20—800	[573]	1970
GdC_2	$+2,8\pm0,3$	20—800	[573]	1970
TbC_2	$+2,8\pm0,3$	20—800	[573]	1970
DyC_2	$+2,9\pm0,3$	20—800	[573]	1970
ErC_2	$+2,2\pm0,3$	20—800	[573]	1970
TmC_2	$+0,30\pm0,03$	20—800	[573]	1970
$TiC_{0,55}$	$\sim0,25$	20—1000	[770]	1970
$TiC_{0,75}$	$\sim0,45$	20—1000	[770]	1970
$TiC_{0,85}$	$\sim0,75$	20—1000	[770]	1970
TiC	$\sim1,8$	20—1000	[770]	1970
$ZrC_{0,6}$	$\sim0,15$	20—1000	[770]	1970
$ZrC_{0,8}$	$\sim0,5$	20—1000	[770]	1970
$ZrC_{0,9}$	$\sim0,9$	20—1000	[770]	1970
ZrC	$\sim1,55$	20—1000	[770]	1970
HfC	$1,42$	300—2000	[771]	1962
$VC_{0,72}$	$\sim0,098$	25—1000	[465]	1974
$VC_{0,77}$	$\sim0,18$	25—1000	[465]	1974
$VC_{0,82}$	$\sim0,46$	25—1000	[465]	1974
$VC_{0,87}$	$1,41$	25—1000	[465]	1974
$NbC_{0,75}$	$\sim0,2$	20—1000	[770]	1970
$NbC_{0,9}$	$\sim0,8$	20—1000	[770]	1970
$NbC_{1,0}$	$\sim1,35$	20—1000	[770]	1970
TaC	$+1,07$	400—2000	[771]	1962
$Cr_{23}C_6$	$+1,72\pm0,11$	0—100	[431]	1962
Cr_7C_3	$+1,06\pm0,05$	0—100	[431]	1962
Cr_3C_2	$+2,33\pm0,04$	0—100	[431]	1962
Mo_2C	$+3,78$	200—800	[771]	1962
W_2C	$+1,95$	200—2000	[771]	1962
WC	$+0,495$	20—1500	[257]	1974
ScN	$+3,8$	20	[14]	1969
TiN	$+2,48$	100—1100	[768]	1958
$TiN_{0,79}$	$0,59$	$(-196)-(+27)$	—*	1974
$TiN_{0,83}$	$0,64$	$(-196)-(+27)$	—*	1974
$TiN_{0,87}$	$0,74$	$(-196)-(+27)$	—*	1974
$TiN_{0,97}$	$1,4$	$(-196)-(+27)$	—*	1974
$ZrN_{0,82}$	$1,9$	$(-196)-(+27)$	—*	1974
$ZrN_{0,85}$	$1,9$	$(-196)-(+27)$	—*	1974
$ZrN_{0,92}$	$1,5$	$(-196)-(+27)$	—*	1974
$ZrN_{0,97}$	$2,0$	$(-196)-(+27)$	—*	1974
$HfN_{1,00}$	$1,5$	$(-196)-(+27)$	—*	1974
$VN_{0,71}$	$0,07$	$(-196)-(+27)$	—*	1974

*L. K. Shvedova, see footnote on p. 229.

Phase	Resistance coefficient, $deg^{-1} \cdot 10^3$	Temp. range, °C	Source	Year
1	2	3	4	5
$VN_{0.80}$	1.1	$(-196)-(+27)$	—*	1974
$VN_{0.85}$	1.4	$(-196)-(+27)$	—*	1974
$VN_{0.87}$	1.7	$(-196)-(+27)$	—*	1974
$VN_{0.93}$	1.7	$(-196)-(+27)$	—*	1974
$VN_{0.96}$	1.7	$(-196)-(+27)$	—*	1974
$NbN_{0.75}$	1.2	$(-196)-(+27)$	—*	1974
$NbN_{0.80}$	1.4	$(-196)-(+27)$	—*	1974
$NbN_{0.89}$	1.2	$(-196)-(+27)$	—*	1974
$NbN_{0.95}$	1.6	$(-196)-(+27)$	—*	1974
$NbN_{1.00}$	3.1	$(-196)-(+27)$	—*	1974
$TaN_{1.00}$	0.1	$(-196)-(+27)$	—*	1974
TiAl	$+1.49$	20—1000	[470]	1962
$TiAl_3$	$+4.78$	20—1000	[470]	1962
V_4Al_{23}	$+3.01$	—	[18]	1965
Ti_5Si_3	$+0.86$	20—120	[594]	1958
TiSi	$+4.13$	20—120	[594]	1958
$TiSi_2$	$+6.3$	20—2000	[594]	1958
ZrSi	$+3.52$	20—120	[594]	1958
$ZrSi_2$	$+1.30$	20—120	[594]	1958
V_3Si	$+0.563$	20—200	[596]	1963
VSi_2	$+3.51$	20—120	[594]	1958
V_5Si_3	$+1.24$	20—200	[596]	1963
$TaSi_2$	$+3.32$	20—120	[594]	1958
$CrSi_2$	$+2.93$	20—120	[594]	1958
$MoSi_2$	$+6.38$	20—120	[594]	1958
WSi_2	$+2.91$	20—120	[594]	1958
MnSi	0.758	20—200	[596]	1963
Fe_3Si	1.41	20—200	[596]	1963
FeSi	0.511	20—200	[596]	1963
Co_3Si	0.85	20—200	[596]	1963
$CoSi_2$	2.48	20—200	[596]	1963
Ni_3Si	1.80	20—200	[596]	1963
Ni_2Si	3.24	20—200	[596]	1963
SmP	4.1	20—800	[598]	1970
CeS	$+0.5$	0—100	[33]	1972
B_4C	$+0.032$	1000—1450	[610]	1961
SiC	$+0.264$	900—1500	[610]	1961
BN	$-20930/T^2$**	—	[612]	1960
Si_3N_4	$-6570/T^2$	350—700	[612]	1960
	$-22\,670/T^2$	700—1000	[612]	1960

*L. K. Shvedova, see footnote on p. 229.
** Calculated from width of forbidden band.

Phase	Transition temp. into supercond. state, T_k, K	Source	Year	Phase	Transition temp. into supercond. state, T_k, K	Source	Year
1	2	3	4	1	2	3	4
$ThBe_{13}$	<1.15	[355]	1972	PrB_6[*1]	<0.35	[245]	1969
$TiBe_2$	<1.02	[354]	1963	NdB_6[*1]	<0.35	[245]	1969
$TiBe_{12}$	<1.15	[355]	1972	SmB_6	<1,28	[245]	1969
$TiBe_{13}$	<1,15	[355]	1972	EuB_6[*2]	<0.35	[245]	1969
Zr_2Be_{17}	<1,15	[355]	1972	GdB_6[*1]	<0.35	[245]	1969
$ZrBe_{13}$	<1.15	[355]	1972	TbB_6[*1]	<0.35	[245]	1969
$ZrBe_{16}$	<1.15	[355]	1972	DyB_6[*1]	<0.35	[245]	1969
Hf_2Be_{17}	<1.15	[355]	1972	HoB_6[*1]	<0.35	[245]	1969
$HfBe_{13}$	<1.15	[355]	1972	HoB_{12}[*1]	<0.35	[245]	1969
Nb_3Be_2	2,3	[355]	1972	ErB_{12}[*1]	<0.35	[245]	1969
$NbBe_2$	2,15	[355]	1972	TmB_{12}[*1]	<0.35	[245]	1969
$NbBe_3$	<1,15	[355]	1972	YbB_6	<1,28	[245]	1969
Nb_2Be_{17}	1,47	[355]	1972	LuB_{12}	0.48	[245]	1969
Ta_3Be_2	1.0	[355]	1972	ThB	<1,20	[354]	1963
Mo_3Be	<1,15	[355]	1972	ThB_2	<1,77	[354]	1963
$MoBe_{22}$	2.51	[245]	1969	ThB_6	0,74	[245]	1969
W_5Be_{21}	<1.15	[355]	1972	TiB	<1,20	[354]	1963
WBe_{22}	4.12	[245]	1969	TiB_2	<1,28	[354]	1963
$MnBe_{12}$	<1.15	[355]	1972	ZrB [*3]	3,4	[245]	1969
$TcBe$	5.21	[245]	1969	ZrB_2	<1.80	[354]	1963
$ReBe_{22}$	9.65	[245]	1969	ZrB_{12}	6.0	[355]	1972
$FeBe_{11}$	<1.15	[355]	1972	HfB [*3]	3.1	[245]	1969
$CoBe_5$	<1,15	[355]	1972	V_2B	<1,20	[354]	1963
$CoBe_{12}$	<1,15	[355]	1972	V_3B_2	<0,1	[355]	1972
$RuBe_2$	1.35	[355]	1972	VB	<1,20	[354]	1963
Ru_3Be_{17}	<1.15	[355]	1972	Nb_3B_2	<0,1	[355]	1972
$RhBe_2$	1,37	[355]	1972	NbB	8.25	[354]	1963
$PdBe_5$	<0,30	[354]	1963	Nb_3B_4	<1,28	[354]	1963
$OsBe_2$	3,07	[355]	1972	NbB_2	<1,0	[355]	1972
CaB_6	<1,28	[245]	1969	$NbB_{2.5}$	3,0—6,4	[355]	1972
BaB_6	<1,28	[354]	1963	Ta_2B [*4]	5,4	[368]	1969
ScB_2	<1,30	[245]	1969	Ta_2B [*5]	3,12	[354]	1963
ScB_4	<1,34	[245]	1969	Ta_3B_2	<0,1	[355]	1972
ScB_{12}	0,39	[245]	1969	TaB	4.0	[245]	1969
YB_6	6,5—7,1	[245]	1969	Ta_3B_4	<1,28	[354]	1963
YB_{12}	4,7	[245]	1969	TaB_2	<1,28	[354]	1963
LaB_6	5,7	[245]	1969	Cr_2B	<1,20	[354]	1963
CeB_6[*1]	<0,35	[245]	1969	CrB	<1,28	[354]	1963

[*1] Antiferromagnetic.

[*2] Ferromagnetic.

[*3] Cubic + extension lines.

[*4] Traces of γ-TaB.

[*5] Indeterminate composition.

Phase	Transition temp. into supercond. state, T_k, K	Source	Year	Phase	Transition temp. into supercond. state, T_k, K	Source	Year
1	2	3	4	1	2	3	4
CrB_2	<1,28	[354]	1963	UC	<1,20	[354]	1963
Mo_2B	5,3	[368]	1969	$TiC_{0.46}$	3,32	[370]	1970
MoB	<1,28	[354]	1963	$TiC_{0.52}$	3,42	[370]	1970
MoB_2	<1,0	[355]	1972	TiC	<1.20	[354]	1963
Mo_2B_5	5,2—8,1	[355]	1972	ZrC	<1,20	[354]	1963
W_2B	3,2	[368]	1969	HfC	<1,23	[354]	1963
WB	<1,28	[354]	1963	V_2C	<1,20	[354]	1963
W_2B_5	<1,28	[354]	1963	VC	<1,20	[354]	1963
Re_2B	2,80	[354]	1963	$NbC_{0.48}$	<1,6	[355]	1972
Ru_7B_3	2,58	[354]	1963	NbC	11,1	[355]	1972
Ru_2B	<1,20	[354]	1963	$TaC_{0.47}$	<1,6	[355]	1972
Rh_7B_3	<0.30	[354]	1963	TaC	10,1	[355]	1972
Rh_2B	<1,00	[354]	1963	Cr_4C	<1,20	[354]	1963
RhB	<1,28	[354]	1963	Cr_7C_3	<1,20	[354]	1963
OsB_2	<1,02	[354]	1963	Cr_3C_2	<1,20	[354]	1963
IrB	>1,28	[354]	1963	Mo_2C	4,0	[355]	1972
PtB	<1,28	[354]	1963	MoC [*1]	14,3	[355]	1972
Sc_4C_3	<1,0	[355]	1972	MoC [*2]	6,5	[245]	1969
$ScC_{0.96}$	<1,38	[245]	1969	W_2C [*2]	2,74	[354]	1963
Y_3C	<1,15	[355]	1972	W_2C [*1]	5,2	[354]	1963
$YC_{0.92}$	<1,38	[245]	1969	WC [*1]	10,0	[355]	1972
Y_2C_3	6,0—11,5	[355]	1972	WC [*2]	1—1,1	[371]	1968
YC_2	3,88	[355]	1972	TcC	3,85	[245]	1969
La_2C_3	5,9—11,0	[355]	1972	Fe_3C	<1,30	[354]	1963
LaC_2	1,61	[355]	1972	ReC [*3]	<1,6	[113]	1971
CeC_2	<1,28	[354]	1963	RuC	<1,90	[354]	1963
PrC_2	<2.0	[355]	1972	RhC	<1,28	[354]	1963
NdC_2	<2,0	[355]	1972	ScN	<1,40	[354]	1963
SmC_2	<2,0	[355]	1972	YN	<1,4	[245]	1969
GdC_2	<2,0	[355]	1972	LaN	1,35	[372]	1965
TbC_2	<2,0	[355]	1972	$CeN_{0.87}$	<1,80	[354]	1969
DyC_2	<2,0	[355]	1972	$PrN_{0.98}$	<1,38	[245]	1963
HoC_2	<2,0	[355]	1972	Th_3N_4	<1,20	[354]	1963
ErC_2	<2,0	[355]	1972	UN	<1,20	[354]	1963
TmC_2	<2,0	[355]	1972	TiN	5,60	[354]	1963
YbC_2	<2,0	[355]	1972	ZrN	9,8	[245]	1969
LuC_2	3,33	[355]	1972	HfN	6,2	[354]	1963
ThC	9 ± 1	[369]	1964	V_5N_2	<1,20	[354]	1963
ThC	<1,20	[354]	1963	VN	8,20	[354]	1963
				Nb_2N	9,5	[355]	1972
				Nb_4N_3	7.20	[354]	1963

[*1]Cub.

[*2]Hex.

[*3]Synthesized when $\rho \geq 60$ kbar.

SUPERCONDUCTIVITY (continued)

Phase	Transition temp. into supercond. state, T_k, K	Source	Year	Phase	Transition temp. into supercond. state, T_k, K	Source	Year
1	2	3	4	1	2	3	4
$NbN_{0.92}$ [*1]	16,30	[355]	1972	Zr_3Al	0,73	[354]	1963
NbN [*2]	$<1,94$	[354]	1963	$ZrAl_2$	$\sim0,3$	[354]	1963
Ta_2N	$<1,20$	[354]	1963	$ZrAl_3$	$<1,02$	[354]	1963
TaN [*1]	$6,5 \pm$	[355]	1972	V_3Al	11,65	[355]	1972
	$\pm 0,5$			Nb_3Al	18,52—	[355]	1972
TaN [*2]	$<4,2$	[355]	1972		18,9		
CrN	$<1,28$	[354]	1963	Nb_2Al	8,5—	[245]	1969
Mo_2N	5,0	[355]	1972		13,5		
MoN	12,0	[355]	1972	Ta_3Al	$<1,02$	[354]	1963
W_2N	$<1,28$	[354]	1963	Mo_3Al	0.58	[354]	1963
$WN_{0.97}$ [*1]	$<1,38$	[245]	1969	$MoAl_5$	$<1,15$	[355]	1972
$ReN_{0.34}$	4—5	[354]	1963	$MoAl_{12}$	$<1,02$	[354]	1963
$CaAl_2$	$<1,02$	[354]	1963	$Re_{24}Al_5$	3,35	[245]	1969
$CaAl_4$	$<1,02$	[354]	1963	$ReAl_6$	1,85	[355]	1972
Sc_3Al	$<1,1$	[245]	1969	$ReAl_{12}$	$<1,15$	[355]	1972
$ScAl_2$	$<1,02$	[245]	1969	$OsAl$	$<1,02$	[354]	1963
Y_3Al	$<1,1$	[245]	1969	Os_2Al_3	$<1,15$	[355]	1972
Y_2Al	$<1,15$	[355]	1972	$OsAl_2$	$<1,15$	[355]	1972
Y_3Al	$<1,15$	[355]	1972	$OsAl_3$	5,9	[354]	1963
YAl	$<1,15$	[355]	1972	$PtAl$	$<0,34$	[245]	1969
YAl_2	$<0,34$	[245]	1969	$PtAl_2$	0,48—	[245]	1969
La_3Al	6,16	[355]	1972		0,55		
$LaAl$	$<0,33$	[245]	1969	$CaSi_2$ [*3]	1,58	[355]	1972
$LaAl_2$	3,237	[355]	1972	$CaSi_2$ [*4]	$<0,32$	[355]	1972
$LaAl_4$	$<1,15$	[355]	1972	Sr_2Si_3	$\sim0,55$	[355]	1972
$CeAl_2$	$<0,34$	[245]	1969	$SrSi_2$	$<0,32$	[355]	1972
$YbAl_3$	0.94	[355]	1972	$ScSi_2$	$<1,00$	[354]	1963
Lu_3Al	<1.1	[245]	1969	YSi	$<1,15$	[355]	1972
Lu_2Al	$<1,02$	[354]	1963	$YSi_{1.90}$ [*3]	$<0,1$	[355]	1972
$LuAl_2$	$<1,02$	[245]	1969	YSi_2 [*5]	<1.0	[354]	1963
Th_3Al	$<0,30$	[354]	1963	YSi_2 [*2]	$<0,30$	[354]	1963
Th_3Al_2	2,6	[355]	1972	$LaSi_2$ [*3]	2,3	[355]	1972
$ThAl_2$	$<0,30$	[354]	1963	$CeSi_2$ [*3]	$<1,00$	[354]	1963
$ThAl_3$	0,75	[354]	1963	$NdSi_2$ [*3]	$<1,00$	[354]	1963
UAl_2	$<1,12$	[354]	1963	Th_3Si_2	$<0,1$	[355]	1972
UAl_3	$<0,07$	[355]	1972	$\alpha\text{-}ThSi_2$	3.2	[245]	1969
$PuAl$	$<1,50$	[354]	1963	$\beta\text{-}ThSi_2$	2.4	[245]	1969
$TiAl_3$	$<1,02$	[354]	1963	U_3Si	$<1,10$	[354]	1963

[*1]Cub.

[*2]Hex.

[*3]Tetr.

[*4]Rhombohed.

[*5]Rhomb.

Phase	Transition temp. into supercond. state, T_k, K	Source	Year	Phase	Transition temp. into supercond. state, T_k, K	Source	Year
1	2	3	4	1	2	3	4
U_3Si_2	<0,1	[355]	1972	Ta_5Si_2	<1,20	[354]	1963
USi_2	<0,30	[354]	1963	Ta_3Si_2	<1,20	[354]	1963
Ti_5Si_3	<1,20	[354]	1963	$TaSi$	4,25— 4,38	[354]	1963
$TiSi$	<1,20	[354]	1963				
$TiSi_2$	<1,20	[354]	1963	$TaSi_2$	<1,20	[354]	1963
Zr_4Si	<1,20	[354]	1963	Cr_3Si	<0,015	[355]	1972
Zr_3Si	<4,2	[373]	1965	Cr_3Si_2	<1,20	[354]	1963
Zr_2Si	<1,20	[354]	1963	$CrSi$	<0,015	[355]	1972
Zr_5Si_3	<1,1	[354]	1963	$CrSi_2$	<1,20	[354]	1963
Zr_3Si_2	<0,1	[355]	1972	Mo_3Si	1,3	[245]	1969
Zr_4Si_3	<1,20	[354]	1963	$MoSi_{0.7}$	1,34	[354]	1963
Zr_6Si_5	<1,20	[354]	1963	Mo_3Si_2	<1,20	[354]	1963
$ZrSi$	<1,20	[354]	1963	$MoSi_2$	<1,20	[354]	1963
$ZrSi_2$	<1,02	[354]	1963	$WSi_{0.7}$	2,84	[354]	1963
Hf_3Si_2	<0,1	[355]	1972	W_3Si_2	2,80	[245]	1969
$HfSi_2$	<1,02	[354]	1963	WSi_2	<1,20	[354]	1963
V_3Si [*1]	16.85	[355]	1972	$ReSi_2$	<1,15	[355]	1972
V_3Si [*2]	17,1	[374]	1969	$FeSi$	<1,28	[354]	1963
V_5Si_3	<0,30	[354]	1963	$CoSi_2$	1,22	[354]	1963
V_6Si_5	<1.5	[167]	1971	$NiSi$	<1,90	[354]	1963
VSi_2	<1,20	[354]	1963	$NiSi_2$	<1,00	[354]	1963
Nb_2Si	<1,20	[354]	1963	$RhSi$	<0,30	[354]	1963
Nb_5Si_3	<1,02	[354]	1963	$PdSi$	0,93	[354]	1963
Nb_5Si_3	8,1	[375]	1969	$OsSi$	<0,30	[354]	1963
Nb_3Si_2	<1,20	[354]	1963	$IrSi$	<1,02	[354]	1963
$NbSi_2$	<1,20	[354]	1963	$IrSi_3$	<1,02	[354]	1963
Ta_5Si	<1,20	[354]	1963	$PtSi$	0,88	[354]	1963
Ta_3Si	<4,2	[373]	1965	LaP	<1,68	[245]	1969

[*1]Monocrystal.

[*2]Polycrystal.

Phase	Transition temp. into supercond. state, T_k, K	Source	Year	Phase	Transition temp. into supercond. state, T_k, K	Source	Year
1	2	3	4	1	2	3	4
$PrP_{0.97}$	7	[27]	1973	La_3S_4	8,25	[355]	1972
V_3P	<1.00	[354]	1963	La_2S_3	<1.25	[245]	1969
VP	<1.01	[245]	1969	CeS	<1.06	[354]	1963
Cr_3P	<1.01	[245]	1969	Ce_3S_4	<1.28	[354]	1963
CrP	<1.01	[245]	1969	Pr_2S_3	<1.68	[245]	1969
Mo_3P	5.31	[245]	1969	Nd_2S_3	<1.68	[245]	1969
MoP	<1.01	[245]	1969	V_3S	<1.15	[355]	1969
W_3P	2.26	[245]	1969	NbS	<1.28	[354]	1963
WP	<1.01	[245]	1969	NbS_2*3	6,1—6,8	[245]	1969
MnP	<0.01	[245]	1969	NbS_2*4	5.0—5.5	[245]	1969
Fe_2P	<0.97	[245]	1969	TaS_2	1,82—1,99	[355]	1972
FeP	<0.97	[245]	1969	MoS_2	<1.28	[354]	1963
Co_2P	<0.97	[245]	1969	W_2S	<1.30	[354]	1963
Ni_3P	<1.01	[245]	1969	FeS	<1.90	[354]	1963
Ni_2P	<1.01	[245]	1969	NiS	<1.28	[354]	1963
Ru_2P	<0.35	[245]	1969	RuS_2	<0.32	[245]	1969
RuP	<0.35	[245]	1969	$Rh_{17}S_{15}$	5,8	[354]	1963
Rh_2P	1.3	[245]	1969	Pd_4S	<0.32	[245]	1969
Rh_5P_4	1.22	[354]	1963	$Pd_{2.2}S$ *5	1,63	[245]	1969
$Pd_{3.0-3.2}P$	$\sim0.35—0.7$	[245]	1969	PdS	<0.35	[245]	1969
Pd_7P_3*1	1.00	[245]	1969	IrS	<0.32	[245]	1969
Pd_7P_3*2	0.70	[245]	1969	$IrS_{2.6}$	<0.32	[245]	1969
Pd_5P_2	<1.1	[354]	1963	B_4C	<1.28	[354]	1963
Pd_5P_3	<1.1	[354]	1963	BN	<1.28	[354]	1963
Ir_2P	<0.35	[245]	1969	SiC	<1.28	[354]	1963
IrP	<0.35	[245]	1969	Al_4C_3	<1.38	[245]	1969
$Pt_{20}P_7$	<0.35	[245]	1969	AlN	1,55	[245]	1969
LaS	<1.25	[245]	1969				

*1High temperature.

*2Low temperature.

*3Double-layer type.

*4Triple-layer type.

*5Tempered.

Phase	Coef. thermo-emf, $\mu V/$ deg	Source	Year	Notes
1	2	3	4	5
Be_5B	$+1,6$	[3]	1975	
Be_2B	$+8,8$	[3]	1975	
CaB_6	$-32,0$	[563]	1961	
SrB_6	$-30,3$	[563]	1961	
BaB_6	$-26,2$	[563]	1961	
ScB_2	$-7,7$	[3]	1975	
YB_4	$-7,2$	—*	1974	
YB_6	$-0,5$	[563]	1961	
YB_{12}	$-3,8\pm0,5$	—**	1970	
LaB_6	$+0,1$	[563]	1961	
CeB_6	$+2,8$	[563]	1961	
PrB_6	$-0,6$	[563]	1961	
NdB_6	$+0,4$	[563]	1961	
SmB_6	$+7,6$	[563]	1961	
EuB_6	$-17,7$	[563]	1961	See [252]
GdB_4	$-6,9$	—*	1974	
GdB_6	$+0,1$	[563]	1961	
TbB_4	$-6,5$	—*	1974	
TbB_6	$-1,1$	[563]	1961	
TbB_{12}	$-4,5\pm0,5$	—**	1970	
DyB_4	$-12,0$	—*	1974	
DyB_{12}	$-2,2\pm0,3$	—**	1970	
HoB_4	$-10,0$	—*	1974	
HoB_{12}	$-2,7\pm0,3$	—**	1970	
ErB_4	$-8,1$	—*	1974	
ErB_{12}	$-0,10\pm0,01$	—**	1970	
TmB_4	$-10,0$	—*	1974	
TmB_{12}	$-0,5\pm0,1$	—**	1970	
YbB_6	$-25,5$	[563]	1961	See [252]
YbB_{12}	$-3,8\pm0,4$	[59]	1971	
LuB_{12}	$-3,6\pm0,4$	—**	1970	
ThB_6	$-0,6$	[563]	1961	
TiB_2	$-2,9$	[459]	1971	
ZrB_2	$-1,2$	[459]	1971	
ZrB_{12}	$-0,9\pm0,1$	—**	1970	
HfB_2	$-1,0$	[459]	1971	
VB_2	$-5,8$	[459]	1971	
Nb_3B_2	$-9,7$	[3]	1975	
NbB	$-8,0$	[3]	1975	
Nb_3B_4	$-5,4$	[3]	1975	
NbB_2	$-3,4$	[459]	1971	
TaB_2	$-2,8$	[459]	1971	
Cr_4B	$-7,7\pm0,7$	[431]	1962	

*E. N. Severyanina, Preparation and Investigation of the Physical Properties of Rare-Earth Metal Tetraborides [in Russian], Author's abstract, Cand. Dissertation, Kiev (1974).

**V. V. Odintsov, Preparation and Physical Properties of Metallic Dodecaborides with the UB_{12} Structural Type [in Russian], Author's abstract, Cand. Dissertation, Kiev (1970).

THERMOELECTRICAL PROPERTIES (continued)

Phase	Source	Coef. thermo-emf, $\mu V/$ deg	Year	Notes
1	2	3	4	5
Cr_2B	−6,9	[3]	1975	
Cr_5B_3	−6,2	[3]	1975	
CrB	−4,7	[3]	1975	
Cr_3B_4	−6,7	[3]	1975	
CrB_2	−3,0	[3]	1975	
Cr_2B_5	−1,6±0,03	[431]	1962	
Mo_2B_5	−9,4	[3]	1975	
W_2B_5	−6,6	[3]	1975	
Mn_4B	+10,0±0,5	[433]	1968	
Mn_2B	+8,2±0,3	[433]	1968	
MnB	+6,7±0,1	[433]	1968	
Mn_3B_4	+4,8±0,3	[433]	1968	
MnB_2	+3,9±0,1	[433]	1968	
Fe_2B	−28	[434]	1971	
FeB	−14,2	[434]	1971	See [569]
Co_3B	−40	[435]	1972	
Co_2B	−27	[435]	1972	
CoB	−51	[435]	1972	See [569]
Ni_3B	−10,8	[435]	1972	
Ni_2B	−1,6	[435]	1972	
NiB	+2,3	[435]	1972	
Rh_7B_3	+0,3	[570]	1971	
$RhB_{\sim1.1}$	+0,6	[570]	1971	
Pd_3B	−7,2	[571]	1972	
Pd_5B_2	+2,0	[571]	1972	
YC	−34,6	[769]	1966	
Y_2C_3	−11,1±0,5	[573]	1970	
YC_2	−4,9±0,5	[573]	1970	
La_2C_3	−1,9±0,5	[573]	1970	
LaC_2	−1,8±0,5	[573]	1970	
Ce_2C_3	−2,2±0,5	[573]	1970	
CeC_2	−0,2±0,5	[573]	1970	
PrC_2	−2,4±0,5	[573]	1970	
Nd_2C_3	−2,2±0,5	[573]	1970	
NdC_2	−2,1±0,5	[573]	1970	
SmC_2	−0,3±0,5	[573]	1970	
GdC_2	−1,8±0,5	[573]	1970	
TbC_2	−1,8±0,5	[573]	1970	
DyC_2	−2,1±0,5	[573]	1970	
ErC_2	−2,8±0,5	[573]	1970	
TmC_2	−4,2±0,5	[573]	1970	
ThC	∼−15	[369]	1964	
UC	∼+50	[369]	1964	
U_2C_3	<0	[576]	1965	In the range 4-250K
PuC	∼+10	[369]	1964	
$TiC_{0.91}$	∼−4,5	[772]	1973	
$TiC_{0.83}$	∼−2,5	[772]	1973	
$TiC_{0.69}$	∼−1,5	[772]	1973	
$TiC_{0.46}$	∼+0,5	[772]	1973	See [578, 773]
$ZrC_{0.903}$	−6,3±0,3	[438]	1964	
$ZrC_{0.820}$	−3,7±0,3	[438]	1964	
$ZrC_{0.756}$	−2,5±0,6	[438]	1964	

Phase	Coef. thermo- emf, $\mu V/$ deg	Source	Year	Notes
1	2	3	4	5
$ZrC_{0.718}$	-0.7 ± 0.2	[438]	1964	
$ZrC_{0.628}$	$+0.1\pm0.3$	[438]	1964	
HfC	-11.7	[582]	1960	
$VC_{0.90}$	-0.5 ± 0.1	[443]	1970	
$VC_{0.85}$	-2.9 ± 0.5	[443]	1970	Cub.; see [578,
$VC_{0.79}$	-5.0 ± 0.5	[443]	1970	580]
$VC_{0.76}$	-5.4 ± 0.6	[443]	1970	
$VC_{0.42}$	$+2.0\pm0.5$	[443]	1970	Hex.
$VC_{0.39}$	$+1.3\pm0.5$	[443]	1970	
$NbC_{0.908}$	-5.5 ± 0.3	[438]	1964	See [578, 580]
$NbC_{0.855}$	-5.8 ± 0.6	[438]	1964	
$NbC_{0.808}$	-3.4 ± 0.4	[438]	1964	
$NbC_{0.759}$	-2.1 ± 0.1	[438]	1964	
$NbC_{0.710}$	-1.9 ± 0.1	[438]	1964	
$TaC_{0.98}$	-11.5	[580]	1971	See [578]
$TaC_{0.80}$	-17.8	[580]	1971	
$TaC_{0.74}$	-19.5	[580]	1971	
$Cr_{23}C_6$	$+2.76\pm0.02$	[431]	1961	
Cr_7C_3	-7.1 ± 0.3	[431]	1961	
Cr_3C_2	-6.7 ± 0.5	[431]	1961	
Mo_2C	-1.9	[582]	1960	
WC	-23.3	[582]	1960	
W_2C	-8.17 ± 0.02	[582]	1960	
NdN	<0	[492]	1965	In the range 10-90K
$TiN_{0.79}$	-5.4	—*	1974	
$TiN_{0.83}$	-6.0	—*	1974	
$TiN_{0.87}$	-6.8	—*	1974	
$TiN_{0.97}$	-7.7	—*	1974	
$ZrN_{0.82}$	-4.9	—*	1974	
$ZrN_{0.85}$	-5.2	—*	1974	
$ZrN_{0.92}$	-5.5	—*	1974	
$ZrN_{0.97}$	-5.9	—*	1974	
$HfN_{1.00}$	-2.9	—*	1974	
V_3N	-5.3 ± 1.2	[444]	1962	
$VN_{0.80}$	$+1.5$	—*	1974	
$VN_{0.85}$	$+0.5$	—*	1974	
$VN_{0.87}$	-0.4	—*	1974	
$VN_{0.93}$	-2.0	—*	1974	
$VN_{0.96}$	-5.0	—*	1974	
Nb_2N	-4.6 ± 0.7	[444]	1962	
$NbN_{0.75}$	-4.2	—*	1974	
$NbN_{0.80}$	-4.7	—*	1974	
$NbN_{0.89}$	-0.4	—*	1974	
$NbN_{0.95}$	-1.5	—*	1974	
NbN	-2.24	[444]	1962	

*L. K. Shvedova, Investigation of the Physical Properties of Transition Metal Nitrides in Groups IV–V [in Russian], Author's abstract, Cand. Dissertation, Kiev (1974).

Phase	Coef. thermo-emf, $\mu V/$deg	Source	Year	Notes
1	2	3	4	5
Ta_2N	$-2,17\pm0,4$	[444]	1962	
$TaN_{1.0}$	$-1,0$	—*	1974	
Cr_2N	$-0,52\pm0,2$	[431]	1962	
CrN	$-92,0\pm4,0$	[431]	1962	
Mo_2N	$+2.18\pm0.5$	[444]	1962	
$TiAl$	$+5,5$	[18]	1965	See [470]
$TiAl_3$	$+4.5$	[18]	1965	See [470]
V_5Al_8	0	[590]	1964	
VAl_3	$-5,5$	[590]	1964	
VAl_6	$+4.64$	[590]	1964	
VAl_{11}	$-0,3$	[590]	1964	
$MnAl_3$	-70	[18]	1965	
β-$NiAl$	~-5	[591]	1969	
Mg_2Si	$600-2000$	[260]	1959	
$MgSi_2$	$180-240$	[260]	1959	
$BaSi_2$	$+600$	[449]	1963	
Sc_5Si_3	$-4,8$	—*	1975	
$ScSi$	$-5,2$	—*	1975	
$ScSi_{1.66}$	$-6,4$	—*	1975	
Y_5Si_3	$-14,0$	[472]	1974	
YSi	$-2,5$	[472]	1974	
YSi_2	$-1,0$	[472]	1974	
La_5Si_3	$-6,0$	—*	1975	
$LaSi$	$-4,0$	—*	1975	
$LaSi_2$	$-5,0$	—*	1975	
Ce_5Si_3	$-1,7$	—*	1975	
$CeSi$	$+3.0$	—*	1975	
$CeSi_2$	$-2,4$	—*	1975	
Pr_5Si_3	$-3,9$	—*	1975	
$PrSi$	$-6,5$	—*	1975	
$PrSi_2$	$-5,8$	—*	1975	
Sm_5Si_3	$-2,3$	—*	1975	
$SmSi$	$+6,2$	—*	1975	
$SmSi_2$	$-3,2$	—*	1975	
Ti_5Si_3	$+2,3$	[632]	1960	
$TiSi$	$+2,4$	[632]	1960	
$TiSi_2$	$+5,2$	[632]	1960	
$ZrSi_2$	$+14,7$	[632]	1960	
V_3Si; V_5Si_3	~0	[596]	1963	$\alpha < 0$
VSi_2	$+10,5$	[632]	1960	See [596]
$NbSi_2$	$+14,4$	[632]	1960	
$TaSi_2$	$+14,0$	[632]	1960	
Cr_3Si	$+16.6$	[260]	1959	
Cr_2Si	$-4,0$	[448]	1958	
$CrSi$	$+5,0$	[448]	1958	

*V. I. Lazorenko, Investigation of the Physical Properties of Rare-Earth Silicides of the Cerium Subgroup [in Russian], Author's abstract, Cand. Dissertation, Kiev (1975). See [297, 593, 632].

Phase	Coef. thermo-emf, $\mu V/$deg	Source	Year	Notes
1	2	3	4	5
Cr_5Si_3	$+0,6$	[260]	1959	
$CrSi_2$	$+86,0$	[632]	1960	
Mo_3Si	$-1,0$	[632]	1960	
Mo_5Si_3	$+2,0$	[632]	1960	
$MoSi_2$	$-3,0$	[632]	1960	
WSi_2	$+0,2$	[632]	1960	
Mn_3Si	$+18,0$	[448]	1958	
Mn_5Si_3	$+14,0$	[448]	1958	
$MnSi$	$+102$	[448]	1958	
$MnSi_2$	$+46,0$	[448]	1958	
Re_3Si	~ 0	[596]	1963	
$ReSi_2$	$+174$	[632]	1960	
$ReSi$	$\sim +35$	[596]	1963	
Fe_3Si	$-2,0$	[448]	1958	
$FeSi$	$+9$	[22]	1971	
$FeSi_2$	$\sim +25$	[596]	1963	$100°$ C
Co_3Si	-10	[596]	1963	
Co_2Si	$-8,0$	[448]	1958	
$CoSi$	$-46,0$	[448]	1958	
$CoSi_2$	$-8,0$	[448]	1958	
$CoSi_3$	$+14,0$	[448]	1958	
Ni_3Si	$-2,0$	[448]	1958	
Ni_5Si_2	$-2,0$	[448]	1958	
Ni_2Si_3	$+9,0$	[448]	1958	
$NiSi$	$+8,0$	[448]	1958	
$NiSi_2$	$+7,0$	[448]	1958	
SmP	$+29,9$	[598]	1970	
UP_2	$\sim +20$	[599]	1969	Monocrystal, $\parallel c$
UP_2	$\sim +37,8$	[599]	1969	Monocrystal, $\perp c$
UP_2	$\sim +20$	[599]	1969	Polycrystal
TiP	$-11,1$	[29]	1961	
FeP_2	<0	[600]	1971	Monocrystal
La_2S_3	$+345$	[33]	1972	
CeS	$+10,0$	[558]	1961	
Ce_2S_3	$+430$	[558]	1961	
Sm_2S_3	<0	[262]	1965	$20-1000°$ C
$ThS_{1.5}$	$+100$	[33]	1972	
$ThS_{1.7}$	$+200$	[33]	1972	
Ti_2S_3	-17	[264]	1964	
TiS_2	$+200$	[264]	1964	
TiS_3	$+600$	[264]	1964	
V_3S_4	$+13,1$	[608]	1967	
$NbS_{1.6}$	$+5,1$	[243]	1966	
Cr_7S_8 Cr_5S_6 Cr_3S_4	-20	[244]	1967	
Cr_2S_3	-70	[244]	1967	Trigon.
Cr_2S_3	-300	[244]	1967	Rhombohed.
Co_3S_4	$+4,8$	[608]	1967	
B_4C	$+80$	[610]	1960	

Phase	Work function, eV	Richardson's Constant, $A/(cm^2 \cdot deg^2)$	Secondary emission coef	Source	Year	Notes
1	2	3	4	5	6	7
CaB_6	2.86	2.6	—	[356]	1970	
SrB_6	2.67	0.14	—	[356]	1970	See [775]
BaB_6	3.45	16	—	[356]	1970	See [775]
ScB_2	2.90	4.6 ± 2.1	—	[774]	1962	
ScB_6	2.96	4.6	0.58	[356]	1970	
YB_4	2.08	$4.47 \cdot 10^{-2}$	—	[477]	1968	
YB_6	3.45	1.77	—	[477]	1968	
YB_{12}	4.6	—	—	[356]	1970	
LaB_6	2.68	73	0.95	[356]	1970	See [775]
LaB_{12}	2.16	—	—	[498]	1958	
CeB_6	2.93	~ 580	0.68	[356]	1970	
CeB_{12}	2.20	—	—	[498]	1958	
PrB_6	3.46	~ 300	—	[356]	1970	
NdB_6	3.97	~ 420	—	[356]	1970	
SmB_6	4.4	—	—	[356]	1970	
EuB_6	4.9	—	—	[776]	1959	
GdB_4	1.45	10^{-3}	—	[777]	1968	
GdB_6	2.06	0.84	0.8	[356]	1970	
TbB_6	3.26	120	0.74	[356]	1970	
DyB_6	3.53	25.1	0.8	[356]	1970	
HoB_6	3.42	13.9	0.7	[356]	1970	
ErB_6	3.37	9.9	0.7	[356]	1970	
TmB_6	$2.75 + 3.3 \cdot 10^{-4}\,T$	—	—	[356]	1970	1100—1800 K
YbB_6	3.13	2.5	—	[356]	1970	
LuB_6	3.0	0.36	0.8	[356]	1970	
ThB_6	2.92	0.3	—	[356]	1970	
UB_2	$3.3 + 0.2 \cdot 10^{-4}\,T$	100	—	[380]	1964	
UB_4	$3.4 - 0.8 \cdot 10^{-4}\,T$	300	—	[380]	1964	
UB_{12}	$2.89 + 2.38 \times \times\, 10^{-4}\,T$	—	—	[356]	1970	
TiB_2	$4.08 + 8 \cdot 10^{-5}\,T$	—	—	[779]	1973	1300—2100 K
ZrB_2	$3.69 + 1.8 \cdot 10^{-4}\,T$	—	—	[779]	1973	1300—2100 K
HfB_2	$3.54 + 2.1 \cdot 10^{-4}\,T$	—	—	[779]	1973	1300—2100 K
VB_2	$3.98 + 1.8 \cdot 10^{-4}\,T$	—	—	[779]	1973	1300—2100 K
Nb_3B_2	$3.35 + 1.0 \cdot 10^{-4}\,T$	—	—	[779]	1973	1300—1850 K
NbB	$3.58 + 1.0 \cdot 10^{-4}\,T$	—	—	[779]	1973	1300—2000 K
Nb_3B_4	$3.88 + 8 \cdot 10^{-5}\,T$	—	—	[779]	1973	1300—2000 K
NbB_2	$4.03 + 1.6 \cdot 10^{-4}\,T$	—	—	[779]	1973	1300—2100 K
TaB_2	$4.06 + 1.4 \cdot 10^{-4}\,T$	—	—	[779]	1973	1300—2100 K
Cr_2B	$2.46 + 4.0 \cdot 10^{-4}\,T$	—	—	[779]	1973	1300—1460 K
Cr_5B_3	$2.72 + 8 \cdot 10^{-5}\,T$	—	—	[779]	1973	1300—1540 K
CrB	$3.02 + 4.6 \cdot 10^{-5}\,T$	—	—	[779]	1973	1300—1680 K
Cr_3B_4	$3.12 + 8 \cdot 10^{-5}\,T$	—	—	[779]	1973	1300—1800 K
CrB_2	$3.18 + 1.0 \cdot 10^{-4}\,T$	—	—	[779]	1973	1300—2000 K
Mo_2B_5	$3.70 + 1.0 \cdot 10^{-4}\,T$	—	—	[779]	1973	1300—2000 K
W_2B_5	$3.79 + 1.2 \cdot 10^{-4}\,T$	—	—	[779]	1973	1300—2100 K

Phase	Work function, eV	Richardson's Constant, $A/(cm^2 \cdot deg^2)$	Secondary emission coef.	Source	Year	Notes
1	2	3	4	5	6	7
MnB_2	4,14	—	—	[780]	1957	
Fe_2B	3,3	—	—	[434]	1971	
FeB	3,6	—	—	[434]	1971	
CeC_2	2.49	—	—	[356]	1970	
ThC_2	3,5	~550	—	[624]	1960	
UC	3.4	—	—	[356]	1970	
UC_2	3,9	—	—	[356]	1970	
PuC	3,4	404	—	[356]	1970	
$TiC_{0.89}$	$4,39 - 1,7 \cdot 10^{-4} T$	—	4.33	[782]	1971	1400—2100 K; See [783]
$TiC_{0.83}$	$4,28 - 1.9 \cdot 10^{-4} T$	—	4.22	[782]	1971	
$TiC_{0.79}$	$4.12 - 2,6 \cdot 10^{-4} T$	—	4,04	[782]	1971	
$TiC_{0.73}$	$4,0 - 2.8 \cdot 10^{-4} T$	—	3,91	[782]	1971	
ZrC	2,18	0,31	—	[356]	1970	
HfC	2.04	10^{-5}	—	[356]	1970	
$VC_{0.72}$	~3.99	—	—	[465]	1974	
$VC_{0.77}$	~4.02	—	—	[465]	1974	
$VC_{0.82}$	~4.05	—	—	[465]	1974	
$VC_{0.87}$	~4,08	—	—	[465]	1974	
$NbC_{0.99}$	$4.25 - 1,9 \cdot 10^{-4} T$	—	4,19	[782]	1971	1400—2100 K
$NbC_{0.91}$	$4.30 - 1.7 \cdot 10^{-4} T$	—	4.25	[782]	1971	
$NbC_{0.88}$	$4.37 - 1,8 \cdot 10^{-4} T$	—	4,31	[782]	1971	
$NbC_{0.81}$	$4.45 - 1.7 \cdot 10^{-4} T$	—	4,4	[782]	1971	
$NbC_{0.72}$	$4,32 - 1,8 \cdot 10^{-4} T$	—	4.26	[782]	1971	
TaC	3,14	0.30	—	[356]	1970	
MoC	3,80	—	—	[356]	1970	
Mo_2C	4.74	—	—	[356]	1970	
WC	3,6	2.7	—	[356]	1970	
W_2C	4.58	190	—	[356]	1970	
TiN	3.75	—	—	[784]	1963	
ZrN	3.90	—	—	[784]	1963	
HfN	3,85—3.90	—	—	[356]	1970	
VN	$2,75 + 5 \cdot 10^{-4} T$	—	—	[785]	1968	1100—2000 K
NbN	3.92	—	—	[784]	1963	See [785]
TaN	$1.35 + 1.67 \cdot 10^{-3} T$	—	—	[785]	1968	1100—2000 K
ThN	3,1	—	—	[356]	1970	
UN	$3,1 + 2,1 \cdot 10^{-4} T$	—	—	[356]	1970	
Zr_5Al_3	~3.0	—	—	[356]	1970	1550 K
$ZrAl_3$	~3.31	—	—	[356]	1970	1400 K
V_5Al_8	$4.25 - 4.5 \cdot 10^{-4} T$	—	—	[356]	1970	1300—1750 K
$NbAl_3$	3.42	—	—	[356]	1970	
$TaAl_3$	~3,0	—	—	[356]	1970	1500 K
YSi_2	$3,26 - 7.5 \cdot 10^{-4} T$	—	—	[356]	1970	
USi_2	$3,01 + 2,1 \cdot 10^{-4} T$	—	—	[356]	1970	
USi_3	$3,22 + 1,1 \cdot 10^{-4} T$	—	—	[356]	1970	
$TiSi_2$	3,95	—	—	[356]	1970	
$ZrSi_2$	3,95	—	—	[356]	1970	
$NbSi_2$	$4,34 - 5.25 \cdot 10^{-4} T$	—	—	[356]	1970	

Phase	Work function, eV	Richardson's Constant, $A/(cm^2 \cdot deg^2)$	Secondary emission coef.	Source	Year	Notes
1	2	3	4	5	6	7
$TaSi_2$	4,71	—	—	[356]	1970	
Cr_2Si	$2,35 + 6,33 \cdot 10^{-4}\, T$	—	—	[356]	1970	
Cr_3Si	$3.49 - 5,8 \cdot 10^{-5}\, T$	—	—	[356]	1970	
$CrSi_2$	$3,78 - 1,2 \cdot 10^{-4}\, T$	—	—	[356]	1970	
$MoSi_2$	4,73	—	—	[356]	1970	
WSi_2	4,62	—	—	[356]	1970	
$ReSi_2$	$4.02 - 2,67 \cdot 10^{-4}\, T$	—	—	[356]	1970	
VSi_2	~ 3.2	—	—	[356]	1970	1600 K
BaS	2,6	—	—	[356]	1970	
LaS	$0,58 + 2,1 \cdot 10^{-3}\, T$	—	—	[786]	1963	
La_2S_3	$0,73 + 2,02 \cdot 10^{-3}\, T$	—	—	[786]	1963	
CeS	$0,79 + 1,86 \cdot 10^{-3}\, T$	—	—	[786]	1963	
Ce_2S_3	$0,92 + 1,78 \cdot 10^{-3}\, T$	—	—	[786]	1963	
PrS	3,10	—	—	[787]	1964	1200 K
PrS	3,72	—	—	[787]	1964	1500 K
PrS	3,90	—	—	[787]	1964	1700 K
Pr_2S_3	3,08	—	—	[787]	1964	1200 K
Pr_2S_3	3,63	—	—	[787]	1964	1500 K
Pr_2S_3	3,84	—	—	[787]	1964	1700 K
NdS	3,09	—	—	[787]	1964	1200 K
NdS	3,69	—	—	[787]	1964	1500 K
NdS	3,89	—	—	[787]	1964	1700 K
Nd_2S_3	3,14	—	—	[787]	1964	1200 K
Nd_2S_3	3,64	—	—	[787]	1964	1500 K
Nd_2S_3	3,93	—	—	[787]	1964	1700 K
GdS	$0.80 + 2,19 \cdot 10^{-3}\, T$	—	—	[603]	1968	1250—1550 K
Gd_2S_3	$0.65 + 1,87 \cdot 10^{-3}\, T$	—	—	[603]	1958	1200—1450 K
ThS	3,4	100	—	[356]	1970	
SiC	$4.37 + 1,2 \cdot 10^{-4}\, T$	—	—	[781]	1970	Polycrystal 1500—2100 K
SiC	4,30	—	—	[781]	1970	Monocrystal Face [0001]

HALL CONSTANT

Phase	Hall constant, $cm^3/K \cdot 10^4$	Source	Year
1	2	3	4
CaB_6	−91,0	[563]	1961
SrB_6	−76,3	[563]	1961
BaB_6	−57.5	[563]	1961

Phase	Hall constant, $cm^3/K \cdot 10^4$	Source	Year
1	2	3	4
YB_2	-3.05	[3]	1975
YB_4	-21.3	[3]	1975
YB_6	-4.56	[563]	1961
YB_{12}	-4.9 ± 0.4	—*	1970
LaB_3	10.8	[3]	1975
LaB_6	-4.96	[563]	1961
CeB_6	-4.18	[563]	1961
PrB_6	-4.33	[563]	1961
NdB_6	-4.39	[563]	1961
SmB_6	$+1.54$	[563]	1961
EuB_6	-50.2	[563]	1961
GdB_6	-4.39	[563]	1961
TbB_6	-4.57	[563]	1961
TbB_{12}	-4.2 ± 0.3	—*	1970
DyB_{12}	-4.6 ± 0.4	—*	1970
HoB_{12}	-5.5 ± 0.4	—*	1970
ErB_{12}	-4.5 ± 0.4	—*	1970
TmB_{12}	-4.7 ± 0.4	—*	1970
YbB_6	-83.6	[563]	1961
YbB_{12}	-8.4 ± 0.7	—*	1970
LuB_{12}	-4.8 ± 0.4	—*	1970
ThB_6	-2.19	[563]	1961
UB_{12}	-0.24 ± 0.02	[458]	1971
TiB_2	-19.6	[459]	1971
ZrB_2	-19.0	[459]	1971
ZrB_{12}	-2.6 ± 0.2	—*	1970
HfB_2	-18.0	[459]	1971
VB_2	-0.82	[459]	1971
Nb_3B_2	-0.46	[3]	1975
NbB	-0.60	[3]	1975
Nb_3B_4	-1.1	[3]	1975
NbB_2	-1.5	[459]	1971
TaB_2	-2.1	[459]	1971
Cr_4B	-1.17 ± 0.05	[431]	1962
Cr_2B	-0.8	[3]	1975
Cr_5B_3	-0.9	[3]	1975
CrB	-1.0	[3]	1975
Cr_3B_4	-1.0	[3]	1975
CrB_2	-1.2	[3]	1975
Cr_2B_5	-0.60 ± 0.05	[431]	1962
Mo_2B_5	-6.6	[3]	1975
W_2B_5	-6.9	[3]	1975
CoB	-104	[435]	1972
Ni_3B	0	[435]	1972
Ni_2B	$+0.53$	[435]	1972
NiB	$+0.63$	[435]	1972
Y_2C_3	-34 ± 1	[573]	1970
YC_2	-11 ± 1	[573]	1970
La_2C_3	-0.8 ± 0.1	[573]	1970
LaC_2	-5.3 ± 0.2	[573]	1970

*V. V. Odintsov, Preparation and Physical Properties of Dodecacarbides with UB_{12} Structure Types [in Russian], Author's abstract, Cand. Dissertation, Kiev (1970).

HALL CONSTANT (continued)

Phase	Hall constant, $cm^3/K \cdot 10^4$	Source	Year
1	2	3	4
Ce_2C_3	-1.4 ± 0.04	[573]	1970
CeC_2	-2.8 ± 0.1	[573]	1970
PrC_2	-5.40 ± 0.2	[573]	1970
Nd_2C_3	$+12.8 \pm 0.4$	[573]	1970
NdC_2	-4.5 ± 0.2	[573]	1970
SmC_2	-4.4 ± 0.2	[573]	1970
GdC_2	-3.0 ± 0.1	[573]	1970
TbC_2	-2.6 ± 0.1	[573]	1970
DyC_2	-3.0 ± 0.1	[573]	1970
ErC_2	-3.2 ± 0.1	[573]	1970
TmC_2	$+136 \pm 4$	[573]	1970
$TiC_{0.91}$	~-9.5	[772]	1973
$TiC_{0.83}$	~-7.5	[772]	1973
$TiC_{0.69}$	~-6	[772]	1973
$TiC_{0.46}$	~-5	[772]	1973
$ZrC_{0.97}$	-10.4	[773]	1970
$ZrC_{0.79}$	-3.7	[773]	1970
$ZrC_{0.58}$	-0.53	[773]	1970
HfC	-12.4	[582]	1960
$VC_{0.72}$	~-0.49	[465]*2	1974
$VC_{0.77}$	~-0.57	[465]*2	1974
$VC_{0.82}$	~-0.62	[465]*2	1974
$VC_{0.87}$	~-0.67	[465]*2	1974
$VC_{0.39}$ *1	-3.82 ± 0.18	[443]	1970
$VC_{0.40}$ *1	-4.86 ± 0.20	[443]	1970
$VC_{0.42}$ *1	-5.28 ± 0.15	[443]	1970
$NbC_{1.0}$	-1.05	[580]	1971
$NbC_{0.84}$	-0.62	[580]	1971
$NbC_{0.80}$	-0.49	[580]	1971
$NbC_{0.74}$	-0.54	[580]	1971
$TaC_{0.98}$	-1.11	[580]	1971
$TaC_{0.80}$	-0.60	[580]	1971
$TaC_{0.74}$	-1.88	[580]	1971
$Cr_{23}C_6$	$+1.2 \pm 0.2$	[431]	1962
Cr_7C_3	-0.38 ± 0.03	[431]	1962
Cr_3C_2	-0.47 ± 0.03	[431]	1962
Mo_2C	-0.85	[582]	1960
W_2C	-13.1 ± 0.7	[582]	1960
WC	-21.8 ± 0.3	[582]	1960
α-WC	-4.0 *4	[583]	1971
	-3.9 *5	[583]	1971
$TiN_{0.79}$	-0.45	—*3	1974
$TiN_{0.83}$	-0.46	—*3	1974

*1Hexagonal.

*2See [443, 580].

*3L. K. Shvedova, Investigation of the Physical Properties of Transition Metal Nitrides in Groups IV-V [in Russian], Author's abstract, Cand. Dissertation, Kiev (1974).

*4Monocrystal H ∥ c.

*5Monocrystal H ⊥ c.

253

Phase	Hall constant, $cm^3/K \cdot 10^4$	Source	Year
1	2	3	4
$TiN_{0.87}$	−0.48	—*1	1974
$TiN_{0.97}$	−0,55	—*1	1974
$ZrN_{0.82}$	−1,30	—*1	1974
$ZrN_{0.85}$	−1,37	—*1	1974
$ZrN_{0.92}$	−1,41	—*1	1974
$ZrN_{0.97}$	−1,44	—*1	1974
$HfN_{1.0}$	−4,00	—*1	1974
V_3N	+0,9 ± 0,1	[444]	1962
$VN_{0.71}$	+0,4	—*1	1974
$VN_{0.80}$	−0,35	—*1	1974
$VN_{0.85}$	−0,37	—*1	1974
$VN_{0.87}$	−0,42	—*1	1974
$VN_{0.93}$	−0,43	—*1	1974
$VN_{0.96}$	−0,46	—*1	1974
Nb_2N	+1,9 ± 0,4	[444]	1962
$NbN_{0.75}$	−0,76	—*1	1974
$NbN_{0.80}$	−0,74	—*1	1974
$NbN_{0.89}$	−0,28	—*1	1974
$NbN_{0.95}$	−0,40	—*1	1974
NbN	+0,52 ± 0,19	[444]	1962
Ta_2N	−0,46 ± 0,1	[444]	1962
TaN	−0,53	—*1	1974
Cr_2N	−0,7 ± 0,1	[431]	1962
CrN	−264 ± 25	[431]	1962
Mo_2N	+2,83 ± 1,2	[444]	1962
Ti_6Al	+0,68	[18]	1965
Ti_3Al	+3.89	[18]	1965
TiAl	+0.30	[18]	1965
V_5Al_8	+2,3	[590]	1964
VAl_3	−47,4	[590]	1964
VAl_6	−2,1	[590]	1964
VAl_{11}	−1,86	[590]	1964
β-NiAl	~+0,65	[591]	1969
Mg_2Si	−1.68 exp (0.064/2 KT)*2	[260]	1959
Mg_2Si	−11,2 exp (0.0072/2KT)*3	[260]	1959
Sc_5Si_3	+2,8	—*4	1975
ScSi	+3.0	—*4	1975
$ScSi_{1.66}$	+3,2	—*4	1975
Y_5Si_3	+1.8	—*4	1975
YSi	+2.0	—*4	1975
YSi_2	+6,1	—*4	1975
La_5Si_3	−1,0	—*4	1975
LaSi	−1.2	—*4	1975

*1L. K. Shvedova, see footnote on p. 253.

*2Intrinsic conductivity (160 < T < 300 K).

*3Impurity (intrinsic) conductivity (T < 160 K).

*4V. I. Lazorenko, Investigations of the Physical Properties of Rare-Earth Metal Silicides of the Cerium Subgroup [in Russian], Author's abstract, Cand. Dissertation, Kiev (1975); see [593, 472].

Phase	Hall constant, $cm^3/K \cdot 10^4$	Source	Year
1	2	3	4
$LaSi_2$	−1,5	—*4	1975
Ce_5Si_3	+1,4	—*4	1975
CeSi	+1,0	—*	1975
$CeSi_2$	+1,2	—*	1975
Pr_5Si_3	+1,0	—*	1975
PrSi	+1,2	—*	1975
$PrSi_2$	+2,5	—*	1975
Sm_5Si_3	−1,9	—*	1975
SmSi	−2,0	—*	1975
$SmSi_2$	−2,1	—*	1975
Ti_5Si_3	−0,27	[633]	1960
TiSi	−0,43	[633]	1960
$TiSi_2$	−0,63	[633]	1960
Zr_5Si_3	−3,75	[633]	1960
$ZrSi_2$	−1,46	[633]	1960
V_3Si	−0,17	[633]	1960
V_5Si_3	−1,0	[633]	1960
VSi_2	−1,95	[633]	1960
$Ta_{4,5}Si$	−2,46	[633]	1960
Ta_2Si	−4,76	[633]	1960
Ta_5Si_3	−4,54	[633]	1960
$TaSi_2$	−0,88	[633]	1960
Cr_3Si	+0,49	[633]	1960
Cr_5Si_3	−0,51	[633]	1960
CrSi	−0,46	[633]	1960
$CrSi_2$	+66,5	[633]	1960
Mo_3Si	−0.26	[633]	1960
Mo_5Si_3	−0,42	[633]	1960
$MoSi_2$	+12,7	[633]	1960
W_3Si	+1,19	[633]	1960
W_5Si_3	−0,3	[633]	1960
WSi_2	+841	[633]	1960
Mn_5Si_3	14	[788]	1965
Re_3Si	+1,79	[633]	1960
ReSi	+65,41	[633]	1960
$ReSi_2$	+8700	[633]	1960
FeSi	−4	[667]	1970
$β-FeSi_2$	−0,3	[633]	1960
CoSi	−1,73	[633]	1960
$CoSi_2$	+2,53	[633]	1960
Ni_2Si	+0,35	[633]	1960
$NiSi_2$	+3,77	[633]	1960
UP_2	+22,5	[599]	1969
TiP	−3,0	[558]	1961
FeP_2**	~−10	[600]	1971
CeS	+2000	[33]	1972
$NbS_{1.6}$	+18.2	[243]	1966

*V. I. Lazorenko, see footnote on p. 254.
**Monocrystal.

FORBIDDEN BAND WIDTHS OF SEMICONDUCTOR REFRACTORY COMPOUNDS

Phase	Forbidden band width, F_0, eV	Source	Year	Notes
1	2	3	4	5
CaB_6	0.20	[564]	1970	See [789]
SrB_6	0.38	[789]	1970	
BaB_6	0.15	[564]	1970	See [789]
SmB_6	0.0023	[566]	1969	See [790]
EuB_6	0.3	[252]	1973	
YbB_6	0.14	[252]	1973	
$MnAl_3$	0.58	[18]	1965	
Mg_2Si	0.75—0,77	[260]	1959	
Ca_2Si	1.9	[260]	1959	
$BaSi_2$	0,48	[449]	1963	
$LaSi_2$	0,19	[449]	1963	
$CrSi_2$	1.3	[449]	1963	
$ReSi_2$	0,12	[449]	1963	
$FeSi$	0.04—0.05	[667]	1970	See [452]
$FeSi_2$	0.5—0.8	[449]	1963	
LaP	0.54	[791]	1970	Gradual change in E_0 in REM monophosphides series
LuP	1.30	[791]	1970	
FeP_2	0.37	[600]	1971	
La_2S_3	2.77	[792]	1970	
Ce_2S_3	2,00	[792]	1970	See [262]
Pr_2S_3	2,60	[792]	1970	
Nd_2S_3	2.65	[792]	1970	See [601]
SmS	0,22	[262]	1965	
Sm_2S_3	2,96	[262]	1965	See [792]
EuS	2.1	[602]	1965	See [792]
Gd_2S_3	2.55	[792]	1970	
YbS	0,40	[605]	1970	
PuS	0.24	[607]	1969	
Pu_3S_4	0,50	[607]	1969	
TiS_2	—	[237]	1967	Degenerated semiconductor
ZrS_2	1.68	[237]	1967	
HfS_2	1,96	[237]	1967	
Nb_2S_3	0,05—0,1	[243]	1966	
B_4C	1.64	[793]	1961	
SiC	1,5—3,5	[11]	1968	
α-BN	4,6	[14]	1969	
β-BN	3	[14]	1969	
AlN	4,26	[249]	1964	
Si_3N_4	3,9	[612]	1960	

Phase	Molar magnetic susceptibility, $\chi \cdot 10^6$	Temp., K	Effective magnetic moment, μ_B	Curie–Weiss constant, θ_p, K	Source	Year	Notes
1	2	3	4	5	6	7	8
GdBe$_{13}$	—	>19	7,82	35	[643]	1973	
TbBe$_{13}$	—	>14	9.7	14	[643]	1973	
DyBe$_{13}$	—	>9	10.5	1	[643]	1973	
CoBe	840	4,2	—	—	[644]	1973	
NiBe	Weak paramagnetic				[644]	1973	
CaB$_6$	813	—	1.41	—	[3]	1975	
SrB$_6$	0	—	0	—	[3]	1975	
BaB$_6$	1520	—	1.92	—	[3]	1975	
ScB$_2$	~100	77—625	—	—	[56]	1970	Slightly dependent on temp.
ScB$_{12}$	−65	77—625	—	—	[56]	1970	Independent of temp.
YB$_2$	Diamagnetic				[3]	1975	See [645]
LaB$_6$	60	293—673	0	—	[3]	1975	
CeB$_6$	—	80—300	2,49	−76	[645]	1967	
PrB$_6$	—	80—300	3,59	−68	[645]	1967	
NdB$_6$	—	80—300	3,54	−42	[645]	1967	
SmB$_6$	—	80	~1,5	—	[645]	1967	Deviates from Curie–Weiss law. Magnetic orderedness absent up to 0.35 K [566].
	—	300	~2,3	—	[645]	1967	
EuB$_6$	—	80—300	8,1	9	[645]	1967	
GdB$_4$	—	82—300	8,07	−50	[646]	1967	
GdB$_6$	—	80—300	8,01	−55	[645]	1967	
TbB$_6$	—	80—300	9,43	−35	[645]	1967	
DyB$_4$	—	82—300	10.4	−21	[646]	1967	
DyB$_6$	—	80—300	10,63	−21	[645]	1967	
HoB$_4$	—	82—300	10.8	−21	[646]	1967	
ErB$_4$	—	82—300	9,52	−11	[646]	1967	
TmB$_4$	—	82—300	7.68	−23	[646]	1967	
YbB$_6$	—	623—1033	4.58	−2	[647]	1968	See [645]
UB$_2$	550	300	—	—	[3]	1975	
UB$_4$	1390	300	—	—	[3]	1975	
UB$_{12}$	70	300	—	—	[3]	1975	
TiB$_2$	31,3	293	—	—	[648]	1974	See [483]
ZrB$_2$	−67,7	293	—	—	[648]	1974	
HfB$_2$	−4,0	293	—	—	[648]	1974	
VB	Paramagnetic				[647]	1968	
VB$_2$	34.1	293	—	—	[648]	1974	See [483]
NbB$_2$	8,0	293	—	—	[648]	1974	
TaB$_2$	−64,8	293	—	—	[648]	1974	

Phase	Molar magnetic susceptibility, $\chi \cdot 10^6$	Temp., K	Effective magnetic moment, μ_B	Curie—Weiss constant, θ_p, K	Source	Year	Notes
1	2	3	4	5	6	7	8
CrB		Paramagnetic			[647]	1968	
CrB_2	390	293	—	—	[648]	1974	Above the Néel point (88 K) the Curie—Weiss law is not obeyed; see [647, 649, 483].
Mo_2B_5	61,5	293	—	—	[648]	1974	
W_2B_5	506,0	293	—	—	[648]	1974	
Mn_2B		Paramagnetic			[486]	1970	
MnB	—	>578	2,71	575	[647]	1968	
Mn_3B_4		Antiferromagnetic —		543	[647]	1968	See [650]
MnB_2	—	>157	2.30	148	[647]	1968	
FeB	—	>598	1,84	625	[647]	1968	
CoB		Paramagnetic			[647]	1968	
Ni_3B	264	293	—	—	[435]	1972	
Ni_2B	94	293	—	—	[435]	1972	
NiB	—5.5	293	—	—	[3]	1975	
LaC_2	905	293	—	—	[257]	1974	
CeC_2	—	70—400	2.19	—61	[257]	1974	
PrC_2	—	70—400	3,15	—5,2	[257]	1974	
NdC_2	—	70—400	3,53	—40	[257]	1974	
SmC_2	—	70—400	2.85	—139	[257]	1974	
GdC_2	—	70—400	7,59	41,3	[257]	1974	
TbC_2	—	70—400	9,57	—91,2	[257]	1974	
DyC_2	—	70—400	10.53	—68,9	[257]	1974	
HoC_2	—	70—400	10.47	—25,6	[257]	1974	
ErC_2	—	70—400	8.75	14,7	[257]	1974	
YbC_2	—	70—400	3.69	—388	[257]	1974	
UC	~800	300	—	—	[575]	1972	
U_2C_3	—	>59	2,10	—170	[576]	1965	
PuC	~1000	270	—	—	[369]	1964	See [647]
$NpC_{0.95}$	—	>317	3,22	—	[588]	1968	See [651]
$TiC_{0.95}$	~5	293	—	—	[652]	1962	
$TiC_{0.90}$	~20	293	—	—	[652]	1962	
$TiC_{0.80}$	~60	293	—	—	[652]	1962	See [257, 319]
$TiC_{0.70}$	~100	293	—	—	[652]	1962	
$ZrC_{0.95}$	~—20	293	—	—	[652]	1962	
$ZrC_{0.90}$	~—10	293	—	—	[652]	1962	
$ZrC_{0.80}$	~10	293	—	—	[652]	1962	
$ZrC_{0.70}$	~30	293	—	—	[652]	1962	
$HfC_{0.95}$	~—30	293	—	—	[652]	1962	
$HfC_{0.90}$	~—20	293	—	—	[652]	1962	See [257, 319]
$HfC_{0.80}$	~0	293	—	—	[652]	1962	
$HfC_{0.70}$	~20	293	—	—	[652]	1962	

MAGNETIC PROPERTIES (continued)

Phase	Molar magnetic susceptibility, $\chi \cdot 10^6$	Temp., K	Effective magnetic moment, μ_B	Curie—Weiss constant, θ_p, K	Source	Year	Notes
1	2	3	4	5	6	7	8
V_2C	176.8	293	—	—	[653]	1970	
$VC_{0.88}$	~25	293	—	—	[652]	1962	
$VC_{0.80}$	~50	293	—	—	[652]	1962	See [257, 319]
$VC_{0.70}$	~80	293	—	—	[652]	1962	
Nb_2C	38.5	293	—	—	[652]	1970	
$NbC_{0.95}$	~18	293	—	—	[652]	1962	
$NbC_{0.90}$	~10	293	—	—	[652]	1962	
$NbC_{0.80}$	~—5	293	—	—	[652]	1962	
$NbC_{0.70}$	~5	293	—	—	[652]	1962	See [257, 319]
$TaC_{0.95}$	~10	293	—	—	[652]	1962	
$TaC_{0.90}$	~0	293	—	—	[652]	1962	
$TaC_{0.80}$	~—15	293	—	—	[652]	1962	
$TaC_{0.70}$	~—5	293	—	—	[652]	1962	
Mo_2C	25.3	293	—	—	[653]	1970	
Fe_3C	—	>483	3.89	233	[647]	1968	See [486]
Fe_2C	—	>520	5.55	246	[647]	1968	
LaN	60	295	—	—	[14]	1969	
CeN	—	300	2,75	—85	[587]	1965	
PrN	—	—	3,7	0	[647]	1968	See [586]
NdN	—	>4	4,0	24	[647]	1968	See [586, 654]
SmN	1125	295	—	—	[586]	1963	
EuN	—	—	~5.0	—200	[647]	1968	
GdN	—	—	8,15	69	[647]	1968	
TbN	—	>42	10,0	34	[647]	1968	
DyN	—	>26	10,9	20	[647]	1968	See [586]
HoN	—	>18	10,7	12	[647]	1968	
ErN	—	>5	9,4	4	[647]	1968	
TmN	—	—	7,60	—18	[647]	1968	
YbN	7250	295	—	—	[586]	1963	See [655]
LuN	Weak paramagnetic				[586]	1963	χ does not depend on temp.
ThN	40	293	—	—	[364]	1972	Does not depend on temp.
UN	—	85—300	3,08	—310	[656]	1965	See [575, 657]
α-U_2N_3	—	96—300	1,92	42	[656]	1965	See [657]
β-U_2N_3	—	—	1,90	—	[656]	1965	
NpN	—	>82	2,13	—	[588]	1968	
TiN	37	293	—	—	[652]	1962	
ZrN	22	293	—	—	[652]	1962	
VN	130	293	—	—	[319]	1974	
NbN	30	293	—	—	[652]	1962	
TaN	25	293	—	—	[652]	1962	
Ta_3N_5	Diamagnetic				[17]	1974	
CrN	—	77	2.41	—	[658]	1969	

Phase	Molar magnetic susceptibility, $\chi \cdot 10^6$	Temp., K	Effective magnetic moment, μ_B	Curie—Weiss constant, θ_p, K	Source	Year	Notes
1	2	3	4	5	6	7	8
$Mo_{16}N_7$	Paramagnetic				[120]	1970	Above 175 K χ does not depend on temp., below it the Curie—Weiss law is obeyed
Mn_5N_2	—	—	3,94	—1070	[647]	1968	
CeAl	—	—	2.34	4	[659]	1974	
$CeAl_3$	—	—	2,63	—4,6	[659]	1974	
PrAl	—	—	—	11	[659]	1974	
$PrAl_3$	—	—	3,74	—14	[659]	1974	
NdAl	—	—	—	—4	[659]	1974	
$NdAl_2$	—	—	3,1	61	[659]	1974	
$NdAl_3$	—	—	4,11	5	[659]	1974	
Gd_3Al_2	—	275—385	8,17	280	[660]	1966	
GdAl	—	82—385	5,69	260	[660]	1966	
$GdAl_2$	—	171—385	7,88	171,5	[660]	1966	See [659]
$GdAl_3$	—	82—385	8,35	—99,5	[660]	1966	
Tb_3Al_2	—	>190	10,1	125	[661]	1971	
TbAl	—	—	10,11	24	[659]	1974	
$TbAl_2$	—	—	9,82	108	[659]	1974	
$TbAl_3$	—	—	10,0	—64	[659]	1974	
Dy_3Al_2	—	>100	11,1	95 ($\parallel c$) 20 ($\perp c$)	[661]	1971	
DyAl	—	—	11.15	17	[659]	1974	
$DyAl_2$	—	—	9.7	68	[659]	1974	
$DyAl_3$	—	—	10,85	—51	[659]	1974	
Ho_3Al_2	—	—	10.9	10	[659]	1974	
HoAl	—	—	11,25	22	[659]	1974	See [662]
$HoAl_3$	—	—	10.89	—26	[659]	1974	
Er_3Al_2	—	—	9,6	—3	[659]	1974	
ErAl	—	—	10.05	28	[659]	1974	
$ErAl_2$	—	—	9,56	16	[659]	1974	
$ErAl_3$	—	—	9,87	6	[659]	1974	
Tm_3Al_2	—	—	7,8	—10	[659]	1974	
TmAl	—	—	7,70	—2	[659]	1974	
$TmAl_3$	—	—	7,88	—19	[659]	1974	
$YbAl_3$	—	—	4.62	—300	[659]	1974	
V_5Al_8	1.566	300	—	—	[18]	1965	
VAl_3	0.25	300	—	—	[18]	1965	
V_4Al_{23}	0.42	300	—	—	[18]	1965	
VAl_6	0.83	300	—	—	[18]	1965	
V_7Al_{45}	1,266	300	—	—	[18]	1965	
VAl_{10}	0,75	300	—	—	[18]	1965	
VAl_{11}	0,95	300	—	—	[18]	1965	
$CrAl_4$	1,866	300	—	—	[18]	1965	

Phase	Molar magnetic susceptibility, $\chi \cdot 10^6$	Temp., K	Effective magnetic moment, μ_B	Curie—Weiss constant, θ_p, K	Source	Year	Notes
1	2	3	4	5	6	7	8
Cr_2Al_{11}	1,113	300	—	—	[18]	1965	
$CrAl_7$	0,5	300	—	—	[18]	1965	
MnAl	—	—	3,17	~646	[663]	1968	
MnAl	7,2	292	—	—	[18]	1965	
Mn_4Al_{11}	6,2	293	—	—	[18]	1965	
Mn_3Al_{10}	7,2	293	—	—	[18]	1965	
$MnAl_4$	5,6	290	—	—	[18]	1965	
$MnAl_6$	0,766	300	—	—	[18]	1965	
FeAl	—	4	—	—	[663]	1968	No magnetic moment observed
$FeAl_2$	5,0	290	—	—	[18]	1965	
Fe_2Al_5	16,0	288	—	—	[18]	1965	
Fe_4Al_{13}	6,7	288	—	—	[18]	1965	
$CoAl_3$	0,866	300	—	—	[18]	1965	
Co_4Al_{13}	0,8	300	—	—	[18]	1965	
Co_2Al_9	1,6	300	—	—	[18]	1965	
Ni_2Al_3	1,2	300	—	—	[18]	1965	
$NiAl_3$	0,2	300	—	—	[18]	1965	
$LaSi_2$ $CeSi_2$ $NdSi_2$	—	—	—	—	[664]	1958	Magnetic order not revealed
Gd_5Si_4	—	—	8,15	349	[659]	1974	
$GdSi_2$	—	50—300	~7,8	~—28	[665]	1967	
Tb_5Si_4	—	—	9,31	216	[659]	1974	
$TbSi_2$	—	140—300	~9,8	~—18	[665]	1967	
Dy_5Si_4	—	—	10,3	233	[659]	1974	
$DySi_2$	—	20—300	~10,5	~—18	[655]	1967	
Ho_5Si_4	—	—	11,1	69	[659]	1974	
HoSi	—	300	10,64	—	[666]	1971	Metamagnetic at 4.2°K
$HoSi_2$	—	20—300	~10,4	~—20	[665]	1967	
Er_5Si_4	—	—	9,9	20	[659]	1974	
ErSi	—	300	9,38	—	[666]	1971	Metamagnetic at 4.2°K
$ErSi_2$	—	4,2—300	~9,5	—	[665]	1967	
TmSi	—	300	7,45	—	[666]	1971	Metamagnetic at 4.2°K
Ti_5Si_3	810	298	—	—	[594]	1958	
TiSi	55	298	—	—	[594]	1958	
$TiSi_2$	129	298	—	—	[594]	1958	See [473]
ZrSi	—67	298	—	—	[594]	1958	
$ZrSi_2$	—103	298	—	—	[594]	1958	
V_3Si	800	300	—	—	[22]	1971	
VSi_2	161	298	—	—	[594]	1958	
V_5Si_3	Paramagnetic				[22]	1971	
$NbSi_2$	—37	298	—	—	[594]	1958	

MAGNETIC PROPERTIES (continued)

Phase	Molar, magnetic susceptibility, $\chi \cdot 10^6$	Temp., K	Effective magnetic moment, μ_B	Curie–Weiss constant, θ_p, K	Source	Year	Notes
1	2	3	4	5	6	7	8
$TaSi_2$	—40	298	—	—	[594]	1958	
Cr_3Si	704	298	—	—	[594]	1958	See [22]
Cr_5Si_3	892	298	—	—	[594]	1958	
CrSi	—	—	3,01	—1990	[647]	1968	
$CrSi_2$	41	298	—	—	[594]	1958	
$MoSi_2$	—36,5	298	—	—	[594]	1958	
WSi_2	—82	298	—	—	[594]	1958	
Mn_3Si	—	300—1100	—	—	[22]	1971	Complex relationship between χ and temp.
Mn_5Si_3	—	293—873	3,9	—	[22]	1971	
MnSi	—	30—300	1,4	—	[22]	1971	
Mn_nSi_{2n-x}	Paramagnetic				[22]	1971	Paramagnetism in higher Mn silicides strongly depends on their composition and temp.
Fe_5Si_3	—	>381	1,135	—	[23]	1966	
FeSi	—	>443	2,55	—107	[647]	1968	See [667]
α-$FeSi_2$	~140	1373	—	—	[668]	1970	See [22]
β-$FeSi_2$	~0	873	—	—	[668]	1970	See [22]
Co_2Si	935	293	—	—	[22]	1971	Curie–Weiss law not obeyed
CoSi	—	—	3,08	—180	[647]	1968	
$CoSi_2$	Pauli paramagnetic				[22]	1971	
Ni_2Si	43,5	293	—	—	[22]	1971	Not dependent on temp.
NiSi	Diamagnetic				[22]	1971	
$Ni_{1,04}Si_{1,93}$	Pauli paramagnetic				[22]	1971	
CeP	—	>10	2,56	5	[647]	1968	See [655]
PrP	—	—	3,77	—4	[647]	1968	See [655]
NdP	—	—	—	11	[659]	1974	
SmP EuP	Van Vleck paramagnetic				[655]	1967	
GdP	—	>15	7,7	0	[647]	1968	
TbP	—	>8	9,2	1	[647]	1968	
DyP	—	—	9,9	6	[647]	1968	
HoP	—	—	10,2	4,2	[647]	1968	
ErP	—	>4	9,3	0	[647]	1968	
TmP	—	—	—	—2	[655]	1967	
YbP	—	—	—	—55	[659]	1974	
UP	—	124—300	3,56	3	[669]	1969	
U_3P_4	—	140—400	2,75	138	[669]	1969	
UP_2	—	Up to 203	2.0	—	[599]	1969	

Phase	Molar, magnetic susceptibility, $\chi \cdot 10^5$	Temp., K	Effective magnetic moment, μ_B	Curie–Weiss constant, θ_p, K	Source	Year	Notes
1	2	3	4	5	6	7	8
Cr_3P	—	>294	2.5	—	[599]	1969	
	Pauli paramagnetic				[647]	1968	
MnP	—	>291,5	2,92	312	[647]	1968	
Fe_3P	—	—	1,94	781	[647]	1968	
Fe_2P	—	—	3,2	478	[647]	1968	See [663]
FeP	—	1,5	—	—	[670]	1967	Magnetic order not revealed; see [647, 663]
FeP_2	Diamagnetic				[600]	1971	Monocrystal; polycrystal is weak paramagnetic; see [647]
Ni_3P	Pauli paramagnetic				[647]	1968	
YS	100	293	—	—	[33]	1972	
Y_5S_7	39,3	293	—	—	[33]	1972	
Y_2S_3	83,4	293	—	—	[33]	1972	
LaS	281	293	—	—	[33]	1972	
La_3S_4	27,2	293	—	—	[221]	1956	
La_2S_3	—18,5	293	—	—	[33]	1972	
LaS_2	—50	293	—	—	[33]	1972	
CeS	2110	293	—	—	[33]	1972	
Ce_3S_4	2125	293	—	—	[221]	1956	
Ce_2S_3	2540	293	—	—	[33]	1972	
CeS_2	2290	293	—	—	[33]	1972	
Pr_2S_3	5385	293	—	—	[33]	1972	
NdS	4370	293	—	—	[33]	1972	
Nd_3S_4	4849	293	—	—	[221]	1956	
Nd_2S_3	5650	293	—	—	[33]	1972	
SmS	4970	293	—	—	[33]	1972	
Sm_3S_4	2350	293	—	—	[221]	1956	
Sm_2S_3	1020	293	1.55	—	[221]	1956	
EuS	—	16—300	8,05	16	[398]	1967	See [263]
Eu_3S_4	11500	298	—	—	[263]	1959	
$Eu_2S_{3,81}$	5800	298	—	—	[263]	1959	
GdS	21740	298	7,52	—	[603]	1968	
γ-Gd_2S_3	59990	298	8,07	—	[603]	1968	
γ-Dy_2S_3	45700	298	—	—	[33]	1972	
δ-Er_2S_3	38600	298	—	—	[33]	1972	
YbS	1450	298	—	—	[33]	1972	
Yb_3S_4	4740	298	—	—	[33]	1972	
Yb_2S_3	7130	298	—	—	[33]	1972	
ThS	Diamagnetic				[550]	1960	
Th_2S_3	Diamagnetic				[550]	1960	
US	5667	293	2,22	185	[234]	1964	See [669]

Phase	Molar magnetic susceptibility, $\chi \cdot 10^6$	Temp., K	Effective magnetic moment, μ_B	Curie—Weiss constant, θ_p, K	Source	Year	Notes
1	2	3	4	5	6	7	8
U_2S_3	2924	293	2.53	27	[234]	1964	Deviates from Curie—Weiss law
U_3S_5	4709	293	3,42	—20	[234]	1964	
α-US_2	3155	293	2,83	—30	[234]	1964	
β-US_2	3470	298	—	—	[33]	1972	
US_3	3470	293	3,42	—150	[234]	1964	
TiS	187	293	—	—	[264]	1964	See [663]
Ti_2S_3	178	293	—	—	[264]	1964	
TiS_2	120	293	—	—	[264]	1964	
TiS_3	—13	293	—	—	[264]	1964	
$V_{0,875}S$	—		3.76	—3120	[663]	1968	
Cr_7S_8	—	285—445	5,62	—1657	[244]	1967	At 450, 590, and 875°K anomalies appear on $\chi(T)$ curve
Cr_7S_8	—	475—575	5,17	—1338	[244]	1967	
Cr_7S_8	—	600—1200	5,04	—1244	[244]	1967	
Cr_5S_6	—	470—580	4,33	—677	[244]	1967	At 470, 600, and 820°K anomalies appear on $\chi(T)$ curve
Cr_5S_6	—	700—820; >930	4,83	—1020	[244]	1967	
Cr_3S_4	—	275—1100	4,24	—565	[244]	1967	See [671]
Cr_2S_3	—	210—1100	4,13	—525	[244]	1967	Trig.
Cr_2S_3	—	245—940	4,10	—610	[244]	1967	Rhombohed.
Cr_2S_5	—	100—390	4,17	—428	[317]	1967	
MoS_2	Diamagnetic				[672]	1973	
α-MnS	—	—	5,6	—403	[663]	1968	
β-MnS	—	—	5,82	—975	[663]	1968	
MnS_2	—	—	6,30	—	[663]	1968	
ReS	43,5	293	—	—	[265]	1963	
ReS_2	—87,5	293	—	—	[265]	1963	
FeS	—	—	5,25	—1140	[663]	1968	
Fe_7S_8	—	—	5,93	—	[663]	1968	
FeS_2 (pyrites)	Diamagnetic				[609]	1967	
FeS_2 (marcasite)	"				[609]	1967	
CoS	—	—	1,7	—645	[663]	1968	
Co_3S_4	Pauli paramagnetic				[608]	1967	
CoS_2	—	—	1,85	—	[663]	1968	
NiS	—	—	2,66	—3000	[663]	1968	
NiS_2	—	—	3,19	—1500	[663]	1968	
α-BN	3.35	293	—	—	[673]	1970	Powder
	0,3	293	—	—	[673]	1970	$\chi_\parallel$

MAGNETIC PROPERTIES (continued)

Phase	Molar magnetic susceptibility, $\chi \cdot 10^6$	Temp., K	Effective magnetic moment, μ_B	Curie–Weiss constant, Θ_p, K	Source	Year	Notes
1	2	3	4	5	6	7	8
α-BN	0,50	293	—	—	[673]	1970	$\chi_\perp$
α-BN		Diamagnetic			[674]	1970	
β-BN	0,3	293	—	—	[673]	1970	
SiC		Diamagnetic			[11]	1968	
Si_3N_4	56—101	293	0.43—0.49	—	[14]	1969	

CURIE AND NÉEL TEMPERATURES

Phase	T_C, K	T_N, K	Source	Year	Notes
1	2	3	4	5	6
$GdBe_{13}$	—	19	[643]	1973	
$TbBe_{13}$	—	14	[643]	1973	
$DyBe_{13}$	—	9	[643]	1973	
$CrBe_{12}$	50	—	[675]	1973	
$FeBe_2$	643	—	[676]	1973	
CeB_6	344	—	[677]	1957	
PrB_6	~0	—	[677]	1957	
NdB_6	455	—	[677]	1957	
GdB_6	60	—	[677]	1957	
GdB_6	—	13,5	[647]	1968	
YbB_6	2	—	[677]	1957	
CrB_2	—	88	[678]	1969	See [483, 649]
MnB	578	—	[647]	1968	See [486]
Mn_3B_4	—	392	[647]	1968	
MnB_2	140	—	[678]	1969	See [486, 649, 650]
Fe_3B	791	—	[679]	1964	
Fe_2B	1015	—	[647]	1968	
FeB	598	—	[647]	1968	
Co_3B	747	—	[647]	1968	
Co_2B	433	—	[647]	1968	
CoB	477	—	[663]	1968	
CeC_2	—	33	[647]	1968	
PrC_2	—	15	[647]	1968	
NdC_2	—	29	[647]	1968	
TbC_2	—	66	[647]	1968	
HoC_2	—	26	[647]	1968	
U_2C_3	—	59	[576]	1965	

Phase	T_C, K	T_N, K	Source	Year	Notes
1	2	3	4	5	6
PuC	—	60	[680]	1965	
$NpC_{0.95}$	190	317	[588]	1968	See [651]
Cr_7C_3	193	—	[681]	1966	
Fe_3C	487	—	[682]	1970	
Fe_5C_2	521	—	[111]	1967	
$\varepsilon\text{-}Fe_2C$	653	—	[647]	1968	
$\chi\text{-}Fe_2C$	520	—	[647]	1968	
CeN	—	15	[655]	1967	
NdN	24	—	[655]	1967	Ferrimagnetic; see [492]
GdN	69	—	[683]	1964	
TbN	34	—	[655]	1967	Ferrimagnetic
DyN	20	—	[655]	1967	Ferrimagnetic
HoN	12	—	[655]	1967	Ferrimagnetic
ErN	4	—	[655]	1967	Ferrimagnetic
UN	—	52	[575]	1972	Antiferromagnetic ordering on ferromagnetic planes (001)
$\beta\text{-}U_2N_3$	186	—	[657]	1962	
NpN	82	—	[588]	1968	
CrN	—	273	[663]	1968	
Mn_4N	738	—	[647]	1968	
$Mn_4N_{0.92}$	750	—	[647]	1968	
$Mn_4N_{0.8}$	778	—	[647]	1968	
Mn_3N_2	1070	—	[677]	1957	
Mn_2N	—	~301	[647]	1968	
Fe_4N	761	—	[647]	1968	See [111]
Fe_3N	567	—	[111]	1967	
$\varepsilon\text{-}Fe_2N$	103	—	[684]	1968	Hex.; $Fe_{2.09}N$
$\zeta\text{-}Fe_2N$	83	—	[684]	1968	Distorted hex.; $Fe_{2.01}N$
CeAl	—	9	[659]	1974	
PrAl	—	20	[659]	1974	
$PrAl_2$	33	—	[659]	1974	
NdAl	—	29	[659]	1974	
$NdAl_2$	65	—	[659]	1974	
$SmAl_2$	123	—	[659]	1974	
Gd_3Al_2	275	—	[660]	1966	
GdAl	—	42	[659]	1975	
$GdAl_2$	171	—	[660]	1966	
$GdAl_3$	—	17	[659]	1974	
Tb_3Al_2	190	—	[661]	1971	
TbAl	—	72	[659]	1974	
$TbAl_2$	114	—	[659]	1974	
$TbAl_3$	—	21	[659]	1974	
Dy_3Al_2	100	—	[661]	1971	See [659]
DyAl	—	20	[659]	1974	
$DyAl_2$	58	—	[659]	1974	
$DyAl_3$	—	23	[659]	1974	
Ho_3Al_2	33	—	[659]	1974	
HoAl	26	—	[662]	1968	
$HoAl_2$	27	—	[659]	1974	
$HoAl_3$	—	9	[659]	1974	

Phase	T_C, K	T_N, K	Source	Year	Notes
1	2	3	4	5	6
Er_3Al_2	—	9	[659]	1974	
ErAl	—	13	[659]	1974	
$ErAl_2$	14,5	—	[659]	1974	
$ErAl_3$	21	—	[659]	1974	
Tm_3Al_2	—	3	[659]	1974	
TmAl	—	10	[659]	1974	
MnAl	646	—	[663]	1968	
Fe_3Al	773	—	[663]	1968	
$FeAl_3$	103	—	[18]	1965	
$PrSi_2$	10,5	—	[664]	1958	
$GdSi_2$	—	27	[665]	1967	
$TbSi_2$	—	17	[665]	1967	
$DySi_2$	—	17	[665]	1967	
HoSi	—	25	[666]	1971	
$HoSi_2$	—	18	[665]	1967	
ErSi	—	10	[666]	1971	
TmSi	—	10	[666]	1971	
Mn_2Si	5	—	[677]	1957	
Mn_5Si_3	—	68	[663]	1968	
MnSi	38	—	[663]	1968	
Fe_3Si	823	—	[663]	1968	
Fe_5Si_3	381	—	[23]	1966	
FeSi	—	~443	[663]	1968	
α-FeSi	798	—	[23]	1966	
$CoSi_2$	170	—	[677]	1957	
CeP	—	9	[655]	1967	
NdP	11	—	[659]	1974	Ferrimagnetic (?)
NdP	—	8	[655]	1967	Antiferromagnetic (?)
GdP	—	15	[655]	1967	
TbP	—	8	[655]	1967	See [653]
DyP	8	—	[655]	1967	Ferrimagnetic (?)
HoP	7	—	[655]	1967	Ferrimagnetic
ErP	—	4	[655]	1967	See [663]
UP	—	124	[669]	1969	
U_3P_4	138	—	[656]	1967	
UP_2	—	203	[599]	1969	
Mn_3P	—	115	[647]	1968	
MnP	291,5	—	[647]	1968	See [663]
Fe_3P	686	—	[679]	1964	See [663]
Fe_2P	266	—	[663]	1968	
FeP	215	—	[663]	1968	
FeP_2	—	~250	[663]	1968	
Ce_2S_3	57	—	[677]	1957	
EuS	15	—	[398]	1967	
US	185	—	[669]	1969	
PuS	—	4,5	[607]	1969	
Pu_3S_4	—	10	[607]	1969	
α-Pu_2S_3	—	7	[607]	1969	
PuS_2	—	15	[607]	1969	
$V_{0.875}S$	—	1040	[663]	1968	
Cr_7S_8	—	136	[244]	1967	

CURIE AND NÉEL TEMPERATURES (continued)

Phase	T_C, K	T_N, K	Source	Year	Notes
1	2	3	4	5	6
Cr_5S_6	305	165	[244]	1967	Ferrimagnetic in 170-305 K range
Cr_3S_4	—	230	[244]	1967	See [671].
Cr_2S_3	120	—	[244]	1967	Rhombohedral; ferrimagnetic
	110	—	[244]	1967	Trigonal; ferrimagnetic.
α-MnS	—	130	[663]	1968	
β-MnS	—	160	[663]	1968	At 100 K magnetic cores change symmetry
MnS_2	—	<77	[663]	1968	
$Fe_{0,9}S$	—	~480	[663]	1968	Conversion: antiferro-ferri
$Fe_{0,9}S$	~535	—	[663]	1968	Conversion: ferri-para.
$Fe_{0,9}S$	—	~600	[663]	1968	At about 580 K conversion para-antiferro
$Fe_{1-x}S$	—	600	[663]	1968	$0.06 \leqslant x \leqslant 0,09$
FeS	—	600	[663]	1968	
Fe_7S_8	578	—	[663]	1968	
CoS	—	358	[663]	1968	
CoS_2	110	—	[663]	1968	
NiS	—	150	[663]	1968	

ELECTRON SPECIFIC HEAT

Phase	Coef. electron specific heat, mJ/(mole·deg^2)	Source	Year	Notes
1	2	3	4	5
VBe_{12}	1,38	[675]	1973	
$CrBe_{12}$	1,79	[675]	1973	
$FeBe_2$	4,32	[676]	1973	
ScB_2	2.2	[483]	1969	
LaB_6	2,6	[484]	1969	
YbB_6	~0	[484]	1969	
ThB_6	4,8	[484]	1969	
TiB	2,96	[368]	1969	
TiB_2	1,40	[368]	1969	Traces of TiB_2

Phase	Coef. electron specific heat, mJ/(mole·deg^2)	Source	Year	Notes
1	2	3	4	5
	1,08	[483]	1969	
ZrB$_2$	0,93	[368]	1969	
HfB$_2$	1,00	[368]	1969	
VB$_2$	4,84	[483]	1969	
NbB	1,39	[368]	1969	Traces Nb$_3$B$_2$
Ta$_2$B	5.76	[368]	1969	Traces γ-TaB
CrB$_2$	13,6	[483]	1969	
Mo$_2$B	8,52	[368]	1969	
MoB	2,32	[368]	1969	
MoB$_2$	3,38	[368]	1969	
W$_2$B	5,20	[368]	1969	
WB	2.12	[368]	1969	
Mn$_2$B	13,76	[486]	1970	
MnB	5,48	[486]	1970	
MnB$_2$	2,80	[486]	1970	
MnB$_2$	4,45	[483]	1969	
Fe$_2$B	10,64	[801]	1968	
FeB	10,33	[801]	1968	
Co$_3$B	21,15	[487]	1968	
Co$_2$B	13,40	[487]	1968	
CoB	0,66	[487]	1968	
ScC	6	[319]	1974	
UC	20,3	[588]	1968	See [488]
ThC	2,92	[802]	1971	
TiC$_{0.5}$	2,8	[803]	1970	
TiC$_{0.9}$	1,285	[803]	1970	
TiC	0,51	[803]	1970	
ZrC	0,75	[319]	1974	
HfC	0,75	[319]	1974	
V$_2$C	2,26	[653]	1970	
VC$_{0.88}$	3.15	[319]	1974	
VC$_{0.84}$	3.0	[319]	1974	
VC$_{0.83}$	2.8	[319]	1974	
NbC$_{0.48}$	1,57	[489]	1968	ε-Fe$_2$N; see [653]
NbC$_{0.5}$	0,839	[319]	1974	Rhomb.
NbC$_{0.98}$	2,83	[489]	1968	
NbC$_{0.91}$	2,52	[489]	1968	
NbC$_{0.86}$	2,22	[489]	1968	
NbC$_{0.83}$	2,15	[489]	1968	
NbC$_{0.77}$	2,11	[489]	1968	
TaC$_{0.47}$	1,20	[489]	1968	$C6$
TaC	3,2	[319]	1974	
TaC$_{0.95}$	2,87	[489]	1968	
TaC$_{0.93}$	2,68	[489]	1968	
TaC$_{0.83}$	2,11	[489]	1968	

Phase	Coef. electron specific heat, mJ/(mole $\cdot$ deg^2)	Source	Year	Notes
1	2	3	4	5
TaC$_{0.78}$	2.05	[489]	1968	
α-MoC$_{0.69}$	4.40	[319]	1974	$B1$
η-MoC$_{0.64}$	3.79	[319]	1974	Hex.
α-Mo$_2$C	3,41	[319]	1974	Rhomb.
β-Mo$_2$C	2,94	[319]	1974	L_3'
WC	0.79	[319]	1974	Hex.
Fe$_3$C	6.25	[490]	1969	
LaN	3,5	[372]	1965	
ThN	3,12	[364]	1972	
UN	48.3	[588]	1968	
TiN	2,5—3,3	[319]	1974	
ZrN	2.67	[319]	1974	
HfN	2,73	[319]	1974	
VN	4,5—8,6	[319]	1974	
NbN$_x$	4,08—4.56	[319]	1974	
NbN$_{0.91}$	2,64	[319]	1974	$B1$
NbN$_{0.84}$	3,01	[319]	1974	
V$_{3.15}$Si	23,4	[374]	1969	
V$_{3.05}$Si	39	[374]	1969	
V$_{2.97}$Si	46,8	[374]	1969	
V$_{2.5}$Si	46,8	[374]	1969	
Co$_{1.015}$Si	3,32	[24]	1972	
Co$_{1.008}$Si	2,63	[24]	1972	
CoSi	1,30	[24]	1972	
CoSi$_{1.015}$	0,46	[24]	1972	
NdS	3.56	[397]	1972	

DIELECTRIC PROPERTIES

Phase	Frequency, cps	Temp., °C	Dielectric constant	Dissipation factor	Source	Year
1	2	3	4	5	6	7
La_2S_3	10^3	—	19.6	—	—*	1969
Pr_2S_3	10^3	—	14.2	—	—*	1969
Nd_2S_3	10^3	—	16.0	—	—*	1969
Sm_2S_3	10^3	—	18,0	—	—*	1969
TiN	—	—	7,0	—	[794]	1971
BN	10^2	10	4,15	0,00103	[14]	1969
	10^2	330	4,4	0.032		
	10^2	500	9,0	1.0 (470° C)		
	10^4	10	4,15	0,00042		
	10^4	330	—	0,0043		
	10^4	500	4,5	0.1 (470° C)		
	10^6	330	—	0,0012		
	10^8	10	4,15	0,000095		
	10^{10}	10	—	0,0003		
	10^{10}	330	—	0,0004		
	10^{10}	470	—	0,005		
Si_3N_4	—	18	9,4	—	[14]	1969
AlN	10^5	—	~30	—	[249]	1964
	10^6	—	26	—		
	10^7	—	~27	—		

*S. A. Kutolin, R. P. Samoilova, et al., Information Reference Sheet No. 000789, Semiconductor Instrument Series [in Russian] (1969).

Chapter IV

OPTICAL PROPERTIES

COLOR OF SOME REFRACTORY COMPOUNDS

Phase	Color in dispersed state (powder)	Phase	Color in dispersed state (powder)
1	2	1	2
Be_2B	Gray with rose shade	DyB_6	Blue
BeB_2	Dark gray	DyB_{100}	Black
$BeB6$	Brick red	HoB_4	Gray-brown
MgB_2	Dark brown	HoB_6	Blue
MgB_6	Dark brown	HoB_{100}	Black
MgB_{12}	Dark brown	ErB_4	Gray-brown
CaB_6	Black	ErB_6	Blue
SrB_6	Black with green shade	ErB_{100}	Black
BaB_6	Black with violet shade	TmB_4	Gray-brown
ScB_2	Gray	TmB_6	Blue
YB_4	Gray-brown	TmB_{100}	Black
YB_6	Blue-violet	YbB_6	Black
LaB_6	Purple-violet	YbB_{100}	Black
CeB_4	Gray-brown	LuB_6	Blue
CeB_6	Blue-violet	LuB_{100}	Black
PrB_4	Gray-brown	ThB_6	Red-violet
PrB_6	Blue	UB_4	Gray-steel
NdB_6	Blue	UB_{12}	Black
SmB_4	Gray-brown	TiB_2	Gray
SmB_6	Blue	ZrB_2	Gray
SmB_{100}	Black	HfB_2	Gray
EuB_6	Dark gray	VB_2	Gray
GdB_4	Gray-brown	NbB_2	Gray
GdB_6	Blue	TaB_2	Gray
GdB_{100}	Black	CrB_2	Gray
TbB_4	Gray-brown	Mo_2B_5	Light gray
TbB_6	Blue	W_2B_5	Light gray
TbB_{100}	Black	MnB	Red-brown
DyB_4	Gray-brown	MnB_2	Red-brown
		Fe_2B	Gray

Phase	Color in dispersed state (powder)	Phase	Color in dispersed state (powder)
1	2	1	2
FeB	Gray	ZrN	Light yellow with greenish shade
Be$_2$C	Reddish		
YC	Gold	HfN	Yellow-brown
YC$_2$	Yellow	V$_3$N	Gray-brown
LaC$_2$	Yellow	VN	Gray-brown
CeC$_2$	Red-yellow	Nb$_2$N	Gray
ThC	Yellow-gray	NbN	Light gray with yellow shade
ThC$_2$	Yellow-gray		
UC	Gray	Ta$_2$N	Black
TiC	Light gray	TaN	Gray with bluish shade
ZrC	Gray	Cr$_2$N	Dark gray
HfC	Gray	CrN	Black
VC	Gray	Mo$_2$N	Dark gray
NbC	Light brown	W$_2$N	Black
TaC	Gold-brown	WN	Brownish
Cr$_{23}$C$_6$	Gray	Mn$_2$N	Gray-blue
Cr$_7$C$_3$	Gray	MnN	Black
Cr$_3$C$_2$	Gray	Re$_3$N	Gray
Mo$_2$C	Dark gray	Co$_3$N	Gray-black
W$_2$C	Gray	Di$_3$N	Dark gray
WC	Gray	Silicides	Gray
Be$_3$N$_2$	Colorless or gray	Be$_3$P$_2$	Yellow
Mg$_3$N$_2$	Yellow-gray	Mg$_3$P$_2$	Colorless with yellow shade
α-Ca$_3$N$_2$	Brownish		
β-Ca$_3$N$_2$	Black	Ca$_3$P$_2$	Brownish red
γ-Ca$_3$N$_2$	Yellow-gold	BaP$_2$	Dark gray
Sr$_2$N	Black	Th$_3$P$_4$	Light steel
Sr$_3$N$_2$	Black	V$_3$P	Gray-black
Ba$_2$N	Black	VP	Gray-black
Ba$_3$N$_2$	Black	VP$_2$	Black
ScN	Dark blue	BeS	From white to gray
LaN	Black	MgS	Gray-white
CeN	Black	CaS	White
PrN	Black	SrS	White
NdN	Black	BaS	Gray white
SmN	Black	ScS	Gold-yellow
TiN	Yellow-bronze	Sc$_2$S$_3$	Yellow

Phase	Color in dispersed state (powder)	Phase	Color in dispersed state (powder)
1	2	1	2
YS	Ruby-red	α-Dy_2S_3	Red brown
Y_5S_7	Yellow	γ-Dy_2S_3	Black
Y_2S_3	Yellow	δ-Dy_2S_3	Green
YS_2	Brown-violet	DyS_2	Red-brown
LaS	Gold-yellow	Ho_2S_3	Yellow-gray
La_3S_4	Blue-black	ErS	Red violet
α-La_2S_3	Dark-red	Er_5S_7	Black
β-La_2S_3	Olive	Er_2S_3	Light brown
γ-La_2S_3	Yellow	Tm_2S_3	Yellow
LaS_2	Brown-yellow	Yb_2S_3	Yellow
CeS	Brass yellow to bronze	Lu_2S_3	Light yellow
Ce_3S_4	Black	ThS	Silvery
γ-Ce_2S_3	Red	Th_2S_3	Brown
CeS_2	Black-brown	Th_4S_7	Red
PrS	Dark gold with green shades	ThS_2	Brown violet, Purple
Pr_3S_4	Blue-black	$ThS_{2.5}$	Brown red
γ-Pr_2S_3	Dark brown	US	Gray
NdS	Gold	U_2S_3	Black
Nd_3S_4	Black	α-US_2	Gray black
α-Nd_2S_3	Jet black	β-US_2	Gray black
β-Nd_2S_3	Ruby	γ-US_2	Black with bluish tint
γ-Nd_2S_3	Brownish with yellow-green shades	Ti_2S	Dark brown
		TiS	Dark brown
Nd_4S_7	Brownish	Ti_3S_4	Black
NdS_2	Gold brown	Ti_2S_3	Black
SmS	Black	$TiS_{1.95}$	Dark bronze
Sm_3S_4	Black	TiS_3	Black
γ-Sm_2S_3	Brown	Zr_2S	Black
EuS	Black	ZrS	Black
Eu_3S_4	Black	Zr_2S_3	Black
GdS	Yellow	ZrS_2	Brownish
α-Gd_2S_3	Brown-black	ZrS_3	Orange-red
γ-Gd_2S_3	Brown	HfS	Black
GdS_2	Brown-violet	Hf_2S_3	Light brown
Tb_2S_3	Brown	HfS_2	Brownish
DyS	Red-violet	HfS_3	Bright-orange
Dy_5S_7	Black	VS	Black

COLOR OF SOME REFRACTORY COMPOUNDS (continued)

Phase	Color in dispersed state (powder)	Phase	Color in dispersed state (powder)
1	2	1	2
V_2S_3	Black with bluish shades	Fe_2S_3	Black
VS_5	Black	FeS_2	Light yellow
TaS_2	Black with green shades	α-CoS	Black
Mo_2S_3	Gray	β-CoS	Gray
MoS_2	Dark gray	CoS_2	Gray-black
WS_2	Dark gray	Ni_3S_2	Yellow-bronze
α-MnS	Green	α-NiS	Black
β-MnS	Red	β-NiS	Black
γ-MnS	Red	γ-NiS	Yellow
ReS	Black	Ni_3S_4	Silver white
ReS_3	Black	NiS_2	Gray
Re_2S_7	Dark brown	RuS_2	Gray
FeS	Dark brown	PdS_2	Gray black
		AlN	Light gray

EMISSION COEFFICIENTS (WAVELENGTH $\lambda = 0.665$ mm)

Phase	Emission coefficient	Temp., °C	Source	Year	Notes
1	2	3	4	5	6
$TiBe_2$	0.73	800—1800	[425]	1962	
$CrBe_2$	0,76	800—1800	[425]	1962	
$ReBe_2$	0,80	800—1600	[425]	1962	
CaB_6	0.75	800—1800	[425]	1962	
SrB_6	0.79	800—1800	[425]	1962	
BaB_6	0.84	800—1600	[425]	1962	
ScB_2	0,89	800—1800	[425]	1962	
YB_6	0,66—0,70	800—1700	[425]	1962	
YB_{12}	0,70	1000	—*	1970	
LaB_6	0,82	800—1700	[425]	1962	
CeB_6	0.72—0,77	800—1800	[425]	1962	
PrB_6	0,76—0,79	800—1900	[425]	1962	
NdB_6	0,51—0,47	800—1600	[425]	1962	

*V. V. Odintsov, Preparation and Physical Properties of Dodecaborides of Metals with UB_{12} Structural Types [in Russian], Author's abstract, Cand. Dissertation, Kiev (1970).

Phase	Emission coefficient	Temp., °C	Source	Year	Notes
1	2	3	4	5	6
SmB_6	0.77	900—1700	[426]	1960	
EuB_6	0,83	800—1800	[425]	1962	
GdB_6	0,66—0,60	800—1800	[425]	1962	
TbB_6	0.74	900—1800	[425]	1962	
TbB_{12}	0.65	1000	—*	1970	
DyB_6	0,8	1600	[425]	1962	
DyB_{12}	0.79	1000	—*	1970	
HoB_6	0.7	1600	[425]	1962	
HoB_{12}	0,83	1000	—*	1970	
ErB_6	0.7	1600	[425]	1962	
ErB_{12}	0,74	1000	—*	1970	
TmB_6	0,57—0,78	800—1900	[425]	1962	
TmB_{12}	0.74	1000	—*	1970	
YbB_6	0,73—0,75	800—1700	[425]	1962	
YbB_{12}	0.69	1000	—*	1970	
LuB_6	0,7	1600	[425]	1962	
LuB_{12}	0,70	1000	—*	1970	
ThB_6	0,69—0,70	—	[277]	1956	λ = 685 μ
UB_{12}	0,77	800—1900	[425]	1962	
TiB_2	0.71	800—1700	[425]	1962	
ZrB_2	0,89—0,91	800—1700	[425]	1962	
ZrB_{12}	0,76	1000	—*	1970	
HfB_2	0,89—0,92	800—1700	[425]	1962	
VB_2	0,72—0,76	800—1700	[425]	1962	
NbB_2	0.77	800—2000	[425]	1962	
TaB_2	0.70	—	[425]	1962	
CrB_2	0.72	800—1700	[425]	1962	
Mo_2B_5	0,80—0,76	800—1700	[425]	1962	
W_2B_5	0.83	800—1900	[425]	1962	
Co_3B	0.82—0,87	800—2000	[425]	1962	
YC	0,81	800—1800	[92]	1962	
Y_2C_3	0,73—0,91	800—1800	[92]	1962	See [327]
YC_2	0,87—0,68	1100—2000	[92]	1962	
TiC	0,90	800—1700	[425]	1962	
TiC	0,70	1500—2400	[427]	1973	λ = 0.65 μ
ZrC	0,75—0,79	800—2000	[425]	1962	
HfC	0.77	800—1600	[425]	1962	
NbC	0.85	800—1800	[425]	1962	
NbC	0,75	1500—2400	[427]	1973	λ = 0.65 μ
TaC	0,62—0,85	800—1700	[425]	1962	
Cr_7C_3	0,92	800—1400	[425]	1962	
Cr_3C_2	0,62—0,80	800—1500	[425]	1962	
Mo_2C	0,71	800—1500	[425]	1962	
W_2C	0,78	800—1800	[425]	1962	
WC	0,73—0,69	800—1700	[425]	1962	
ScN	0,79—0,87	800—1800	[425]	1962	
TiN	0.82—0,79	800—1700	[425]	1962	
TiN	0.60	1500—2200	[427]	1973	λ = 0.65 μ
ZrN	0.43—0,76	800—1800	[425]	1962	
HfN	0.84	800—1900	[425]	1962	
V_3N	0.82	800—1600	[425]	1962	
VN	0,77	800—1800	[425]	1962	
Nb_2N	0.82	800—1700	[425]	1962	

EMISSION COEFFICIENTS (WAVELENGTH λ = 0.665 mm) (continued)

Phase	Emission coefficients	Temp., °C	Source	Year	Notes
1	2	3	4	5	6
NbN	0.83	800—1700	[425]	1962	
Ta_2N	0.83	800—1700	[425]	1962	
TaN	0,79	800—1700	[425]	1962	
Cr_2N	0.69	800—1700	[425]	1962	
CrN	0,66—0,40	1200—2000	[425]	1962	
Mg_2Si	0.67—0.69	800—1000	[425]	1962	
$GdSi_2$	0,80—0,83	800—1600	[425]	1962	
$TiSi_2$	0.82	800—1700	[425]	1962	
Ti_5Si_3	0.74	800—1700	[425]	1962	
$ZrSi_2$	0.72	800—1800	[425]	1962	
VSi_2	0,73—0,89	800—1600	[425]	1962	
$NbSi_2$	0,80	800—1700	[425]	1962	
$TaSi_2$	0.74	800—1800	[425]	1962	
CrSi	0.80	800—1800	[425]	1962	
Cr_3Si_2	0.79	800—1700	[425]	1962	
$CrSi_2$	0.79	800—1600	[425]	1962	
Mo_3Si	0.77	800—1700	[425]	1962	
Mo_5Si_3	0,75	800—1700	[425]	1962	
$MoSi_2$	0,75—0,79	800—2000	[425]	1962	
Mn_3Si	0,68—0,78	800—1100	[425]	1962	
$MnSi_2$	0,70—0.83	800—1200	[425]	1962	
$ReSi_2$	0,70—0,89	800—1400	[425]	1962	
CoSi	0,67—0,86	800—1300	[425]	1962	
NiSi	0,67—0,82	800—1200	[425]	1962	
TiP	0.83	800—1300	[425]	1962	
La_2S_3	0,79	800—1500	[425]	1962	
Ce_2S_3	0.78—0,91	800—1800	[425]	1962	
Pr_2S_3	0,69	800—1300	[425]	1962	
Nd_2S_3	0.68	800—1900	[425]	1962	
B_4C	0,85	800—1500	[425]	1962	
SiC	0,68	1100—1500	[425]	1962	
α-BN	0,64—0,62	800—1700	[425]	1962	
Si_3N_4	0.77	800—1600	[425]	1962	Mixture of α and β-phases
AlN	0,85	800—1400	[425]	1962	Vacuum
AlN	0.80	800—2000	[425]	1962	Argon

INFRARED ABSORPTION SPECTRA *

Phase	Wavelength, mm	Intensity	Phase	Wavelength, mm	Intensity
Mo_2B	7.2	Weak	AlN	14,0	Average
Mg_3N_2	4.8	Average	B_4C	9,5	"
Mg_3N_2	7.1	"	B_4C	12,9	Weak
Mg_3N_2	15,2	"	SiC	12.0	"
AlN	8,45	Weak	BN	7,28	Strong
AlN	9.46	"	BN	12.3	Average

*Ref. 428.

Chapter V

MECHANICAL PROPERTIES

TENSILE STRENGTH

Phase	σ_B, kg/mm^2	Temp., °C	Porosity, %	Source	Year
1	2	3	4	5	6
ZrB$_2$ Be$_2$C	183 9,14—9,83	— 20	— —	[3] [11]	1975 1968
TiC	~6,5 ~5,4 ~3,5 ~0,5	0 1000 2000 2700	20	[734]	1968
ZrC	~7,6 ~8,4 ~6 ~0,5	0 1000 2000 2750	30	[734]	1968
NbC	~6,7 ~6,8 ~6,4 ~0,5	0 1000 2000 2950	30	[734]	1968
TaC	~10 ~11,9 ~10,5 ~1	0 1500 2500 3200	20	[734]	1968
TaC	~8 ~10,1 ~7 ~1	0 2000 2500 3200	30	[734]	1968
Cr$_3$C$_2$	~5,0 * ~3,2 * ~3,5 ** ~1,7 **	900 1000 900 1000	—	[257]	1974

*Long-term strength for 10 h.
**Long-term strength for 100 h.

Phase	σ_B, kg/mm^2	Temp., °C	Porosity, %	Source	Year
1	2	3	4	5	6
WC	35	20	—	[257]	1974
TiSi$_2$	15	20	—	[260]	1959
MoSi$_2$	28	980	—	[260]	1959
	29,4	1200			
Ni$_2$Si	0,6	20	~0	[735]	1960
	11,2	600			
	14,2	650			
	5.9	750			
NiSi	0,6	20	~0	[735]	1960
	0,8	500			
	2,0	550			
	1,1	650			
	0,53	750			
U$_3$Si	70	25	—	[260]	1959
B$_4$C	7,3	25	~0	[538]	1956
SiC	~2,8 [*1]	900	—	[11]	1968
	~2,4 [*1]	1000			
	~2,3 [*2]	900			
	~1,5 [*2]	1000			
BN	11,12 [*3]	25	4—5	[14]	1969
	10,60 [*3]	350			
	2,70 [*3]	700			
	1,53 [*3]	1000			
	5,10 [*4]	25			
	4,90 [*4]	350			
	1,33 [*4]	700			
	0,76 [*4]	1000			
Si$_3$N$_4$	1.5—2.75	20	20—25	[14]	1969
AlN	27	25	—	[14]	1969
	18.95	1000			
	12.7	1400			

[*1]For 10 h.
[*2]For 100 h.
[*3]Parallel to the hot-pressing direction.
[*4]Perpendicular to the hot-pressing direction.

Phase	$\sigma_{bend},$ kg/mm^2	Temp., °C	Porosity, %	Source	Year
1	2	3	4	5	6
TiBe$_{12}$	7,8 2,5 1,9	1260 1370 1510	—	[2]	1966
Zr$_2$Be$_{17}$	17,5 16,1 28,0 28,0 24,6	20 982 1260 1370 1510	—	[2]	1966
ZrBe$_{13}$	17,5 12,6 28,0 26,0 17,5	20 982 1260 1370 1510	—	[2]	1966
Nb$_2$Be$_{17}$	21,7 24,5 49,1 44,2 25,2	20 982 1260 1370 1510	—	[2]	1966
NbBe$_{12}$	13,4 10,5 31,5 31,5 13.4	20 982 1260 1370 1510	—	[2]	1966
Ta$_2$Be$_{17}$	21,0 28,1 54,7 39,3 24,6	20 982 1260 1370 1510	—	[2]	1966
TaBe$_{12}$	21,7 37,2 30,2 18,2	20 1260 1370 1510	—	[2]	1966
MoBe$_{12}$	29,3 8,3	1260 1510	—	[2]	1966
CaB$_6$ LaB$_6$ SmB$_6$ YB$_4$	14,1 12,9 15,1 2,90	20 — — —	∼0 — — 22—26	[3] [3] [3] [6]	1975 1975 1975 1966

Phase	σ_{bend}, kg/mm^2	Temp., °C	Porosity, %	Source	Year
1	2	3	4	5	6
YB$_6$	2.70	—	22—26	[6]	1966
YB$_{12}$	1.65	—	22—26	[6]	1966
EuB$_6$	18.7	—	—	[3]	1975
GdB$_4$	6.75	—	30—32	[6]	1966
GdB$_6$	3.20	—	30—32	[6]	1966
GdB$_6$	21.1	20	8.5	[737]	1960
ThB$_4$	14.06	—	—	[3]	1975
UB$_4$	42.18	—	—	[3]	1975
TiB$_2$	24.5	20	1.0	[3]	1975
ZrB$_2$	9.3 9.6 6.6 3.4 2.1 2.4 0.8 1.0 0.7	20 800 1000 1100 1200 1300 1500 1670 1750	22—24	[736]	1962
CrB$_2$	62.0	20	—	[3]	1975
Mo$_2$B MoB Mo$_2$B$_5$	17.53—35.1	20	10—35	[3]	1975
TiC	1.5 2.5 0.8 0.6 1.4 0.8 4.0 10.4 5.7 3.6 1.3	20 800 1000 1200 1400 1600 1800 1900 2000 2200 2450	21—25	[736]	1962
ZrC	7.51 8—10	1000 1220	2.2	[257]	1974
Mo$_2$C	5.0 4.8 14.8 21.4 11.7	20 1000 1300 1600 1800	26—28	[736]	1962

BENDING STRENGTH (continued)

Phase	σ_{bend}, kg/mm^2	Temp., °C	Porosity, %	Source	Year
1	2	3	4	5	6
WC	3,0 6,3 1,6 6,9 13,5	20 1000 1500 1800 2000	14—16	[736]	1962
WC	35	20	—	[625]	1957
LaSi$_2$	27,2	20	—	[737]	1960
NdSi$_2$	6,18	20	—	[737]	1960
GdSi$_2$	4,45	20	—	[737]	1960
DySi$_2$	6,9	20	—	[737]	1960
TiSi$_2$	21,0	20	—	[260]	1959
MoSi$_2$	35,1 21 10,6 6,0	20 980 1040 1100	—	[260]	1959
B$_4$C	34,0 24,6 20,9 19,5	20 870 1093 1316	—	[11]	1968
SiC	16,9 17,6 12,6	25 1200 1500	4	[611]	1961
Si$_3$N$_4$	16,0 15,2 14,5 14,7	20 600 900 1200	32,6 30,6 30,4 22,0	[14]	1969
Si$_3$N$_4$	3,2 4,7	1550 1600	25—30	[738]	1974

COMPRESSIVE STRENGTH

Phase	σ_{comp}, kg/mm^2	Temp., °C	Porosity, %	Source	Year	Notes
1	2	3	4	5	6	7
ZrBe$_{13}$	133 105 49	20 871 1371	—	[2]	1966	
NbBe$_{12}$	140 91 56	20 871 1371	—	[2]	1966	
TaBe$_{12}$	~105 133	20 871	—	[2]	1966	
TiB$_2$	135,0 22,7 25,8 18,3 11,0	20 1000 1200 1400 1600	~0	[740]	1960	
ZrB$_2$	158,7 30,6 24,1 24,4 47,1	20 1000 1200 1400 1600	~0	[740]	1960	
CrB$_2$	127,9 86,8 40,2 58,1	20 1000 1200 1400	~0	[740]	1960	
Be$_2$C	73,9	20	~0	[11]	1968	
UC	30,1 ± 4	20	20—25	[724]	1959	Parallel to applied pressure
	12,6 ± 2,2	20	20—25	[724]	1959	Perpendicular to applied pressure
TiC	138,0 87,5 51,0 35,0 23,0 31,0 16,4 9,45	20 1000 1200 1400 1600 1800 2000 2200	~0	[740]	1960	See [739]

Phase	σ_{comp}, kg/mm^2	Temp., °C	Porosity, %	Source	Year	Notes
1	2	3	4	5	6	7
ZrC	83,4 49,7 26,4	20 1000 1200	~0	— *	1969	
VC	62	20	—	[256]	1957	See [739]
NbC	242,3	20	~0	— *	1969	
Cr$_3$C$_2$	104,8 94,9 57,2 57,1 42,1	20 1000 1100 1200 1400	~0	— *	1969	
WC	360	20	—	[256]	1957	See [625]
WC	272,1 141,0	20 1000	~0	— *	1969	
TiN	129,8	20	3,4	[256]	1957	
ZrN	100	20	~0	[256]	1957	
U$_3$Si	35,0 5,5	600 800	—	[417]	1960	
TiSi$_2$	117,9 39,7 10,5 5,5	20 1000 1100 1200	~0	— *	1969	
MoSi$_2$	113,0 40,5 35,0 39,0 4,5	20 1000 1200 1400 1600	~0	[740]	1960	

*L. I. Struk, Investigation of Pressing Process and Some Physical Properties of Refractory Compounds [in Russian], Author's abstract, Cand. Dissertation, Kiev (1969).

Phase	σ_{comp}, kg/mm^2	Temp., °C	Porosity, %	Source	Year	Notes
1	2	3	4	5	6	7
WSi$_2$	126,9 59,5	20 1000	~0	— *	1969	
CoSi	3.8 6.3 34.0	20 500 750	~0	[735]	1960	
CoSi$_2$	10.0 15,2 60,0	20 500 750	~0	[735]	1960	
Ni$_2$Si	31,6 57.9 76.0	20 500 600	~0	[735]	1960	
NiSi	15,8 46,7 50,7 62,5	20 500 600 750	~0	[735]	1960	
B$_4$C SiC BN Si$_3$N$_4$	180 150 24—32 13,5	20 25 20 —	— 4 — 25—30	[11] [611] [14] [738]	1968 1961 1969 1974	

*L. I. Struk, See footnote on p. 284.

MODULUS OF ELASTICITY

Phase	Modulus of elasticity, kg/mm^2	Temp., °C	Source	Year	Notes
1	2	3	4	5	6
Zr$_2$Be$_{17}$	3 420 2 860 1 870 1 430 715	20 871 1260 1371 1510	[2]	1966	

Phase	Modulus of elasticity, kg/mm²	Temp., °C	Source	Year	Notes
1	2	3	4	5	6
$ZrBe_{13}$	3 360 2 860 1 440—2 860 1 430 715	20 871 1260 1371 1510	[2]	1966	
Nb_2Be_{17}	3 070 2 860 1 580 1 780	20 871 1260 1370	[2]	1966	
$NbBe_{12}$	3 360 2 860 2 150 1 780 1 440	20 871 1260 1370 1510	[2]	1966	
Ta_2Be_{17}	3 570 2 860 1 225 1 430 715	20 871 1260 1370 1510	[2]	1966	
$TaBe_{12}$	1 930 2 370 1 725 2 430 715	20 871 1260 1370 1510	[2]	1966	
$MoBe_{12}$	1 070 860 71	1260 1370 1510	[2]	1966	
CaB_6	46 000	20	[741]	1958	
BaB_6	39 300	20	[741]	1958	
LaB_6	48 800	20	[741]	1958	
CeB_6	38 600	20	[741]	1958	
ThB_4	15 120	20	[742]	1961	
UB_4	45 000	20	[742]	1961	
TiB_2	54 000	20	[743]	1961	See [430, 485]
ZrB_2	35 000	20	[741]	1958	See [430]
VB_2	27 300	20	[63]	1961	See [430]
NbB_2	65 000	20	[744]	1968	
TaB_2	26 200	20	[743]	1961	See [430, 744]
Mo_2B_5	68 500	20	[744]	1968	
CrB_2	21 500	20	[741]	1958	See [470]
W_2B_5	79 000	20	[744]	1968	
Fe_2B	29 000	20	[434]	1971	
FeB	35 000	20	[434]	1971	

Phase	Modulus of elasticity, kg/mm²	Temp., °C	Source	Year	Notes
1	2	3	4	5	6
Be₂C	32 000	20	[11]	1968	
	32 000	540			
	24 600	830			
	21 050	1100			
UC	22 500	20	[745]	1972	
TiC	46 000	20	[743]	1961	See [746]
ZrC	35 500	20	[741]	1958	See [491]
HfC	35 900	20	[743]	1961	
VC	43 000	20	[743]	1961	
NbC	34 500	20	[11]	1968	See [491, 744]
TaC	29 100	20	[11]	1968	See [747]
Cr₃C₂	38 000	20	[743]	1961	
Mo₂C	54 400	20	[741]	1958	See [485]
W₂C	42 800	20	[11]	1968	
WC	71 000	20	[743]	1961	
UN	26 700	20	[745]	1972	
TiN	25 600	20	[741]	1958	
NbN	49 300	20	[744]	1968	Cub.
ε-TaN	58 700	20	[744]	1968	
Mg₂Si	5 430	20	[260]	1959	
U₃Si	19 300	20	[447]	1960	
TiSi₂	26 400	20	[743]	1961	
ZrSi₂	26 800	20	[743]	1961	
Mo₃Si	30 000	20	[743]	1961	See [750]
MoSi₂	43 000	20	[743]	1961	See [750]
SiC	39 400	20	[11]	1968	See [749]
	39 300	200			
	38 900	400			
	38 300	600			
	37 850	800			
	37 000	1000			
	36 700	1100			
	36 200	1200			
	35 600	1250			
	35 000	1300			
	33 000	1350			
α-BN	8 650 *	25	[248]	1979	
	6 150 *	350			
	1 080 *	700			
	1 160 *	1000			
	3 440 **	25			
	2 430 **	350			
	360 **	700			

*Parallel to hot-pressing direction.
**Perpendicular to hot-pressing direction.

MODULUS OF ELASTICITY (continued)

Phase	Modulus of elasticity, kg/mm^2	Temp., °C	Source	Year	Notes
1	2	3	4	5	6
Si_3N_4	4 700	20	[248]	1969	
	4 860	300			
	4 830	550			
	4 760	850			
	4 720	950			
	4 600	1100			
AlN	35 050	25	[14]	1969	See [249]
	30 000—33 000	20	[748]	1970	Whiskers

IMPACT STRENGTH

Phase	Impact strength, kg·m/mm^2	Source	Year	Notes
$ZrBe_{13}$	0.115	[2]	1966	20° C
$ZrBe_{13}$	13,8	[2]	1966	1095° C
$NbBe_{12}$	0.115	[2]	1966	20° C
$NbBe_{12}$	27,6	[2]	1966	1260° C
TiC	9,9	[447]	1960	
$MoSi_2$	1,1	[447]	1960	
SiC	1,12—1,59	[14]	1969	
Si_3N_4	0,77—1,02	[14]	1969	

HARDNESS ON THE MINERALOGICAL SCALE

Phase	Hardness number(tentative scale)	Source	Year	Notes
1	2	3	4	5
UB_4	8	[124]	1962	
UB_{12}	8	[124]	1962	
TiB_2	>9	[751]	1968	
TiB_{12}	9	[751]	1968	
ZrB_2	~8	[751]	1968	Scratches corundum and SiC
VB	8—9	[751]	1968	See [124]
VB_2	8—9	[625]	1957	

Phase	Hardness number (tentative scale)	Source	Year	Notes
1	2	3	4	5
VB_4	>8	[751]	1968	Scratches quartz and topaz
VB_{12}	>8	[751]	1968	Scratches quartz and topaz
NbB_2	8	[124]	1962	
Cr_5B_3, Cr_3B_2	9	[751]	1968	
CrB	8.5	[751]	1968	See [124]
CrB_2	9	[751]	1968	
Mo_2B	8—9	[751]	1968	
MoB	8	[751]	1968	
WB	9	[751]	1968	
WB_2	9	[751]	1968	
Mn_3B_4	8	[254]	1960	
Be_2C	9	[538]	1956	
UC_2	7	[538]	1956	
TiC	8—9	[751]	1968	
ZrC	8—9	[751]	1968	
VC	>9	[751]	1968	
NbC	>9	[625]	1957	Scratches corundum
TaC	9	[751]	1968	
$Cr_{23}C_6$	>9	[751]	1968	Scratches corundum
Cr_3C_2	~7	[538]	1956	
Mo_2C	~7	[538]	1956	
MoC	7—8	[751]	1968	
W_3C	9—10	[124]	1962	
W_2C	9—10	[751]	1968	
WC	>9	[625]	1957	
Fe_3C	7—8	[538]	1968	
ScN	7—8	[751]	1968	
AlN	9—10	[49]	1964	
TiN	9—10	[625]	1957	
ZrN	8	[625]	1957	
VN	9—10	[751]	1968	
NbN	>8	[625]	1957	
TaN	>8	[625]	1957	
$ZrSi_2$	~6	[625]	1957	Hardness of glass
VSi_2	6—7	[625]	1957	Scratches glass
Cr_3Si	>6	[625]	1957	Scratches glass
Cr_2Si	~9	[625]	1957	Scratches quartz and corundum
Cr_3Si_2	>6	[625]	1957	Scratches glass
BeS	7.5	[751]	1968	
MoS_2	1	[752]	1965	

ROCKWELL HARDNESS

Phase	HRA, kg/mm^2	Source	Year	Phase	HRA, kg/mm^2	Source	Year
1	2	3	4	1	2	3	4
LaB_6	83	[751]	1968	Cr_7C_3	67	[751]	1968
GdB_6	86	[751]	1968	Cr_3C_2	86,5—	[704]	1966
TiB_2	86	[751]	1968		89		
ZrB	69—72	[751]	1968	Mo_2C	74	[751]	1968
ZrB_2	84	[751]	1968	W_2C	80	[751]	1968
ZrB_{12}	92,5	[751]	1968	WC	81	[751]	1968
VB_2	83	[751]	1968	TiN	75	[751]	1968
CrB	89—91	[751]	1968	ZrN	84	[751]	1968
CrB_2	84	[751]	1968	NbN	86	[751]	1968
Mo_2B	90	[751]	1968	CrN	78	[751]	1968
α-MoB	90	[751]	1968	YSi_2	32	[751]	1968
β-MoB	90	[751]	1968	$LaSi_2$	31	[737]	1960
MoB_2	90	[751]	1968	$GdSi_2$	80	[735]	1960
Mo_2B_5	90	[751]	1968	$EuSi_2$	80	[735]	1960
Co_2B	90	[751]	1968	$DySi_2$	80	[737]	1960
Co_2B_5	82	[751]	1968	U_3Si	~23	[447]	1960
La_2C_3	~77	[751]	1968	$TiSi_2$	81	[751]	1968
TiC	92,5—93,5	[751]	1968	V_3Si	78	[751]	1968
				V_5Si_3	79	[751]	1968
ZrC	87	[751]	1968	VSi_2	32	[751]	1968
HfC	84	[751]	1968	Cr_3Si	85	[751]	1968
NbC	83	[751]	1968	$CrSi$	82	[751]	1968
TaC	82	[751]	1968	Mo_5Si_3	74	[751]	1968
$Cr_{23}C_6$	83	[751]	1968	$MoSi_2$	74	[751]	1968

VICKERS HARDNESS

Phase	HV, kg/mm^2	Source	Year	Notes
1	2	3	4	5
$ZrBe_{13}$	1650	[751]	1968	
$NbBe_{12}$	1200	[751]	1968	
Nb_2Be_{17}	1000	[751]	1968	
Ta_2Be_{17}	1120	[751]	1968	
$CrBe_2$	1288	[751]	1968	
$MoBe_{12}$	950	[751]	1968	
SmB_6	1391 $\pm$ 159	[737]	1960	
YbB_6	1538 $\pm$ 33	[737]	1960	
UC	700 $\pm$ 150	[724]	1959	Porosity 20.4%
TiC	3200	[751]	1968	See [753, 754, 746]

VICKERS HARDNESS (continued)

Phase	HV, kg/mm^2	Source	Year	Notes
1	2	3	4	5
HfC	3202—2533	[751]	1968	
WC	2200	[751]	1968	
Nb$_3$Al	790—800	[751]	1968	
Nb$_2$Al	850—870	[751]	1968	
NbAl	490—510	[751]	1968	
CrAl$_7$	42	[18]	1965	
CrAl$_{11}$	540	[751]	1968	
MnAl$_4$	740	[751]	1968	
FeAl$_3$	730	[751]	1968	
FeAl$_5$	1000	[751]	1968	
Co$_2$Al$_5$	820	[751]	1968	
Co$_2$Al$_9$	660	[751]	1968	
Ni$_2$Al$_3$	1000	[751]	1968	
NiAl$_3$	720	[751]	1968	
Zr$_2$Si	1180—1280	[751]	1968	
Zr$_5$Si$_3$	1280—1390	[751]	1968	
ZrSi	1020—1180	[751]	1968	
ZrSi$_2$	830—980	[751]	1968	
Nb$_4$Si	470—550	[712]	1956	
Nb$_5$Si$_3$	400—600	[712]	1956	
NbSi$_2$	660—700	[712]	1956	
Ta$_{4.5}$Si	1000—1200	[751]	1968	
Ta$_2$Si	1200—1500	[751]	1968	p = 40 kg; 30 sec
Ta$_5$Si$_3$	1200—1500	[751]	1968	
TaSi$_2$	1000—1200	[751]	1968	
Cr$_3$Si	900—980	[751]	1968	
Cr$_3$Si$_2$	1050—1200	[751]	1968	
CrSi	950—1050	[751]	1968	
CrSi$_2$	880—1100	[751]	1968	
Mo$_3$Si	1320—1550	[260]	1959	
Mo$_5$Si$_3$	1200—1320	[260]	1959	
MoSi$_2$	1320—1550	[260]	1959	
W$_3$Si$_2$	770	[751]	1968	p = 100 g

MICROHARDNESS

Phase	Hμ, kg/mm^2	Load, g	Source	Year	Notes
1	2	3	4	5	6
MnBe$_2$	1413	—	[46]	1959	
CoBe	443	—	[46]	1959	
Ru$_2$Be$_{17}$	900	—	[49]	1971	
Rh$_2$Be$_{17}$	1200	—	[49]	1971	
Os$_2$Be$_{17}$	940	—	[49]	1971	
Be$_5$B	623	—	[51]	1973	

Phase	$H\mu$, kg/mm^2	Load, g	Source	Year	Notes
1	2	3	4	5	6
Be$_2$B	890	—	[51]	1973	
BeB$_2$	3180	—	[51]	1973	
BeB$_6$	2577	—	[51]	1973	
CaB$_6$	2700 ± 220	30	[274]	1956	
SrB$_6$	2920 ± 90	30	[269]	1961	
BaB$_6$	3000 ± 290	30	[274]	1956	
ScB$_2$	1780 ± 276	200	[270]	1960	
YB$_4$	2850 ± 150	—	[6]	1966	
YB$_6$	2575 ± 100	—	[6]	1966	
YB$_{12}$	2500 ± 150	—	[6]	1966	
LaB$_6$	2770 ± 60	30	[274]	1956	
CeB$_6$	3140 ± 190	30	[274]	1956	
PrB$_4$	1930	—	— *	1974	
PrB$_6$	2470	100	[751]	1968	
NdB$_4$	1953	—	— *	1974	
NdB$_6$	2540 ± 170	70	[751]	1968	
SmB$_6$	2500 ± 300	100	[275]	1959	
EuB$_6$	2660	100	[751]	1968	
GdB$_4$	1900 ± 50	—	[6]	1966	
GdB$_6$	1850 ± 50	—	[6]	1966	
TbB$_4$	1897	—	— *	1974	
TbB$_6$	2300	100	[751]	1968	
TbB$_{12}$	2600 ± 100	100	— **	1970	
DyB$_4$	1896	—	— *	1974	
DyB$_{12}$	2400 ± 100	100	— **	1970	
HoB$_4$	1684	—	— *	1974	
HoB$_{12}$	2700 ± 100	100	— **	1970	
ErB$_4$	1754	—	— *	1974	
ErB$_{12}$	2800 ± 100	100	— **	1970	
TmB$_4$	1768	—	— *	1974	
TmB$_{12}$	3000 ± 100	100	— **	1970	
YbB$_6$	2660	100	[751]	1968	
YbB$_{12}$	3300 ± 100	30	— **	1970	
LuB$_{12}$	2900 ± 100	100	—**	1970	
ThB$_6$	1740 ± 123	20	[277]	1956	
ThB$_{76}$	2310	—	[62]	1972	
UB$_2$	1510	—	[291]	1968	
UB$_4$	>2500	—	[291]	1968	
UB$_{12}$	>2000	—	[291]	1968	
TiB	2700—2800	30	[726]	1959	
TiB$_2$	3370 ± 60	30	[755]	1952	
ZrB	3500—3600	30	[726]	1959	
ZrB$_2$	2252 ± 22	30	[755]	1952	See [65]
ZrB$_{12}$	2750—2850	—	[65]	1970	
HfB$_2$	2900 ± 500	30	[751]	1968	See [66]

*E. N. Severyanina, Producing and Investigating the Physical Properties of Rare-Earth Metal Tetraborides [in Russian], Cand. Dissertation, Kiev (1974).

**V. V. Odintsov, Preparation and Physical Properties of Metal Dodecaborides with the UB$_{12}$ Structural Type [in Russian], Cand. Dissertation, Kiev (1970).

Phase	$H\mu$, kg/mm^2	Load, g	Source	Year	Notes
1	2	3	4	5	6
V_3B_2	2280	50	[751]	1968	
V_3B_4	2350	50	[751]	1968	
VB_2	2800 ± 13	30	[755]	1952	
Nb_3B_2	2290	50	[751]	1968	
NbB	2195	50	[751]	1968	
Nb_3B_4	2290	30	[751]	1968	
NbB_2	2600	30	[751]	1968	
Ta_3B_2	2770	50	[751]	1968	
TaB	3130	50	[751]	1968	
Ta_3B_4	3350	50	[751]	1968	
TaB_2	2500 ± 42	30	[751]	1968	
Cr_4B	1240 ± 60	50	[751]	1968	
Cr_2B	1350 ± 100	50	[751]	1968	
Cr_5B_3	1420—1520	50	[3]	1975	
CrB	1200—1300	100	[751]	1968	
Cr_3B_4	1400—1500	100	[751]	1968	
CrB_2	2100 ± 80	50	[751]	1968	
Mo_2B	2500	50	[751]	1968	
α-MoB	2350	50	[751]	1968	
β-MoB	2500	50	[751]	1968	
MoB_2	1200	50	[751]	1968	
Mo_2B_5	2350	50	[751]	1968	
W_2B	2420 ± 120	50	[756]	1957	
WB	3700	50	[756]	1957	
WB_2	2660 ± 12	30	[755]	1952	
W_2B_5	2663 ± 12	30	[741]	1958	
WB_4	4000 ± 200	50	[8]	1974	See [70]
Mn_4B	1050 ± 50	—	[9]	1967	
Mn_2B	1800 ± 50	—	[9]	1967	
MnB	2050 ± 50	—	[9]	1967	
Mn_3B_4	2000 ± 50	—	[9]	1967	
MnB_2	1700 ± 50	—	[9]	1967	
MnB_4	3600 ± 100	—	[9]	1967	
ReB_2	3100	—	[280]	1968	
Fe_2B	1340 ± 50	—	[3]	1975	
FeB	1650 ± 50	—	[3]	1975	
Co_3B	1150	50	[757]	1959	
Co_2B	1150	50	[757]	1959	
CoB	1150	50	[757]	1959	
CoB_2	2575	50	[757]	1959	
Ni_3B	1190	—	[758]	1967	
Ni_2B	1430	—	[758]	1967	
Ni_4B_3	1486	—	[758]	1967	
NiB	1546	—	[758]	1967	
Ru_7B_3	1131	—	— *	1971	
$Ru_{11}B_8$	1307	—	— *	1971	
$RuB_{\sim 1.1}$	1409	—	— *	1971	
Ru_2B_3	1523	—	— *	1971	

*V. A. Kosenko, Investigation of the Conditions for Preparation and Some Properties of Platinoid Borides [in Russian], Cand. Dissertation, Kiev (1971).

MICROHARDNESS (continued)

Phase	$H\mu$, kg/mm²	Load, g	Source	Year	Notes
1	2	3	4	5	6
RuB_2	2263	—	— *	1971	
Rh_7B_3	777	50	[570]	1971	
$RhB_{\sim 1.1}$	1213	50	[570]	1971	
Pd_3B	470	—	[571]	1972	
Pd_5B_2	595	—	[571]	1972	
$OsB_{1.2}$	1642	—	— *	1971	
$OsB_{1.6}$	1878	—	— *	1971	
OsB_2	2899	—	— *	1971	
$IrB_{1.15}$	1652	—	[572]	1972	
$PtB_{1.1}$	940	—	— *	1971	
Be_2C	2690	—	[751]	1968	
ScC	2720	20	[751]	1968	
YC	120	5	[759]	1961	
Y_2C_3	900 ± 160	—	[92]	1962	
YC_2	400 ± 25	400	[288]	1967	
LaC_2	215 ± 20	400	[288]	1967	
CeC_2	219 ± 7	400	[288]	1967	
PrC_2	166 ± 12	400	[288]	1967	
NdC_2	213 ± 13	400	[288]	1967	
SmC_2	224 ± 23	400	[288]	1967	
GdC_2	281 ± 31	400	[288]	1967	
TbC_2	370 ± 9	400	[288]	1967	
DyC_2	392 ± 16	400	[288]	1967	
HoC_2	611 ± 40	400	[288]	1967	
ErC_2	660 ± 65	400	[288]	1967	
TmC_2	648 ± 72	400	[288]	1967	
YbC_2	168 ± 3	400	[288]	1967	
LuC_2	721 ± 29	400	[288]	1967	
UC	923 ± 56	50	[320]	1969	
$TiC_{0.96}$	3170 ± 170	50	— **	1971	See [760, 761]
$TiC_{0.89}$	2950 ± 140	50	— **	1971	
$TiC_{0.79}$	2570 ± 120	50	— **	1971	
$TiC_{0.72}$	2310 ± 90	50	— **	1971	
$TiC_{0.62}$	1930 ± 110	50	— **	1971	
$ZrC_{0.97}$	2950 ± 120	50	— **	1971	
$ZrC_{0.90}$	2720 ± 140	50	— **	1971	
$ZrC_{0.80}$	2430 ± 100	50	— **	1971	
$ZrC_{0.70}$	2080 ± 120	50	— **	1971	
$ZrC_{0.65}$	1870 ± 110	50	— **	1971	
$HfC_{0.98}$	2830 ± 140	50	— **	1971	
$HfC_{0.91}$	2590 ± 120	50	— **	1971	
$HfC_{0.80}$	2280 ± 90	50	— **	1971	
$HfC_{0.71}$	2000 ± 130	50	— **	1971	
$HfC_{0.62}$	1810 ± 110	50	— **	1971	

*V. A. Kosenko, see footnote on p. 293.

**V. Ya. Naumenko, Producing and Studying Some Properties of Transition Metal Borides in Groups IV-V in Their Homogeneity Regions [in Russian], Cand. Dissertation, Kiev (1971).

Phase	$H\mu$, kg/mm^2	Load, g	Source	Year	Notes
1	2	3	4	5	6
$VC_{0.41}$**	1610 ± 80	50	[466]	1975	
$VC_{0.42}$**	1680 ± 60	50	[466]	1975	
$VC_{0.44}$**	1850 ± 50	50	[466]	1975	
$VC_{0.46}$**	1910 ± 60	50	[466]	1975	
$VC_{0.49}$**	2020 ± 70	50	[466]	1975	
$VC_{0.87}$	2480 ± 110	50	[465]	1974	
$VC_{0.82}$	2290 ± 90	50	[465]	1974	
$VC_{0.77}$	2170 ± 120	50	[465]	1974	
$VC_{0.72}$	2010 ± 150	50	[465]	1974	
$NbC_{0.42}$**	1890 ± 110	50	[466]	1975	
$NbC_{0.45}$**	1960 ± 80	50	[466]	1975	
$NbC_{0.47}$**	2000 ± 60	50	[466]	1975	
$NbC_{0.49}$**	2060 ± 60	50	[466]	1975	
$NbC_{0.50}$**	2080 ± 50	50	[466]	1975	
$NbC_{0.99}$	2170 ± 130	50	— *	1971	See [761]
$NbC_{0.92}$	2410 ± 90	50	— *	1971	
$NbC_{0.82}$	2620 ± 100	50	— *	1971	
$NbC_{0.76}$	2560 ± 80	50	— *	1971	
$NbC_{0.72}$	2440 ± 60	50	— *	1971	
Ta_2C	1714 ± 159	30	[741]	1958	
$TaC_{0.99}$	1720 ± 140	50	— *	1971	
$TaC_{0.91}$	2150 ± 110	50	— *	1971	
$TaC_{0.85}$	2380 ± 80	50	— *	1971	
$TaC_{0.82}$	2310 ± 60	50	— *	1971	
$TaC_{0.73}$	2170 ± 100	50	— *	1971	
$Cr_{23}C_6$	1663	100	[751]	1968	
Cr_7C_3	1882	200	[751]	1968	
Cr_3C_2	1800	150	[751]	1968	Anisotropy inherent [762-764]
Mo_2C	1499 ± 130	30	[741]	1958	
MoC	1500	100	[751]	1968	
WC	1716	50	[751]	1968	
$TiN_{0.79}$	1670	100	— ***	1974	
$TiN_{0.83}$	1730	100	— ***	1974	
$TiN_{0.87}$	1800	100	— ***	1974	
$TiN_{0.97}$	2050	100	— ***	1974	
$ZrN_{0.85}$	1480	100	— ***	1974	
$ZrN_{0.97}$	1670	100	— ***	1974	
$HfN_{1.0}$	1600	100	— ***	1974	
V_3N	1900 ± 102	50	[765]	1961	
$VN_{0.87}$	1060	100	— ***	1974	
$VN_{0.93}$	1230	100	— ***	1974	
$VN_{0.96}$	1310	100	— ***	1974	
Nb_2N	1720 ± 100	50	[763]	1961	
$NbN_{0.75}$	1780	50	[751]	1968	
$NbN_{0.97}$	1525 ± 136	50	[751]	1968	

*V. Ya. Naumenko, see footnote on p. 294.

**Hexagonal.

***L. K. Shedova, Investigation of the Physical Properties of Transition Metal Nitrides in Groups IV-V [in Russian], Cand. Dissertation, Kiev (1974).

MICROHARDNESS (continued)

Phase	Hµ, kg/mm²	Load, g	Source	Year	Notes
1	2	3	4	5	6
NbN	1461	100	[751]	1968	
Ta$_2$N	1220 ± 120	50	[751]	1968	
TaN	2416	90	[751]	1968	
Cr$_2$N	1571 ± 49	50	[751]	1968	
CrN	1093 ± 93	50	[751]	1968	
Mo$_2$N	630 ± 86	20	[751]	1968	
CaAl$_4$	200	—	[18]	1965	
SrAl$_4$	160	—	[18]	1965	
BaAl$_4$	280	—	[18]	1965	
ThAl$_3$	520	—	[18]	1965	
ScAl$_2$	765	50	[122]	1964	
ScAl$_3$	380	50	[122]	1964	
TiAl	180	—	[18]	1965	
TiAl$_3$	680	—	[18]	1965	} See [470]
Zr$_3$Al	445	—	[18]	1965	
Zr$_5$Al$_3$	580	—	[18]	1965	
Zr$_3$Al$_2$	475	—	[18]	1965	
Zr$_2$Al	580	—	[18]	1965	
ZrAl$_3$	560	—	[18]	1965	
VAl$_3$	395	—	[766]	1954	
NbAl$_3$	375	—	[766]	1954	
TaAl$_3$	440—450	—	[18]	1965	
CrAl$_{11}$	710	—	[18]	1965	
CrAl$_7$	316	—	[766]	1954	
MoAl$_5$	366	—	[766]	1954	
MnAl$_4$	560	—	[18]	1965	
Fe$_2$Al$_5$	720—900	—	[18]	1965	
FeAl$_3$	650—800	—	[18]	1965	
CoAl	530	—	[751]	1968	
CoAl$_9$	450—740	—	[18]	1965	
Ni$_2$Al$_3$	720	—	[751]	1968	
NiAl$_3$	770	—	[751]	1968	
Mg$_2$Si	457	—	[751]	1968	
BaSi$_2$	930 ± 50	50	[449]	1963	
Sc$_5$Si$_3$	973 ± 13 *	50—100	[767]	1965	
ScSi	1005 ± 6 *	50—100	[767]	1965	
ScSi$_{1.66}$	897 ± 4 *	50—100	[767]	1965	
YSi	1106 ± 20 *	40—50	[767]	1965	
YSi$_{1.66}$	845 ± 12 *	40—50	[767]	1965	
YSi$_2$	804 ± 9 *	40—50	[767]	1965	
La$_5$Si$_3$	592 ± 5 *	30—80	[767]	1965	
La$_3$Si$_2$	525 ± 13 *	30—80	[767]	1965	
LaSi	632 ± 18,3 *	30—80	[767]	1965	
LaSi$_2$	476 ± 14 *	30—80	[767]	1965	
Ce$_5$Si$_3$	597 ± 3 *	40—100	[767]	1965	
Ce$_3$Si$_2$	460 ± 4 *	40—100	[767]	1965	
CeSi	628 ± 15 *	40—100	[767]	1965	
CeSi$_2$	520 ± 20 *	40—100	[767]	1965	

*Values are given in the load ranges for which the microhardness alters immaterially.

Phase	$H\mu$, kg/mm^2	Load, g	Source	Year	Notes
1	2	3	4	5	6
$PrSi_2$	875 ± 9 *	50—100	[767]	1965	
$NdSi_2$	698 ± 11 *	40—100	[767]	1965	
$ThSi_2$	1120	100	[737]	1960	
Ti_5Si_3	986	100	[751]	1968	
$TiSi$	1039	100	[751]	1968	
$TiSi_2$	892	50	[260]	1959	
Zr_2Si	1180—1280	50	[260]	1959	
Zr_5Si_3	1280—1390	50	[260]	1959	
$ZrSi$	1020—1180	50	[260]	1959	
$ZrSi_2$	1063	50	[260]	1959	
$HfSi_2$	930	50	[751]	1968	
V_3Si	1430—1560	50	[712]	1956	
V_5Si_3	1350—1510	50	[712]	1956	
VSi_2	890—960	50	[712]	1956	
Nb_4Si	690—820	100	[260]	1959	
Nb_5Si_3	700	100	[260]	1959	
$NbSi_2$	1050	50	[751]	1968	
$TaSi_2$	1407	50	[260]	1959	
Cr_3Si	1005	50	[751]	1968	
Cr_3Si_2	1280	50	[751]	1968	
$CrSi$	1005	50	[751]	1968	
$CrSi_2$	1131	50	[260]	1959	
$CrSi_2$	798	50	[751]	1968	Annealed
$CrSi_2$	704	50	[751]	1968	Cast
Mo_3Si	1310	100	[751]	1968	
Mo_5Si_3	1170	100	[751]	1968	
$MoSi_2$	1200	50	[260]	1959	
$MoSi_2$	707	50	[751]	1968	Cast
$MoSi_2$	735	50	[751]	1968	Annealed
W_3Si_2	770	50	[751]	1968	
WSi_2	1074	50	[260]	1959	
$ReSi_2$	1500 ± 40	50	[751]	1968	
$FeSi_2$	1074	50	[260]	1959	
$CoSi$	1000	—	[735]	1960	20° C
$CoSi$	300	—	[735]	1960	500° C
$CoSi$	115	—	[735]	1960	1000° C
$CoSi_2$	552	—	[735]	1960	20° C
$CoSi_2$	322	—	[735]	1960	500° C
$CoSi_2$	77	—	[735]	1960	1000° C
Ni_3Si	400	—	[260]	1959	
Ni_2Si	440	—	[735]	1960	20° C
Ni_2Si	320	—	[735]	1960	500° C
Ni_2Si	120	—	[735]	1960	750° C
$NiSi$	400	—	[735]	1960	20° C
$NiSi$	256	—	[735]	1960	500° C
ζ-$NiSi_2$	1560 ± 30	50	[181]	1970	Rhomb.
$NiSi_2$	1019	50	[260]	1959	
$OsSi_2$	1950	—	[182]	1970	

*See footnote on p. 296.

MICROHARDNESS (continued)

Phase	Hμ, kg/mm^2	Load, g	Source	Year	Notes
1	2	3	4	5	6
LaP	158 ± 14	—	[751]	1968	
TiP	1300	100	[751]	1968	
LaS	677 ± 61	—	[751]	1968	
La$_2$S$_3$	360 ± 44	—	[751]	1968	
CeS	683 ± 54	—	[751]	1968	
Ce$_2$S$_3$	403 ± 36	—	[751]	1968	
Nd$_2$S$_3$	330 ± 50	—	[751]	1968	
ThS	363 ± 40	30	[751]	1968	
Th$_2$S$_3$	227 ± 28	30	[751]	1968	
α-TiS	580 ± 65	—	[752]	1965	
Ti$_2$S$_3$	200 ± 23	—	[752]	1965	
ZrS$_2$	95 ± 17	—	[752]	1965	
Nb$_2$S$_3$	40 ± 12	—	[752]	1965	
α-TaS$_2$	35 ± 10	—	[752]	1965	
Cr$_2$S$_3$	480 ± 80	—	[752]	1965	
MoS$_2$	26 ± 5	—	[752]	1965	
FeS	380 ± 28	—	[752]	1965	

COMPRESSIBILITY

Phase	Compress. factor, $\Delta V/V_0$	Temp., °C	Source	Year
TiC	$4.72 \cdot 10^{-7}\, p - 2.16 \cdot 10^{-12}\, p^2$	30	[257]	1974
TiC	$4.78 \cdot 10^{-7}\, p - 2.19 \cdot 10^{-12}\, p^2$	75	[257]	1974
TiN	$3.32 \cdot 10^{-7}\, p - 2.13 \cdot 10^{-12}\, p^2$	30	[14]	1969
TiN	$3.51 \cdot 10^{-7}\, p - 2.13 \cdot 10^{-12}\, p^2$	75	[14]	1969
BN	$34 \cdot 10^{-7}\, p - 54 \cdot 10^{-12}\, p^2$	45	[14]	1969

Chapter VI

CHEMICAL PROPERTIES

RESISTANCE OF POWDERED REFRACTORY COMPOUNDS TO
ACIDS AND ALKALIS*

Phase	Reagent	Temp., °C	Process time, h	Insoluble residue, %	Source	Year
1	2	3	4	5	6	7
MgB_2	H_2O HCl (1,19) HNO_3 (1,43) H_2SO_4 (1,84)	20	—	CS** CS CS CS	[124]	1959
MgB_{12}	HCl (1,19) HCl (1,19) HNO_3 (1,43) HNO_3 (1,43) H_2SO_4 (1,84) H_2SO_4 (1,84) HNO_3 (1,43) + + H_2O_2 (30%)	20 Boiling 20 Boiling 20 Boiling "	—	ND ND ND ND ND ND ND CS	[124]	1959
Mg_3B_2	H_2O HCl (1,19) HNO_3 (1,43) H_2SO_4 (1,84)	20	—	CS	[124]	1959
CaB_6	HCl (1,19) HCl (1,19) HCl (1,19) H_2SO_4 (1,84) HNO_3 (1,42) HNO_3 (1,42) HNO_3 (1,42) $NaOH$ (50%) $NaOH$ (50%) $NaOH$ (50%) Na_2CO_3(50%) Na_2CO_3 (50%) Na_2CO_3 (50%)	20	1 2 240 1—240 1 2 24 1 2 240 1 2 240	99,5 99,5 98,5 ND 8,5 2,7 CS 97,8 97,8 97,4 99,7 99,2 99,5	[524]	1961

*Key to abbreviations: CS, complete solution; PS, partial solution; CSH, complete solution with hydrolysis; SMC, solution of most of the compound with the formation of salt residues; ND, not dissolved.

** Dissolves slowly.

Phase	Reagent	Temp., °C	Process time, h	Insoluble residue, %	Source	Year
1	2	3	4	5	6	7
SrB_6	HCl (1.19)	20	1	99.3	[524]	1961
	HCl (1,19)		24	98.6		
	HCl (1,19)		240	98,5		
	H_2SO_4 (1,84)		1—240	ND		
	HNO_3 (1,42)		1	1.5		
	HNO_3 (1,42)		2	CS		
	NaOH (50%)		1	99,2		
	NaOH (50%)		2	98,8		
	NaOH (50%)		240	98,1		
	Na_2CO_3 (50%)		24	98,7		
	Na_2CO_3 (50%)		240	98.5		
SrB_6	H_2O	20	—	ND	[124]	1959
		Boiling	—	ND		
BaB_6	HCl (1,19)	20	1	99.2	[524]	1961
	HCl (1.19)		2	98.9		
	HCl (1,19)		24	98.4		
	HCl (1,19)		240	97.0		
	H_2SO_4 (1,84)		1—240	ND		
	HNO_3 (1,42)		1	6,0		
	HNO_3 (1.42)		2	1.2		
	NNO_3 (1.42)		24	CS		
	NaOH (50%)		1	98.8		
	NaOH (50%)		24	98.1		
	NaOH (50%)		240	98.1		
	Na_2CO_3 (50%)		1	99,4		
	Na_2CO_3 (50%)		24	98.6		
	Na_2CO_3 (50%)		240	98,6		
ScB_2	HCl (1,19)	80	1	25,0	[526]	1965
	H_2SO_4 (1.84)	300	1	32,8		
	HNO_3 (1,4)	120	1	22.5		
	HCl(1 : 1)	100	1	21,4		
	H_2SO_4 (1 : 1)	150	1	34,5		
	HNO_3 (1 : 1)	110	1	15,8		
	HCl (1 : 2)	20	1	0,3		
	HCl (1 : 2)	20	1.5	0.2		
	HCl (1 : 2)	20	2	0.5		
ScB_{12}	HCl (1.19)	80	5	99,2	[526]	1965
	H_2SO_4 (1.84)	300	4	86,4		
	HNO_3 (1,4)	120	1	19,2		
	HCl (1 : 1)	100	5	98,7		
	H_2SO_4 (1 : 1)	150	4	89.4		
	HNO_3 (1 : 1)	110	1	17,8		

Phase	Reagent	Temp., °C	Process time, h	Insoluble residue, %	Source	Year
1	2	3	4	5	6	7
ScB_{12}	HCl (1 : 2) HCl (1 : 2) HCl (1 : 2)	20 20 20	1 1.5 2	99.5 99.6 99.4	[526]	1965
YB_4	HNO_3 (1 : 1) HCl (1 : 1) H_2SO_4 (1 : 1) NaOH (15%)	20 Boiling " "	— 0.5 10 min —	CS ND CS ND	[6]	1966
YB_6	HNO_3 (1 : 1) HCl (1 : 1) H_2SO_4 (1 : 1) NaOH (15%)	20 Boiling " "	— 0.5 10 min —	CS ND CS ND	[6]	1966
YB_{12}	HNO_3 (1 : 1) HCl (1 : 1) H_2SO_4 (1 : 1) NaOH (15%)	Boiling " " "	0.5 0.5 0.5 —	SMC ND ND ND	[6]	1966
LaB_6	H_2SO_4 (1,84) H_2SO_4 (1 : 1) H_2SO_4 (1 : 5) H_2SO_4 (1 : 4)+ citric acid (50%) H_2SO_4 (1 : 4) + + oxalic acid (25%) H_2SO_4 (1 : 4) + + 0.2 N Trilon-B HCl (1,19) HCl (1 : 1) HCl (1 : 5) HCl (1,19) + citric acid (50%) HCl (1,19) + + 0.2 N Trilon-B HNO_3 (1,34) HNO_3 (1 : 1) HNO_3 (1 : 5) NaOH (10%) NaOH (40%)	280 250 120 125 210 128 115 110 105 110 110 120 110 103 102 105	1	CS 88,7 98.7 98,6 99,0 98,9 96,2 98,0 98,6 99,6 98,7 CS CS CS 98,5 99,7	[527]	1970
CeB_6	H_2SO_4 (1,84) H_2SO_4 (1 : 1) H_2SO_4 (1 : 5)	280 250 120	1	CS 89,6 95,5	[527]	1970

Phase	Reagent	Temp., °C	Process time, h	Insoluble residue, %	Source	Year
1	2	3	4	5	6	7
CeB_6	H_2SO_4 (1 : 4) + + citric acid (50%)	125	1	98,0	[527]	1970
	H_2SO_4 (1 : 4) + + oxalic acid (25%)	210		98,2		
	H_2SO_4 (1 : 4) + + 0.2 N Trilon-B	128		97.5		
	HCl (1,19)	115		95,0		
	HCl (1 : 1)	110		95,2		
	HCl (1 : 5)	105		98,5		
	HCl (1,19) + citric acid (50%)	110		98,1		
	HCl (1,19)+0,2 N Trilon-B	110		98.0		
	HNO_3 (1,34)	120		CS		
	HNO_3 (1 : 1)	110		CS		
	HNO_3 (1 : 5)	103		CS		
	NaOH (10%)	102		98,9		
	NaOH (40%)	105		99,6		
PrB_6	H_2SO_4 (1.84)	280	1	CS	[527]	1970
	H_2SO_4 (1 : 1)	250		89.4		
	H_2SO_4 (1 : 5)	120		98.2		
	H_2SO_4 (1 : 4) + + citric acid (50%)	125		98,7		
	H_2SO_4 (1 : 4) + + oxalic acid (25%)	210		99,6		
	H_2SO_4 (1 : 4) + + 0,2 N Trilon-B	128		98,1		
	HCl (1,19)	115		95,6		
	HCl (1 : 1)	110		90.8		
	HCl (1 : 5)	105		98.2		
	HCl (1,19) + citric acid (50%)	110		99,2		
	HCl (1,19) + + 0.2 N Trilon-B	110		98,9		
	HNO_3 (1,34)	120		CS		
	HNO_3 (1 : 1)	110		CS		
	HNO_3 (1 : 5)	103		CS		
	NaOH (10%)	102		99.3		
	NaOH (40%)	105		99,9		

Phase	Reagent	Temp., °C	Process time, h	Insoluble residue, %	Source	Year
1	2	3	4	5	6	7
NdB_6	H_2SO_4 (1,84)	280	1	CS	[527]	1970
	H_2SO_4 (1 : 1)	250		96,0		
	H_2SO_4 (1 : 5)	120		98,6		
	H_2SO_4 (1 : 4) + + citric acid (50%)	125		99,0		
	H_2SO_4 (1 : 4) + + oxalic acid (25%)	210		99,5		
	H_2SO_4 (1 : 4) + + 0.2 N Trilon-B	128		99,3		
	HCl (1,19)	115		95,3		
	HCl (1 : 1)	110		94,6		
	HCl (1 : 5)	105		98,9		
	HCl (1,19) + citric acid (50%)	110		98,9		
	HCl (1,19) + + 0.2 N Trilon-B	110		99,2		
	HNO_3 (1.34)	120		CS		
	HNO_3 (1 : 1)	110		CS		
	HNO_3 (1 : 5)	103		CS		
	NaOH (10%)	102		99,2		
	NaOH (40%)	105		98.9		
SmB_6	HCl (1 : 1)	Slight heating	5 min	78—80	[482]	1961
	HNO_3 (1 : 1)	Same	5 min	CS		
	H_2SO_4 (1 : 1)	" "	2	77		
	3 h HCl (1 : 1) + + 1 h HNO_3 (1,43)	20	5 min	CS		
	H_2SO_4 (1 : 1) + + HNO_3 (1,43)	Slight heating	5 min	CS		
	NaOH (15%)	106	1	99,4		
	2 h NaOH + 1 h H_2O_2	103	1	99,6		
GdB_4	HNO_3 (1 : 1)	20	10 min	CS	[6]	1966
	HCl	Boiling	—	ND		
	H_2SO_4	"	—	ND		
	NaOH (15%)	"	—	ND		

Phase	Reagent	Temp., °C	Process time, h	Insoluble residue, %	Source	Year
1	2	3	4	5	6	7
GdB$_6$	HCl (1 : 1)	103	20	91—93	[482]	1961
	HNO$_3$ (1 : 1)	103	5 min	CS		
	H$_2$SO$_4$ (1 : 1)	103	2	87		
	H$_2$SO$_4$ (1 : 1) + + HNO$_3$ (1,43)	Slight heating	5 min	CS		
	NaOH (15%)	106	1	99.4		
	2 h NaOH + 1 h H$_2$O$_2$	103	1	99.9		
ThB$_4$	H$_2$O	20	—	ND	[124]	1959
	H$_2$O	Boiling	—	ND		
	HNO$_3$ (1 : 1)	20	—	CS		
	HCl (1,19)	20	—	CS		
	H$_2$SO$_4$ (1,82)	Boiling	—	CS		
ThB$_6$	H$_2$O	20	—	ND	[124]	1959
	H$_2$O	Boiling	—	ND		
	HCl (1,19)	"	—	ND		
	HNO$_3$ (1,43)	20	—	ND		
	H$_2$SO$_4$ (1,82)	Boiling	—	ND		
UB$_2$	H$_2$SO$_4$ (1,82)	Boiling	—	ND	[124]	1959
	HNO$_3$ (1,43)	20	—	CS		
	HCl (1,19)	Boiling	—	ND		
	HF (1,15)	20	—	CS		
	NaOH (50%)	Boiling	—	ND		
UB$_4$	HCl (1,19)	20	—	CS	[124]	1959
	HCl (1,19)	Boiling	—	CS		
	HNO$_3$ (1,43)	20	—	CS		
	HNO$_3$ (1,43)	Boiling	—	CS		
	H$_2$SO$_4$ (1,82)	20	—	ND		
	H$_2$SO$_4$ (1,82)	Boiling	—	CS		
	HF (1,15)	"	—	CS		
	H$_2$O$_2$ (30%)	20	—	CS		
	Na$_2$O$_2$ (20%)	20	—	CS		
UB$_{12}$	HCl (1,19)	20	—	ND	[124]	1959
	HCl (1,19)	Boiling	—	ND		
	HNO$_3$ (1,43)	20	—	CS		
	HNO$_3$ (1,43)	Boiling	—	CS		
	H$_2$SO$_4$ (1,82)	20	—	ND		

Phase	Reagent	Temp., °C	Process time, h	Insoluble residue %	Source	Year
1	2	3	4	5	6	7
UB_{12}	H_2SO_4 (1.82)	Boiling	—	ND	[124]	1959
	HF (1.15)	20	—	ND		
	HF (1.15)	Boiling	—	ND		
	H_2O_2 (30%)	20	—	ND		
	Na_2O_2 (20%)	20	—	ND		
TiB_2	H_2SO_4 (1.84)	Boiling	96	5.5 *	[528]	1960
	H_2SO_4 (1.84)	"	1	43.3 *	[528]	1960
	H_2SO_4 (1.84)	20	24	89	[328]	1959
	H_2SO_4 (1.84)	Boiling	2	58	[328]	1959
	H_2SO_4 (1 : 10)	20	168	45.7 *	[528]	1960
	H_2SO_4 (1 : 4)	20	24	96	[328]	1959
	H_2SO_4 (1 : 4)	Boiling	2	68	[328]	1959
	H_3PO_4 (1 : 3)	20	24	98	[328]	1959
	H_3PO_4 (1 : 3)	Boiling	2	65	[328]	1959
	H_3PO_4 (1.21)	20	24	90	[328]	1959
	H_3PO_4 (1.21)	Boiling	2	SMC	[328]	1959
	HNO_3 (1 : 10)	20	96	97.5 *	[528]	1960
	HNO_3 (1 : 10)	Boiling	1	95.5 *	[528]	1960
	HNO_3 (1 : 1)	20	24	31	[328]	1959
	HNO_3 (1 : 1)	Boiling	2	CSH	[328]	1959
	HNO_3 (1.43)	20	24	97 *	[528]	1960
	HNO_3 (1.43)	Boiling	2	CSH	[328]	1959
	HNO_3 (1.43) + + HF (1.15)	"	15 min	1	[328]	1959
	3 h HCl (1.19) + + 1 h HNO_3 (1.43)	20	24	9	[328]	1959
	3 h HCl (1.19) + + 1 h HNO_3 (1.43)	Boiling	2	SMC	[328]	1959
	$HClO_4$ (1 : 3)	20	2	87	[328]	1959
	$HClO_4$ (1 : 3)	Boiling	24	28	[328]	1959
	$HClO_4$ (1.35)	20	24	30	[328]	1959
	$HClO_4$ (1.35)	Boiling	15 min	CS	[328]	1959
	$H_2C_2O_4$ (1 : 3)	20	24	89	[328]	1959
	$H_2C_2O_4$ (1 : 3)	Boiling	2	CSH	[328]	1959
	$H_2C_2O_4$ (sat.)	20	24	94	[328]	1959
	$H_2C_2O_4$ (sat.)	Boiling	2	51	[328]	1959
	HF (1 : 10)	20	27	15.6 *	[528]	1960
	HF (1.15)	20	96	16.6 *	[528]	1960
	HF (1.15)	Boiling	2	64	[328]	1959
	HCl (1.19)	"	2	58	[328]	1959
	HCl (1.19)	20	24	12 *	[528]	1960
	HCl (1.19)	Boiling	1	5.5 *	[528]	1960

*Particle size 200μ.

Phase	Reagent	Temp., °C	Process time, h	Insoluble residue, %	Source	Year
1	2	3	4	5	6	7
TiB_2	HCl (1 : 10)	20	96	3,9 *	[528]	1960
	HCl (1 : 10)	Boiling	1	12 *	[528]	1960
	HCl (1 : 2)	"	0.5	94 *	[528]	1960
	HCl (1 : 2)	20	24	93,5	[328]	1959
	HCl (1 : 2)	Boiling	2	61	[328]	1959
	3 h HCl (1 : 1) + + 1 h HNO_3 (1 : 1)	20	24	30	[328]	1959
	3 h HCl (1 : 1) + + 1 h HNO_3 (1 : 1)	Boiling	2	SMC	[328]	1959
	3 h $C_2H_2O_4$ (sat.) + 1 h H_2O_2 (30%) + 1 h HNO_3 (1,43)	20	24	1	[328]	1959
	3 h $C_2H_2O_4$ (sat.) + 1 h H_2O_2 (30%) + 1 h HNO_3 (1.43)	Boiling	2	6	[328]	1959
	3 h $C_2H_2O_4$ + + 2 h H_2SO_4 (1,84)	20	24	87	[328]	1959
	3 h $C_2H_2O_4$ (sat.) + 2 h H_2SO_4 (1,84)	Boiling	2	50	[328]	1959
	7 h HCl (1,19) + + 3 h bromine water	11	2	35	[328]	1959
	7 h $HClO_4$ (1,35) + 3 h HCl (1,19)	20	24	27	[328]	1959
	7 h $HClO_4$ (1,35) + 3 h HCl (1,19)	Boiling	2	SMC	[328]	1959
	1 h H_2SO_4 (1,84) + 4 h H_3PO_4 + 2 h H_2O	20	24	91	[328]	1959
	1 h H_2SO_4 (1,84) + 4 h H_3PO_4 + 2 h H_2O	Boiling	2	48	[328]	1959
	10 h H_2SO_4 (1,84) + 1 h K_2SO_4	"	2	6	[328]	1959
	7 h H_2SO_4 (1,84) + 3 h HNO_3 (1,43)	"	2	1	[328]	1959

*Particle size 200 μ.

Phase	Reagent	Temp., °C	Process time, h	Insoluble residue, %	Source	Year
1	2	3	4	5	6	7
TiB$_2$	30 ml H$_2$O$_2$ (1 : 3) + 5 drops H$_2$SO$_4$ (1,84)	Boiling	1	PS*	[525]	1961
	30 ml H$_2$O$_2$ (1 : 3) + 10 drops HNO$_3$ (1,43)	"	1	PS*	[525]	1961
	KNO$_3$ — 1% H$_2$SO$_4$ soln.	"	1	PS*	[525]	1961
	HCl (1 : 1) + + H$_2$C$_2$O$_4$	"	1	71 *	[525]	1961
	H$_2$C$_2$O$_4$ (6%)	"	1	81 *	[525]	1961
	HCl (1 : 1) + Trilon-B	"	1	84 *	[525]	1961
	HCl (1 : 1) + + C$_6$H$_8$O$_7$	"	1	83 *	[525]	1961
	NaOH (30%)	20	24	92	[328]	1959
	NaOH (30%)	Boiling	2	CS	[328]	1959
	NaOH (10%)	"	2	CS	[328]	1959
ZrB$_2$	HCl (1,19)	Boiling	2	6	[328]	1959
	HCl (1,19)	"	1	25,4 **	[528]	1960
	HCl (1,19)	20	24	2 **	[528]	1960
	HCl (1 : 1)	20	24	93	[328]	1959
	HCl (1 : 1)	Boiling	2	7	[328]	1959
	HCl (1 : 2)	"	0,5	20,8 **	[528]	1960
	HCl (1 : 2)	"	1,5	47,4 **	[528]	1960
	HCl (1 : 2)	"	2,5	27,7 **	[528]	1960
	HCl (1 : 2)	"	4	3,02 **	[528]	1960
	HCl (1 : 10)	"	1	7 **	[528]	1960
	HCl (1 : 10)	20	16	25.4 **	[528]	1960
	H$_2$O	20	24	5,75 **	[528]	1960
	H$_2$O	Boiling	1	0.94 **	[528]	1960
	HNO$_3$ (1.43)	20	24	12	[328]	1959
	HNO$_3$ (1,43)	Boiling	2	CS	[328]	1959
	HNO$_3$ (1 : 1)	20	24	23	[328]	1959
	HNO$_3$ (1 : 1)	Boiling	2	4	[328]	1959
	HNO$_3$ (1 : 10)	20	0,5	CS**	[528]	1960
	HNO$_3$ (1 : 10)	Boiling	1	14 **	[528]	1960
	HNO$_3$ (1,43)	20	96	74,5 **	[528]	1960
	HNO$_3$ (1.43)	Boiling	1	93,1 **	[528]	1960
	H$_2$SO$_4$ (1,84)	20	24	65	[328]	1959
	H$_2$SO$_4$ (1,84)	Boiling	2	1	[328]	1959
	H$_2$SO$_4$ (1 : 4)	20	24	51	[328]	1959
	H$_2$SO$_4$ (1 : 4)	Boiling	2	5	[328]	1959
	H$_2$SO$_4$ (1 : 10)	20	0,5	CS**	[528]	1960
	H$_2$SO$_4$ (1 : 10)	Boiling	1	27 **	[528]	1960

*TiO$_2$ precipitated.

**Particle size 200 μ.

Phase	Reagent	Temp., °C	Process time, h	Insoluble residue, %	Source	Year
1	2	3	4	5	6	7
ZrB_2	H_2SO_4 (1,84)	20	96	3,99 *	[528]	1960
	H_2SO_4 (1,84)	Boiling	1	CS*	[528]	1960
	H_3PO_4 (1,21)	20	24	63	[328]	1959
	H_3PO_4 (1,21)	Boiling	2	SMC	[328]	1959
	H_3PO_4 (1 : 3)	20	24	89	[328]	1959
	H_3PO_4 (1 : 3)	Boiling	2	SMC	[328]	1959
	$HClO_4$ (1,35)	20	24	10	[328]	1959
	$HClO_4$ (1,35)	Boiling	25	4	[328]	1959
	$HClO_4$ (1 : 3)	20	24	71	[328]	1959
	$HClO_4$ (1 : 3)	Boiling	2	48	[328]	1959
	$H_2C_2O_4$ (sat.)	20	24	55	[328]	1959
	$H_2C_2O_4$ (sat.)	Boiling	2	5	[328]	1959
	$H_2C_2O_4$ (6%)	"	1	67	[525]	1961
	$H_2C_2O_4$ (1 : 3)	20	24	38	[328]	1959
	$H_2C_2O_4$ (1 : 3)	Boiling	2	SMC	[328]	1959
	HF (1,15)	"	2	25	[328]	1959
	HF (1,15)	20	24	84,4	[528]	1960
	HF (1 : 10)	20	24	77	[528]	1960
	HF (1 : 10)	Boiling	1	86,2	[528]	1960
	3 h HCl (1,19) + + 1 h HNO₃ (1,43)	20	24	7	[328]	1959
	3 h HCl (1,19) + + 1 h HNO₃ (1,43)	Boiling	2	6	[328]	1959
	3 h HCl (1 : 1) + + 1 h HNO₃ (1 : 1)	20	24	16	[328]	1959
	3 h $H_2C_2O_4$ (sat.) + 1 h H_2O_2 (30%) + 1 h HNO₃ (1,43)	20	24	6	[328]	1959
	3 h $H_2C_2O_4$ (sat.) + 2 h H_2SO_4 (1,84)	20	24	59	[328]	1959
	3 h $H_2C_2O_4$ (sat.) + 2 h H_2SO_4	Boiling	2	10	[328]	1959
	7 h HCl (1,19) + + 3 bromine water	Boiling	2	18	[328]	1959
	7 h $HClO_4$ (1,35) + + 3 h HCl (1,19)	20	24	90	[328]	1959
	7 h $HClO_4$ (1,35) + 3 h HCl (1,19)	Boiling	2	8	[328]	1959

*Particle size 200 μ.

Phase	Reagent	Temp., °C	Process time, h	Insoluble residue, %	Source	Year
1	2	3	4	5	6	7
ZrB_2	1 h H_2SO_4 (1.84) + 4 h H_3PO_4 (1.21) + + 2 h H_2O	20	24	SMC	[328]	1959
	1 h H_2SO_4 (1.84) + 4 h H_3PO_4 (1.21) + + 2h H_2O	Boiling	2	SMC	[328]	1959
	H_2SO_4 (1.84) + + K_2SO_4	"	2	6	[328]	1959
	H_2SO_4 (1.84) + + $K_2S_2O_8$	"	2	7	[328]	1959
	7 h H_2SO_4 (1.84) + 3 h HNO_3 (1.43)	"	15 min	4	[328]	1959
	HNO_3 (1.43) + + HF (1.15)	"	15 min	4	[328]	1959
	HCl (1 : 1) + + $H_2C_2O_4$	"	1	71	[529]	1961
	HCl (1 : 1) + + $C_6H_8O_7$	"	1	76	[529]	1961
	HCl (1 : 1) + Trilon-B	"	1	75	[529]	1961
	KNO_3, 1% H_2SO_4 soln.	"	1	52	[529]	1961
	30 ml H_2O_2 (1 : 3) + 5 drops H_2SO_4 (1.84)	Boiling	1	97	[529]	1961
	30 ml H_2O_2 (1 : 3) + 10 drops HNO_3 (1.43)	"	1	ND	[529]	1961
	HaOH (30%)	20	24	CS	[328]	1959
	NaOH (30%)	Boiling	2	CS	[328]	1959
	NaOH (10%)	"	2	98	[328]	1959
HfB_2	H_2SO_4 (1.84)	20	24	91.58	[530]	1968
		Boiling	2	1.87		
	H_2SO_4 (1 : 4)	20	24	96.07		
		Boiling	2	3.84		
	HNO_3 (1.40)	20	24	47.87		
		Boiling	2	2.98		
	HNO_3 (1 : 1)	20	24	54.35		
		Boiling	2	3.24		
	HCl (1.19)	20	24	97.10		
		Boiling	2	6.37		
	HCl (1 : 1)	20	24	96.60		
		Boiling	2	5.65		
	$HClO_4$ (1.65)	20	24	100.00		
		Boiling	2	2.54		

Phase	Reagent	Temp., °C	Process time, h	Insoluble residue, %	Source	Year
1	2	3	4	5	6	7
HfB_2	$HClO_4$ (1 : 3)	20	24	95,12	[530]	1968
		Boiling	2	36,85		
	$H_2C_2O_4$ (sat.)	20	24	93,94		
		Boiling	2	5,05		
	H_3PO_4 (1 : 4)	20	24	97,51		
		Boiling	2	100,00		
	$3HCl$ (1,19) + + HNO_3 (1,40)	20	24	5,95		
	$3HCl$ (1,19) + + HNO_3 (1,40)	Boiling	2	2,31		
	$3HCl$ (1 : 1) + + HNO_3 (1 : 1)	20	24	28,81		
		Boiling	2	2,93		
	$3H_2C_2O_4$ (sat.) + H_2O_2 (30%) + HNO_3 (1,4)	20	24	3,36		
		Boiling	2	2.57		
	$3H_2C_2O_4$ (sat.) + $2H_2SO_4$ (1.84)	20	24	96,32		
		Boiling	2	2,94		
	7 h HCl (1.19) + + 3 h bromine water	20	24	29,05		
		Boiling	2	3,13		
	7 h $HClO_4$(1,65) + + 3 h HCl (1,19)	20	24	91,24		
		Boiling	2	3,00		
	H_2SO_4 (1,84) + + $4H_3PO_4$ (1.42) + + $2H_2O$	20	24	93,43		
		Boiling	2	5,43		
	5 ml H_2SO_4 (1.84) + 5 g K_2SO_4	20	24	99,21		
		Boiling	2	0		
	5 ml H_2SO_4 (1,84) + 5 g $K_2S_2O_8$	20	24	33,17		
		Boiling	2	0		
	$7H_2SO_4$ (1.84) + + $3HNO_3$ (1,40)	20	24	2,09		
		Boiling	2	1,87		
VB_2	HCl (1.19)	20	24	63	[328]	1959
		Boiling	2	3		
	HCl (1 : 1)	20	24	62		
	HCl (1 : 1)	Boiling	2	10		

Phase	Reagent	Temp., °C	Process time, h	Insoluble residue, %	Source	Year
1	2	3	4	5	6	7
VB_2	HNO_3 (1,43)	20	24	1	[328]	1959
	HNO_3 (1.43)	Boiling	2	2		
	HNO_3 (1 : 1)	20	24	3		
	HNO_3 (1 : 1)	Boiling	2	2		
	H_2SO_4 (1,84)	20	24	49		
	H_2SO_4 (1,84)	Boiling	2	13		
	H_2SO_4 (1 : 4)	20	24	60		
	H_2SO_4 (1 : 4)	Boiling	2	7		
	H_3PO_4 (1,21)	20	24	66		
	H_3PO_4 (1,21)	Boiling	2	SMC		
	H_3PO_4 (1 : 3)	20	24	62		
	H_3PO_4 (1 : 3)	Boiling	2	24		
	$HClO_4$ (1,35)	20	24	4		
	$HClO_4$ (1,35)	Boiling	2	0		
	$HClO_4$ (1 : 3)	20	24	47		
	$HClO_4$ (1 : 3)	Boiling	2	2		
	$H_2C_2O_4$ (sat.)	20	24	60		
	$H_2C_2O_4$ (sat.)	Boiling	2	17		
	$H_2C_2O_4$ (6%)	"	1	29	[525]	1961
	$H_2C_2O_4$ (1 : 3)	20	24	24	[328]	1959
	$H_2C_2O_4$ (1 : 3)	Boiling	2	37	[328]	1959
	HF (1,15)	"	2	13	[328]	1959
	30 ml H_2O_2 (1 : 3) + 5 drops H_2SO_4 (1,82)	"	1	SMC	[525]	1961
	30 ml H_2O_2 (1 : 3) + 10 dropsHNO$_3$ (1,43)	"	1	27	[525]	1961
	KNO_3, 1% H_2SO_4 soln.	Boiling	1	28	[525]	1961
	HCl (1 : 1) + $H_2C_2O_4$	"	1	34	[525]	1961
	HCl (1 : 1) + $C_6H_8O_7$	"	1	36	[525]	1961
	HCl (1 : 1) + Trilon-B	20	24	64	[328]	1959
	NaOH (30%)	Boiling	2	61	[328]	1959
	NaOH (10%)	20	24	98	[328]	1959
NbB_2	HCl (1,19)	Boiling	2	91	[328]	1959
	HCl (1,19)	20	24	99		
	HCl (1 : 1)	Boiling	2	95		
	HCl (1 : 1)	20	24	94		
	HNO_3 (1,4)	Boiling	2	100		
	HNO_3 (1 : 1)	20	24	99		
	HNO_3 (1 : 1)	Boiling	2	100		
	H_2SO_4 (1,84)	20	24	100		
	H_2SO_4 (1,84)	Boiling	2	3		

Phase	Reagent	Temp., °C	Process time, h	Insoluble residue, %	Source	Year
1	2	3	4	5	6	7
NbB_2	H_2SO_4 (1 : 4)	20	24	100	[328]	1959
	H_2SO_4 (1 : 4)	Boiling	2	22		
	H_3PO_4 (1.35)	20	24	100		
	H_3PO_4 (1.35)	Boiling	2	PS		
	H_3PO_4 (1 : 3)	20	24	100		
	H_3PO_4 (1 : 3)	Boiling	2	24		
	$HClO_4$ (1.35)	20	24	98		
	$HClO_4$ (1.35)	Boiling	2	98		
	$HClO_4$ (1 : 3)	20	24	98		
	$H_2C_2O_4$ (sat.)	20	24	97		
	$H_2C_2O_4$ (sat.)	Boiling	2	50		
	$H_2C_2O_4$ (6%)	"	1	94	[525]	1961
	$H_2C_2O_4$ (1 : 3)	20	24	93	[328]	1959
	$H_2C_2O_4$ (1 : 3)	Boiling	2	98		
	HF (1.15)	"	2	44		
	3 h HCl (1. 19) + + 1 h HNO_3 (1.43)	20	24	71		
	3 h HCl (1,19) + + 1 h HNO_3 (1.43)	Boiling	2	80		
	3 h HCl (1 : 1) + + 1 h HNO_3 (1:1)	20	24	96		
	3 h HCl (1 : 1) + + 1 h HNO_3 (1:1)	Boiling	2	CS		
	3 h $H_2C_2O_4$ (sat.) + 1 h H_2O_2 (30%) + 1 h HNO_3 (1.43)	20	24	5		
	3 h $H_2C_2O_4$ (sat.) + 1 h H_2O_2 (30%) + 1 h HNO_3 (1.43)	Boiling	2	26		
	H_2O_2 (1 : 2)	20	24	28,44	[332]	1968
		Boiling	2	11,80		
	HCl + KIO_3	20	24	ND		
		Boiling	2	ND		
	HCl + KIO_4	20	24	ND		
		Boiling	2	ND		
	HNO_3 + KI	20	24	ND		
		Boiling	2	ND		
	HNO_3 + KJO_3	20	24	99,64		
		Boiling	2	99,55		
	3 h. $H_2C_2O_4$ (sat.) + 2 h H_2SO_4 (1,84)	20	24	86	[328]	1959
	3 h $H_2C_2O_4$ (sat.) + 2 h H_2SO_4 (1,84)	Boiling	2	86		

Phase	Reagent	Temp., °C	Process time, h	Insoluble residue, %	Source	Year
1	2	3	4	5	6	7
NbB_2	7 h HCl (1,19) + + 3 h bromine water	Boiling	2	58	[328]	1959
	7 h $HClO_4$ (1.35) + 3 h HCl (1,19)	20	24	73		
	7 h $HClO_4$ (1.35) + 3 h HCl (1.19)	Boiling	2	60		
	1 h H_2SO_4 (1.84) + 4 h H_3PO_4 (1.21) + + 2 h H_2O	"	2	10		
	1 h H_2SO_4 (1.84) + 4 h H_3PO_4 (1.21) + + 2 h H_2O	20	15 min	SMC		
	H_2SO_4 (1.84) + + K_2SO_4	Boiling	2	3		
	H_2SO_4 (1.84) + + HNO_3 (1,43)	"	2	SMC		
	HNO_3 (1,43) + + HF (1.15)	"	15 min	4		
	30 ml H_2O_2 (1 : 3) + 5 drops H_2SO_4 (1,82)	"	1	PS	[525]	1961
	30 ml H_2O_2 (1 : 3) + 10 drops HNO_3 (1,43)	"	1	PS		
	KNO_3, 1% H_2SO_4 soln.	"	1	98		
	HCl (1 : 1) + + $H_2C_2O_4$	"	1	95		
	HCl (1 : 1) + $C_6H_8O_7$	"	1	98		
	NaOH (30%)	20	24	95	[328]	1959
	NaOH (30%)	Boiling	2	CS		
	NaOH (10%)	"	2	95		
TaB_2	HCl (1,19)	20	24	100	[328]	1959
	HCl (1,19)	Boiling	2	99		
	HCl (1 : 1)	20	24	100		
	HCl (1 : 1)	Boiling	2	98		
	HNO_3 (1.43)	20	24	100		
	HNO_3 (1.43)	Boiling	2	100		
	HNO_3 (1 : 1)	20	24	100		
	HNO_3 (1 : 1)	Boiling	2	100		
	H_2SO_4 (1,84)	20	24	99		
	H_2SO_4 (1,84)	Boiling	2	3		

Phase	Reagent	Temp., °C	Process time, h	Insoluble residue, %	Source	Year
1	2	3	4	5	6	7
TaB$_2$	H$_2$SO$_4$ (1 : 4)	20	24	100	[328]	1959
	H$_2$SO$_4$ (1 : 4)	Boiling	2	99		
	H$_3$PO$_4$ (1,21)	20	24	100		
	H$_3$PO$_4$ (1 : 3)	20	24	100		
	H$_3$PO$_4$ (1 : 3)	Boiling	2	100		
	HClO$_4$ (1,35)	20	24	100		
	HClO$_4$ (1,35)	Boiling	2	100		
	HClO$_4$ (1 : 3)	20	24	100		
	HClO$_4$ (1 : 3)	Boiling	2	99		
	H$_2$C$_2$O$_4$ (sat.)	20	24	100		
	H$_2$C$_2$O$_4$ (sat.)	Boiling	2	94		
	H$_2$C$_2$O$_4$ (1 : 3)	20	24	99		
	H$_2$C$_2$O$_4$ (1 : 3)	Boiling	2	99		
	HF (1,15)	"	2	20		
	3 h HCl (1,19) + + 1 h HNO$_3$ (1,43)	20	24	99		
	3 h HCl (1,19) + + 1 h HNO$_3$(1,43)	Boiling	2	SMC		
	3 h HCl (1 : 3) + + 1 h HNO$_3$ (1 : 1)	20	24	99		
	3 h HCl (1 : 3) + + 1 h HNO$_3$ (1:1)	Boiling	2	SMC		
	3 h H$_2$C$_2$O$_4$ (sat.) + 1 h H$_2$O$_2$ (30%) + 1 h HNO$_3$ (1,43)	20	24	32		
	3 h H$_2$C$_2$O$_4$ (sat.) + 1 h H$_2$O$_2$ (30%) + 1 h HNO$_3$ (1,43)	Boiling	2	SMC		
	3 h H$_2$C$_2$O$_4$ (sat.) + 2 h H$_2$SO$_4$ (1,84)	•	2	99		
	3 h H$_2$C$_2$O$_4$ (sat.) + 2 h H$_2$SO$_4$ (1.84)	20	24	99		
	7 h HCl (1,19) + + 3 h bromine water	Boiling	2	99		
	7 h HClO$_4$ (1,35) + 3 h HCl (1,19)	20	24	99		
	7 h HClO$_4$ (1,35) + 3 h HCl (1,19)	Boiling	2	100		

Phase	Reagent	Temp., °C	Process time, h	Insoluble residue, %	Source	Year
1	2	3	4	5	6	7
TaB_2	1 h H_2SO_4 (1,84) $+$ 4 h H_3PO_4 (1,21) $+$ $+$ 2 h H_2O	Boiling	2	23	[328]	1959
	1 h H_2SO_4 (1,84) $+$ 4 h H_3PO_4 (1,21) $+$ $+$ 2 h H_2O	20	24	SMC		
	H_2SO_4 (1,84) $+$ $+$ K_2SO_4	Boiling	2	5		
	KNO_3, 1% H_2SO_4 soln.	"	1	ND	[525]	1961
	30 ml H_2O_2 (1 : 3)+15 drops H_2SO_4 (1,82)	"		PS		
	30 ml H_2O_2 (1 : 3) $+$ 10 drops HNO_3 (1,43)	"		PS		
	H_2O_2 (1 : 2)	20	24	88,62	[332]	1968
		Boiling	2	80,56		
	$HCl + KIO_3$	20	24	99,06		
		Boiling	2	98,88		
	$HCl + KIO_4$	20	24	99,68		
		Boiling	2	99,32		
	$HNO_3 + KI$	20	24	99,18		
		Boiling	2	98,83		
	$HNO_3 + KIO_3$	20	24	98,38		
		Boiling	2	98,27		
	HCl (1 : 1) $+$ $+$ $H_2C_2O_4$	"	1	91	[525]	1961
	HCl (1 : 1) $+$ $+$ $C_6H_8O_7$	"	1	98		
	HCl (1 : 1) $+$ $+$ Trilon-B	"	1	98		
	NaOH (30%)	20	24	CS	[328]	1959
	NaOH (30%)	Boiling	2	CS		
	NaOH (10%)	"	2	45		
CrB_2	HCl (1,19)	20	24	36	[328]	1959
	HCl (1,19)	Boiling	2	3		
	HCl (1 : 1)	20	24	51		
	HCl (1 : 1)	Boiling	2	6 *	[528]	1960
	HCl (1 : 2)	"	0,5	10,2 *		
	HCl (1 : 2)	"	0,5	8,5 *		

*Particle size 200 μ.

Phase	Reagent	Temp., °C	Process time, h	Insoluble residue, %	Source	Year
1	2	3	4	5	6	7
CrB_2	HNO_3 (1,43)	20	24	99	[328]	1959
	HNO_3 (1,43)	Boiling	2	22		
	HNO_3 (1 : 1)	20	24	99		
	HNO_3 (1 : 1)	Boiling	2	41		
	H_2SO_4 (1,84)	20	24	99		
	H_2SO_4 (1 : 4)	20	24	9		
	H_2SO_4 (1 : 4)	Boiling	2	3		
	H_3PO_4 (1.21)	20	24	100		
	H_3PO_4 (1,21)	Boiling	2	SMC		
	H_3PO_4 (1 : 3)	20	24	100		
	H_3PO_4 (1 : 3)	Boiling	2	18		
	$HClO_4$ (1,35)	20	24	96		
	$HClO_4$ (1,35)	Boiling	2	0		
	$HClO_4$ (1 : 3)	20	24	100		
	$HClO_4$ (1 : 3)	Boiling	2	4		
	$H_2C_2O_4$ (sat.)	20	24	44		
	$H_2C_2O_4$ (sat.)	Boiling	2	2		
	$H_2C_2O_4$ (1 : 3)	20	24	97		
	$H_2C_2O_4$ (1 : 3)	Boiling	2	75		
	HF (1,15)	"	2	2		
	3 h HCl (1,19) + + 1 h HNO$_3$ (1,43)	20	24	80		
	3 h HCl (1,19) + + 1 h HNO$_3$ (1,43)	Boiling	2	29		
	3 h HCl (1 : 3) + + 1 h HNO$_3$ (1 : 1)	20	24	95		
	3 h HCl (1 : 3) + + 1 h HNO$_3$ (1 : 1)	Boiling	2	27		
	3 h $H_2C_2O_4$ (sat.) + 1 h H_2O_2 (30%) + 1 h HNO_3 (1,43)	20	24	99		
	3 h $H_2C_2O_4$ (sat.) + 1 h H_2O_2 (30%) + 1 h HNO_3 (1,43)	Boiling	2	89		
	3 h $H_2C_2O_4$ (sat.) + 2 h H_2SO_4 (1,84)	20	24	31		
	3 h $H_2C_2O_4$ (sat.) + 2 h H_2SO_4 (1,84)	Boiling	2	3		
	35 ml HCl (1,19) + 1 ml bromine water	"	2	2		

Phase	Reagent	Temp., °C	Process time, h	Insoluble residue, %	Source	Year
1	2	3	4	5	6	7
CrB_2	7 h $HClO_4$ + 3 h HCl (1.19)	Boiling	2	3	[328]	1959
	7 h $HClO_4$ (1.35) + 3 h HCl (1.19)	20	24	49		
	1 h H_2SO_4 (1.84) + 4 h H_3PO_4 (1.21) + 2 h H_2O	20	24	86		
	1 h H_2SO_4 (1.84) + 4 h H_3PO_4 (1.21) + 2 h H_2O	Boiling	2	6		
	50 ml H_2SO_4 (1.84) + 5 g K_2SO_4	"	2	CS		
	50 ml H_2SO_4 (1.84) + 5 g $K_2S_2O_8$	"	2	CS		
	36 ml H_2SO_4 (1.84) + 215 ml HNO_3 (1.43)	"	2	2		
	HNO_3 (1.43) + HF (1.15)	"	1	4		
	HCl (1 : 1) + $H_2C_2O_4$ (sat.)	»	1	PS	[525]	1961
	30 ml H_2O_2 (1 : 3) + 5 drops H_2SO_4 (1.84)	"	1	97		
	30 ml H_2O_2 (1 : 3) + 10 drops HNO_3	"	1	PS		
	KNO_3, 1% H_2SO_4 soln.	"	1	98		
	NaOH (30%)	20	20	99	[328]	1959
	NaOH (30%)	Boiling	2	88		
	NaOH (10%)	"	2	98		
Mo_2B_5	HCl (1.19)	20	24	95	[328]	1959
	HCl (1.19)	Boiling	2	73		
	HCl (1 : 1)	20	24	94		
	HCl (1 : 1)	Boiling	2	85		
	HNO_3 (1.43)	20	24	9		
	HNO_3 (1.43)	Boiling	2	3		
	HNO_3 (1 : 1)	20	24	9		
	HNO_3 (1 : 1)	Boiling	2	9		
	H_2SO_4 (1.84)	20	24	95		
	H_2SO_4 (1.84)	Boiling	2	7		

Phase	Reagent	Temp., °C	Process time, h	Insoluble residue, %	Source	Year
1	2	3	4	5	6	7
Mo_2B_5	H_2SO_4 (1 : 4)	20	24	97	[328]	1959
	H_2SO_4 (1 : 4)	Boiling	2	65		
	H_3PO_4 (1.21)	20	24	93		
	H_3PO_4 (1.21)	Boiling	2	SMC		
	H_3PO_4 (1 : 3)	20	24	93		
	H_3PO_4 (1 : 3)	Boiling	2	77		
	$HClO_4$ (1,35)	20	24	8		
	$HClO_4$ (1.35)	Boiling	2	9		
	$HClO_4$ (1 : 3)	20	24	90		
	$HClO_4$ (1 : 3)	Boiling	2	16		
	$H_2C_2O_4$ (sat.)	20	24	91		
	$H_2C_2O_4$ (sat.)	Boiling	2	88		
	$H_2C_2O_4$ (1 : 3)	20	24	92		
	$H_2C_2O_4$ (1 : 3)	Boiling	2	88		
	HF (1,15)	"	2	60		
	3 h HCl (1,19 + + 1 h HNO_3 (1,43)	20	24	0.5		
	3 h HCl (1,19) + + 1 h HNO_3 (1,43)	Boiling	2	SMC		
	3 h HCl (1 : 1) + + 1 h HNO_3 (1 : 1)	20	24	0.3		
	3 h HCl (1 : 1) + + 1 h HNO_3 (1 : 1)	Boiling	2	CS		
	3 h $H_2C_2O_4$ (sat.) + 1 h H_2O_2 (30%) + 1 h HNO_3 (1,43)	20	24	0.3		
	H_2O_2 + HNO_3 (1,43)	Boiling	2	SMC		
	3 h $H_2C_2O_4$ (sat.) + 2 h H_2SO_4 (1,84)	20	24	63		
	3 h $H_2C_2O_4$ (sat.) + 2 h H_2SO_4 (1,84)	Boiling	2	63		
	7 h HCl (1,19) + + 3 h bromine water	"	2	23		
	7 h HCl (1,19) + + 3 h bromine water	20	24	61		
	1 h H_2SO_4 (1,84) + 4 h H_3PO_4 (1,21) + + 2 h H_2O	20	24	58		

Phase	Reagent	Temp., °C	Process time, h	Insoluble residue, %	Source	Year
1	2	3	4	5	6	7
Mo_2B_5	1 h H_2SO_4 (1,84) + 4 h H_3PO_4 (1,21) + + 2 h H_2O	Boiling	2	67	[328]	1959
	50 ml H_2SO_4 (1,84) + 5 g K_2SO_4	"	2	8		
	50 ml H_2SO_4 (1,84) + 5 h $K_2S_2O_8$	"	2	1		
	HNO_3 (1,43) + + HF (1,15)	"	1	1		
	NaOH (30%)	20	24	68		
	NaOH (30%)	Boiling	2	67		
	30 ml H_2O_2 (1 : 3) + 5 drops H_2SO_4	"	1	CS	[525]	1961
	30 ml H_2O_2 (1 : 3) + 10 drops HNO_3	"	1	CS		
	KNO_3, 1% H_2SO_4 soln.	"	1	45		
	HCl (1 : 1) + +$H_2C_2O_4$	"	1	75		
	30 ml HCl (1 : 1)+ + $C_6H_8O_7$	"	1	95		
	HCl (1 : 1) + Trilon-B	"	1	96		
				Weight decrease mg/ $(cm^2 \cdot h)$		
α-MoB	KCl (0,1 N)	30	10	$\sim 3,96 \times \times 10^{-2}$	[531]	1961
			100	$\sim 1,25 \times \times 10^{-2}$		
			500	$\sim 0,83 \times \times 10^{-2}$		
β-MoB	KCl (0,1 N.)	30	10	$\sim 2,75 \times \times 10^{-2}$	[531]	1961
			100	$\sim 0,83 \times \times 10^{-2}$		
			500	$\sim 0,5 \times \times 10^{-2}$		
W_2B_5	HCl (1,19)	20	2	96	[328]	1959
	HCl (1,19)	Boiling	24	96		

Phase	Reagent	Temp., °C	Process time, h	Insoluble residue, %	Source	Year
1	2	3.	4	5	6	7
W_2B_5	HCl (1 : 1)	20	2	95	[328]	1959
	HCl (1 : 1)	Boiling	24	97		
	HNO_3 (1,43)	20	24	CSH		
	HNO_3 (1.43)	Boiling	2	9		
	HNO_3 (1 : 1)	20	24	CSH		
	HNO_3 (1 : 1)	Boiling	2	11		
	H_2SO_4 (1.84)	20	24	100		
	H_2SO_4 (1.84)	Boiling	2	2		
	H_2SO_4 (1 : 1)	20	24	96		
	H_2SO_4 (1 : 1)	Boiling	2	97		
	H_3PO_4 (1,21)	20	24	96		
	H_3PO_4 (1,21)	Boiling	2	SMC		
	H_3PO_4 (1 : 3)	20	24	89		
	H_3PO_4 (1 : 3)	Boiling	2	93		
	$HClO_4$ (1.35)	20	24	94		
	$HClO_4$ (1.35)	Boiling	2	3		
	$HClO_4$ (1 : 3)	20	24	96		
	$HClO_4$ (1 : 3)	Boiling	2	100		
	$H_2C_2O_4$ (sat.)	20	24	92		
	$H_2C_2O_4$ (sat.)	Boiling	2	87		
	$H_2C_2O_4$ (1 : 3)	20	24	91		
	$H_2C_2O_4$ (1 : 3)	Boiling	2	88		
	HF (1.15)	"	2	75		
MnB	HCl (1,19)	Boiling	—	CS	[124]	1959
	HCl (1,19)	20	—	CS		
	HNO_3 (1.43)	20	—	CS		
	HNO_3 (1.43)	Boiling	—	CS		
	H_2SO_4 (1.84)	20	—	CS		
	H_2SO_4 (1.84)	Boiling	—	CS		
Mn_3B	H_2O	20	—	ND	[256]	1957
	H_2O	Boiling	—	ND		
	HCl (1.19)	20	—	CS		
	HNO_3 (1.43)	20	—	CS		
	H_2SO_4 (1.84)	20	—	CS		
Mn_4B	H_2O	20	—	ND	[256]	1957
	H_2O	Boiling	—	CS		
	HCl (1.19)	20	—	CS		
	HNO_3 (1.43)	20	—	CS		
	H_2SO_4 (1.84)	20	—	CS		
FeB	HCl (1,19)	20	—	ND	[256]	1957
	HCl (1,19)	Boiling	—	ND		

RESISTANCE OF POWDERED REFRACTORY COMPOUNDS TO ACIDS
AND ALKALIS (continued)

Phase	Reagent	Temp., °C	Process time, h	Insoluble residue, %	Source	Year
1	2	3	4	5	6	7
FeB	HNO_3 (1.43) HNO_3 (1.43) H_2SO_4 (1.84) H_2SO_4 (1.84)	20 Boiling 20 Boiling	— — — —	CS CS ND ND	[256]	1957
Co_2B	HCl (1,19) HCl (1,19) HNO_3 (1.43)	20 Boiling 20	— — —	ND CS CS	[256]	1957
CoB	HCl (1,19)	20 Boiling	—	ND	[256]	1957
NiB	H_2O HNO_3 (1.43) 3 h HCl (1,19) + + 1 h HNO_3 (1.43)	20	—	CS	[124]	1959
BeC_2	H_2O HCl (1 : 1) H_2SO_4 (1 : 1) HNO_3 (1 : 1)	20	—	CS*	[532]	1952
Be_2C	H_2O HCl (1 : 1) H_2SO_4 (1 : 1) HNO_3 (1 : 1)	20	—	CS**	[532]	1952
Be_3C_2	H_2O HCl (1 : 1) HNO_3 (1 : 1) H_2SO_4 (1 : 1)	20	—	CS	[124]	1959
MgC_2	H_2O	20	—	CS	[532]	1952
Mg_2C_3	H_2O	20	—	CS	[532]	1952

*C_2H_2 is given off.
**CH_4 is given off.

Phase	Reagent	Temp., °C	Process time, h	Insoluble residue, %	Source	Year
1	2	3	4	5	6	7
CaC_2	H_2O	20	—	CS*	[533]	1971
	HCl (1,19)	20	—	CS	[538]	1956
SrC_2	H_2O	20	—	CS	[532]	1952
BaC_2	H_2O	20	—	CS	[532]	1952
	HCl (1 : 1)				[538]	1956
ScC	H_2O	20	—	CS	[461]	1971
ScC_x	H_2O; D_2O	20	—	CS**	[534]	1971
ScC_2	H_2O	20	—	CS**	[535]	1970
YC	H_2O	20	5 min	CS	[92]	1961
	HCl (1,19)		1 min	CS		
	HCl (1 : 1)		1 min	CS		
	HNO_3 (1,43)		15 min	PS		
	HNO_3 (1 : 1)		5 min	CS		
	H_2SO_4 (1,84)		15	PS		
	H_2SO_4 (1 : 1)		2	PS		
	NaOH (25%)		15 min	CS		
Y_2C_3	H_2O	20	3 min	CS	[92]	1961
	HCl (1,19)		1 min	CS		
	HCl (1 : 1)		1 min	CS		
	HNO_3 (1,43)		20	PS		
	HNO_3 (1 : 1)		5 min	CS		
	H_2SO_4 (1,84)		20	PS		
	H_2SO_4 (1 : 1)		5 min	CS		
	NaOH (25 %)		10 min	CS		
YC_2	H_2O	20	5 min	CS	[92]	1961
	HCl (1.19)		20	PS		
	HCl (1 : 1)		20	PS		

*C_2H_2 is given off.
**Hydrocarbons are given off.

Phase	Reagent	Temp., °C	Process time, h	Insoluble residue, %	Source	Year
1	2	3	4	5	6	7
YC_2	HNO_3 (1.43) HNO_3 (1 : 1) H_2SO_4 (1.84) H_2SO_4 (1 : 1) NaOH (25%)	20	20 20 20 20 15	ND PS ND PS CS	[92]	1961
La_2C_3	H_2O	20	—	CS*	[461]	1971
LaC_2	H_2O; D_2O	20	—	CS*	[534]	1971
	H_2SO_4 HCl HNO_3 NaOH	20	1	12—15	[537]	1973
	Organic acids	—	1	CS**		
	NaOH (40%)	110	1	CSH		
	Organic solvents	20	1	ND		
Ce_2C_3	H_2O	20	—	CS*	[461]	1971
CeC_2	H_2O; D_2O	20	—	CS*	[534]	1971
	H_2SO_4 HCl HNO_3 NaOH	20	1	12—15	[537]	1973
	Organic acids	—	1	CS**		
	NaOH (40%)	110	1	CSH		
	Organic solvents	20	1	ND		
Pr_2C_3	H_2O	20	—	CS*	[461]	1971
PrC_2	H_2O	20	—	CS*	[461]	1971
	Organic acids	—	1	CS**	[537]	1973
	H_2SO_4 HCl HNO_3 NaOH	20	1	12—15		

*Hydrocarbons given off.
**Hydrocarbons and hydrogen given off.

Phase	Reagent	Temp., °C	Process time, h	Insoluble residue, %	Source	Year
1	2	3	4	5	6	7
PrC_2	NaOH (40%) Organic solvents	110 20	1 1	CSH ND	[537]	1973
NdC_2	H_2O; D_2O H_2SO_4 HCl HNO_3 NaOH	20 20	— 1	CS* 12—15	[534] [537]	1971 1973
NdC_2	Organic acids NaOH (40%) Organic solvents	— 110 20	1 1 1	CS** CSH ND	537	1973
SmC_2	H_2O; D_2O Mineral acids	20	—	CS* CS	[534] [529]	1971 1961
GdC_2	H_2O; D_2O	20	—	CS	[534]	1971
TbC_2	H_2O; D_2O	20	—	CS	[534]	1971
DyC_2	H_2O	20	—	CS	[461]	1971
HoC_2	H_2O; D_2O	20	—	CS	[534]	1971
ErC_2	H_2O	20	—	CS	[461]	1971
TmC_2	H_2O	20	—	CS	[461]	1971
ThC	H_2O HCl (1 : 1) HCl (1 : 1) H_2SO_4 (1 : 1) H_2SO_4 (1 : 1) HNO_3 (1 : 1)	100 20 110 20 135 20	2 1 0.5 0.5 0.15 2	CSH CSH CSH CSH CSH 100	[538]	1956

*Hydrocarbons given off.
**Hydrocarbons and hydrogen given off.

Phase	Reagent	Temp., °C	Process time, h	Insoluble residue, %	Source	Year
1	2	3	4	5	6	7
ThC	HNO_3 (1 : 1)	115	1	CSH	[538]	1956
	Tartaric acid	120	1	CSH		
	NaOH (25%)	20	1	CSH		
	NaOH (25%)	110	0.5	CSH		
ThC_2	H_2O	25	—	CS*	[539]	1967
	Concentrated mineral acids	20	—	PS	[538]	1956
U_2C_3	HCl (1,19)	Boiling	—	CS	[532]	1952
	HNO_3 (1,4)					
	H_2SO_4 (1,84)					
UC_2	HCl (1,19)	Boiling	—	CS	[532]	1952
	HNO_3 (1,4)					
	H_2SO_4 (1.84)					
TiC	HCl (1.19)	20	24	99	[540]	1961
	HCl (1.19)	Boiling	2	100		
	HCl (1 : 1)	20	24	100		
	HCl (1 : 1)	Boiling	2	97		
	HNO_3 (1.43)	20	24	CS		
	HNO_3 (1.43)	Boiling	2	SMC		
	HNO_3 (1 : 1)	20	24	CS		
	HNO_3 (1 : 1)	Boiling	2	CS		
	H_2SO_4 (1.84)	20	24	ND		
	H_2SO_4 (1.84)	Boiling	2	88		
	H_2SO_4 (1 : 4)	20	24	100		
	H_2SO_4 (1 : 4)	Boiling	2	97		
	H_3PO_4 (1.21)	20	24	99		
	H_3PO_4 (1.21)	Boiling	2	98		
	H_3PO_4 (1 : 3)	20	24	98		
	H_3PO_4 (1 : 3)	Boiling	2	99		
	$HClO_4$ (1.35)	20	24	100		
	$HClO_4$ (1.35)	Boiling	2	CS		
	$HClO_4$ (1 : 3)	20	24	100		
	$HClO_4$ (1 : 3)	Boiling	2	CS		
	$H_2C_2O_4$ (sat.)	20	24	100		
	$H_2C_2O_4$ (sat.)	Boiling	2	100		
	3 h HCl (1.19) + + 1 h HNO_3 (1.43)	20	24	4		

*Hydrocarbons given off.

Phase	Reagent	Temp., °C	Process time, h	Insoluble residue, %	Source	Year
1	2	3	4	5	6	7
TiC	3 h HCl (1,19) + + 1 h HNO$_3$ (1,43)	Boiling	2	CS	[540]	1961
	2 h H$_2$SO$_4$ (1,84) + 1 h HNO$_3$ (1.43)	20	24	CS		
	2 h H$_2$SO$_4$ (1,84) + 1 h HNO$_3$ (1,43)	Boiling	2	CS		
	4 h HNO$_3$ (1,4) + + 1 h HF (1.15)	20	24	CS		
	H$_2$SO$_4$ (1,81) + + H$_3$PO$_4$ (1 : 4)	20	24	98		
	H$_2$SO$_4$ (1,84) + + H$_3$PO$_4$ (1 : 3)	20	24	100		
	H$_2$SO$_4$ (1,84) + + H$_3$PO$_4$ (1 : 3)	Boiling	2	100		
	H$_2$SO$_4$ (1,84) + + H$_2$C$_2$O$_4$ (sat.)	20	24	99		
	H$_2$SO$_4$ (1,84) + + H$_2$C$_2$O$_4$ (sat.)	Boiling	2	84		
TiC$_x$	H$_2$SO$_4$ + H$_2$O$_2$	25—80	0,5—24	— *	[541]	1971
ZrC	HNO$_3$ (1 : 1)	20	24	76	[540]	1961
	HNO$_3$ (1 : 1)	Boiling	2	6		
	H$_2$SO$_4$ (1,84)	20	24	97		
	H$_2$SO$_4$ (1,84)	Boiling	2	CS		
	H$_2$SO$_4$ (1 : 4)	20	24	98		
	H$_2$SO$_4$ (1 : 4)	Boiling	2	76		
	H$_3$PO$_4$ (1,21)	20	24	98		
	H$_3$PO$_4$ (1,21)	Boiling	2	SMC		
	H$_3$PO$_4$ (1 : 3)	20	24	96		
	H$_3$PO$_4$ (1 : 3)	Boiling	2	88		
	HClO$_4$ (1.35)	20	24	97		
	HClO$_4$ (1.35)	Boiling	2	2		
	HClO$_4$ (1 : 3)	20	24	99		
	HClO$_4$ (1 : 3)	Boiling	2	84		
	H$_2$C$_2$O$_4$ (sat.)	20	24	98		
	H$_2$C$_2$O$_4$ (sat.)	Boiling	2	92		
	3 h HCl (1.19) + + 1 h HNO$_3$ (1,43)	20	24	14		

*The resistance falls with a reduction in the carbon content.

Phase	Reagent	Temp., °C	Process time, h	Insoluble residue, %	Source	Year
1	2	3	4	5	6	7
ZrC	3 h HCl (1,19) + + 1 h HNO$_3$ (1,43)	Boiling	2	6	[540]	1961
	4 h HNO$_3$ (1,43) + 1 h HF (1,15)	20	24	CS		
	H$_2$SO$_4$ (1,84) + + H$_3$PO$_4$ (1 : 1)	20	2	97		
	H$_2$SO$_4$ (1,84) + + H$_3$PO$_4$ (1 : 1)	Boiling	2	PS		
	H$_2$SO$_4$ (1 : 4) + + H$_3$PO$_4$ (1 : 3)	20	24	PS		
	H$_2$SO$_4$ (1 : 4) + + H$_3$PO$_4$ (1 : 3)	Boiling	2	PS		
	1 h H$_2$SO$_4$ (1.84) + 1 h H$_2$C$_2$O$_4$ (sat.)	20	24	96		
	1 h H$_2$SO$_4$ (1,84) + 1 h H$_2$C$_2$O$_4$ (sat.)	Boiling	2	CS		
	1 h H$_2$SO$_4$ (1 : 4) + 1 h H$_2$C$_2$O$_4$ (sat.)	20	24	96		
	1 h H$_2$SO$_4$ (1 : 4) + 1 h H$_2$C$_2$O$_4$ (sat.)	Boiling	2	91		
	NaOH (10%)	20	24	100		
	NaOH (10%)	Boiling	2	100		
	NaOH (20%)	20	24	100		
	NaOH (20%)	Boiling	2	100		
	4 h NaOH (20%) + 1 h bromine water	20	24	93		
	4 h NaOH (20%) + 1 h bromine water	Boiling	2	87		
	4 h NaOH (20%) + 1 h H$_2$O$_2$ (30%)	20	24	53		
	4 h NaOH (20%) + 1 h H$_2$O$_2$ (30%)	Boiling	2	3		
HfC	HCl (1.19)	20	24	100	[540]	1961
	HCl (1.19)	120	2	100		
	HCl (1 : 1)	20	24	96		
	H$_2$SO$_4$ (1,84)	20	24	100		
	H$_2$SO$_4$ (1,84)	280	2	CS		
	H$_2$SO$_4$ (1 : 4)	116	2	88		
	HNO$_3$ (1,43)	20	24	60		

Phase	Reagent	Temp., °C	Process time, h	Insoluble residue, %	Source	Year
1	2	3	4	5	6	7
HfC	HNO_3 (1,43)	112	2	CS	[540]	1961
	H_3PO_4 (1,21)	20	24	97		
	H_3PO_4 (1,21)	115	2	CS		
	$HClO_4$ (1,35)	20	24	97		
	$HClO_4$ (1,35)	20	2	2		
	$H_2C_2O_4$ (sat.)	20	24	98		
	$H_2C_2O_4$ (sat.)	104	2	98		
	3 h HCl (1,19) + + 1 h HNO_3 (1,43)	20	24	14		
	3 h HCl (1,19) + + 1 h HNO_3 (1,43)	106	2	6		
	H_2SO_4 (1 : 1) + + H_3PO_4 (1 : 1)	20	24	2		
	H_2SO_4 (1 : 1) + + H_3PO_4 (1 : 1)	160	2	CS		
	H_2SO_4 (1.84) + + H_3PO_4 (1,21)	20	24	97		
	H_2SO_4 (1.84) + + H_3PO_4 (1,21)	250	2	SMC		
	NaOH (20%)	110	2	ND		
	NaOH + bromine water	20	24	81		
	NaOH + H_2O_2 (30%)	20	24	53		
	NaOH + H_2O_2 (30%)	110	2	CSH		
	K_3 [Fe (CN)$_6$] (10%) + NaOH (20%)	100	2	37		
	K_3 [Fe (CN)$_6$] (10%) + NaOH (20%)	20	24	83		
VC	H_3PO_4 (1.7)	300—310	1	ND	[542]	1966
	H_2SO_4 (1.84)	200—310	1	— *		
	HNO_3 (1,43)	100—110	1	CS		
	Aqua regia	110	1	CS		
	H_2SO_4 (1 : 1) + + HNO_3 (1,43)	100—110	1	CS		
	HNO_3 (1 : 1) + + NH_4F (5%)	100—110	1	CS		
	HCl (1,19) + + $(NH_4)_2S_2O_8$ (25%)	100—110	1	60,6		

*Decomposition with separation of $V_2(SO_4)_3$.

Phase	Reagent	Temp., °C	Process time, h	Insoluble residue, %	Source	Year
1	2	3	4	5	6	7
VC	HCl (1.19) + H_2O_2 (30%)	100—110	1	78.6	[542]	1966
	HCl (1.19) + bromine soln. (sat.)	100—110	1	45.7		
	H_2SO_4 (1 : 4), (1 : 3) + H_2O_2 (30%)	100—110	1	5.0—15.4		
	H_2SO_4 (1 : 4), (1 : 3) + $(NH_4)_2S_2O_8$ (25%)	100—110	1	27,4		
	H_2O_2 (30%)	100—110	1	3—5		
	$(NH_4)_2S_2O_8$ (25%)	100—110	1	32,5		
	H_2O_2 (30%) + tartaric acid (50%)	100—110	1	64,1		
	H_2O_2 (30%) + citric acid (50%)	100—110	1	62,7		
	H_2O_2 (30%) + oxalic acid	100—110	1	60,4		
	H_2O_2 (30%) + Trilon-B	100—110	1	30,0		
	H_2O_2 (30%) + NH_4F (5%)	100—110	1	28,7		
	Organic acids	—	1	97—98		
	Oxalic acid	—	1	97—98		
	Citric acid	—	1	97—98		
	Oxalic acid	—	1	97—98		
	Acetic acid	—	1	97—98		
NbC	HCl (1.19)	20	24	100	[540]	1961
	HCl (1.19)	115	2	96		
	HCl (1 : 1)	20	24	100		
	HCl (1 : 1)	108	2	99		
	HNO_3 (1.43)	20	24	100		
	HNO_3 (1.43)	120	2	PS		
	HNO_3 (1 : 1)	105	2	PS		
	H_2SO_4 (1.84)	20	24	ND		
	H_2SO_4 (1.84)	275	2	CS		
	H_3PO_4 (1.21)	20	24	100		
	H_3PO_4 (1.21)	120	2	ND		
	$H_2C_2O_4$ (sat.)	104	2	99		

Phase	Reagent	Temp., °C	Process time, h	Insoluble residue, %	Source	Year
1	2	3	4	5	6	7
NbC	H_2O_2 (30%)	Boiling	1—1.5	CS*	[543]	1966
	3 h HCl (1,19) + + 1 h HNO$_3$ (1,43)	20	24	92	[540]	1961
	3 h HCl (1,19) + + 1 h HNO$_3$ (1,43)	105	2	CSH		
	1 h H_2SO_4 (1,84) + 1 h HNO$_3$ (1,43)	20	24	100		
	1 h H_2SO_4 (1,84) + 1 h HNO$_3$ (1,43)	140	2	22		
	1 h H_2SO_4 (1,84) + 1 h H_3PO_4 (1,21)	20	24	91		
	1 h H_2SO_4 (1,84) + 1 h H_3PO_4 (1,21)	240	2	CS		
	1 h H_2SO_4 (1,84) + 1 h $H_2C_2O_4$ (sat.)	20	24	100		
	1 h H_2SO_4 (1,84) + 1 h $H_2C_2O_4$ (sat.)	180	2	95		
	NaOH (20%)	20	24	99		
	NaOH (20%)	110	2	100		
	NaOH + bromine water	20	24	100		
	NaOH + bromine water	105	2	84		
	NaOH (20%) + + H_2O_2 (30%)	20	24	71		
	NaOH (20%) + + H_2O_2 (30%)	112	2	88		
	K_3 [Fe (CN)$_6$] + + NaOH (20%)	20	24	PS		
	K_3 [Fe (CN)$_6$] + + NaOH (20%)	110	2	PS		
TaC	HCl (1,19)	20	24	100	[540]	1961
	HCl (1,19)	120	2	98		
	HCl (1 : 1)	112	2	98		
	HNO$_3$ (1,43)	20	24	100		
	HNO$_3$ (1,43)	114	2	99		
	HNO$_3$ (1 : 1)	105	2	98		

*When mixed with H_2O_2 (30%) and citric, oxalic, tartaric acids, Trilon-B, and ammonium salts, complete solution occurred in 20 min.

Phase	Reagent	Temp., °C	Process time, h	Insoluble residue, %	Source	Year
1	2	3	4	5	6	7
TaC	H_2SO_4 (1.84)	20	24	100	[540]	1961
	H_2SO_4 (1.84)	260	2	0		
	H_2SO_4 (1 : 4)	115	2	93		
	H_3PO_4 (1.21)	20	24	98		
	H_3PO_4 (1.21)	Boiling	2	SMC		
	NaOH (20%)	20	24	99		
	NaOH (20%)	108	2	100		
	NaOH + bromine water	20	24	100		
	NaOH + bromine water	110	2	SMC		
	NaOH (20%) + H_2O_2 (30%)	20	24	62		
	NaOH (20%) + H_2O_2 (30%)	105	2	SMC		
	K_3 [Fe $(CN)_6$] + NaOH (20%)	100	2	57		
	$H_2C_2O_4$ (sat.)	20	24	97		
	$H_2C_2O_4$ (sat.)	105	2	98		
	3 h HCl (1,19) + 1 h HNO_3 (1,43)	20	24	99		
	3 h HCl (1,19) + 1 h HNO_3 (1,43)	115	2	98		
	1 h H_2SO_4 (1,84) + 1 h HNO_3 (1,43)	20	24	91		
	1 h H_2SO_4 (1,84) + 1 h HNO_3 (1,43)	150	2	96		
	1 h H_2SO_4 (1,84) + 1 h H_3PO_4 (1,21)	20	24	98		
	1 h H_2SO_4 (1,84) + 1 h H_3PO_4 (1,21)	180	2	SMC		
	1 h H_2SO_4 (1,84) + 1 h $H_2C_2O_4$ (sat.)	20	24	97		
	1 h H_2SO_4 (1,84) + 1 h $H_2C_2O_4$ (sat.)	—	2	97		
TaC	HF (1,15) + HNO_3 (1,43)	20	—	CS	[538]	1956
Ta_2C	HF (1,15) + HNO_3 (1,43)	20	—	CS	[538]	1956

Phase	Reagent	Temp., °C	Process time, h	Insoluble residue, %	Source	Year
1	2	3	4	5	6	7
Cr_3C_2	H_2O HCl (1.19)	20 Boiling	100 —	97 CS	[540] [538]	1961 1956
Cr_3C_2	H_2SO_4 (1 : 1)	20	48	100	[529]	1961
	H_2SO_4 (1,84)	280	1	CS		
	H_2SO_4 (1 : 1)	136	1	65,1		
	H_2SO_4 (1 : 4)	105	1	95,3		
	1 h H_2SO_4 (1.84) + 1 h HNO_3 (1.43)	120	1	83,5		
	H_2SO_4 (1.84) + + tartaric acid	134	1	74.13		
	H_2SO_4 (1,84) + + CrO_3	120	1	33.2		
	H_2SO_4 (1.84) + + Trilon-B	126	1	10,36		
	H_3PO_4 (1.21)	20	48	100		
	4 h H_3PO_4 (1,21) + 1 h H_2SO_4 (1.84) + + 2 h H_2O	119	1	94.1		
	4 h H_3PO_4 (1.21) + 1 h H_2SO_4 (1.84)	20	48	100		
	3 h HCl (1.19) + + 1 h HNO_3 (1.43)	106	1	90,9		
	3 h HCl (1.19) + + 1 h HNO_3 (1.43)	20	48	98,7		
	$H_2C_2O_4$ (sat.)	100	1	98.5		
	$H_2C_2O_4$ (sat.)	20	48	100		
	Tartaric acid 30% soln.	100	1	100		
	NaOH (50%)	20	48	100		
	NaOH (30%)	110	1	99.8		
	NaOH (20%) + + H_2O_2 (30%)	100	1	95.5		
	NaOH + bromine water	106	1	88,1		
	$K_3[Fe(CN)_6]$, alkaline soln.	100	1	61.5		
	Ethyl alcohol	20	48	99,0		
	Methyl alcohol	20	48	99,4		
	Toluene	20	48	99.5		
	Benzene	20	48	99.4		
	Dichloroethane	20	48	99,6		
	Acetone	20	48	99,6		
	Chloroform	20	48	99.8		

Phase	Reagent	Temp., °C	Process time, h	Insoluble residue, %	Source	Year
1	2	3	4	5	6	7
Cr_7C_3	HCl (1.19)	20	48	92.3	[529]	1961
	HCl (1 : 1)	20	48	99.9		
	HCl (1 : 1)	110	1	3.49		
	H_2SO_4 (1 : 1)	20	48	99.8		
	H_2SO_4 (1,84)	265	1	CS		
	H_2SO_4 (1 : 1)	137	1	1,62		
	H_2SO_4 (1,84) + + HNO_3 (1,43)	125	1	90.6		
	H_3PO_4 (1,21)	20	48	100		
	4 h H_3PO_4 (1,21) + 1 h H_2SO_4 (1,84) + + 2 h H_2O	127	1	3.93		
		20	48	100		
	3 h HCl (1,19) + + 1 h HNO_3 (1,43)	20	48	94,9		
		106	48	93.8		
	HCl (1,19) + + H_2O_2 (30%)	105	1	5,55		
	$H_2C_2O_4$ (sat.)	104	1	95,47		
	$H_2C_2O_4$ (sat.)	20	48	100		
	NaOH (30%)	110	1	96.1		
	NaOH + H_2O_2 (30%)	100	1	96.3		
	NaOH (20%) + + bromine water	102	1	85,9		
	K_3 [Fe (CN)$_6$], alkaline soln.	100	1	53,2		
	Ethyl alcohol	20	48	99,6		
	Methyl alcohol	20	48	99,7		
	Toluene	20	48	99.8		
	Benzene	20	48	99.6		
	Dichloroethane	20	48	99.8		
	Acetone	20	48	99,6		
	Chloroform	20	48	99,8		
Mo_2C	HCl (1,19)	20	24	80	[540]	1961
	HCl (1.19)	Boiling	2	89		
	HCl (1 : 1)	20	24	88		
	HCl (1 : 1)	Boiling	2	83		
	HNO_3 (1.43)	20	24	CS		
	HNO_3 (1,43)	Boiling	2	CS		
	HNO_3 (1 : 1)	20	24	CS		
	HNO_3 (1 : 1)	Boiling	2	CS		
	H_2SO_4 (1,84)	20	24	89		
	H_2SO_4 (1,84)	Boiling	2	CS		
	H_2SO_4 (1 : 4)	20	24	90		
	H_2SO_4 (1 : 4)	Boiling	2	83		
	H_3PO_4 (1,21)	20	24	93		
	H_3PO_4 (1,21)	Boiling	2	76		

Phase	Reagent	Temp., °C	Process time, h	Insoluble residue, %	Source	Year
1	2	3	4	5	6	7
Mo_2C	H_3PO_4 (1 : 3)	20	24	92	[540]	1961
	$HClO_4$ (1,35)	20	24	CS		
	$HClO_4$ (1,35)	Boiling	2	73		
	$HClO_4$ (1 : 3)	20	24	89		
	$HClO_4$ (1 : 3)	Boiling	2	58		
	$H_2C_2O_4$ (sat.)	20	24	89		
	$H_2C_2O_4$ (sat.)	Boiling	2	90		
	3 h HCl (1,19) + + 1 h HNO_3 (1,43)	20	24	CS		
	3 h HCl (1,19) + + 1 h HNO_3 (1,43)	Boiling	2	CS		
	2 h H_2SO_4 (1,84) + 1 h HNO_3 (1,43)	20	24	1		
	2 h H_2SO_4 (1,84) + 1 h HNO_3 (1,43)	Boiling	2	CS		
	4 h HNO_3 (1,4) + + 1 h HF (1,15)	20	24	CS		
	H_2SO_4 (1,84) + + H_3PO_4 (1,21)	20	24	90		
	H_2SO_4 (1,84) + + H_3PO_4 (1,21)	Boiling	2	CS		
	H_2SO_4 (1,84) + + H_3PO_4 (1 : 3)	20	24	92		
	H_2SO_4 (1,84) + + H_3PO_4 (1 : 3)	Boiling	2	88		
	H_2SO_4 (1,84) + + $H_2C_2O_4$ (sat.)	20	24	80		
	H_2SO_4 (1,84) + + $H_2C_2O_4$ (sat.)	Boiling	2	73		
	H_2SO_4 (1 : 4) + + $H_2C_2O_4$ (sat.)	20	24	89		
	H_2SO_4 (1 : 4) + + $H_2C_2O_4$ (sat.)	Boiling	2	88		
	NaOH (20%)	20	24	90		
	NaOH (20%)	Boiling	2	90		
	NaOH (10%)	20	24	90		
	NaOH (10%)	Boiling	2	94		
	4 h NaOH (20%) + 1 h bromine water	20	24	65		
	4 h NaOH (20%) + 1 h	Boiling	2	60		

Phase	Reagent	Temp., °C	Process time, h	Insoluble residue, %	Source	Year
1	2	3	4	5	6	7
Mo_2C	bromine water 4 h NaOH (20%) + 1 h H_2O_2 (30%)	20	24	31	[540]	1961
	4 h NaOH (20%) + 1 h H_2O_2 (30%)	Boiling	2	36		
	1 h $K_3[Fe(CN)_6]$+ + 4 h NaOH (20%)	20	24	69		
	1 h $K_3[Fe(CN)_6]$+ + 4 h NaOH (20%)	Boiling	2	69		
Mo_2C	HNO_3 (1,43) + + HF (1,15)	Boiling	—	CS	[538]	1956
W_2C	HCl	—	—	ND	[529]	1961
	H_2SO_4	—	—	ND		
	HNO_3	—	—	CS		
	HF + HNO_3	—	—	CS		
WC	HCl (1,19)	20	24	97	[540]	1961
	HCl (1,19)	Boiling	2	48		
	HCl (1 : 1)	20	24	96		
	HCl (1 : 1)	Boiling	2	92		
	HNO_3 (1.43)	20	24	63		
	HNO_3 (1.43)	Boiling	2	1		
	HNO_3 (1 : 1)	20	24	72		
	HNO_3 (1 : 1)	Boiling	2	10		
	H_2SO_4 (1,84)	20	24	91		
	H_2SO_4 (1,84)	Boiling	2	1		
	H_2SO_4 (1 : 4)	20	24	96		
	H_2SO_4 (1 : 4)	Boiling	2	95		
	H_3PO_4 (1,21)	20	24	91		
	H_3PO_4 (1,21)	Boiling	2	93		
	H_3PO_4 (1 : 3)	20	24	96		
	H_3PO_4 (1 : 3)	Boiling	2	90		
	$HClO_4$ (1,35)	20	24	98		
	$HClO_4$ (1,35)	Boiling	2	40		
	$HClO_4$ (1 : 3)	20	24	98		
	$HClO_4$ (1 : 3)	Boiling	2	93		
	$H_2C_2O_4$ (sat.)	20	24	95		
	$H_2C_2O_4$ (sat.)	Boiling	2	95		
	3 h HCl (1,19) + + 1 h HNO_3 (1,43)	20	24	28		

Phase	Reagent	Temp., °C	Process time, h	Insoluble residue, %	Source	Year
1	2	3.	4	5	6	7
WC	3 h HCl (1,19) + + 1 h HNO_3 (1,43)	Boiling	2	3	[540]	1961
	2 h H_2SO_4 (1,84) + 1 h HNO_3 (1,43)	20	24	92		
	2 h H_2SO_4 (1,84) + 1 h HNO_3 (1,43)	Boiling	2	42		
	4 h HNO_3 (1.4) + + 1 h HF (1,15)	20	24	60		
	H_2SO_4 (1,84) + + H_3PO_4 (1 : 4)	20	24	96		
	H_2SO_4 (1,84) + + H_3PO_4 (1 : 4)	Boiling	2	CS		
	H_2SO_4 (1,84) + + H_3PO_4 (1 : 3)	20	24	96		
	H_2SO_4 (1,84) + + H_3PO_4 (1 : 3)	Boiling	2	93		
	H_2SO_4 (1,84) + + $H_2C_2O_4$ (sat.)	20	24	95		
	H_2SO_4 (1,84) + + $H_2C_2O_4$ (sat.)	Boiling	2	70		
	H_2SO_4 (1 : 4) + + $H_2C_2O_4$ (sat.)	20	24	94		
	H_2SO_4 (1 : 4) + + $H_2C_2O_4$ (sat.)	Boiling	2	95		
	NaOH (20%)	20	24	97		
	NaOH (20%)	Boiling	2	98		
	NaOH (10%)	20	24	98		
	NaOH (10%)	Boiling	2	98		
	4 h NaOH (20%) + 1 h bromine water	20	24	70		
	4 h NaOH (20%) + 1 h. bromine water	Boiling	2	60		
	4 h NaOH (20%) + 1 h H_2O_2 (30%)	20	24	88		
	4 h NaOH (20%) + 1 h H_2O_2 (30%)	Boiling	2	87		
	1 h K_3 [Fe (CN)$_6$] (10%) + 4 h NaOH (20%)	20	24	68		

Phase	Reagent	Temp., °C	Process time, h	Insoluble residue, %	Source	Year
1	2	3	4	5	6	7
WC	1 h K_3 [Fe (CN)$_6$] (10%) + 4 h NaOH (20%)	Boiling	2	58	[540]	1961
Mn_5C_2	HCl (1,19)	Boiling	—	ND	[11]	1968
Mn_7C_3	HNO_3 (1,43)	20	—	ND	[11]	1968
$Mn_{23}C_6$	HNO_3 (1,4) H_2SO_4 (1,84) H_2SO_4 (1,84) HF (1,15) + + HNO_3 (1,43)	Boiling 20 Boiling 20	— — — —	ND ND ND CS	[11]	1968
Mn_3C	Concentrated mineral acids	20	—	CS	[538]	1956
Fe_2C	HCl (1,19) HNO_3 (1,43)	Boiling "	—	ND CS	[11]	1968
Fe_3C	HCl (1,19) HNO_3 (1,43) Diluted mineral acids	Boiling " 20	— — —	ND CS CS	[11] [11] [538]	1968 1968 1956
Co_2C	HCl (1,19) HNO_3 (1,43) H_2SO_4 (1,84) 3 h HCl (1,19) + + 1 h HNO_3 (1,43) HF (1,15) + + HNO_3 (1,43) HCl (1,19) + + H_2SO_4 (1,84)	Boiling " " 20 20 20	— — — — — —	ND ND ND CS CS CS	[11]	1968
Co_3C	HCl (1,19) HNO_3 (1,43) H_2SO_4 (1,84) 3 h HCl (1,19) + + 1 h HNO_3 (1.43) HF (1,15) + + HNO_3 (1,43) HCl (1,19) + + H_2SO_4 (1,84)	Boiling " " 20 20 20	— — —	ND ND ND CS CS CS	[11]	1968

Phase	Reagent	Temp., °C	Process time, h	Insoluble residue, %	Source	Year
1	2	3	4	5	6	7
Mg_3N_2	H_2O	20	—	ND	[14]	1969
	HCl (1.19)	100		CS		
	HCl (1.19)	Boiling		CS		
	HNO_3 (1,4)	"		CS		
	C_2H_5OH	20		ND		
Ca_3N_2	HCl (1 : 1)	20	—	CS	[14]	1969
	H_2SO_4 (1 : 1)			CS		
	C_2H_5OH			ND		
SrN, Sr_2N	H_2O	20	—	CS	[14]	1969
Sr_3N_2	H_2O	20	—	PS	[14]	1969
Ba_3N_2	H_2O	20	—	ND*	[14]	1969
$ScN_{0.97}$	H_2O	20	24	ND	[544]	1962
	H_2O	Boiling	1	ND		
	H_2O	"	2	ND		
	H_2O	"	4	ND		
	H_2SO_4 (1,84)	"	1	16,89		
	H_2SO_4 (1,84)	"	2	3,84		
	H_2SO_4 (1,84)	"	4	2,82		
	HNO_3 (1,4)	"	1	CS		
	HNO_3 (1.4)	"	2	CS		
	HNO_3 (1.4)	"	4	CS		
	HCl (1.19)	"	1	CS		
	HCl (1.19)	"	2	CS		
	HCl (1.19)	"	4	CS		
	H_2SO_4 (1 : 1)	"	1	0,52		
	H_2SO_4 (1 : 1)	"	2	0,36		
	H_2SO_4 (1 : 1)	"	4	0,42		
	HNO_3 (1 : 1)	"	1	CS		
	HNO_3 (1 : 1)	"	2	CS		
	HNO_3 (1 : 1)	"	4	CS		
	HCl (1 : 1)	"	1	CS		
	HCl (1 : 1)	"	2	CS		
	HCl (1 : 1)	"	4	CS		
	NaOH (10%)	"	1	71,94		
	NaOH (10%)	"	2	71,37		
	NaOH (10%)	"	4	68,93		
	NaOH (20%)	"	1	26,01		

*$Ba(OH)_2$ is precipitated.

Phase	Reagent	Temp., °C	Process time, h	Insoluble residue, %	Source	Year
1	2	3	4	5	6	7
$ScN_{0.97}$	NaOH (20%)	Boiling	2	18,49	[544]	1962
	NaOH (20%)	"	4	9,59		
	NaOH (40%)	"	1	CS		
	NaOH (40%)	"	2	CS		
	NaOH (40%)	"	4	CS		
Th_3N_4	H_2O	20	—	CS*	[532]	1952
UN	H_2O	20	—	CS	[532]	1952
	H_3PO_4 (1,21)			SMC		
U_2N_3, UN_2	HCl (1,19)	20	—	ND	[532]	1952
UN_2	H_2SO_4 (1,82)	20	—	ND	[532]	1952
U_2N_3	HNO_3 (1,43)	20	—	PS	[14]	1969
	NaOH (25%)			ND		
NpN	H_2O	20	—	ND	[14]	1969
	HCl (1 : 1)			CS		
	HCl (1,19)			CS		
TiN	HCl (1 : 1)	20	24	99	[545]	1958
	HCl (1 : 1)	Boiling	2	98		
	HCl (1,19)	20	24	89		
	HCl (1,19)	Boiling	2	98		
	H_2SO_4 (1 : 4)	20	24	98		
	H_2SO_4 (1 : 4)	Boiling	2	95		
	H_2SO_4 (1,84)	20	24	97		
	H_2SO_4 (1,84)	Boiling	2	24		
	HNO_3 (1 : 1)	20	24	11		
	HNO_3 (1 : 1)	Boiling	2	5		
	HNO_3 (1,43)	20	24	10		
	HNO_3 (1,43)	Boiling	2	PS		
	$HClO_4$ (1 : 3)	20	24	98		
	$HClO_4$ (1 : 3)	Boiling	2	94		
	$HClO_4$ (1,35)	20	24	99		
	$HClO_4$ (1,35)	Boiling	2	PS		
	$HClO_4$ (1 : 4)	20	24	97		
	H_3PO_4 (30%)	Boiling	2	2		

*Dissolves to form ThO_2.

Phase	Reagent	Temp., °C	Process time, h	Insoluble residue, %	Source	Year
1	2	3	4	5	6	7
TiN	NaOH (1%)	Boiling	2	PS	[545]	1958
	NaOH (10%)	"	2	PS		
	NaOH (40%)	"	2	PS		
	NaOH + H_2O_2 (1%)	"	2	9		
	NaOH + H_2O_2 (10%)	"	2	16		
	NaOH + H_2O_2 (40%)	"	2	43		
	H_2O_2 (30%)	100	1	39,1	— *	1969
	H_2O	25	24	ND		
	H_2O	100	6	ND		
	HCl (1 : 1) + + $K_2S_2O_8$ (5%)	100	1	94.73		
ZrN	H_2O	25	24	ND	— *	1969
	H_2O	100	6	ND		
	HCl (1 : 1)	50	1	98,14		
	HCl (1 : 1)	95	1	44,9		
	HCl (1,19)	50	1	92,6		
	HCl (1,19)	105	1	5,3		
	H_2SO_4 (1 : 1)	50	1	97,53		
	H_2SO_4 (1 : 1)	95	1	7,50		
	H_2SO_4 (1.84)	50	1	94.16		
	H_2SO_4 (1.84)	95	1	10,1		
	HNO_3 (1 : 1)	50	1	99,19		
	HNO_3 (1 : 1)	95	1	93.66		
	HNO_3 (1,43)	20	24	98	[545]	1958
	HNO_3 (1.43)	Boiling	2	84		
	$HClO_4$ (1 : 3)	20	24	100		
	$HClO_4$ (1 : 3)	Boiling	2	98		
	$HClO_4$ (1,35)	20	24	99		
	$HClO_4$ (1,35)	Boiling	2	98		
	H_3PO_4 (1 : 4)	20	24	SMC		
	H_2O_2 (30%)	Boiling	2	100		
	NaOH (1%)	"	2	100		
	NaOH (10%)	"	2	100		
	NaOH (40%)	"	2	42		
	NaOH + H_2O_2 (1%)	"	2	99		
	NaOH + H_2O_2 (10%)	"	2	87		

*O. P. Kulik, Investigation of the Chemical Properties of some Transition Nitrides of Limiting Composition and in the Homogeneity Region [in Russian], Author's abstract, Cand. Dissertation, Kiev (1969). The chemical resistance falls with a rise in the nitrogen content.

RESISTANCE OF POWDERED REFRACTORY COMPOUNDS TO ACIDS AND ALKALIS (continued)

Phase	Reagent	Temp., °C	Process time, h	Insoluble residue, %	Source	Year
1	2	3	4	5	6	7
ZrN	NaOH + H$_2$O$_2$ (40%)	Boiling	2	48	[545]	1958
	3 h HCl (1,19) + + 1 h HNO$_3$ (1,43)	20	24	82		
	3 h HCl (1,19) + + 1 h HNO$_3$ (1,43)	Boiling	2	25		
	1 h HClO$_4$ (1,35) + 1 h HCl (1.19)	20	24	76		
	1 h HClO$_4$ (1,35) + 1 h HCl (1.19)	Boiling	2	12		
	1 h HNO$_3$ (1,43) + 1 h H$_2$O$_2$	20	24	94		
	1 h HNO$_3$ (1,43) + 1 h H$_2$O$_2$	Boiling	2	65		
	HNO$_3$ (1.43) + + HF (1,15)	"	5 min	CS		
	3 h H$_2$C$_2$O$_4$ (sat.) + 1 h H$_2$SO$_4$ (1,84)	20	24	90		
	2 h H$_2$SO$_4$ + 1 h H$_2$O$_2$	20	24	25		
	1 h H$_2$SO$_4$ (1,84) + 1 h HNO$_3$ (1,43) + + 4 h H$_2$O	20	24	81		
	1 g K$_2$SO$_4$ + 1 ml H$_2$SO$_4$ (1,84)	Boiling	2	CS		
HfN	H$_2$O	25	24	ND	— *	1969
	H$_2$O	100	6	ND		
	HCl (1 : 1)	50	1	95,9		
	HCl (1 : 1)	95	1	17,3		
	HCl (1.19)	50	1	89,65		
	HCl (1.19)	105	1	2,9		
	H$_2$SO$_4$ (1 : 1)	50	1	91,28		
	H$_2$SO$_4$ (1 : 1)	95	1	2,4		
	H$_2$SO$_4$ (1,84)	50	1	94,4		
	H$_2$SO$_4$ (1.84)	95	1	26,3		
	HNO$_3$ (1 : 1)	50	1	97,73		

*See footnote on p. 340.

Phase	Reagent	Temp., °C	Process time, h	Insoluble residue, %	Source	Year
1	2	3	4	5	6	7
HfN	HNO_3 (1 : 1)	95	1	16.7	— *	1969
	HNO_3 (1.4)	50	1	97.3		
	HNO_3 (1.4)	95	1	24.0		
	HCl (1 : 1) + + $K_2S_2O_8$ (5%)	100	1	17.2		
	H_2O_2 (30%)	100	1	ND		
	NaOH (10%)	105	6	CS		
	NaOH (20%)	120	1	CS		
	NaOH (40%)	120	1	CS		
VN	H_2O	25	24	ND	— *	1969
	H_2O	100	6	ND		
	HCl (1 : 1)	50	1	ND		
	HCl (1 : 1)	95	1	ND		
	HCl (1.19)	50	1	ND		
	HCl (1.19)	105	1	ND		
	H_2SO_4 (1 : 1)	95	1	ND		
	H_2SO_4 (1 : 1)	140	1	ND		
	H_2SO_4 (1.84)	95	1	ND		
	H_2SO_4 (1.84)	280	1	83.06		
	HNO_3 (1 : 1)	50	1	38.5		
	HNO_3 (1 : 1)	95	1	2.3		
	HNO_3 (1.4)	50	1	13.5		
	HNO_3 (1.4)	95	1	CS		
	HCl (1 : 1) + + $K_2S_2O_8$ (5%)	100	1	79.2		
	H_2O_2 (30%)	100	1	80.0		
	NaOH (10%)	105	1	99.2		
	NaOH (20%)	120	1	99.2		
	NaOH (40%)	120	1	98.35		
NbN	HCl (1 : 3)	20	24	99	[545]	1958
	HCl (1 : 3)	Boiling	2	94		
	HCl (1.19)	20	24	100		
	HCl (1.19)	Boiling	2	99		
	H_2SO_4 (1 : 4)	20	24	99		
	H_2SO_4 (1 : 4)	Boiling	2	84		
	H_2SO_4 (1.84)	20	24	100		
	H_2SO_4 (1.84)	Boiling	2	0		
	HNO_3 (1 : 1)	20	24	98		
	HNO_3 (1 : 1)	Boiling	2	100		
	HNO_3 (1.43)	20	24	100		
	HNO_3 (1.43)	Boiling	2	100		
	$HClO_4$ (1 : 3)	20	24	100		
	$HClO_4$ (1 : 3)	Boiling	2	100		

*O. P. Kulik, see footnote on p. 340.

Phase	Reagent	Temp., °C	Process time, h	Insoluble residue, %	Source	Year
1	2	3	4	5	6	7
NbN	$HClO_4$ (1,35)	20	24	98	[545]	1958
	$HClO_4$ (1.35)	Boiling	2	100		
	H_2O_2 (30%)	"	2	16		
	NaOH (1%)	"	2	96		
	NaOH (10%)	"	2	87		
	NaOH (40%)	"	2	87		
	NaOH (1%) + $+ H_2O_2$	"	2	17		
	NaOH (10%) + $+ H_2O_2$	"	2	CS		
	NaOH (40%) + $+ H_2O_2$	"	2	CS		
	3 h HCl (1,19) + $+ 1$ h HNO_3 (1,43)	20	24	99		
	3 h HCl (1,19) + $+ 1$ h HNO_3 (1,43)	Boiling	2	99		
	1 h $HClO_4$ (1,35) + 1 h HCl (1,19)	20	24	98		
	1 h $HClO_4$ (1.35) + 1 h HCl (1,19)	Boiling	2	95		
	1 h HNO_3 (1,43) + 1 h H_2O	20	24	26		
	1 h HNO_3 (1,43) + 1 h H_2O	Boiling	2	15		
	HNO_3 (1,43) + $+ HF$ (1,15)	"	5 min	CS		
	10 g K_2SO_4 + $+ 10$ ml H_2SO_4 (1,84)	"	2	CS		
TaN	HCl (1 : 1)	20	24	98	[545]	1958
	HCl (1 : 1)	Boiling	2	99		
	HCl (1.19)	20	24	99		
	HCl (1.19)	Boiling	2	98		
	H_2SO_4 (1 : 4)	20	24	100		
	H_2SO_4 (1 : 4)	Boiling	2	100		
	H_2SO_4 (1,84)	20	24	100		
	H_2SO_4 (1,84)	Boiling	2	77		
	HNO_3 (1 : 1)	20	24	99		
	HNO_3 (1 : 1)	Boiling	2	98		
	HNO_3 (1.43)	20	24	98		
	HNO_3 (1.43)	Boiling	2	98		
	$HClO_4$ (1 : 3)	20	24	100		
	$HClO_4$ (1,35)	20	24	100		
	$HClO_4$ (1,35)	Boiling	2	98		
	H_3PO_4 (1 : 4)	20	24	96		
	H_3PO_4 (1 : 4)	Boiling	2	100		

Phase	Reagent	Temp., °C	Process time, h	Insoluble residue, %	Source	Year
1	2	3	4	5	6	7
TaN	H_2O_2 (30%)	Boiling	2	41	[545]	1958
	NaOH (1%)	"	2	93		
	NaOH (10%)	"	2	PS		
	NaOH (40%)	"	2	PS		
	NaOH (1%) + + H_2O_2	"	2	84		
	NaOH (10%) + + H_2O_2	"	2	39		
	NaOH (40%) + H_2O_2	"	2	CSH		
	3 h HCl (1.19)+ +1 h HNO_3 (1,43)	"	2	100		
	HNO_3 (1,43) + + HF (1,15)	"	5 min	0		
	2 h H_2SO_4 (1,82) + 1 h H_2O_2	"	2	93		
	10 g K_2SO_4 + + 10 ml H_2SO_4 (1,84)	"	5—6	CS		
Cr_2N	HCl (1 : 1)	Boiling	2—3	CS	[546]	1961
	HCl (conc.)			CS		
	H_2SO_4 (1 : 4)			CS		
	H_2SO_4 (conc.)			SMC		
	HNO_3 (1 : 1)			98		
	HNO_3 (conc.)			97		
	$HClO_4$ (conc.)			CS		
	H_2O_2			98		
	HNO_3 + HCl (1 : 3)			98		
	HNO_3 + H_2O_2			99		
	H_2SO_4 + H_2O_2			ND		
	H_2SO_4 + K_2SO_4			SMC		
Cr_2N	H_2O	100	6	ND	—*	1969
	HCl (1 : 1) + + $K_2S_2O_8$ (5%)	100	1	99,63		
	NaOH (10%)	105	1	99.8		
	NaOH (20%)	120	1	98.87		
	NaOH (40%)	120	1	96,98		
CrN	HCl (1 : 1)	Boiling	2—3	96	[546]	1961
	HCl (conc.)			94		
	H_2SO_4 (1 : 4)			CS		
	H_2SO_4 (conc.)			CS		
	HNO_3 (1 : 1)			98		

*O. P. Kulik, see footnote on p. 340.

RESISTANCE OF POWDERED REFRACTORY COMPOUNDS TO ACIDS AND ALKALIS (continued)

Phase	Reagent	Temp., °C	Process time, h	Insoluble residue, %	Source	Year
1	2	3	4	5	6	7
CrN	HNO_3 (conc.) $HClO_4$ (conc.) H_2O_2 $HNO_3 + HCl$ (1:3) $HClO_4 + HCl$ (1:1) $HNO_3 + H_2O_2$ $H_2SO_4 + H_2O_2$ $H_2SO_4 + K_2SO_4$	Boiling	2—3	92 CS ND 91 95 98 98 SMC	[546]	1961
CrN	H_2O HCl (1:1) + + $K_2S_2O_8$ (5%) NaOH (10%) NaOH (20%) NaOH (40%)	100 100 105 120 120	6 1 1 1 1	ND 99,55 99,59 99,11 98,28	—*	1969
W_2N	H_2O HCl (1,19)	20 Boiling	— —	CS** ND	[14]	1969
Mn_3N_2	H_2O HCl (1.19) HNO_3 (1,4) H_2SO_4 (1.82) 3 h HCl (1,19) + + 1 h HNO_3 (1,43)	Boiling	—	ND ND ND ND CS	[14]	1969
Mn_5N_2	HCl (1,19) HNO_3 (1:1) H_2SO_4 (1.82) 3 h HCl (1,19) + + 1 h HNO_3	Boiling	— —	CS CS CS CS	[532]	1952
Mn_2N	HCl (1,19) HNO_3 (1,43) H_2SO_4 (1,82) HF (1,15) + + HNO_3 (1,43) HCl (1,19) + + H_2SO_4 (1,82) NaOH 25%)	Boiling	—	ND ND ND CS** CS** CS	[14]	1969

*O. P. Kulik, see footnote on p. 340.
**NH_3 is given off.

Phase	Reagent	Temp., °C	Process time, h	Insoluble residue, %	Source	Year
1	2	3	4	5	6	7
FeN	H_2O	100	—	CS	[14]	1969
Fe_2N	HCl (1.19)	20	—	CS	[14]	1969
Fe_3N	HNO_3 (1.4) H_2SO_4 (1.82)	20	—	CS	[14]	1969
Co_2N	HCl (1 : 1) HCl (1 : 1) H_2SO_4 (1.82) HCl (1.19) HNO_3 (1.4)	20 Boiling 20 20 20	—	CS	[14]	1969
Co_3N	HCl (1 : 1) HCl (1 : 1) HNO_3 (1 : 1) HNO_3 (1 : 1) HCl (1.19) HNO_3 (1.43) H_2SO_4 (1.82) H_2SO_4 (1.82)	20 Boiling 20 Boiling 20 20 20 Boiling	—	CS	[14]	1969
Ni_3N	HCl (1 : 1) HNO_3 (1 : 1) H_2SO_4 (1 : 1) NaOH (25%)	20	—	CS CS CS ND	[14]	1969
Mg_2Si	Tartaric acid (50%) Citric acid (50%) Oxalic acid (50%) Acetic acid (10%) Trilon-B (0.2-N) Trilon-B (0.02-N) Trilon-B (0.05-N) H_2O_2 (3%) Rochelle salt (10%)	20 20 20 20 20 100 20 20 20	0.16 0.16 2—4 0.16 0.5 1 24 48 48	CS* CS* CS* CS CS PS PS ND PS	[547]	1969

*Spurting with the formation of solid reaction products.

RESISTANCE OF POWDERED REFRACTORY COMPOUNDS TO ACIDS AND ALKALIS (continued)

Phase	Reagent	Temp., °C	Process time, h	Insoluble residue, %	Source	Year
1	2	3	4	5	6	7
Mg_2Si	Bromine water (sat.)	20	48	PS	[547]	1969
	NH_4F (10%)	20	48	ND		
	$(NH_4)_2S_2O_8$ (12,5%)	20	0,16	CS*		
	H_2O	20	48	ND		
	H_2O	100	1	ND		
$MgSi_2$	HCl (1 : 1)	20	—	CS	[260]	1959
$CaSi$, $CaSi_2$	H_2SO_4 (1 : 1)	20	—		[260]	1959
	HCl (1.19)					
	HNO_3 (1,43)					
	H_2SO_4 (1,82)					
$SrSi_2$, $BaSi_2$	H_2O	Boiling	—	CS	[260]	1959
$BaSi_2$	HCl (1 : 1)	20	—	CS	[256]	1957
	HCl (1.19)					
Silicides	Mineral acids	—	—	—**	[548]	1970
	H_2O	—	—	—***	[548]	1970
$ThSi_2$	HCl (1,19)	20	—	SMC	[549]	1956
	HJ (1,47)					
	HF (1,15)					
	3 h HCl (1,19) + + 1 h HNO_3 (1,43)					
$ThSi$, $ThSi_2$	NaOH (20%) H_2O_2	20	—	ND	[549]	1956
$ThSi_2$	HCl (1,19).	20	—	ND	[549]	1956
$ThSi_2$	HNO_3 (1,43)	20	—	ND	[550]	1955
	H_2SO_4 (1,82)	20	—	ND		
	H_2SO_4 (1,1)	20	—	SMC		

*See footnote on p. 346.
**Instantaneous decomposition with decrepitation.
***Practically unresistant.

Phase	Reagent	Temp., °C	Process time, h	Insoluble residue, %	Source	Year
1	2	3	4	5	6	7
$NpSi_2$	H_2O	20 Boiling	—	ND	[550]	1955
$TiSi_2$	HCl (1,19)	20	—	SMC	[529]	1961
	HCl (1,19)	Boiling	2	99.7		
	HCl (1,19)	"	1	ND		
	HCl (1 : 1)	"	2	99,8		
	HCl (1 : 1)	"	1	ND		
	H_2SO_4 (1,84)	"	2	99,6		
	H_2SO_4 (1,84)	"	3,5	ND		
	H_2SO_4 (1 : 1)	"	2	99,6		
	H_2SO_4 (1 : 1)	"	3,5	ND		
	H_2SO_4 (1 : 10)	"	2	99,8		
	H_2SO_4 (1 : 10)	"	3,5	ND		
	H_3PO_4 (1,21)	"	2	99,7		
	HF (1,15)	"	2,5	SMC		
	$KHSO_4$	"	1	ND		
	HF (1,15) + + HNO_3 (1,43)	"	2	CS		
	HCl (1,19) + + HNO_3 (1,43)	"	2	99,5		
	4 h H_3PO_4 (1,21) + + 1 h H_2SO_4 (1,82) + 2 h H_2O	"	2	SMC*		
	$H_2C_2O_4$ (sat.) + + H_2O	"	2	86,4		
	1 h $H_2C_2O_4$ (sat.) + 2 h H_2SO_4 (1,82)	"	2	85,5		
	NaOH (1%)	"	2	ND		
	NaOH (20%)	"	30 min	CS		
	Na_2O_2 (30%)	"	15 min	CS		
$ZrSi_2$	HCl (1,19)	Boiling	1	ND	[529]	1961
	HCl (1 : 1)		1	ND		
	H_2SO_4 (1,84)		3	ND		
	H_2SO_4 (1 : 1)		3	ND		
	H_2SO_4 (1 : 10)		3	ND		
	H_3PO_4 (1,21)		2	99,9		
	HF (1,15)		2	SMC		
	$KHSO_4$		2	ND		
	HF (1,15) + HNO_3 (1.43)		2	CS		
	$KHSO_4$ + KHF_2		15 min	CS		

*Insoluble residue appears upon evaporation of solution prior to the start of the appearance of SO_3 vapor.

Phase	Reagent	Temp., °C	Process time, h	Insoluble residue, %	Source	Year
1	2	3	4	5	6	7
$ZrSi_2$	$KHF_2 + H_2SO_4$ (1,84)	Boiling	4	CS	[529]	1961
	$KHSO_4 + $ $+ H_2SO_4 + $ $+ SiOCl_2$		7	ND		
	NaOH (1%)		—	ND		
	NaOH (20%)		30 min	CS		
	Na_2O_2 (30%)		15 min	CS		
VSi_2	HCl (1 : 1)	Boiling	1	ND	[529]	1961
	HCl (1.19)		1	ND		
	H_2SO_4 (1,84)		3	ND		
	H_2SO_4 (1 : 1)		3	ND		
	H_2SO_4 (1 : 10)		3	ND		
	H_3PO_4 (1.21)		2	99.5		
	HF (1,15)		3	SMC		
	HF (1,15) + HNO_3 (1,43)		2	CS		
	4 h H_3PO_4 + 1 h H_2SO_4 + 2 h H_2O		—	CS*		
	$KHSO_4 + KHF_2$		1,5 min	CS		
	$KHF_2 + H_2SO_4$ (1,82)		3	CS		
	$KHSO_4 + $ $+ H_2SO_4 + CrO_3$		5	ND		
	$KHSO_4 + $ $+ H_2SO_4 + $ $+ CrO_2Cl_2$		4,5	ND		
	$KHSO_4 + $ $+ H_2SO_4 + $ $+ SiOCl_2$		3	ND		
	NaOH (1%)		—	ND		
	NaOH (20%)		30 min	CS		
	Na_2CO_3		1	ND		
	Na_2O_2 (30%)		15 min	CS		
$NbSi_2$	HCl (1 : 1)	Boiling	1	ND	[529]	1961
	HCl (1.19)		1	ND		
	H_2SO_4 (1.84)		3	ND		
	H_2SO_4 (1 : 1)		3	ND		
	H_2SO_4 (1 : 10)		3	ND		
	HF (1.15)		1	SMC		
	HF (1,15) + HNO_3 (1.43)		2	CS		

*See footnote on p. 348.

Phase	Reagent	Temp., °C	Process time, h	Insoluble residue, %	Source	Year
1	2	3	4	5	6	7
$NbSi_2$	HCl (1,19) + HNO_3 (1,43)	Boiling	2	95,4	[529]	1961
	4 h H_3PO_4 + 1 h H_2SO_4 + 2 h H_2O		—	CS*		
	$KHSO_4$ + H_2SO_4 + $SiOCl_2$		2	96,5		
	$NaOH$ (1%)		5,5	ND		
	$NaOH$ (20%)		—	ND		
	Na_2O_2 (30%)		15 min	CS		
$TaSi_2$	HF 1.15) + HNO_3 (1,43)	Boiling	2	CS	[529]	1961
	HCl (1,19) + HNO_3 (1,43)		2	95,5		
	4 h H_3PO_4 (1.21) + 1 h H_2SO_4 (1,84) + 2 h H_2O		—	CS*		
	$H_2C_2O_4$ + H_2O		2	96,5		
	1 h $H_2C_2O_4$ + 2 h H_2O + 2 h H_2SO_4 (1.84)		2	96,6		
	$KHSO_4$ + H_2SO_4 + $SiOCl_2$		10	ND		
	$NaOH$ (1%)		—	ND		
	$NaOH$ (20%)		30 min	CS		
	Na_2O_2 (30%)		15 min	CS		
$CrSi_2$	HCl (1,19)	Boiling	2	44,5	[529]	1961
	HF 1.15)		1	PS		
	HF + HNO_3		2	CS		
	HCl + HNO_3		2	91,6		
	4 h H_3PO_4 + 1 h H_2SO_4 + 2 h H_2O		—	CS*		
	$H_2C_2O_4$ + H_2O		2	41,4		
	$H_2C_2O_4$ + H_2O		2	62,8		
	$NaOH$ (1%)		—	ND		
	$NaOH$ (20%)		30 min	CS		
	Na_2O_2 (30%)		15 min	CS		
Mo_3Si	HNO_3 (5%)	70	1	27	[341]	1968
	HNO_3 (10%)			CS		

*See footnote on p. 348.

Phase	Reagent	Temp., °C	Process time, h	Insoluble residue, %	Source	Year
1	2	3	4	5	6	7
Mo_5Si_3	HCl (conc.)	20	50	ND	[552]	1958
	HCl (conc.)	Boiling	2	ND		
	H_2SO_4 (conc.)	"	1	ND		
	H_2SO_4 (1 : 1)	20	50	ND		
	H_2SO_4 (1 : 1)	Boiling	2	ND		
	HNO_3 (conc.)	20	24	CS		
	HNO_3 (conc.)	Boiling	0.5	CS		
	HNO_3 (1 : 5)	20	24	CS		
	HNO_3 (1 : 5)	Boiling	0.5	CS		
	HF	"	1	ND		
	$H_3PO_4 + H_2SO_4 + H_2O$	20	50	ND		
		Boiling	0,5	CS		
	HCl + HNO_3	20	24	CS		
		Boiling	0.5	CS		
	HF + HNO_3	20	0,15	CS		
		Boiling	0.10	CS		
	$H_2C_2O_4 + H_2O_2$	20	24	PS		
		Boiling	0.5	CS		
	$H_2C_2O_4 + H_2O_2 + H_2SO_4$	20	24	CS		
		Boiling	0,5	CS		
	HCl + $KMnO_4$	"	1	ND		
	H_3PO_4 + HF	"	1	PS		
	H_2SO_4 + HF	20	1	PS		
	$H_2C_2O_4 + H_2SO_4$	Boiling	1	ND		
$MoSi_2$	HCl (1,19)	Boiling	2	99,4	[552]	1958
	HCl (1 : 1)	"	2	99,6	[552]	1958
	HNO_3 (5%)	70	1	99,6	[341]	1968
	HNO_3 (10%)	70	1	99,1	[341]	1968
$MoSi_2$	H_2SO_4 (1.84)	Boiling	2	99.2	[552]	1958
	H_2SO_4 (1 : 1)		2	99,8		
	H_3PO_4 (1.21)		2	96,7		
	HF (1.15)		1	PS		
	HCl (1,19)		3	ND		
	$KHSO_4$		1	ND		
	HF (1.15) + HNO_3 (1,43)		2	CS		
	HCl (1,19) + HNO_3 (1,43)		2	99,0		
	4 h H_3PO_4 (1.21) + 1 h H_2SO_4 (1.84) + 2 h H_2O		—	CS*		
	$H_2C_2O_4 + H_2O$		2	99,2		

*An insoluble residue appears upon evaporation prior to the start of the appearance of SO_3 vapors.

Phase	Reagent	Temp., °C	Process time, h	Insoluble residue, %	Source	Year
1	2	3	4	5	6	7
$MoSi_2$	1 h $H_2C_2O_4$ + 2 h H_2SO_4 (1,84)	Boiling	2	95,2	[552]	1958
	$KHSO_4$ + KHF_2		1,2	CS		
	KHF_2 + H_2SO_4 (1,84)		3	PS		
	$KHSO_4$ + + H_2SO_4 + CuO		5	ND		
	$KHSO_4$ + + H_2SO_4 + + CrO_2Cl_2		4,5	ND		
	$KHSO_4$ + + H_2SO_4 + + $SiOCl_2$		9,5	ND		
	NaOH (1%)		—	ND		
	NaOH (20%)		30 min	CS		
	Na_2CO_3 (20%)		1	ND		
	Na_2O_2 (30%)		15 min	CS		
W_5Si_3	HNO_3 (2%)	70	1	91,9	[341]	1968
	HNO_3 (5%)		1	85,9		
	HNO_3 (10%)		1	84,9		
	HNO_3 (15%)		1	85,7		
	HNO_3 (20%)		1	87,9		
	NaOH (10 g/liter)		20 min	96,4		
	NaOH (20 g/liter)		20 min	96,3		
	NaOH (50 g/liter)		20 min	97,0		
	NaOH (100 g/liter)		20 min	96,7		
	NaOH (200 g/liter)		20 min	97,2		
WSi_2	HNO_3 (2%)	70	1	96,1	[341]	1968
	HNO_3 (5%)			95,2		
	HNO_3 (10%)			94,6		
	HNO_3 (15%)			94,9		
	HNO_3 (20%)			95,9		
WSi_2	HF (1,15)	Boiling	2,5	CS	[529]	1961
	HJ (1,47)		3	ND		
	$KHSO_4$		1	PS		
	HF (1,15) + HNO_3 (1,43)		2	CS		
	$H_2C_2O_4$ + H_2O		2	93,4		
	$KHSO_4$ + KHF_2		20 min			
	KHF_2 + H_2SO_4 (1,82)		3,5			
	$KHSO_4$ + + H_2SO_4 + CuO		5	ND		
	$KHSO_4$ + + H_2SO_4 + + CrO_2Cl_2		4,5	ND		

Phase	Reagent	Temp., °C	Process time, h	Insoluble residue, %	Source	Year
1	2	3	4	5	6	7
WSi_2	$KHSO_4 +$ $+ H_2SO_4 +$ $+ SiOCl_2$	Boiling	3	ND	[529]	1961
	NaOH (20%)		30 min	CS		
	Na_2CO_3 (20%)		1	ND		
	Na_2O_2 (30%)		15 min	ND		
	HCl (1.19)		2	99.2		
	HCl (1 : 1)		1.5	ND		
	H_2SO_4 (1,82)		4	ND		
	H_2SO_4 (1 : 1)		4	ND		
	H_2SO_4 (1 : 10)		4	ND		
Mn_2Si, MnSi, $MnSi_2$	HCl (1.19) HNO_3 (1.43)	20	—	ND	[260]	1959
$MnSi_2$	H_2SO_4 (1,82)	20	—	ND	[260]	1959
	HF (1.15)			CS		
	3 h HCl (1,19) $+$ $+ 1$ h HNO_3 (1.43)			CS		
	HF (1,15) $+ HNO_3$ (1.43)			CS		
Fe_2Si, FeSi	HCl (1,19)	20	—	CS	[260]	1959
FeSi	HNO_3 (1,43)	20	—	CS	[260]	1959
	H_2SO_4 (1,82)	20	—	CS		
$FeSi_2$	HCl (36%)	115	1	96	[553]	1973
	HCl (19%)	110		97		
	H_2SO_4 (98%)	280		99		
	H_2SO_4 (20%)	120		99		
	H_3PO_4 (88%)	155		94		
	H_3PO_4 (29%)	120		ND		
	HF (48%)	—		CS		
	HNO_3 (65%) $+$ $+ HF$ (48%)	—		CS		
	H_3PO_4 (88%) $+$ $+ HCl$ (36%)	—		98		
	H_2SO_4 (20%) $+$ $+ HCl$ (36%)	—		96		
	H_2SO_4 (20%) $+$ $+ (NH_4)_2S_2O_8$ (25%)	120		ND		
	KOH (1%)	100		96		
	KOH (5%)	100		99		
	KOH (10%)	102		96		
	NH_4F (10%)	100		22		

RESISTANCE OF POWDERED REFRACTORY COMPOUNDS TO ACIDS AND ALKALIS (continued)

Phase	Reagent	Temp., °C	Process time, h	Insoluble residue, %	Source	Year
1	2	3	4	5	6	7
$FeSi_2$	NH_4F (10%) + + HCl (36%)	100	1	CS	[553]	1973
	NH_4F (10%) + + H_3PO_4 (88%)	120		CS		
	NH_4F (10%) + + H_2SO_4 (20%)	120		CS		
	H_2O					
	H_2O_2					
	Acetic acid					
	Citric acid					
	Tartaric acid	—	—	ND		
	Oxalic acid					
	$(NH_4)_2S_2O_8$					
	Trilon-B					
$CoSi$; Co_2Si	HCl (1.19)	20	—	CS	[260]	1959
$CoSi_2$	HCl (36%)	115	1	97	[553]	1973
	HCl (19%)	110		95		
	H_2SO_4 (98%)	280		99		
	H_2SO_4 (20%)	120		96		
	H_3PO_4 (88%)	155		CS		
	HF (48%)	—		CS		
	HNO_3 (65%)	120		98		
	HNO_3 (65%) + + HF (48%)	—		CS		
	HNO_3 (65%) + + 3HCl (36%)	—		96		
	H_2SO_4 (20%) + + HCl (36%)	—		98		
	H_2SO_4 (20%) + + $(NH_4)_2S_2O_8$ (25%)	120		ND		
	KOH (1%)	100		96		
	KOH (5%)	100		95		
	KOH (10%)	102		93		
	KOH (20%)	105		96		
	NH_4F (10%)	100		36		
	NH_4F (10%) + + HCl (36%)	100		CS		
	NH_4F (10%) + + H_2SO_4 (20%)	120		3		

Phase	Reagent	Temp., °C	Process time, h	Insoluble residue, %	Source	Year
1	2	3	4	5	6	7
$CoSi_2$	H_2O H_2O_2 Acetic acid Citric acid Tartaric acid Oxalic acid $(NH_4)_2S_2O_8$ Trilon-B	—	—	ND	[553]	1973
Ni_2Si	H_2SO_4 (1.82) HCl (1.19)	20	—	CS	[260]	1959
NiSi	HNO_3 (1.43) H_2SO_4 (1.82)	20	—	CS	[260]	1959
ξ-$NiSi_2$	HCl (36%) H_2SO_4 (98%) H_2SO_4 (20%) H_3PO_4 (95%) H_3PO_4 (24%) HF (48%) $H_2C_2O_4$ (10%) Tartaric acid (50%) Citric acid (50%) H_2O HNO_3 (65%) + + 3HI (36%) HNO_3 (65%) + + HF (48%) H_3PO_4 + H_2SO_4 + H_2O (2 : 1 : 2) H_2O_2 (30%) Bromine water (sat.) $(NH_4)_2S_2O_8$ (50%) NH_4F (10%) Trilon-B (sat.) KOH (1%) KOH (5%) KOH (10%) KOH (20%)	Boiling	1	92 73 77 85 35 53 100 98 82 100 84 0 90 99 92 93 71 99,6 80 77 76 73	[181]	1970

Phase	Reagent	Temp., °C	Process time, h	Insoluble residue, %	Source	Year
1	2	3	4	5	6	7
ξ-NiSi$_2$	H_2SO_4 (20%) + + H_2O_2 (30%)	Boiling	1	99,66	[181]	1970
	H_2SO_4 (20%) + + $(NH_4)_2S_2O_8$ (50%)			74		
	HCl (36%) + + NH$_4$F (10%)			14		
	H_2SO_4 (20%) + + NH$_4$F (10%)			86		
	Tartaric acid (50%) + HCl (36%)			91		
	Tartaric acid (50%) + H$_2O_2$ (30%)			100,2		
	Oxalic acid (10%) + + $(NH_4)_2S_2O_8$ (50%)			101		
Mg$_3$P$_2$	H_2O HCl (1 : 1) HCl (1,19) H_2SO_4 (1,82) HNO$_3$ (1,43) HF HF + HNO$_3$ (1 : 1) 3 h HCl (1,19) + + 1 h HNO$_3$ (1,43) KBr + Br$_2$	20	—	CS*	[29]	1961
Ca$_3$P$_2$	H_2O HCl (1 : 1) HCl (1,19) H_2SO_4 (1,82) HNO$_3$ (1,43) HF (40%) HF + HNO$_3$ (1 : 1) 3 h HCl (1,19) + +1 h HNO$_3$ (1,43)	20	—	CS*	[29]	1961
Sr$_3$P$_2$	H_2O HCl (1,19) H_2SO_4 (1,82)	20	—	CS 98 CS	[29]	1961

*Phosphine is evolved.

Phase	Reagent	Temp., °C	Process time, h	Insoluble residue, %	Source	Year
1	2	3	4	5	6	7
LaP	HNO_3 (conc.) HNO_3 (1 : 1) HCl (conc.) HCl (1 : 1) H_2SO_4 (conc.) H_2SO_4 (1 : 1) Aqua regia H_2O NaOH (5%)	—	1	CS CS CS CS Slightly decomposed Same CS ND ND	[554]	1963
CeP	HNO_3 (conc.) HNO_3 (1 : 1) HCl (conc.) HCl (1 : 1) H_2SO_4 (1 : 1) Aqua regia CH_3COOH H_2O H_2O_2 (3%) NaOH (5%) CH_3OH (84%) C_2H_5OH	—	1	CS CS CS CS CS CS CS 95,6 86,0 84,5 90,0 83,4	[555]	1970
PrP	HNO_3 (conc.) HNO_3 (1 : 1) HCl (conc.) HCl (1 : 1) H_2SO_4 (conc.) H_2SO_4 (1 : 1) Aqua regia H_2O H_2O_2 (3%) NaOH (5%) CH_3OH (84%) C_2H_5OH	—	1 1 1 1 48 48 1 1 1 1 1 1	CS CS CS CS 6.5 3.1 CS 98.3 64.0 97.8 90,0 96,9	[555]	1970
NdP	HNO_3 (conc.) HNO_3 (1 : 1) HCl (conc.) HCl (1 : 1) H_2SO_4 (conc.) H_2SO_4 (1 : 1) Aqua regia H_2O NaOH (5%)	—	1	CS CS CS CS Slightly decomposed Same CS ND ND	[556]	1965

Phase	Reagent	Temp., °C	Process time, h	Insoluble residue, %	Source	Year
1	2	3	4	5	6	7
SmP	HNO_3 (1.4) HNO_3 (1 : 1) H_2SO_4 (1 : 1) HCl (1,19) HCl (1 : 1) HF (conc.) Aqua regia Bromine water	20	1	CS*	[191]	1965
SmP	NaOH (30%) NaOH (30%) NaOH (10%) NaOH (10%) H_2O_2 (25%) H_2O_2 (25%) H_2O	20 107 20 102 20 105 20	24 4 24 4 24 4 24	ND 99,6 99,8 99,5 99,7 99,7 99,6	[191]	1965
YbP	HNO_3 (conc.) HNO_3 (1 : 1) HCl (conc.) HCl (1 : 1) H_2SO_4 (conc.) H_2SO_4 (1 : 1) Aqua regia CH_3COOH (conc.) H_2O H_2O_2 (3%) NaOH (5%) CH_3OH C_2H_5OH	—	1	CS CS CS CS 70 CS CS CS ND ND ND ND ND	[555]	1970
TiP	H_2O HCl (1 : 1) HCl (1,19) HNO_3 (1,43) H_2SO_4 (1,82) HF (40%) HF (40%) + + HNO_3 (1 : 1) 3 h HCl (1,19) + +. 1 h HNO_3 (1.43)	Boiling 20 Slight heating	8—10 6 6 6 6 6 — —	100 100 100 100 100 100 CS CS*	[29]	1961

*Phosphine is given off.

Phase	Reagent	Temp., °C	Process time, h	Insoluble residue, %	Source	Year
1	2	3	4	5	6	7
V_3P	H_2O HCl (1 : 1) HCl (1.19) HNO_3 (1.43) H_2SO_4 (1,82) 3 h HCl (1,19) + + 1 h HNO_3 (1.43)	Boiling	—	100 ND ND CS* CS* CS*	[29]	1961
VP	H_2O HCl (1 : 1) HCl (1,19) H_2SO_4 (1,82) HNO_3 (1.43) 3 h HCl (1,19) + + 1 h HNO_3(1,43)	Boiling	—	100 ND ND CS* CS* CS*	[29]	1961
VP_2	H_2O HCl (1 : 1) HCl (1,19) H_2SO_4 (1,82) HNO_3 (1.43) 3 h HCl (1.19) + + 1 h HNO_3 (1.43)	Boiling	—	ND ND ND CS* CS* CS*	[29]	1961
NbP	H_2O HCl (1 : 1) HCl (1.19) H_2SO_4 (1,82) HNO_3 (1.43) 3 h HCl (1,19) + + 1 h HNO_3 (1,43)	Boiling	—	ND ND ND CS* CS* CS*	[29]	1961
TaP	H_2O HCl (1 : 1) HCl (1.19) H_2SO_4 (1.82) HNO_3 (1.43) 3 h HCl (1,19) + + 1 h HNO_3 (1,43)	Boiling	—	ND ND ND CS* CS* CS*	[29]	1961
CrP	H_2O	Boiling	—	ND	[29]	1961

*Phosphine is given off.

Phase	Reagent	Temp., °C	Process time, h	Insoluble residue, %	Source	Year
1	2	3	4	5	6	7
CrP	HCl (1 : 1)	Boiling	—	ND	[557]	1961
	HCl (1,19)			ND		
	H_2SO_4 (1,84)			CS		
	H_2SO_4 (1 : 4)			ND		
	H_2SO_4 (1,84) + + HNO_3 (1,43)			CS		
	H_2SO_4 (1 : 1) + + HNO_3 (1,43)			ND		
	HNO_3 (1,43) + + HF (40%)			ND		
	3 h HCl (1,19) + + 1 h HNO_3 (1,43)			ND		
	H_2SO_4 (1 : 4) + + $(NH_4)_2S_2O_8$			ND		
	HNO_3 (1,43) + + $H_2C_2O_4$ (35%)			ND		
	HNO_3 (1,43) + + H_2O_2 (30%)			ND		
	NaOH (20%) + +bromine water			ND		
	HNO_3 (1,43) + + H_2SO_4 (1 : 1) + + $H_2C_2O_4$ (35%)			ND		
	NaOH (20%) + + H_2O_2 (30%) + + $C_2H_2O_4$ (35%)			ND		
MoP	H_2O	Boiling	—	ND	[29]	1961
	HCl (1 : 1)			ND		
	HCl (1.19)			ND		
	H_2SO_4 (1,82)			ND		
	HNO_3 (1,43)			CS		
MoP_2	H_2O	Boiling	—	ND	[29]	1961
	HCl (1,19)			ND		
	HCl (1 : 1)			ND		
	H_2SO_4 (1.82)			ND		
	HNO_3 (1,43)			CS		
FeP_2	H_2O	Boiling	—	ND	[29]	1961
	HCl (1,19)			ND		
	H_2SO_4 (1.82)			ND		
	3 h HCl (1,19) + + 1 h HNO_3 (1,43)			CS		
Fe_3P	H_2O	Boiling	—	ND	[29]	1961
	HCl (1,19)			ND		
	H_2SO_4 (1,82)			ND		
	3 h HCl (1,19) + + 1 h HNO_3 (1,43)			CS		

Phase	Reagent	Temp., °C	Process time, h	Insoluble residue, %	Source	Year
1	2	3	4	5	6	7
FeP	H_2O HCl (1,19) H_2SO_4 (1.82)	Boiling	—	ND	[29]	1961
U_3P_4	H_2O HCl (1,19) HNO_3 (1.43) H_2SO_4 (1,82) HF (40%) HF $+$ HNO_3 (1 : 1) 3 h HCl (1,19) $+$ $+$ 1 h HNO_3 (1.43)	Boiling	—	PS CS CS CS CS CS CS	[29]	1961
Np_3P_4	H_2O HF $+$ HNO_3 (1 : 1) HCl (1,19)	20	—	ND CS CS	[29]	1961
YS	Diluted organic acids CH_3COOH (1 : 10) J_2 (solution) $KMnO_4$	20	—	CS CS —* —**	[33] [33] [33] [558]	1972 1972 1972 1961
LaS	HCl (1 : 5) HNO_3 (1 : 5) H_2SO_4 (1 : 5) CH_3COOH (1 : 1)	20	—	CS CS CS CS	[558]	1961
La_2S_3	H_2O HCl (1 : 5) HNO_3 (1 : 5) H_2SO_4 (1 : 5) H_3PO_4 (1.21) CH_3COOH (1 : 1) $H_6C_4O_6$ (50%) NaOH (20%)	100 20 20 20 20 20 100 100	1 — — — — — — 1	99,9 CS CS CS CS CS CS ND	[558]	1961
CeS	HCl (1 : 5) HNO_3 (1 : 5) H_2SO_4 (1 : 5) CH_3COOH (1 : 1) H_2O	20 20 20 20 100	— — — — 1	CS CS CS CS 100	[558]	1961

---

*Oxidized with iodine solution.

**Oxidized with $KMnO_4$ solution.

Phase	Reagent	Temp., °C	Process time, h	Insoluble residue, %	Source	Year
1	2	3	4	5	6	7
Ce_2S_3	HCl (1 : 5) HNO_3 (1 : 5) H_2SO_4 (1 : 5) H_3PO_4 (1.21) CH_3COOH (1 : 1) $H_6C_4O_6$ (50%) NaOH (20%) H_2O_2 (30%)	20 20 20 20 20 100 100 100	— — — — — — 1 1	CS CS CS CS CS CS ND 21,0	[558]	1961
PrS	HCl (1 : 1) HNO_3 (1,43) H_2SO_4 (1.82)	20	—	CS	[33]	1972
Pr_3S_4	Diluted inorganic acids and CH_3COOH	20	—	CS	[33]	1972
Pr_2S_3	HCl (1 : 5) HNO_3 (1 : 5) H_2SO_4 (1 : 5) CH_3COOH (1 : 1)	20	—	CS	[558]	1961
NdS	HCl (1 : 1) HNO_3 (1,43) H_2SO_4 (1,82)	20	—	CS	[33]	1972
Nd_3S_4	Diluted inorganic acids and CH_3COOH	20	—	CS	[33]	1972
Nd_2S_3	HCl (1 : 5) HNO_3 (1 : 5) H_2SO_4 (1 : 5) CH_3COOH (1 : 1)	20	—	CS	[558]	1961
SmS	HCl (1 : 1) HNO_3 (1,43) H_2SO_4 (1.82)	20	—	CS	[33]	1972
Sm_3S_4	Diluted inorganic acids and CH_3COOH	20	—	CS	[33]	1972

RESISTANCE OF POWDERED REFRACTORY COMPOUNDS TO ACIDS
AND ALKALIS (continued)

Phase	Reagent	Temp., °C	Process time, h	Insoluble residue, %	Source	Year
1	2	3	4	5	6	7
ThS	HCl (1 : 1) HNO$_3$ (1,43)	20	—	CS	[550]	1955
Th$_2$S$_3$	HCl (1 : 1) HNO$_3$ (1.43)	20	—	CS	[550]	1955
ThS$_{1,7}$	H$_2$O H$_2$SO$_4$ (1 : 1) HCl (1,19) H$_3$PO$_4$ (1 : 1) H$_2$C$_2$O$_4$ (sat.) CH$_3$COOH NaOH (20%)	Boiling	0,5	99.8 CSH 8,5 20,5 24.2 18,6 20,5	[559]	1971
ThS$_2$	H$_2$O H$_2$SO$_4$ (1 : 1) HCl (1,19) H$_3$PO$_4$ (1 : 1) H$_2$C$_2$O$_4$ (sat.) CH$_3$COOH NaOH (20%)	Boiling	0,5	99,8 CSH CS 1,8 20,1 14,3 7,38	[559]	1971
US	Inorganic acids	20	—	CS	[33]	1972
	Alkalis and NH$_4$OH (soln.)	20		ND		
U$_2$S$_3$	Inorganic acids	20	—	CS	[33]	1972
	CH$_3$COOH (1 : 1)	100		ND		
U$_3$S$_5$	Inorganic acids	20	—	ND	[33]	1972
	CH$_3$COOH (1 : 1)	100				
US$_3$	Inorganic acids	20	—	CS	[33]	1972
	CH$_3$COOH					
Ti$_2$S$_3$	H$_2$O H$_2$O H$_2$SO$_4$ (1.84)	20 Boiling 20	24 0,5 24	ND 99.8 10,2	[559]	1971

Phase	Reagent	Temp., °C	Process time, h	Insoluble residue, %	Source	Year
1	2	3	4	5	6	7
Ti_2S_3	H_2SO_4 (1 : 1)	20	24	CS*	[559]	1971
	H_2SO_4 (1 : 1)	136	0,5	CSH		
	HCl (1,19)	20	24	95.4		
	HCl (1,19)	112	0.5	36,2		
	HCl (1 : 1)	20	24	CS*		
	HNO_3 (1,49)	20	24	CS		
	HNO_3 (1 : 1)	20	24	CS		
	H_3PO_4 (1.21)	20	24	93,5		
	H_3PO_4 (1 : 1)	110	0.5	88,6		
	H_2O_2 (30%)	20	24	CS		
	$H_2C_2O_4$ (sat.)	20	24	97,7		
	$H_2C_2O_4$ (sat.)	100	0.5	50,2		
	CH_3COOH	20	24	98,5		
	CH_3COOH	100	0,5	43,05		
	NaOH (10%)	20	24	90,5		
	NaOH (20%)	100	0,5	30,5		
	NaOH (40%)	20	24	75,6		
Zr_2S_3	H_2O	20	24	ND	[559]	1971
	H_2O	100	0,5	99,8		
	H_2SO_4 (1,84)	20	24	7,4		
	H_2SO_4 (1 : 1)	20	24	CS**		
	H_2SO_4 (1 : 1)	136	0.5	CSH		
	HCl (1,19)	20	24	40,4		
	HCl (1,19)	112	0.5	18,5		
	HCl (1 : 1)	20	24	98,0		
	HNO_3 (1,49)	20	24	CS		
	HNO_3 (1 : 1)	20	24	CS		
	H_3PO_4 (1,21)	20	24	CS		
	H_2O_2 (30%)	20	24	CS		
	$H_2C_2O_4$ (sat.)	20	24	90.2		
	$H_2C_2O_4$ (sat.)	100	0,5	38,3		
	CH_3COOH	20	24	99,8		
	CH_3COOH	100	0,5	21,7		
	NaOH (10%)	20	24	84,3		
	NaOH (20%)	100	0,5	22,8		
	NaOH (40%)	20	24	53,6		
ZrS_2	H_2O	20	24	ND	[559]	1971
	H_2SO_4 (1,84)			CS		
	H_2SO_4 (1 : 1)			CS**		
	HCl (1,19)			18,35		
	HCl (1 : 1)			9,25		
	HNO_3 (1,49)			CS		

*$TiO_2 \cdot n H_2O$ is formed.

**$ZrO_2 \cdot n H_2O$ is formed.

RESISTANCE OF POWDERED REFRACTORY COMPOUNDS TO ACIDS
AND ALKALIS (continued)

Phase	Reagent	Temp., °C	Process time, h	Insoluble residue, %	Source	Year
1	2	3	4	5	6	7
ZrS_2	HNO_3 (1 : 1) H_3PO_4 (1,21) H_2O_2 (30%) $H_2C_2O_4$ (sat.) CH_3COOH NaOH (10%) NaOH (40%)	20	24	CS CS CS CS ND 80,5 2,5	[559]	1971
HfS_2	H_2O H_2SO_4 (1.84) H_2SO_4 (1 : 1) HCl (1.19) HCl (1 : 1) HNO_3 (1.49) HNO_3 (1 : 1) H_3PO_4 (1,21) H_2O_2 (30%) $H_2C_2O_4$ (sat.) CH_3COOH NaOH (10%) NaOH (40%)	20	24	ND CS CS* CS CS CS CS CS CS CS ND 74.2 CS	[559]	1971
$NbS_{1.6}$	H_2SO_4 (1.84) H_2SO_4 (1 : 1) HCl (1.19) HCl (1 : 1) HNO_3 (1.4) HNO_3 (1 : 1) H_3PO_4 (1.21) H_2O_2 (30%) $H_2C_2O_4$ (6%) Bromine water H_2O NaOH (40%) NaOH (10%)	Boiling	1	CS 88.4 97.8 98.0 CS CS 93.4 CS 97.8 PS ND 73.6 90.3	[36]	1965
$1s\text{-}TaS_2$ (α)	H_2SO_4 (1,84) H_2SO_4 (1 : 1) HCl (1.19) HCl (1 : 1)	Boiling	1	CS 99.6 98,0 98.7	[36]	1965

*$HfO_2 \cdot nH_2O$ is formed.

Phase	Reagent	Temp., °C	Process time, h	Insoluble residue, %	Source	Year
1	2	3	4	5	6	7
$1s$-TaS_2 (α)	HNO_3 (1.4)	Boiling	1	CS*	[36]	1965
	HNO_3 (1 : 1)			CS		
	H_3PO_4 (1.21)			1.4		
	H_2O_2 (30%)			CS		
	$H_2C_2O_4$ (6%)			79.6		
	Bromine water			PS*		
	H_2O			ND		
	NaOH (40%)			97.0		
	NaOH (10%)			36.0		
B_4C	HCl (1.19)	20	24	98	[560]	1959
	HCl (1.19)	115	1	98		
	HCl (1 : 1)	105	30 min	97.8		
	HCl (1 : 1)	105	2	97.8		
	H_2SO_4 (1.82)	20	24	98		
	H_2SO_4 (1.82)	130	1	98		
	H_2SO_4 (1 : 1)	130	30 min	98		
	H_2SO_4 (1 : 1)	130	2	97.7		
	H_2SO_4 (1 : 1)	130	4	98		
	HNO_3 (1.43)	20	24	97		
	HNO_3 (1.43)	130	1	97		
	HNO_3 (1 : 1)	105	30 min	96.9		
	HNO_3 (1 : 1)	105	1	96.5		
	HNO_3 (1 : 1)	105	2	96.1		
	$HClO_4$ (1.35)	110	4	96.9		
	$HClO_4$ (1.35)	20	24	98		
	$HClO_4$ (1.35)	115	1	98		
	3 h HCl (1.19) + + 1 h HNO_3 (1.43)	20	24	97		
	H_2SO_4 (1.82) + + HNO_3 (1.43)	230	4	91.2		
	NaOH (50%)	20	40	98.3		
	NaOH (25%)	20	40	99.2		
	NaOH (12%)	20	40	99		
	NaOH (6%)	20	40	98.6		
	NaOH (3%)	20	40	98.8		
	NaOH (1%)	20	40	99		
	NaOH (1%)	100	1	99		
	NaOH (1%)	100	2	98.5		
	NaOH (25%)	100	2	99		
	NaOH (12%)	100	2	98.5		
	HF (1.15) + + HNO_3 (1.43)	180	2	90.8		
	$HClO_4$ (1 : 1)	115	30 min	98		
	$HClO_4$ (1 : 1)	115	1	98		
	$HClO_4$ (1 : 1)	115	2	96.7		
	NaOH (10%) + + H_2O_2 (10%)	100	1	98		

*Tantalum hydroxide forms and sulfur separates.

Phase	Reagent	Temp., °C	Process time, h	Insoluble residue, %	Source	Year
1	2	3	4	5	6	7
B_4C	NaOH (10%) + + H_2O_2 (10%)	100	2	96	[560]	1959
	NaOH (10%) + + Br_2	100	1	99.6		
SiC	HCl (1 : 1)	Boiling	—	100	[529]	1961
	HCl (1.19)	"	1	100		
	HNO_3 (1 : 1)	"	1	100		
	HNO_3 (1.43)	"	1	100		
	HF (1.15)	"	1	100		
	HNO_3 (1.43) + + HF (1.15)	"	1	100		
	H_3PO_4 (1.21)	230	1	PS		
BN	H_2SO_4 (1.82)	Boiling	6—10	CS	[529]	1961
	H_2SO_4 (1.82)	20	—	—		
	H_2SO_4 (20%)	20	—	10,7		
	H_3PO_4 (1.21)	20	—	1,3		
	HNO_3 (1.43)	20	—	8,9		
	HF (1.15)	20	—	17,5		
	NaOH (20%)	Boiling	15—20 min	CS		
	NaOH (20%)	20	—	8,9		
	CCl_4	20	—	1,3		
	C_2H_5OH (95%)	20	—	14,6		
	CH_3COOH	20	—	13,0		
Si_3N_4	HCl (20%)	Boiling	500	ND	[529]	1961
	HNO_3 (65%)	"	500	ND		
	HNO_3 (65%)	Fuming	500	ND		
	H_2SO_4 (10%)	70	500	ND		
	H_2SO_4 (77%)	20	500	ND		
	H_2SO_4 (85%)	20	500	ND		
	H_3PO_4 (1.21)	20	500	ND		
	$H_4P_2O_7$	20	500	ND		
	HF (1.15)	Boiling	192	13,9		
	NaOH (20%)	20	500	ND		
	NaOH (50%)	Boiling	115	ND		
	H_2SO_4 + $CuSO_4$ + + $KHSO_4$ (conc.)	"	500	ND		
	HF (1.15) + + HNO_3 (1.43)	"	68	56		

RESISTANCE OF POWDERED REFRACTORY COMPOUNDS TO ACIDS AND ALKALIS (continued)

Phase	Reagent	Temp., °C	Process time, h	Insoluble residue, %	Source	Year
1	2	3	4	5	6	7
AlN	H_2O	20	24	36,71	[544]	1962
	H_2O	Boiling	1	42,34		
	H_2O	"	2	41,86		
	H_2O	"	4	41,00		
	H_2SO_4 (1,84)	"	1	93,60		
	H_2SO_4 (1,84)	"	2	90.04		
	H_2SO_4 (1,84)	"	4	86.03		
	HNO_3 (1,4)	"	1	32,48		
	HNO_3 (1,4)	"	2	6,37		
	HNO_3 (1,4)	"	4	2.11		
	HCl (1,19)	"	1	52,50		
	HCl (1,19)	"	2	25,03		
	HCl (1,19)	"	4	10,12		
	H_2SO_4 (1 : 1)	"	1	0,95		
	H_2SO_4 (1 : 1)	"	2	0.52		
	H_2SO_4 (1 : 1)	"	4	0.45		
	HNO_3 (1 : 1)	"	1	18.49		
	HNO_3 (1 : 1)	"	2	4.88		
	HNO_3 (1 : 1)	"	4	CS		
	HCl (1 : 1)	"	1	26,57		
	HCl (1 : 1)	"	2	7,36		
	HCl (1 : 1)	"	4	2.05		
	NaOH (10%)	"	1	0,95		
	NaOH (10%)	"	2	0,52		
	NaOH (10%)	"	4	0.35		
	NaOH (20%)	"	1	CS		
	NaOH (20%)	"	2	CS		
	NaOH (20%)	"	4	CS		
	NaOH (40%)	"	1	CS		
	NaOH (40%)	"	2	CS		
	NaOH (40%)	"	4	CS		

RESISTANCE OF COMPACTED REFRACTORY COMPOUNDS TO ACIDS AND ALKALIS

Phase	Reagent	Temp., °C	Process time, h	Insoluble residue, %	Source	Year
1	2	3	4	5	6	7
TiB_2	HCl (1,19)	20	24	99,9	[528]	1960
	HCl (1,19)		96	99,6		
	HCl (1,19)		240	99,2		

Phase	Reagent	Temp., °C	Process time, h	Insoluble residue, %	Source	Year
1	2	3	4	5	6	7
TiB_2	HNO_3 (1.42)	20	24	93.9	[528]	1960
	HNO_3 (1.42)		96	84.9		
	HNO_3 (1.42)		240	69.5		
	H_2SO_4 (1.84)		24	99.9		
	H_2SO_4 (1.84)		96	99.6		
	H_2SO_4 (1.84)		240	99.0		
ZrB_2	HCl (1.19)	20	24	97.5	[528]	1960
	HCl (1.19)		96	95.1		
	HCl (1.19)		240	89.6		
	HNO_3 (1.42)		24	94.4		
	HNO_3 (1.42)		96	86.4		
	HNO_3 (1.42)		240	66.7		
	H_2SO_4 (1.84)		24	98.8		
	H_2SO_4 (1.84)		96	95.8		
	H_2SO_4 (1.84)		240	94.8		
Cr_3C_2	HCl (1 : 1)	112	1	0.039 *	[529]	1962
	HNO_3 (1 : 1)	112		Not detected		
	H_2SO_4 (1 : 1)	136		0.003 *		
	$H_2C_2O_4$ (sat.)	130		0.015 *		
	3 h HCl + 1 h HNO_3	102		0.15 *		
	4 h H_3PO_4 + 1 h H_2SO_4 + 2 h H_2O	189		Not detected		
	NaOH + bromine water	110		0.12 *		
Cr_7C_3	HCl (1 : 1)	110	1	7.5 *	[529]	1962
	HNO_3 (1 : 1)	112		Not detected		
	H_2SO_4 (1 : 1)	125		26.4 *		
	$H_2C_2O_4$ (sat.)	135		0.036 *		
	3 HCl (1.19) + + 1 h HNO_3 (1.42)	102		0.036 *		
	4 h H_3PO_4 (1.21) + 1 h H_2SO_4 (1.84) + 2 h H_2O	135		Not detected		
	NaOH + bromine water	115		"		
$MoSi_2$	HCl (1.19)	20	24	99.91	[528]	1960
	HCl (1.19)		96	99.84		
	HCl (1.19)		240	99.26		
	HNO_3 (1.42)		24	99.54		

*Corrosion rate, $g/(m^2 \cdot h)$. Specimens hot pressed. P = 3-7%.

Phase	Reagent	Temp., °C	Process time, h	Insoluble residue, %	Source	Year
1	2	3	4	5	6	7
$MoSi_2$	HNO_3 (1,42)	20	96	99.16	[528]	1960
	HNO_3 (1,42)		240	98,44		
	H_2SO_4 (1,84)		24	99,93		
	H_2SO_4 (1,84)		96	99.93		
	H_2SO_4 (1,84)		240	99.93		
B_4C	HCl (1,19)	20	24	99,76	[528]	1960
	HCl (1,19)		96	99,41		
	HCl (1,19)		240	99,35		
	HNO_3 (1,42)		24	99,59		
	HNO_3 (1,42)		96	99,35		
	HNO_3 (1,42)		240	99,35		
	H_2SO_4 (1,84)		24	98,95		
	H_2SO_4 (1,84)		96	98,95		
	H_2SO_4 (1,84)		240	98,43		
SiC *	HCl (20%)	Boiling	1008	0,3	[611]	1961
	HCl (20%)	175	144	—0,3		
	HCl (20%)	200	144	0,9		
	HCl (20%)	225	144	1,5		
	HCl (37%)	Boiling	144	0,0		
	HCl (37%)	200	48	0,0		
	HNO_3 (30%)	Boiling	144	0.6		
	HNO_3 (50%)	"	1008	0,0		
	HNO_3 (50%)	200	144	12.2		
	HNO_3 (70%)	Boiling	144	0,6		
	HNO_3 (70%)	200	144	5,8		
	HNO_3 (70%)	225	144	3,0		
	H_2SO_4 (60%)	Boiling	144	0,0		
	H_2SO_4 (60%)	200	144	—0,9		
	H_2SO_4 (80%)	Boiling	1008	—0,3		
	H_2SO_4 (95%)	"	144	—2,4		
	H_2SO_4 (95%)	200	288	—0,6		
	H_3SO_4 (95%)	225	144	3,66		
	H_3PO_4 (40%)	Boiling	144	0,0		
	H_3PO_4 (60%)	"	1008	0,0		
	H_3PO_4 (60%)	200	144	—0,3		
	H_3PO_4 (85%)	Boiling	144	6,9		
	H_3PO_4 (85%)	200	288	1,5		

*For all determinations of SiC resistance, instead of the insoluble residue we
quote the solution rate (corrosion) mm/min, determined for the process time,
indicated in the table. The rate was determined on specimens of the following
compositions, %: 96.5 SiC; 2.5 Si_{free}; 0.4 C_{free}; 0.4 Al; 0.2 Fe; P = 0.4.

RESISTANCE OF COMPACTED REFRACTORY COMPOUNDS TO ACIDS
AND ALKALIS (continued)

Phase	Reagent	Temp., °C	Process time, h	Insoluble residue, %	Source	Year
1	2	3	4	5	6	7
SiC	H_3PO_4 (85%)	225	144	0.3	[611]	1961
		60	24	960.8		
	40% HF + + 10% HNO_3	60	144	496		
		60	288	369		
		60	432	308		
		60	576	263		
	Na_2SO_4 (10%)	Boiling	144	0.6		
	Na_2SO_4 (10%)	"	288	0.6		
	NaOH (50%)	"	24	6668		
	NaOH (25%)	"	144	224		
	Na_2CO_3 (10%)	"	144	84,2		
	Na_2CO_3 (10%)	"	285	40,6		
	Na_2CO_3 (10%)	"	432	19,2		
	Na_2CO_3 (10%)	"	576	8.5		
	Na_2CO_3 (10%)	"	720	—1.2		
	Na_2CO_3 (10%)	"	864	—9.6		
	Na_2CO_3 (10%)	"	1008	10.2		

RESISTANCE TO OXIDATION

Phase	Temp., °C	Oxidation time, h	Wt. change, mg/cm^2	Source	Year
1	2	3	4	5	6
Be_5B	1000	20	+258	[693]	1960
	1200	14.5	+120		
Be_2B	1000	20	+132	[693]	1960
	1200	14,5	+126		
BeB_2	1000	20	+22	[693]	1960
	1100	20	+34		
	1200	14,5	+99,4		
		20	+48		
BeB_4	1000	20	+30	[693]	1960
	1200	15,5	—2,9		
BeB_6	1000	20	+64	[693]	1960
	1200	14.5	—7,25		

Phase	Temp., °C	Oxidation time, h	Wt. change, mg/cm^2	Source	Year
1	2	3	4	5	6
CaB$_6$	900	0,5 1 2 10	+36% +37% +38% +38%	[524]	1961
SrB$_6$	900	0,5 1 2 10	+42% +42% +43% +43%	[524]	1961
BaB$_6$	900	0,5 1 2 10	+44% +45% +46% +46%	[524]	1961
ScB$_2$, ScB$_{12}$ GdB$_6$	To 600 *1 1000	— 140	— +6	[526] [708]	1965 1959
TiB$_2^{*2}$	450 500 550 600 700 800 900 1000	1	+0,42 +0,63 +0,63 +1,78 +2,00 +7,36 +20,4 +12,0	[694]	1958
TiB$_2^{*3}$	1000	0,8 2,8 9,3 19 29 40 48 63 82,5 102	+6,8 +10 +19 +25 +20 +24 +28 +29 +32 +30	[695]	1958

*1Resistant in air.
*2P = 1.6%.
*3P = 2-3%.

Phase	Temp., °C	Oxidation time, h	Wt. change, mg/cm^2	Source	Year
1	2	3	4	5	6
TiB$_2^{*1}$	1000	119 147 170	+29 +29 +31	[695]	1958
TiB$_2$	1100	20	+26	[693]	1960
TiB$_2$	1200	2 5 25 50 75 100	+10 +24.5 +38.4 +62.0 +68.1 +73.7	[696]	1960
ZrB$_2$	1000 1100	150 20	+30 +22	[695] [693]	1958 1960
ZrB$_2$	1150	8 16 24 32 48 200	+0.5 +1.2 +2 +3 +3 +4	[254]	1960
HfB$_2$	590 *2	—	—	[698]	1971
NbB$_2^{*3}$	450 500 550 600 700 800 900 1000	1	+0.25 +0.99 +1.74 +1.86 +4.99 +16.2 +28.6 +32.5	[694]	1958
TaB$_2$	700 800 900	1 2 1 1 2	+1.24 +1.81 +1.69 +2.52 +3.34	[254]	1960

*1 p = 2-3 %.
*2 Oxidation commenced in air.
*3 p = 1.4 %.

Phase	Temp., °C	Oxidation time, h	Wt. change, mg/cm^2	Source	Year
1	2	3	4	5	6
Cr_4B	900	0,5—1.5	~0.2	[699]	1965
		2—2.5	~3.0		
	1000	0,5—3,0	~3.5		
Cr_3B_2	600	0.5	~3.5	[699]	1965
		3,5	~4.5		
	700	0,5	~6,0		
		3,5	~6,5		
	800	0,5—2	~8		
	900	0.5	~9		
		1	~10		
		1.5—2,5	~10.5		
		3—3,5	~11		
	1000	0,5	~15.5		
		1	~19.5		
		2	~23,5		
		3	~25,5		
CrB	900	0.5	~0.2	[699]	1965
		1,5—2	~0.3		
		2,5—4	~0.4		
	1000	0,5	~0.55		
		1	~0,65		
		1,5	~0,85		
		2—3,5	~0,9		
		4—5	~1,0		
Cr_3B_4	600	0,5	~0.5	[699]	1965
		1	~0,75		
		1,5	~1		
		2—3	~1.5		
	700	0,5—1,5	~2.75		
		2	~3,25		
		2,5—3	~3,75		
	900	0,5—3,5	~4.5		
CrB_2	700	0,5—2,5	~4,3	[699]	1965
	900	0,5	~4,75		
		1	~5,3		
		2	~5,5		
		3—3,5	~6,5		
		4—4,5	~6,6		
	1000	0,5	~7,25		
		1—3,5	~7,75		

Phase	Temp., °C	Oxidation time, h	Wt. change, mg/cm^2	Source	Year
1	2	3	4	5	6
CrB$_2$	1100	5	~2	[700]	1974
	1200	1	~5		
		1,5	~6,5		
		2	~8		
		2,5	~9		
		3,0	~18		
		3,5	~22		
		4	~28		
		4,5	~30		
		5	~31		
	1240	1,5	~5,5		
		2	~6,5		
		2,5	~7		
		3	~7,5		
		3,5	~27,5		
		4	~40		
		4,5	~42,5		
		5	~45		
Mo$_2$B$_5$	To 800	Not oxidized		[700]	1974
	800	1—5	~1		
	900	1—5	~1,5		
	1000	1	~1,5		
		2—5	~2,5		
	1100	0,5	~3		
		1	~6		
		1,5	~10		
		2	~14		
		2,5	~16		
		3	~17,5		
		4	~19,5		
		5	~20,5		
	1200	0,5	~8,5		
		1	~17		
		1,5	~20		
		2	~22		
		2,5	~22,5		
		3—3,5	~23		
		4	~23,5		
		5	~26		
W$_2$B$_5$	To 800	Not oxidized		[700]	1974
	800	1—5	~2		
	900	1—5	~2,5		
	1000	1,5	~2,5		
		2	~3		
		2,5—5	~4		

RESISTANCE TO OXIDATION (continued)

Phase	Temp., °C	Oxidation time, h	Wt. change, mg/cm^2	Source	Year
1	2	3	4	5	6
W$_2$B$_5$	1100	1	~3	[700]	1974
		2	~5		
		3	~6		
		4	~7		
		5	~8		
	1200	1	~6		
		2	~8		
		3	~10		
		5	~12.5		
YC	20	1	+3.33	[92]	1962
		5	+21,65		
		15	+123		
		20	+138		
		40	+156		
		50	+162		
Y$_2$C$_3$	20	1	+2.51	[92]	1962
		5	+10,65		
		15	+48		
		30	+81		
		40	+97,2		
		50	+112,5		
YC$_2$	20	1	+3,52	[92]	1962
		5	+10		
		20	+19,6		
		50	+33		
YC$_2$	20	150	~120 *	[93]	1964
LaC$_2$	20	150	~60 *	[93]	1964
CeC$_2$	20	150	~180 *	[93]	1964
PrC$_2$	20	150	~30 *	[93]	1964
UC	350	0.5	2,5 **	[697]	1968
	400	0,5	5,8 **		
	450	0,5	23,8 **		

*Acetylene given off.
**Oxidation in oxygen.

Phase	Temp., °C	Oxidation time, h	Wt. change, mg/cm^2	Source	Year
1	2	3	4	5	6
TiC	To 600		Not oxidized	[701]	1972
	700	1—5	~3		
	800	1	~3,5		
		2—5	~4		
	900	1—5	~1		
	1000	1—5	~1,5		
	1100	1	~7		
		3	~13.5		
		5	~18.5		
	1200	0,5	~7,5		
		1	~12.5		
		2	~26.5		
		3	~34.5		
		3,5—5	~36		
ZrC	To 700	Slightly oxidized		[702]	1973
	800	5	~22,5		
	900	5	~25		
	1000	5	~20		
	1100	1,5	~22,5		
		2	~30		
		3	~40		
		5	~55		
	1200	1,5	~25		
		2	~32		
		3	~42,5		
		4	~51		
		5	~60		
HfC	To 700	Practically not oxidized		[702]	1973
	800	1	~8		
		3	~13		
		5	~16		
	900	0.5	~12		
		1	~22		
		2	~70		
		4	~85		
	1000	0.5	~20		
		1	~38		
		3	~105		
		5	~142		
	1100	0.5	~35		
		1	~65		
		2	~112		
		2,5—5	~130		

Phase	Temp., °C	Oxidation time, h	Wt. change, mg/cm^2	Source	Year
1	2	3	4	5	6
HfC	1200	0,5 1 2 3—3,5	~40 ~65 ~115 ~145	[702]	1973
VC	900	—	—	[697]	1968
NbC	450 600 1100—1400	1 2 1 Active oxidation commences	+1,39 +4,96 +11,7 —	[697]	1968
TaC	800 900 1100—1400	1 2 1 2 Active oxidation commences	+0,493 +1,29 +10,0 +39,4 —	[697]	1968
$Cr_{23}C_6$	800—1100	1—2	0	[705]	1961
Cr_7C_3	800 900 1000	1 2 3 4 1 2 3 4 1 2 3	+8,7 +12,1 +12,8 +12,8 +28,7 +35,4 +42,1 +47 +69,9 +116,9 +142,5	[705]	1961
Cr_3C_2	800—1100	1—4	0	[705] *	1961

See [703, 704].

Phase	Temp., °C	Oxidation time, h	Wt. change, mg/cm^2	Source	Year
1	2	3	4	5	6
Mo$_2$C	600—800	Strong oxidation		[703]	1973
	900	1	~—40		
		2	~—80		
	1000	0.5	~—100		
		1	~—270		
	1100	0,5	~—170		
	1200	0,5	~—250		
WC	500—700	Substantial oxidation		[703]	1973
	800—1200 *1	Complete oxidation			
ScN	600	Oxidation commences		[544]	1962
UN	355	0—15	0,5	[706]	1964
	406	0—1	2,8		
		1—5	2,4		
		5—10	2,3		
		10—15	2,5		
	454	0—1	8		
		1—10	7,7		
		10—15	7,5		
	482	0—1	22		
		1—5	20		
TiN	700	1	+16	[468] *1	1955
	800		+17		
	900		+18		
	1000		+25		
ZrN	580 *2	3	—	[707]	1970
HfN	650 *2	1	—	[707]	1970
VN	400 *2	1	—	[707]	1970
CrN	900 *2	1	—	[707]	1970

*1At 1100°C the oxidation process slows down because of sintering.
*2See [697, 707].
*3Oxidation commences.

Phase	Temp., °C	Oxidation time, h	Wt. change, mg/cm^2	Source	Year
1	2	3	4	5	6
TiAl [*1]	600	8	~-15	[470]	1962
		12	~-10		
	800	8	~0.9		
		16	~1.1		
		24	~1.1		
	1150	2	~10		
		32	~50		
TiAl$_3$[*2]	600	12	~0	[470]	1962
	800	8	~1.6		
		16	~2.5		
		24	~3		
	1150	8	~50		
		16	~120		
BaSi$_2$	1300	1.0	$+4.9$	—[*2]	1961
		2.3	$+8.25$		
		4.5	$+10.6$		
		6.5	$+13.0$		
		10.5	$+15.2$		
		14.7	$+16.7$		
LaSi$_2$	1300	0.5	$+12.7$	—[*3]	1961
		1.0	$+16.2$ [*3]		
DySi$_2$	1000	0.5	$+2.60$	[708]	1959
		1.0	$+4.34$		
		18	$+21.4$		
		42	$+35.8$		
		66	$+44.5$		
		96	$+52.1$		
		398	$+80.0$		
		782	$+90.8$		
U$_3$Si	260	1	-50	[447]	1960
	345	1	$+1000$		

[*1]At 600-800°C a thin oxide film forms on the specimens, which at 1150°C thickens rapidly, and for TiAl is easily removed in layers; for TiAl$_3$ the oxide film remains on the specimen, but inadequately protects it from further oxidation.

[*2]V. S. Neshpor, Investigation of the Conditions for Obtaining, and Some Physical Properties of, Transition Metal Silicides [in Russian], Author's abstract, Cand. Dissertation, Kiev (1961).

[*3]Specimen destroyed.

Phase	Temp., °C	Oxidation time, h	Wt. change, mg/cm^2	Source	Year
1	2	3	4	5	6
TiSi$_2$	500—700	—	—*1	[709]	1974
	800	1	~1.5		
		2—6	~2		
	900	1	~2		
		2	~2.5		
		3	~3		
		4	~3.5		
		5	~4		
		6	~4.5		
	1000	1	~2.5		
		2	~3		
		3	~4		
		4—6	~5		
	1100	0,5	~11		
		1	~13		
		2—6	~14		
	1200	0,5	~14		
		1	~16		
		2	~18		
		3	~20		
		4	~21.5		
		5—6	~23		
Zr$_2$Si	1100	4	+29.0	[260]	1959
Zr$_5$Si$_3$	1100	4	+26.9	[260]	1959
ZrSi	1100	4	+3,0	[260]	1959
	1200		+42		
ZrSi$_2$	500	Not oxidized		[709]	1974
	600	1—6	~—0,5		
	700	1—6	~—1.0		
	900	0.5	~1.5		
	900	1—6	~3		
	1000	1	~1.5		
		2	~2		
		3	~2.5		
		4	~3.5		
		5	~4		
		6	~4,5		

*1Oxidation rate insignificant.

Phase	Temp., °C	Oxidation time, h	Wt. change, mg/cm²	Source	Year
1	2	3	4	5	6
ZrSi₂	1100	1 2 3 4 5 6	~2.5 ~3.5 ~4 ~4.5 ~5 ~5.5	[709]	1974
	1200	1 2 3 4 5 6	~1 ~2.5 ~6.5 ~7.5 ~8 ~8.5		
HfSi₂	500—600	Not oxidized		[709]	1974
	700	1—6	~7		
	800	1 2 3—6	~10 ~15 ~20		
	900	1 2 3	~15 ~20 ~25		
	900	4 5 6	~30 ~33 ~35		
	1000	1 2 3 4 5 6	~20 ~30 ~35 ~40 ~45 ~50		
	1100	1 2 3 4 5 6	~37.5 ~55 ~70 ~75 ~80 ~85		
	1200	1 2 3 4 5 6	~95 ~120 ~125 ~135 ~142.5 ~150		
V₃Si	1250	1	—63	[712]	1956
V₅Si₃	1250 1400	1	—7.7 —600	[712]	1956

Phase	Temp., °C	Oxidation time, h	Wt. change, mg/cm^2	Source	Year
1	2	3	4	5	6
VSi$_2$	500	1—6	$\sim$—0.5	[71]	1974
	600	1—6	$\sim$—0.75		
	700	1—6	$\sim$—1.0		
	800	1	$\sim$—1.5		
		2—6	$\sim$—1.75		
	900	1—2	$\sim$—0.5		
		3	$\sim$+0.25		
		4—6	$\sim$+0.5		
	1000	0.5	$\sim$1.3		
		1.5	$\sim$1.5		
		3	$\sim$1.75		
		4—6	$\sim$2		
	1100	0.5	$\sim$1.5		
		1	$\sim$1.75		
		2	$\sim$2		
		3	$\sim$2.25		
		4—6	$\sim$2.5		
	1200	0.5	$\sim$0		
		1	$\sim$1.3		
		2	$\sim$3.0		
		3	$\sim$3.5		
		4	$\sim$3.75		
		5	$\sim$4		
		6	$\sim$4.25		
NbSi$_2$	500	2—6	$\sim$40	[710]	1974
	600	1	$\sim$50		
		2	$\sim$70		
		3	$\sim$90		
		4	$\sim$100		
		5	$\sim$120		
	700	1	$\sim$80		
		2	$\sim$130		
		3	$\sim$180		
		4	$\sim$230		
		5	$\sim$280		
		6	$\sim$320		
	800	1	$\sim$180		
		2	$\sim$330		
		3	$\sim$430		
		4	$\sim$520		
		5	$\sim$600		
		6	$\sim$650		
	900	1	$\sim$290		
		2	$\sim$350		
		3—6	$\sim$380		

Phase	Temp., °C	Oxidation time, h	Wt. change, mg/cm^2	Source	Year
1	2	3	4	5	6
NbSi$_2$	1000	1	$\sim$70	[710]	1974
		2	$\sim$90		
		3	$\sim$100		
		4—6	$\sim$110		
	1100	6	$\sim$25		
	1200	6	$\sim$40		
Ta$_{4.5}$Si	1500	1	$+240$	[260]	1959
Ta$_5$Si$_3$	1500	1	$+125$	[260]	1959
TaSi$_2$	500	1—6	$\sim$5	[710]	1974
	600	1—6	$\sim$10		
	700	1	$\sim$20		
		2	$\sim$30		
		3	$\sim$35		
		4—6	$\sim$40		
	800	1	$\sim$60		
		2	$\sim$100		
		3	$\sim$110		
		4—6	$\sim$115		
	900	1	$\sim$100		
		2	$\sim$150		
		3—6	$\sim$180		
	1000	1	$\sim$185		
		2—6	$\sim$200		
	1100	1—2	$\sim$10		
		3	$\sim$15		
		4	$\sim$25		
		5	$\sim$50		
		6	$\sim$75		
	1200	1—2	$\sim$15		
		3	$\sim$20		
		4	$\sim$60		
		5	$\sim$105		
		6	$\sim$110		
Cr$_3$Si	1260	100	$+7,3$	[260]	1959
	1300	4	$+11.4$		
Cr$_5$Si$_3$	1300	4	$+12.5$	[260]	1959

Phase	Temp., °C	Oxidation time, h	Wt. change, mg/cm^2	Source	Year
1	2	3	4	5	6
CrSi$_2$	700 900 1000 1100 1200	3	0.40 * 0.77 1.03 2.25 3.56	[711]	1974
Mo$_3$Si	1500	4	—812	[260]	1959
Mo$_5$Si$_3$	1500	4	—67	[260]	1959
MoSi$_2$	1100	20	+1.4	[693]	1960
MoSi$_2$	1150	8 16 24 32 48 200	+2 +4 +5 +5 +6 +8	[260]	1959
MoSi$_2$	1200	4 20 50 75 100	+0.3 +0.6 +1.9 +2.1 +2.1	[696]	1960
MoSi$_2$	1500	4	+1.3	[260]	1959
W$_3$Si	1500	4	—445	[260]	1959
W$_5$Si$_3$	1500	4	—205	[260]	1959
WSi$_2$	1200 1500	4	—17 —23	[260]	1959
MnSi$_2$	1200	0.5 1.0 2.0 4.0 6.0	+2.5 +3.6 +5.4 +6.8 +7.5	—**	1961

*As the oxygen pressure rises from 0.1 to 7.6 torr, the oxidation rate rises, and with a further rise to 760 torr, it falls slightly.

**V. S. Neshpor, see footnote on p. 380.

Phase	Temp., °C	Oxidation time, h	Wt. change, mg/cm^2	Source	Year
1	2	3	4	5	6
ReSi$_2$	1400	0,5 1,0 3,0 4,3 6,0 8,3	+3,2 +7,15 +7,3 +7,2 +7,2 +7,3	—*	1961
FeSi$_2$	1200	0,5 1,0 2,0 4,0 6,0	—0,5 +0,4 +1,8 +2,7 +3,3	—*	1961
CeP B$_4$C	500 ** 1100	— 20	— —0,8	[713] [693]	1970 1960
B$_4$C	1200	5 25 50 100	—1,11 —3,88 —8,1 —11,3	[696]	1960
SiC	1400	50 100 200 500 1200	—5,2% +8,2% +9,2% +16,1% +20,7%	[697]	1968
BN	700 1000	2 10 30 60 2 10 30 60	—0,014 —0,062 —0,138 —0,235 —0,35 —0,85 —4,8 —10.0	[14]	1969
Si$_3$N$_4$ AlN	1200 800 *	80 —	+5 —	[14] [544]	1969 1962

*V. S. Neshpor, see footnote on p. 380.

**Oxidation commences in air.

Phase	Temp., °C	Chlorination time, h	Wt. change, mg/cm^2	Source	Year	Notes
1	2	3	4	5	6	7
TiC	400—700	—	—	[256]	1957	Powder easily chlorinated to form chlorides, chlorates, and sulfochlorides
ZrC, NbC, TaC	700—800	—	—	[256]	1957	Powder easily decomposes
VC	300—500	—	—	[256]	1957	—
Cr$_3$C$_2$	To 900—1000	—	—	[256]	1957	Resistant powder
Mo$_2$C	1000—1200	—	—	[256]	1957	Powder decomposes
MoC	900—1000	—	—	[256]	1957	Powder decomposes to form MoCl$_6$ and C
W$_2$C	400	—	—	[256]	1957	Powder decomposes to form MoCl$_6$ and C
WC	To 500—700	—	—	[256]	1957	Stable
TiSi$_2$, ZrSi$_2$, HfSi$_2$	900	—	—	[256]	1957	Decomposes
MoSi$_2$	1000	2	—1500	[260]	1959	Compact
WSi$_2$	200—300	—	—	[260]	1959	Decomposes to form WCl$_6$

Phase	Temp., °C	Chlori-nation time, h	Wt. change, mg/cm²	Source	Year	Notes
1	2	3	4	5	6	7
SiC *	200	6	0.0	[611]	1961	
	200	24	8,4			
	200	48	0,9			
	200	72	0,6			
	200	120	0,3			
	200	192	0,3			
	200 **	6	0,0			
	200 **	24	0,0			
	200 **	48	—1,2			
	200 **	120	—0,3			
	300	6	13,5			
	300	24	3,9			
	300	48	1,2			
	300	120	1,8			
	300	192	0,9			
	300	264	0,9			
	300	336	0,9			
	400	6	21,7			
	400	24	8,7			
	400	48	4,5			
	400	120	2,7			
	400	192	2,1			
	400	264	2,1			
	400	336	2,7			
	400 **	48	7,2			
	400 **	120	4,2			
	400 **	192	3,0			
	400 **	264	1.8			
	500	6	24			
	500	24	3 040			
	500	72	1 142			
	500	144	580			
	500	216	387			
	500	288	290			
	500	360	290			
	500	432	193			
	600	6	4 840			
	600	24	4 444			
	600	120	1 080			
	600	192	700			

*For all determinations of the chlorine resistance of SiC, instead of weight change we quote the destruction (corrosion) rate, mm/min, determined during chlorination as indicated in the table. The resistance was determined on specimens of the following composition, %: 96.5 SiC; 2.5 Si_{free}; 0.4 C_{free}; 0.4 Al; 0.2 Fe. P = 4%.

**The resistance was measured in moist chlorine.

Phase	Temp., °C	Chlorination time, h	Wt. change, mg/cm²	Source	Year	Notes
1	2	3	4	5	6	7
SiC *	600 600 600 600 600 ** 600 ** 600 ** 600 ** 800 1000	264 336 408 480 48 120 192 264 6 6	533 433 378 340 322 133 85 63 22 600 32 300	[611]	1961	
B₄C	<1000	—	—	[254]	1960	Reacts to form BCl₃ and C
BN	700 700 700 1000 1000	3 20 40 3 20	— —0.25 —0.55 —2.7 —17.0	[254]	1960	Compact
Si₃N₄	Up to 900	500	—	[14]	1969	Compact, stable
Si₃N₄	350—240	2	—0.7			Powder decomposes
AlN	Up to 900	—	—	[256]	1957	Reacts to form AlCl₃

**See footnote on p. 388.

Chapter VII

REFRACTORY PROPERTIES

WETTABILITY BY MOLTEN METALS

Phase	Wetting metal	Temp., °C	Medium	Contact angle, θ°	Source	Year
1	2	3	4	5	6	7
TiB$_2$	Cu	1100	Argon	143	[10]	1973
	Cu	1200	"	137		
	Cu	1120	Vacuum	142		
	Al	900—1100	Argon	98		
	Al	1200	"	60		
	Al	1000	Vacuum	114		
	Ga	800	"	115		
	In	500	Argon	124		
	Ge	1000	Vacuum	127		
	Ge	1100	Argon	70		
	Sn	250—600	"	114		
	Sn	350	Vacuum	132		
	Pb	350—800	Argon	106		
	Pb	450	Vacuum	130		
	Bi	320	Argon	141		
	Bi	370	Vacuum	134		
	Fe	1450—1550	Argon	100		
	Fe	Tmelt	"	118		
	Co	1500—1600	"	100—64		
	Co	Tmelt	"	25		
	Ni	1480—1600	"	83—72		
	Ni	Tmelt	"	64		
	Ni	1480	Helium	39		
	Ni	1500	Vacuum	0		
	Cd	450	"	122		
	Si	1450	Argon	34		
	Mn	Tmelt	"	75		
ZrB$_2$	Cu	1100	Argon	135	[10]	1973
	Cu	1400	"	36		
	Al	900—1100	"	106		
	Al	1200	"	103		
	Al	1000	Vacuum	107		
	Ga	800	"	117		
	In	500	Argon	114		
	Ge	1000	"	102		
	Sn	250—600	"	110		
	Sn	350	Vacuum	127		
	Fe	1450—1550	Argon	105		
	Fe	Tmelt	"	122		

WETTABILITY BY MOLTEN METALS (continued)

Phase	Wetting metal	Temp., °C	Medium	Contact angle, θ°	Source	Year
1	2	3	4	5	6	7
ZrB_2	Co	1500—1600	Argon	88—81	[10]	1973
	Co	1500	Vacuum	39		
	Ni	1480—1600	Argon	89—78		
	Ni	1500	Vacuum	55		
	Cd	450	"	136		
	Si	T melt	Argon	44		
	Pb	350	Vacuum	110		
	Bi	370	"	127		
	Mn	T melt	Argon	56		
HfB_2	Al	900—1100	Argon	134	[10]	1973
	Al	1200	"	119		
	In	500	Vacuum	114		
	Ge	1000	Argon	141		
	Fe	1450—1550	"	100		
	Co	1500—1600	"	54—10		
	Ni	1480—1600	"	102—99		
VB_2	Ge	1000	Argon	60	[10]	1973
	Co	1500—1600	"	54—10		
	Cu	1100—1400	"	150—114		
NbB_2	Cu	1100	Argon	109	[10]	1973
	Cu	1200	"	107		
	Al	900—1100	"	125		
	Al	1200	"	25		
	Ga	400	"	103		
	In	500	Vacuum	133		
	Ge	1000	Argon	66		
	Sn	250—600	"	102		
	Pb	350—800	"	125		
	Bi	320	"	110		
	Fe	1450—1550	"	30—0		
	Co	1500—1600	"	46—22		
	Ni	1480—1600	"	24		
TaB_2	Cu	1100—1400	Argon	77—47	[10]	1973
	Ag	1500	"	118		
	Al	900—1100	"	138		
	Al	1200	"	73		
	In	500	"	117		
	Fe	1450—1550	"	0		
	Co	1500—1600	"	26—0		
	Ni	1500	"	43		

WETTABILITY BY MOLTEN METALS (continued)

Phase	Wetting metal	Temp., °C	Medium	Contact angle, θ°	Source	Year
1	2	3	4	5	6	7
CrB_2	Cu	1100	Argon	26	[10]	1973
	Cu	1200	"	15		
	Al	900—1100	"	96		
	Al	1200	"	36		
	Al	1100	Vacuum	58		
	Ga	400	Argon	128		
	In	500	"	78		
	Ge	1000	"	126		
	Ge	1100	"	77		
	Sn	250—600	"	100		
	Sn	350	Vacuum	147		
	Pb	350—800	Argon	124		
	Pb	450	Vacuum	135		
	Fe	1450—1550	Argon	0		
	Fe	Tmelt	"	25		
	Co	1500—1600	"	28—0		
	Co	Tmelt	"	76		
	Ni	1480—1600	"	23—31		
	Ni	Tmelt	"	40		
	Cd	450	Vacuum	112		
	Si	Tmelt	Argon	44		
	Bi	370	Vacuum	135		
	Mn	Tmelt	Argon	0		
Mo_2B_5	Cu	1100	Argon	104	[10]	1973
	Cu	1100	Vacuum	127		
	Al	900—1100	Argon	134		
	Al	1200	"	88		
	Al	1000	Vacuum	—*		
	Ge	1000	Argon	28		
	Sn	250—600	"	100		
	Sn	350	Vacuum	135		
	Fe	1450—1550	Argon	0		
	Co	1500—1600	"	46—22		
	Co	Tmelt	"	85		
	Ni	1480—1600	"	0		
	Ni	Tmelt	"	75		
	Cd	450	Vacuum	140		
	Si	Tmelt	Argon	60		
	Pb	450	Vacuum	120		
	Bi	370	"	133		
	Mn	Tmelt	Argon	0		
W_2B_5	Cu	1100	Argon	101	[10]	1973
	Cu	1200	"	93		
	Cu	1100	Vacuum	60		
	Al	1000	"	—*		

*Reacts.

WETTABILITY BY MOLTEN METALS (continued)

Phase	Wetting metal	Temp., °C	Medium	Contact angle, $\theta°$	Source	Year
1	2	3	4	5	6	7
W_2B_5	Si	T_{melt}	Argon	75	[10]	1973
	In	500	"	130		
	Ge	1000	"	128		
	Sn	250—600	"	100		
	Sn	350	Vacuum	105		
	Bi	320	Argon	128		
	Bi	370	Vacuum	115		
	Pb	450	"	102		
	Fe	1450—1550	Argon	40—0		
	Fe	T_{melt}	"	—*		
	Co	T_{melt}	"	97		
	Co	1500—1600	"	128—94		
	Ni	T_{melt}	"	84		
	Cd	450	Vacuum	138		
	Mn	T_{melt}	Argon	22		
Be_2C	Si	1450	Hydrogen	54	[687]	1954
	Si	1450	Helium	63		
	Ni	1500	Vacuum	92		
	Ni	1500	Hydrogen	90		
	Ni	1500	Helium	75		
$TiC_{0.99}$	Cu	1100	Vacuum	$\sim$105	[688]	1971
$TiC_{0.88}$				$\sim$100		
$TiC_{0.75}$				$\sim$90		
$TiC_{0.6}$				$\sim$50		
TiC	Ag	980	Vacuum	108	[10]	1973
	Zn	620	Argon	120	[10]	1973
	Al	1000	Vacuum	149	[10]	1973
	Al	700	Argon	118	[10]	1973
	Ga	800	Vacuum	147	[10]	1973
	In	250	"	145	[10]	1973
	Tl	400	"	127	[10]	1973
	Si	1500	"	32	[10]	1973
	Si	T_{melt}	Argon	68	[10]	1973
	Ge	1000	Vacuum	133	[10]	1973
	Sn	300	"	148	[10]	1973
	Pb	400	"	143	[10]	1973
	Pb	660	"	120	[10]	1973
	Sb	700	"	145	[10]	1973
	Bi	320	"	145	[10]	1973
	Mg	1300	"	50	[10]	1973
	Mg	1300	Argon	15	[10]	1973
	Mn	T_{melt}	"	68	[10]	1973

*Reacts.

Phase	Wetting metal	Temp., °C	Medium	Contact angle, θ°	Source	Year
1	2	3	4	5	6	7
TiC	Fe	1550	Vacuum	41	[687]	1954
	Fe	1550	Hydrogen	39	[687]	1954
	Fe	1550	Helium	36	[687]	1954
	Fe	1550	Argon	125	[10]	1973
	Co	1500	Hydrogen	36	[687]	1954
	Co	1500	Helium	39	[687]	1954
	Co	1500	Vacuum	5	[687]	1954
	Co	1450	"	30	[10]	1973
	Co	1450	Argon	16	[10]	1973
	Ni	1450	Vacuum	30	[687]	1954
	Ni	1500	"	38	[689]	1962
	Ni	2200	"	20	[690]	1963
	Ni	1450	Argon	25	[10]	1973
	Ni	1500	"	0	[10]	1973
	Ni	1450	Hydrogen	17	[687]	1954
	Ni	1450	Helium	32	[687]	1954
$ZrC_{1.0}$	Cu	1100	Vacuum	$\sim$140	[688]	1971
$ZrC_{0.88}$				$\sim$115		
$ZrC_{0.72}$				$\sim$80		
$ZrC_{0.6}$				$\sim$40		
ZrC	Cu	1100—1500	Argon	140—118	[10]	1973
	Au	—	Vacuum	151	[688]	1971
	Al	900—1000	"	150	[10]	1973
	Ga	800	"	134	[10]	1973
	In	250	"	143	[10]	1973
	Tl	400	Vacuum	128	[10]	1973
	Si	1500	"	22	[10]	1973
	Si	T melt	Argon	57	[10]	1973
	Ge	1000	Vacuum	135	[10]	1973
	Sn	300	"	150	[10]	1973
	Pb	400	"	147	[10]	1973
	Sb	700	"	110	[10]	1973
	Bi	320	"	141	[10]	1973
	Mn	T melt	Argon	70	[10]	1973
	Fe	1490	Vacuum	45	[421]	1965
	Fe	1550	"	49	[10]	1973
	Fe	1550	Argon	140	[10]	1973
	Fe	T melt	"	50	[10]	1973
	Co	1420	Vacuum	36	[421]	1965
	Co	1500	Argon	15	[10]	1973
	Co	T melt	"	86	[10]	1973
	Ni	1380	Vacuum	24	[421]	1965
	Ni	1450	Argon	32	[10]	1973
	Ni	T melt	"	43	[10]	1973

Phase	Wetting metal	Temp., °C	Medium	Contact angle, θ°	Source	Year
1	2	3	4	5	6	7
HfC	Cu	1100—1200	Vacuum	134—131,5	[421]	1965
	Al	900—1000	"	148—146	[10]	1973
	Ga	800	"	147	[10]	1973
	In	250	"	147	[10]	1973
	Tl	400	"	132	[10]	1973
	Si	1500	"	23	[10]	1973
	Ge	1000	"	140	[10]	1973
	Sn	300	"	153	[10]	1973
	Pb	400	"	150	[10]	1973
	Sb	700	"	130	[10]	1973
	Bi	320	"	148	[10]	1973
	Fe	1490	"	45	[421]	1965
	Fe	1550	"	52	[10]	1973
	Fe	1550	Argon	148	[10]	1973
	Co	1420	Vacuum	40	[421]	1965
	Co	1550	"	36	[10]	1973
	Co	1550	Argon	40	[10]	1973
	Ni	1380	Vacuum	23	[421]	1965
	Ni	1450	"	28	[10]	1973
	Ni	1450	Argon	28	[10]	1973
$VC_{0,88}$ $VC_{0,83}$	Cu	1100—1200	Vacuum	50—40 38—25	[421]	1965
VC	Na	200—500	Argon	181—157	[10]	1973
	Al	900	Vacuum	130	[10]	1973
	Ga	800	"	120	[10]	1973
	In	250	"	119	[10]	1973
	Tl	400	"	111	[10]	1973
	Si	1500	"	0	[10]	1973
	Ge	1000	"	121	[10]	1973
	Sn	300	"	130	[10]	1973
	Pb	400	"	130	[10]	1973
	Sb	700	"	113	[10]	1973
	Bi	320	"	103	[10]	1973
	Fe	1490	"	20	[421]	1965
	Fe	1550	"	0	[10]	1973
	Fe	1550	Argon	0	[10]	1973
	Co	1420	Vacuum	13	[421]	1965
	Co	1500	"	0	[10]	1973
	Co	1500	Argon	0	[10]	1973
	Ni	1380	Vacuum	17	[421]	1965
	Ni	1450	"	0	[10]	1973
	Ni	1450	Argon	0	[10]	1973
$NbC_{1.0}$ $NbC_{0.82}$ $NbC_{0.75}$ $NbC_{0.66}$	Cu	1100	Vacuum	~40 ~85 ~90 ~70	[688]	1971

Phase	Wetting metal	Temp., °C	Medium	Contact angle, θ°	Source	Year
1	2	3	4	5	6	7
NbC	Cu	1100—1200	Vacuum	70—48	[421]	1965
	Au	—	"	60	[688]	1971
	Al	900—1000	"	136—134	[10]	1973
	Ga	800	"	108	[10]	1973
	In	250	"	151	[10]	1973
	Tl	400	"	121	[10]	1973
	Si	1500	"	0	[10]	1973
	Ge	1000	"	148	[10]	1973
	Sn	300	"	130	[10]	1973
	Pb	400	"	148	[10]	1973
	Sb	700	"	111	[10]	1973
	Bi	320	"	135	[10]	1973
	Fe	1490	"	25	[421]	1965
	Fe	1550	"	0	[10]	1973
	Fe	1550	Argon	0	[10]	1973
	Co	1420	Vacuum	14	[421]	1965
	Co	1500	"	0	[10]	1973
	Co	1500	Argon	0	[10]	1973
	Ni	1380	Vacuum	18	[421]	1965
	Ni	1450	"	0	[10]	1973
	Ni	1450	Argon	0	[10]	1973
$TaC_{1.0}$ $TaC_{0.9}$ $TaC_{0.78}$	Cu	1100	Vacuum	$\sim$60 $\sim$80 $\sim$100	[688]	1971
TaC	Cu	1100—1200	Vacuum	78—60	[421]	1965
	Al	900—1000	"	145—146	[10]	1973
	Ga	800	"	130	[10]	1973
	In	250	"	154	[10]	1973
	Tl	400	"	138	[10]	1973
	Si	1500	"	0	[10]	1973
	Ge	1000	"	128	[10]	1973
	Sn	300	"	140	[10]	1973
	Pb	400	"	130	[10]	1973
	Sb	700	"	119	[10]	1973
	Bi	320	"	135	[10]	1973
	Fe	1490	"	23	[421]	1965
	Fe	1550	"	0	[10]	1973
	Fe	1550	Argon	0	[10]	1973
	Co	1420	Vacuum	13	[421]	1965
	Co	1500	"	0	[10]	1973
	Co	1500	Argon	0	[10]	1973
	Ni	1380	Vacuum	16	[421]	1965
	Ni	1400	"	0	[10]	1973
	Ni	1400	Argon	0	[10]	1973
Cr_3C_2	Cu	1100—1200	Vacuum	47—44	[421]	1965
	Al	900—1000	"	120—121	[10]	1973
	Ga	800	"	120	[10]	1973

WETTABILITY BY MOLTEN METALS (continued)

Phase	Wetting metal	Temp., °C	Medium	Contact angle, $\theta°$	Source	Year
1	2	3	4	5	6	7
Cr_3C_2	In	250	Vacuum	143	[10]	1973
	Tl	400	"	133	[10]	1973
	Si	1500	"	0	[10]	1973
	Si	Tmelt	Argon	0	[10]	1973
	Ge	1000	Vacuum	121	[10]	1973
	Sn	300	"	120	[10]	1973
	Pb	400	"	124	[10]	1973
	Sb	700	"	106	[10]	1973
	Bi	320	"	102	[10]	1973
	Mn	Tmelt	Argon	0	[10]	1973
	Fe	1490	Vacuum	~0	[421]	1965
	Fe	1550	Argon	0	[10]	1973
	Co	1420	Vacuum	~0	[421]	1965
	Co	1450	Argon	0	[10]	1973
	Ni	1380	Vacuum	~0	[421]	1965
	Ni	1400	Argon	0	[10]	1973
Mo_2C	Cu	1100—1200	Vacuum	18—~0	[421]	1965
	Al	900—1000	"	118	[10]	1973
	Ga	800	"	118	[10]	1973
	In	250	"	150	[10]	1973
	Tl	400	"	130	[10]	1973
	Si	1500	"	0	[10]	1973
	Si	Tmelt	Argon	0	[10]	1973
	Ge	1000	Vacuum	76	[10]	1973
	Sn	300	"	140	[10]	1973
	Pb	400	"	141	[10]	1973
	Sb	700	"	141	[10]	1973
	Bi	320	"	105	[10]	1973
	Mn	Tmelt	Argon	0	[10]	1973
	Fe	1490	Vacuum	~0	[421]	1965
	Fe	1550	Argon	0	[10]	1973
	Co	1420	Vacuum	~0	[421]	1965
	Co	1450	Argon	0	[10]	1973
	Ni	1380	Vacuum	~0	[421]	1965
	Ni	1400	Argon	0	[10]	1973
WC	Cu	1100—1200	Vacuum	20—7	[421]	1965
	Al	900—1000	"	135—133	[10]	1973
	Ga	800	"	122	[10]	1973
	In	250	"	148	[10]	1973
	Si	1500	"	0	[10]	1973
	Ge	1000	"	63	[10]	1973
	Sn	300	"	141	[10]	1973
	Pb	400	"	145	[10]	1973
	Sb	700	"	98	[10]	1973
	Bi	320	"	144	[10]	1973
	Fe	1490	"	~0	[421]	1965
	Fe	1550	Argon	0	[10]	1973

Phase	Wetting metal	Temp., °C	Medium	Contact angle, θ°	Source	Year
1	2	3	4	5	6	7
WC	Co	1420	Vacuum	~0	[421]	1965
	Co	1500	Argon	0	[10]	1973
	Ni	1380	Vacuum	~0	[421]	1965
	Ni	1450	Argon	0	[10]	1973
TiN	Cu	1100	Vacuum	180	[10]	1973
	Cu	1180	"	126	[691]	1966
	Cu	1560	Ammonia	148	[10]	1973
	Cu	1130	Argon	136	[10]	1973
	Al	850—1000	Vacuum	147	[10]	1973
	Al	900	Argon	135	[10]	1973
	Cd	450	Vacuum	139	[10]	1973
	Si	$\sim T_{melt}$	"	102	[10]	1973
	Pb	450	"	102	[10]	1973
	Sn	350	"	140	[10]	1973
	Bi	370	Argon	147	[10]	1973
	Mn	$\sim T_{melt}$	"	74	[10]	1973
	Fe	1550	Vacuum	~100	[10]	1973
	Fe	$\sim T_{melt}$	Ammonia	91	[10]	1973
	Fe	1550	Argon	132	[10]	1973
	Co	1550	Vacuum	104	[10]	1973
	Co	$\sim T_{melt}$	Argon	84	[10]	1973
	Ni	1550	Vacuum	~70	[10]	1973
	Ni	$\sim T_{melt}$	Argon	113	[10]	1973
	Ni	1450—1500	Nitrogen	110	[692]	1974
ZrN	Cu	1100	Vacuum	138	[10]	1973
	Cu	1180	"	148	[691]	1966
	Cu	1500	Ammonia	146	[10]	1973
	Au	—	Vacuum	140	[688]	1971
	Al	850—1000	"	160	[10]	1973
	Al	900	Argon	167	[10]	1973
	Cd	450	Vacuum	120	[10]	1973
	Si	$\sim T_{melt}$	Argon	45	[10]	1973
	Sn	350	Vacuum	130	[10]	1973
	Pb	450	"	105	[10]	1973
	Bi	370	"	120	[10]	1973
	Mn	$\sim T_{melt}$	Argon	76	[10]	1973
	Fe	1550	Vacuum	110	[10]	1973
	Fe	1550	Ammonia	140	[10]	1973
	Co	1500—1600	Vacuum	7	[691]	1966
	Ni	1550	"	72	[10]	1973
NbN	Cu	1180	Vacuum	130	[691]	1966
	Cu	1500	Ammonia	150	[10]	1973
	Al	900	Vacuum	156	[691]	1966
	Fe Co Ni	1500—1600	"	—*	[691]	1966

*Reacts.

Phase	Wetting metal	Temp., °C	Medium	Contact angle, $\theta°$	Source	Year
1	2	3	4	5	6	7
Cr_2N	Cu	1180	Vacuum	36	[691]	1966
	Al	900	"	—[1]		
	Fe ⎫					
	Co ⎬	1500—1600	"	—[1]		
	Ni ⎭					
CrN	Cu	1180	Vacuum	—[2]	[691]	1966
	Al	900	"	149		
B_4C	Zn	540—620	Vacuum	121,5—119	[254]	1960
	Cu	995—1090	"	130—17		
	Al	600—670	"	117—118		
	Pb	225—395	"	121—113,5		
	Fe	1780	Helium	—[1]		
	Co	1780	"	>90		
	Ni	1780	"	>90		
BN	Cu	1100	Vacuum	146	[10]	1973
	Al	850—1000	"	142		
	Sn	250	"	150		
	Cd	450	"	130		
	Si	$\sim T$melt	Argon	126		
	Pb	450	Vacuum	145		
	Bi	370	"	134		
	Mn	$\sim T$melt	Argon	119		
	Fe	$\sim T$melt	"	112		
	Co	$\sim T$melt	"	118		
	Ni	1100	"	134		
AlN	Cu	1100	Vacuum	100	[10]	1973
	Cu	1180	"	155	[691]	1966
	Al	850—1000	"	138	[10]	1973
	Sn	250	"	100	[10]	1973
	Bi	250	"	98	[10]	1973
	Fe ⎫					
	Co ⎬	1500—1600	"	—[3]	[691]	1966
	Ni ⎭					

[1]Reacts.
[2]Diffusion of Cu into CrN.
[3]Does not wet.

WETTABILITY BY MOLTEN METALS (continued)

Phase	Wetting metal	Temp., °C	Medium	Contact angle, θ°	Source	Year
1	2	3	4	5	6	7
Si_3N_4	Cu	1100	Vacuum	60	[10]	1973
	Cd	450	"	129		
	Si	T_{melt}	"	134		
	Sn	250	"	~135		
	Sn	450	"	126		
	Pb	450	"	105		
	Bi	370	Argon	130		
	Mn	~T_{melt}	"	74		
	Fe	~T_{melt}	"	90		
	Co	~T_{melt}	"	90		
	Ni	~T_{melt}	"	90		
	Ni	1550	Vacuum	120		

RESISTANCE TO MOLTEN SALTS, ALKALIS, AND OXIDES

Phase	Fusion composition	Temp., °C	Type of reaction	Source	Year	Notes
1	2	3	4	5	6	7
MgB_2	Na_2CO_3 NaOH	800 550	Decomposes	[254]	1960	
CaB_6	KOH	800	Decomposes	[254]	1960	
SrB_6 and other alkaline-earth borides	K_2CO_3 $KHSO_4$ PbO_2 KNO_3	800	Decomposes	[254]	1960	
TiB_2	NaOH Na_2CO_3 $KHSO_4$ PbO_2 Na_2O_2	550 800 200—300 900 750	Decomposes " " Reacts vigorously Same	[641]	1954	
ZrB_2	NaOH $KHSO_4$ PbO_2 Na_2O_2	550 800 300 750	Decomposes	[641]	1954	

Phase	Fusion composition	Temp., °C	Type of reaction	Source	Year	Notes
1	2	3	4	5	6	7
VB_2	NaOH K_2CO_3 Na_2O_2	550 800—900 750	Decomposes	[641]	1954	Instantly
NbB_2	NaOH Na_2CO_3	550 800	Decomposes	[641]	1954	
TaB_2	NaOH Na_2CO_3 $KHSO_4$ Na_2O_2	550 800 200—300 750	Decomposes	[641]	1954	Very rapidly
CrB_2	NaOH	550	Decomposes	[641]	1954	
MoB_2	NaOH	550	Decomposes	[641]	1954	
WB_2	Na_2CO_3 KNO_3	800 350	Decomposes	[641]	1954	
Ni_3B	NaOH + Na_2O_2	650—700	Decomposes	[641]	1954	
UB_2	NaOH	550	Decomposes	[641]	1954	
UB_4, UB_{12}	PbO_2 Na_2O_2	900 750	Decomposes	[641]	1954	
TiC	NaCl + $CaCl_2$ + Cl + Na NaCl + $BaCl_2$	700 1100	Reacts actively Reacts weakly	[640]	1965	Rapid solution —
ZrC	NaCl + $BaCl_2$	1100	Reacts weakly	[640]	1965	
Cr_3C_2	NaCl + $CaCl_2$ + Cl + Na NaCl + $BaCl_2$	700 1100	Reacts actively Does not react	[640]	1965	Rapid solution —

Phase	Fusion composition	Temp., °C	Type of reaction	Source	Year	Notes
1	2	3	4	5	6	7
WC	NaCl + + BaCl$_2$	1100	Reacts weakly	[640]	1965	
AlN	KOH	400	Decomposes	[14]	1969	
TiN	NaOH	650	Decomposes	[14]	1969	
ZrN	NaOH	650	Decomposes	[14]	1969	Reacts slowly, gives off NH$_3$
VN	NaOH	650	Decomposes	[14]	1969	
Nb$_2$N	NaOH	650	Decomposes	[14]	1969	
NbN	NaOH	350	Decomposes	[14]	1969	Gives off N$_2$
WN	Na$_2$CO$_3$	800	Decomposes	[256]	1957	Gives off NH$_3$, forms Na$_2$WO$_4$
LaSi$_2$	NaOH	650—700	Decomposes	[642]	1954	In 20 min
ThSi$_2$	NaOH	550	Decomposes	[642]	1954	Very easily
TiSi$_2$, ZrSi$_2$, HfSi$_2$	NaOH	650—700	Decomposes	[642]	1954	In 20 min
HfSi$_2$	Na$_2$CO$_3$ Na$_2$B$_4$O$_7$ KHSO$_4$	650—700 650—700 200—300	Decomposes " Resistant	[642]	1954	In 20 min
VSi$_2$, NbSi$_2$, TaSi$_2$	NaOH + + Na$_2$CO$_3$	650—700	Resistant	[642]	1954	
CrSi, Cr$_3$Si$_2$, Cr$_2$Si, CrSi$_2$	K$_2$GO$_3$ + + NaNO$_3$	650—700	Decomposes	[260]	1959	Forms silicates and chromates

RESISTANCE TO MOLTEN SALTS, ALKALIS, AND OXIDES
(continued)

Phase	Fusion composition	Temp., °C	Type of reaction	Source	Year	Notes
1	2	3	4	5	6	7
MoSi$_2$	K$_2$CO$_3$ + + KNO$_3$	650—700	Reacts actively	[642]	1954	
	NaCl + + CaCl$_2$ + + Cl + Na	700		[640]	1965	
	NaOH	400—500	Same	[260]	1959	
	NaCl + + BaCl$_2$	1100	Does not react	[640]	1965	
WSi$_2$	K$_2$CO$_3$ + + KNO$_3$	650—700	Decomposes	[642]	1954	
	NaCl + + CaCl$_2$ + + Cl + Na	700	Reacts actively	[640]	1965	
MnSi, MnSi$_2$	K$_2$CO$_3$ + + KNO$_3$	650—700	Decomposes	[642]	1954	
FeSi, CoSi, NiSi	NaOH	550	Decomposes	[642]	1954	
FeSi	K$_2$CO$_3$	800—900	Decomposes	[642]	1954	
B$_4$C	BaCO$_3$	600—700	Decomposes	[685]	1961	
	Na$_2$CO$_3$ + + NaNO$_3$			[685]	1961	
	NaOH + + NaNO$_3$			[642]	1954	
	CaO, MgO			[685]	1961	
SiC	Na$_2$B$_4$O$_7$	750	Decomposes	[685]	1961	
	Na$_2$CO$_3$	800	"	[686]	1938	
	K$_2$Cr$_2$O$_7$	400	"	[686]	1938	
	PbCrO$_4$	850	"	[686]	1938	
	Na$_2$SO$_4$	900	"	[686]	1938	
	NaOH	550	"	[686]	1938	
	Na$_2$O$_2$	750	"	[686]	1938	
	CuO	800	"	[686]	1938	
	CaO	800	"	[686]	1938	
	MgO	1000	"	[686]	1938	
	SiO$_2$	2000—2500	SiO$_2$ + + SiC = = Si + CO$_2$	[686]	1938	

Phase	Fusion composition	Temp., °C	Type of reaction	Source	Year	Notes
1	2	3	4	5	6	7
SiC	PbO_2	900	Decomposes	[686]	1938	
	Cr_2O_3	1370	"	[686]	1938	
	MnO	1360	"	[686]	1938	
	FeO	1360	"	[686]	1938	
	NiO	1300	"	[686]	1938	
	NaOH	350	1 725 *	[611]	1961	
	NaOH	500	33 100 *	[611]	1961	
	Na_2CO_3	900	>360 *	[611]	1961	
	LiCl	900	1 038 *	[611]	1961	
	NaCl	900	9.3 *	[611]	1961	
	KCl	900	322 *	[611]	1961	
	$MgCl_2$	900	152 *	[611]	1961	
	$CaCl_2$	900	3 000 *	[611]	1961	
	LiF	900	2 360 *	[611]	1961	
BN	Sb_2O_3	—	Reacts	[248]	1969	
	Cr_2O_7	—	Same			
	MoO_3	—	"			
	AsO_3	—	"			
	K_2CO_3	800—900	Decomposes			
Si_3N_4	NaOH	400—500	Decomposes	[248]	1969	
	Na_2O_2	—	Partially decomposes			
	$PbCrO_4$, PbO_2, PbO	—	Decomposes			
	NaCl + + KCl	900	Decomposes after 144 h			
	$NaB (SiO_3)_2$ + V_2O_5	1100	Decomposes after 4 h			
	NaF + + ZrF_4	850	Decomposes after 100 h			

*Corrosion rate, mm/min.

Phase	Melt	Temp., °C	Contact time, h	Medium	Type of reaction	Concentration of component compound entering molten metal, %	Source	Year
1	2	3	4	5	6	7	8	9
TiB_2	Zn	550	80	Air	NR	Ti not detected	[637]	1962
	Zn	940	240	"	NR	—	[637]	1962
	Cd	450	10	"	NR	Ti trace	[621]	1960
	Cd	450	40	"	NR	0.026 Ti	[621]	1960
	Cd	450	80	"	NR	—	[637]	1962
	Al	1000	0,2	Argon	NR	—	[637]	1962
	Si	1550	0.3	"	AR	—	[637]	1962
	Sn	350	10	Air	NR	Ti trace	[621]	1960
	Sn	350	40	"	NR	0,01 Ti	[621]	1960
	Sn	350	80	"	NR	—	[637]	1962
	Pb	450	10	"	NR	Ti trace	[621]	1960
	Pb	450	40	"	NR	0.06 Ti	[621]	1960
	Pb	450	80	"	NR	—	[637]	1962
	Bi	375	10	"	NR	Ti trace	[621]	1960
	Bi	375	40	"	NR	0.05 Ti	[621]	1960
	Bi	375	80	"	NR	—	[637]	1960
	Cr	1900	0,3	$CO + N_2$	AR	—	[637]	1962
	Co	1550	0,3	$CO + N_2$	R	—	[637]	1962
	Ni	1500	0,3	$CO + N_2$	R	—	[637]	1962
	Carbon steel	1600	0.1	$CO + N_2$	R	—	[621]	1960
	Iron	1600	0.1	$CO + N_2$	R	—	[621]	1960
	Cryolite	1050	19.5	—	NR	—	[637]	1962
ZrB_2	Zn	550	80	—	NR	Zr trace	[637]	1962
	Zn	940	180	—	NR	—		
	Cu	450	80	—	NR	Zr residue		
	Al	1000	0.2	Argon	NR	—		
	Si	1550	0.2	»	WR	—		
	Sn	350	80	Air	NR	Zr trace		
	Pb	450	80	—	NR	Zr trace		
	Bi	375	80	—	NR	Zr residue		
	Cr	1900	0.2	$CO + N_2$	WR	—		
	Co	1550	0.2	$CO + N_2$	R	—		
	Ni	1500	0.3	$CO + N_2$	R	—		

*Key: (NR) no reaction; (AR) active reaction; (R) reacts; (WR) weak reaction.

Phase	Melt	Temp., °C	Contact time, h	Medium	Type of reaction	Concentration of component compound entering molten metal, %	Source	Year
1	2	3	4	5	6	7	8	9
ZrB_2	Carbon steel	1620	2	$CO + N_2$	NR	—	[637]	1962
	Iron	1520	12	$CO + N_2$	NR	—		
	Brass	900	86	Air	NR	—		
	Basic slag	1520	12	$CO + N_2$	NR	—		
	Acid slag	1520	12	$CO + N_2$	NR	—		
	Cryolite	1050	20	—	WR	—		
CrB_2	Cd	450	10	Air	NR	Cr trace	[621]	1960
	Cd	450	40	"	NR	< 0.01	[621]	1960
	Cd	450	80	"	NR	Cr trace	[637]	1962
	Al	1000	0.2	Argon	WR	—	[637]	1962
	Si	1550	0.2	"	AR	—	[637]	1962
	Pb	450	10	Air	NR	Cr trace	[621]	1960
	Pb	450	40	"	NR	0.01 Cr	[621]	1960
	Pb	450	80	"	NR	Cr trace	[637]	1962
	Bi	375	80	"	NR	Cr trace	[637]	1962
	Zn	940	132	"	NR	—	[637]	1962
	KhVG steel	1620	0.1	$CO + N_2$	NR	—	[621]	1960
	Iron	1520	0.1	$CO + N_2$	NR	—	[621]	1960
	Basic slag	1520	0.1	$CO + N_2$	R	—	[621]	1960
	Acid slag	1520	—	$CO + N_2$	R	—	[621]	1960
	Cryolite	1050	20	—	WR	—	[637]	1962
W_2B_5	Zn	940	168	—	NR	—	[637]	1962
	Cryolite	1050	8	—	R			
MnB	Zn	600	—	—	NR	—	[638]	1955
NiB	Zn	600	—	—	AR	—	[638]	1955
TiC	Zn	550	10	—	NR	0.02 Ti	[637]	1962
	Cd	450	10	—	NR	0.01 Ti	[621]	1960
	Al	1000	0.1	—	WR	—	[637]	1962
	Si	1500	0.1	—	R	—	[637]	1962

Phase	Melt	Temp., °C	Contact time, h	Medium	Type of reaction	Concentration of component compound entering molten metal, %	Source	Year
1	2	3	4	5	6	7	8	9
TiC	Sn	350	10	Air	NR	—	[621]	1960
	Pb	450	10	"	NR	0,01 Ti	[621]	1960
	Bi	375	10	"	NR	0.018 Ti	[621]	1960
	Co	1550	0,2	CO + N_2	R	—	[637]	1962
	Ni	1500	0,3	CO + N_2	R	—	[637]	1962
	Carbon steel	1620	0.3	Air	R	—	[621]	1960
	Iron	1520	0.3	"	R	—	[621]	1960
	Basic slag	1520	0,1	"	NR	—	[621]	1960
	Acid slag	1520	0,1	"	NR	—	[621]	1960
	Cryolite	1050	8	"	WR	—	[637]	1962
ZrC	Zn	550	6	Argon	NR	0,02 Zr	[637]	1962
	Cd	450	10	Air	NR	0.01 Zr		
	Al	1000	0.2	Argon	R	—		
	Si	1500	0.2	"	R	—		
	Sn	350	10	Air	NR	Zr not detected		
	Pb	450	10	"	NR	Zr trace		
	Bi	375	10	"	NR	Zr trace		
	Cr	1900	0,2	CO + N_2	NR	—		
	Co	1550	0,2	CO + N_2	R	—		
	Ni	1500	0,2	CO + N_2	R	—		
	Cryolite	1050	20	Air	WR	—		
Cr_3C_2	Zn	940	24	Argon	R	—	[637]	1962
Mo_2C	Zn	940	168	—	WR	—	[637]	1962
WC	Zn	940	144	—	WR	—	[637]	1962
	Cryolite	1050	3.5	—	R			
TiN	Cd	450	10	Air	WR	0.20 Ti	[621]	1960
	Cd	450	40	"	WR	0.07 Ti		
	Sn	350	10	"	NR	Ti not detected		
	Sn	350	40	"	WR	0,26 Ti		

Phase	Melt	Temp., °C	Contact time, h	Medium	Type of reaction	Concentration of component compound entering molten metal, %	Source	Year
1	2	3	4	5	6	7	8	9
TiN	Pb	450	10	Air	WR	0,04 Ti	[621]	1960
	Pb	450	40	"	WR	0.20 Ti		
	Bi	375	10	"	NR	Ti trace		
	Carbon steel	1620	0.1	$CO + N_2$	WR	—		
	Iron	1520	0.1	$CO + N_2$	NR	—		
	Basic slag	1520	0.3	$CO + N_2$	NR	—		
	Acid slag	1520	0.1	$CO + N_2$	NR	—		
TiN	Cryolite	1050	36	—	WR	—	[637]	1962
ZrN	Cryolite	1050	8	—	NR	—	[637]	1962
$ThSi_2$	Cu	1130	0.5	—	NR	—	[639]	1959
	Ni	1500	0,5	—	AR	Complex silicide		
$TiSi_2$	Cu	1130	0.5	—	AR	Two-phase structure in cooled melt	[639]	1959
	Ni	1500	0.5	—	AR	Ti, Si in solid solution		
Ti_5Si_3	Cu	1130	1	—	NR	—	[639]	1959
	Ni	1500	1	—	NR	—		
ZrSi	Cu	1130	1	—	NR	—	[639]	1959
	Ni	1500	1	—	NR	—		

Phase	Melt	Temp., °C	Contact time, h	Medium	Type of reaction	Concentration of component compound entering molten metal, %	Source	Year
1	2	3	4	5	6	7	8	9
ZrSi$_2$	Cu	1130	0.5	—	WR	Two-phase structure in cooled melt	[639]	1959
	Ni	1500	0,5	—	AR	Complex silicide with low m.p.		
TaSi$_2$	Ni	1500	0,5	—	AR	Fine new-phase dispersions	[639]	1959
MoSi$_2$	Cu	1130	0.5	—	WR	Mo, Si not detected	[639]	1959
	Na, Cu, Fe, Ag	—	—	Vacuum	NR	—	[637]	1962
	Zn	940	204	"	NR	—	[637]	1962
	Al	1000	5	"	R	—	[637]	1962
	Si	1550	0.1	"	R	—	[637]	1962
	Sn, Pb	1000	5	"	NR	—	[637]	1962
	Ni	1500	0.5	—	AR	Single-phase structure	[639]	1959
WSi$_2$	Ni	1500	0.5	—	AR	New phase	[639]	1959
BaS	Ce	1150	0,1	—	NR	—	[33]	1972
	Ce	1300	0,5		R			
	U	1400	0.2		R			
	U	1500	—		R			
	U	1900	—		R			

Phase	Melt	Temp., °C	Contact time, h	Medium	Type of reaction	Concentration of component compound entering molten metal, %	Source	Year
1	2	3	4	5	6	7	8	9
CeS	Zn	500	0,1	Vacuum	NR	—	[33]	1972
	Zn	700	0,1		NR			
	Mg	900	0,1		NR			
	Al	1500	0,2		WR			
	Ti	1500	0,2		NR			
	Sn	1200	0,1		NR			
	Th	1825	0,1		NR			
	Ce	1500	0,3		NR			
	Bi	1400	0,1		NR			
	Bi	1500	0,2		NR			
	Bi	1400	0,5		NR			
	Pt	1900	0,1		WR			
	Pt	1900	0,2		R			
ThS	Ce	1500	0,3	Vacuum	NR	—	[33]	1972
	Th	1825	0,1					
	Mg	900	0,1					
	Al	1500	0,2					
	Fe	1500	0,2					
Th_2S_3	Ce	1500	0,3	—	WR	—	[33]	1972
	Th	1825	0,1					
Th_4S_7	U	1300	0,1	—	NR	—	[33]	1972
	U	1475	0,5		NR			
	Ce	1500	0,3		WR			
SiC	Al	700	72	—	970.5 *	—	[611]	1961
	Al	900	24		30 *			
	Al	900	72		WR			
	Sn	400	24		5.4 *			
	Pb	600	24		31,8 *			
	Bi	600	22		~100 *			
	Mg	750	24		546 *			
	Mg	800	24		WR			
	Zn	600	72		14,6 *			

*Corrosion rate, mm/min.

Phase	Melt	Temp., °C	Contact time, h	Medium	Type of reaction	Concentration of component compound entering molten metal, %	Source	Year
1	2	3	4	5	6	7	8	9
Si_3N_4	Cu	1150	7	—	R	—	[248]	1969
	Zn	550	500		NR		[248]	1969
	Mg	750	20		WR		[248]	1969
	Al	800	900		NR		[248]	1969
	Al	1000	950		NR		[248]	1969
	Sn	300	144		NR		[248]	1969
	Fe	—	—		R		[248]	1969
	Iron	1450	2		WR		[637]	1962
	Brass	950	72		NR		[637]	1962
	Cryolite	1050	36		WR		[637]	1962
	$NaNH_2 + NaNO_3$	350	—		NR		[248]	1969
	$NaCl + KCl$	790	—		NR			
	$NaCl + KCl$	900	144		NR			
	NaOH	450	5		NR			
	$NaF + ZrF_4$	800	100		NR			
	$NaB \cdot (SiO_3)_2 + V_2O_5$	1100	4		NR			
BN	Fe	1600	0,5	—	NR	—	[248]	1969
AlN	Copper matte	1300	5	—	AR	—	[640]	1965
	Cu slag	1300	5		WR			
	Ni matte	1250	5		R			
	Ni slag	1250	5		R			
	Basalt	1400	0.5		AR			

RESISTANCE TO REACTION IN THE SOLID PHASE AND WITH NITROGEN

Reacting mixture	Temp., °C	Basic reaction products	Reaction time, h	Source	Year	Notes
1	2	3	4	5	6	7
$Be_5B + 25\% C$	800 900 1000 1300	$Be_5B + C$ $Be_2C + C$ $Be_2C + $ tr. C Be_2C	—	—*	1961	
$Be_2B + $ $+ 25\% C$	800 900 1000 1200 1300	$Be_2B + C$ $Be_2B + C$ $Be_2C + C$ $Be_2C + $ tr. C Be_2C	—	—*	1961	
$BeB_2 + $ $+ 25\% C$	800 900 1000 1200 1300	$BeB_2 + C$ $BeB_2 + C +$ $+ $ tr. Be_2C $Be_2C + C$ $Be_2C + $ tr. C Be_2C	—	—*	1961	
$BeB_4 + $ $+ 25\% C$	800 900 1000 1200 1300	$BeB_4 + C$ $BeB_4 + C$ $BeB_4 + C$ $Be_2C + C +$ $+ $ tr. BeB_4 Be_2C	—	—*	1961	
$MgB_2 + N_2$	<900 950—1000	No reaction Mg_3N_2	2—3	[254]	1960	
$MgB_6 + N_2$	≤1350	No reaction	2—3	[254]	1960	
$MgB_{12} + N_2$	≤1350	No reaction	2—3	[254]	1960	
$Mg_3B_2 + N_2$	900	Mg_3N_2	2—3	[254]	1960	

*G. S. Markevich, Investigation of the Be–B System [in Russian], Author's Dissertation, Leningrad (1961).

Reacting mixture	Temp., °C	Basic reaction products	Reaction time, h	Source	Year	Notes
1	2	3	4	5	6	7
LaB_6 + Ta *	1000—1600	LaB_6 + Ta	2—5	[637]	1962	
	1800	Reacts	2			
	1800	Same	5			
	2000	"	2			
	2000	"	5			
	2100	Reacts strongly	2			
	2100	Same	5			
LaB_6 + Mo *	1200	Reacts weakly	2	[637]	1962	
	1400	Reacts	5			
	1600	Same	5			
	1800	"	2			
	1800	"	5			
	2000	"	2			
	2000	"	5			
	2100	Reacts strongly	2			
	2100	Same	5			
LaB_6 + W *	1400	LaB_6 + W	2—5	[637]	1962	
	1600	Reacts weakly	5			
	1800	Same	2			
	1800	Reacts	5			
	2000	Same	2			
	2000	"	5			
	2100	"	2			
	2100	"	5			
LaB_6 + Mo_2C	1800—2100	LaB_6 + Mo_2C	2—5	[637]	1962	
LaB_6 + $MoSi_2$	1200—1500	$MoSi_2$ + X	—	[629]	1959	
DyB_6 + Ta	1500	TaB_2 + Dy	—	[7]	1973	
UB_2 + B	1100—1200	UB_4	—	[380]	1964	

Note. Here and below the asterisks indicate data for contact interaction in which both or one of the reacting phases are in the compacted state, in contrast to reactions in powder mixtures or gas-on-powder reactions.

Reacting mixture	Temp., °C	Basic reaction products	Reaction time, h	Source	Year	Notes
1	2	3	4	5	6	7
CeB_6, TiB_2, ZrB_2, ThB_6, NbB_2, TaB_2, Cr_2B, Cr_5B_3, CrB, Cr_3B_2, Cr_3B_4, CrB_2, Mo_2B, Mo_2B_5, WB, W_2B_5, Fe_2B, FeB, Co_2B, CoB, Ni_2B	2000—2200	Resistant to carbon	—	[254]	1960	See [716]
CeB_4, Ti_2B, TiB, Ti_2B_5, ZrB, ZrB_{12}, ThB_4, Nb_2B, NbB, Nb_3B_4, Nb_3B_2, Ta_3B_4, Ta_3B_2, Cr_4B, Cr_2B_5, W_2B_5	2000—2200	Not resistant to carbon	—	[254]	1960	See [716]
$TiB_2 + Nb^*$	1300	—	2	[640]	1965	Temperatures at which reactions commence
$TiB_2 + Ta^*$	1600	—	2	[640]	1965	
$TiB_2 + Mo^*$	1400	—	5	[640]	1965	
$TiB_2 + W^*$	>1800	—	5	[640]	1965	
$TiB_2 + MgO^*$	1200	—	2	[640]	1965	
$TiB_2 + ZrO_2^*$	1100	—	2	[640]	1965	
$TiB_2 + Fe$	1250	FeB	—	[718]	1964	
$TiB_2 + Co$	1250	Co_3B	—	[718]	1964	
$TiB_2 + Ni$	1250	Ni_3B	—	[718]	1964	
$TiB_2 + TiSi_2$	1500	Reacts	3	[715]	1974	
$TiB_2 + Ti_5Si_3$	2000	Same	1	[715]	1974	
$TiB_2 + TaSi_2$	2000	Reacts weakly	1	[715]	1974	
$TiB_2 + Ta_5Si_3$	2200	Same	1	[715]	1974	
$TiB_2 + CrSi_2$	1600	Reacts	3	[715]	1974	
$TiB_2 + Cr_5Si_3$	1700	Same	0,75	[715]	1974	
$TiB_2 + WSi_2$	2000	Reacts weakly	1	[715]	1974	
$TiB_2 + W_5Si_3$	2150	Same	2	[715]	1974	

Reacting mixture	Temp., °C	Basic reaction products	Reaction time, h	Source	Year	Notes
1	2	3	4	5	6	7
$TiB_2 + N_2$	700—1200	Reacts weakly	10	[722]	1973	No ternary phases detected in reaction products; see [717, 629]
$TiB_2 + Cr$	1850—2000	$TiB + CrB$	2	[714]	1969	
$TiB_2 + Mo$	1950—2050	$TiB + \beta\text{-}MoB$	2	[714]	1969	
$TiB_2 + W$	1950—2400	$TiB + \beta\text{-}WB$	2	[714]	1969	
$TiB_2 + Re$	1900—2200	$TiB + Re_7B_3$	2	[714]	1969	
$ZrB_2 + Nb^*$	1200	—	5	[640]	1965	These are initial-reaction temperatures
$ZrB_2 + Ta^*$	1200	—	5	[640]	1965	
$ZrB_2 + Mo^*$	1200	—	5	[640]	1965	
$ZrB_2 + W^*$	1200	—	5	[640]	1965	
$ZrB_2 + MgO^*$	1100	—	2	[640]	1965	
$ZrB_2 + ZrO_2^*$	1300	—	2	[640]	1965	
$ZrB_2 + Fe$	1200	FeB	—	[718]	1964	
$ZrB_2 + Co$	1200	Co_3B	—	[718]	1964	
$ZrB_2 + Ni$	1200	Ni_3B	—	[718]	1964	
$ZrB_2 + ZrSi_2$	1700	Reacts	2	[715]	1974	
$ZrB_2 + Zr_5Si_3$	2000	Reacts	1	[715]	1974	
$ZrB_2 + TaSi_2$	2000	Reacts weakly	1	[715]	1974	
$ZrB_2 + Ta_5Si_3$	2200	Same	1	[715]	1974	
$ZrB_2 + CrSi_2$	1600	Reacts	1	[715]	1974	
$ZrB_2 + Cr_5Si_3$	1730	Same	1	[715]	1974	
$ZrB_2 + MoSi_2$	1800	"	3	[715]	1974	
$ZrB_2 + Mo_5Si_3$	2000	"	1	[715]	1974	
$ZrB_2 + WSi_2$	2060	Reacts weakly	1	[715]	1974	
$ZrB_2 + W_5Si_3$	2000	Same	1	[715]	1974	
$ZrB_2 + N_2$	700—1200	"	10	[722]	1973	No ternary phases detected in reaction products; see [629]
$ZrB_2 + Cr$	1800—2000	$ZrB + CrB$	2	[714]	1969	
$ZrB_2 + Mo$	2050	$ZrB + \beta\text{-}MoB$	2	[714]	1969	
$ZrB_2 + W$	2050—2100	$ZrB + \beta\text{-}WB$	2	[714]	1969	
$ZrB_2 + Re$	1950—2200	$ZrB + Re_7B_3$	2	[714]	1969	
$ZrB_2 + ZrO_2$	1000—2000	No reaction	1	[719]	1971	

Reacting mixture	Temp., °C	Basic reaction products	Reaction time, h	Source	Year	Notes
1	2	3	4	5	6	7
$ZrB_2 + ZrN$	—	Limited solid solutions based on ZrN and ZrB_2	—	[720]	1972	No ternary phases
$HfB_2 + HfN$	—	Limited solid solutions based on HfN and HfB_2	—	[720]	1972	No ternary phases
$HfB_2 + HfO_2$	1000—2000	No reaction	1	[719]	1971	
$NbB_2 +$ $+ Nb_2O_5$	1000—1100 1200—1300 1400—1700 1800 1900	$NbB_2 + NbO_2$ $NbO_2 + $ tr. $NbB_2 + NbB$ $NbO_2 + NbB$ $NbB + $ tr. X $Nb_3B_2 +$ $+ $ tr. NbB	1	[719]	1971	
$TaB_2 +$ $+ Ta_2O_5$	1000 1100 1200—1700 1800 1900 2000	No reaction $Ta_2O_5 + Ta_3B_4$ $Ta_2O_5 + TaB$ $TaB + Ta$ tr. $TaB + Ta$ $Ta_2B + Ta$	1	[719]	1971	
$NbB_2 + Ta$	1800—2000	$NbB_2 + Ta$	—	[629]	1959	
TiB_2 (67%)+ $+ B_4C$ (33%)	1600 2150	TiB_2, B_4C TiB_2, B_4C	—	[725]	1952	
ZrB_2 (67%)+ $+ B_4C$ (33%)	2100	ZrB_2, B_4C	—	[725]	1952	
VB_2 (67%)+ $+ B_4C$ (33%)	2000	$VB_2 + B_4C +$ $+ C$	—	[725]	1952	

Reacting mixture	Temp., °C	Basic reaction products	Reaction time, h	Source	Year	Notes
1	2	3	4	5	6	7
TaB_2(67%)+ + B_4C (33%)	2000	$TaB_2 + B_4C$	—	[725]	1952	
CrB (67%) + + B_4C (33%)	1500— 2000 1700— 2100	CrB + B_4C + + C $CrB_2 + B_4C +$ + C	—	[725]	1952	
$TaB_2 + Nb*$	1600	Reacts weakly	5	[637]	1962	
	1800	Reacts	2			
	1800 2000	Same Reacts strongly	5 2			
	2000 2100 2100	Same " "	5 2 5			
$TaB_2 + Ta*$	1600	Reacts weakly	5	[637]	1962	
	1800 2000	Same Reacts	5 2			
	2000	Same	5			
$TaB_2 + Mo*$	1600	Reacts weakly	5	[637]	1962	
	1800 1800	Same Reacts	2 5			
	2000 2000 2100	Same " Reacts strongly	2 5 2			
	2100	Same	5			
$TaB_2 + W*$	1600	Reacts weakly	5	[637]	1962	
	1800 1800	Same Reacts	2 5			

Reacting mixture	Temp., °C	Basic reaction products	Reaction time, h	Source	Year	Notes
1	2	3	4	5	6	7
$TaB_2 + Nb^*$ $TaB_2 + Ta^*$ $TaB_2 + Mo^*$ $TaB_2 + W^*$	1400 1400 1400 1500	— — — —	2 2 2 5	[640]	1965	This is the initial-reaction temperature
$Cr_3B_2 + N_2$	800 900 1000 1180	$Cr_3B_2 + CrN$ $CrN + Cr_2N +$ $+ BN$ $CrN + Cr_2N +$ $+ BN$ $Cr_2N + BN$	—	[721]	1951	
$CrB + N_2$	550 800 900 1000 1180	CrB CrB $CrB + CrN$ $CrB + CrN +$ $+ BN$ $Cr_2N + BN$	—	[721]	1951	
$Cr_3B_4 + N_2$	550 900 1050 1100	Cr_3B_4 Cr_3B_4 $Cr_2N + Cr_3B_4$ $Cr_2N + BN$	—	[721]	1951	
$CrB_2 + N_2$	1000 1180	CrB_2 $Cr_2N + BN$	—	[721]	1951	
$CrB_2 + Cr_2O_3$	1000 1100 1200— 1400 1500 1600 1700	$Cr_2O_3 +$ $+ CrB + Cr_3O$ $Cr_2B + Cr_2O_3$ $Cr_2B +$ $+ tr. Cr_2O_3$ $Cr_2B +$ $+ tr. Cr_2O_3 +$ $+ tr. Cr$ $Cr_2B + Cr$ CrB	1	[719]	1971	
$Mo_2B_5 +$ $+ Nb^*$ $Mo_2B_5 + Ta^*$ $Mo_2B_5 + Mo^*$ $Mo_2B_5 + W^*$	~1300 ~1300 ~1300 ~1200	— — — —	5 5 5 5	[640] [640] [640] [640]	1965 1965 1965 1965	These are the initial-reaction temperatures; see [629]

Reacting mixture	Temp., °C	Basic reaction products	Reaction time, h	Source	Year	Notes
1	2	3	4	5	6	7
Mo_2B_5 + Fe	1120	FeB	—	[718]	1964	No ternary phases detected
Mo_2B_5 + Fe	1700—1750	FeB, MoB_2	—	[718]	1964	
Mo_2B_5 + Ni	—	α-MoB, NiB	—	[718]	1964	
Mo_2B_5 + Co	—	α-MoB, Co_3B	—	[718]	1964	
Mo_2B (67%) + + B_4C (33%)	1600	MoB_2 + B_4C + + C	—	[725]	1952	
MoB (67%) + + B_4C (33%)	1800, 2000	MoB_2 + B_4C + + C	—	[725]	1952	
Mo_2B_5 (67%) + + B_4C (33%)	1800, 2000	MoB + B_4C + + C	—	[725]	1952	
W_2B_5 + WC	2350—2400	WC + W_2B_5	—	[726]	1959	
W_2B_5 + Nb*	~1400	—	5	[640]	1965	These are the initial-reaction temperatures
W_2B_5 + Ta*	~1400	—	5	[640]	1965	
W_2B_5 + Mo*	~1300	—	5	[640]	1965	
W_2B_5 + W*	~1400	—	5	[640]	1965	
W_2B + N_2	700	W_2B	—	[721]	1951	
	800	W_2B				
	850	W_2B + W + + BN				
	900	W_2B + W + + BN				
	1000	W_2B + W + + BN				
	1100	W + BN				
WB + N_2	700	WB	—	[721]	1951	
	750	WB				
	800	WB + BN				
	850	WB + W + + BN				
	900	W + (WB) + + BN				
	1000	W + BN				
	1100	W + BN				

RESISTANCE TO REACTION IN THE SOLID PHASE AND
WITH NITROGEN (continued)

Reacting mixture	Temp., °C	Basic reaction products	Reaction time, h	Source	Year	Notes
1	2	3	4	5	6	7
$W_2B_5 + N_2$	700 750 850 900 1100	WB_2 WB_2 $WB_2 + W +$ $+ (BN)$ $W + (WB_2) +$ $+ BN$ $W + BN$	—	[721]	1951	
$Fe_2B + N_2$	350 400 450 500 550 600 700 770	Fe_2B ζ-phase $+ BN$ $\varepsilon + BN$ $\varepsilon + \gamma' + BN$ $\gamma' + \varepsilon + BN$ α-Fe $+ \gamma' +$ $+ \varepsilon + BN$ α-Fe $+ BN$ α-Fe $+ BN$	—	[721]	1951	
$FeB + N_2$	300 400 550 600 770	FeB $Fe_2B + \zeta +$ $+ BN$ $\gamma' + \varepsilon + BN$ $\gamma' + \varepsilon + \alpha$-Fe$+$ $+ BN$ α-Fe $+ BN$	—	[721]	1951	
$Be_2C + N_2$	1250	$Be_3N_2 + C$	—	[532]	1952	
$UC + Mo$ $UC + W$	1000 1000	Mo_2C WC, W_2C	— —	[510] [510]	1966 1966	
$UC + Zr$ } $UC + Be$ }	950	Reacts by sintering under a pressure of 15 kg/mm^2	12	[724]	1959	
$UC + Si$ $UC + Ni$	1000 1000	$UC + USi_3$ $UC + U_6Ni$ (and other phases in the U—Ni system)	— 0,6	[724] [724]	1959 1959	
(TiC, ZrC, HfC, NbC, TaC) $+ C$	—	Eutectic fusions	—	[723]	1961	

Reacting mixture	Temp., °C	Basic reaction products	Reaction time, h	Source	Year	Notes
1	2	3	4	5	6	7
TiC + Nb*	1600 1800	TiC + Nb Reacts weakly	2 2	[637]	1961	
	1600 1800	TiC + Nb Reacts weakly	5 5			
	2000	Reacts	5			
TiC + Ta*	1600— 1800 2000	TiC + Ta Reacts	2—5 5	[637]	1961	
TiC + Mo*	1600— 2000	TiC + Mo	2—5	[637]	1961	
TiC + 2B ZrC + 2B VC + 2B NbC + 2B TaC + 2B Cr$_3$C$_2$ + 2BN Mo$_2$C + 2B WC + 2B	2400 2600 2000 2250 2400 1650 2300 2450	TiB$_2$ ZrB$_2$ VB$_2$ NbB$_2$ TaB$_2$ Cr$_2$N + Cr MoB W$_2$B$_5$	—	[625]	1957	
TiC + N$_2$ ZrC + N$_2$ HfC + N$_2$ VC + N$_2$ NbC + N$_2$ TaC + N$_2$ Cr$_3$C$_2$ + N$_2$	1400— 1800 1400— 1800 1400— 1800 1100— 1700 1100— 1700 1100— 1800 1100— 1400 1450	Solid solution No reaction Same Cr$_3$ (C, N)$_2$	— — 5—14 1	[339]	1970	$p = 1-$ -300 atm

Reacting mixture	Temp., °C	Basic reaction products	Reaction time, h	Source	Year	Notes
1	2	3	4	5	6	7
$Mo_2C + N_2$	1100—1700	No reaction	5—24	[339]	1970	$p = 1-$ -300 atm
	1100	$Mo_2C + Mo$ (C, N)	14			
	1700	$Mo_2C +$ $+\eta\text{-}MoC_{1-x}$	2			
	1450	$Mo_2C + Mo$ (C, N)	1			
$WC + N_2$	1100—1700	No reaction	—			
$TiC + W^*$	1400—1800	$TiC + W$	2—5	[637]	1962	
	2000	Reacts weakly	2			
	2000	Same	5			
$TiC (56\%) +$ $+ B_4C (44\%)$	1500—2150	$TiB + B_4C$	—	[725]	1952	
$TiC + Nb^*$	1700	—	5	[640]	1965	These are initial-reaction temperatures
$TiC + Ta^*$	1900		5			
$TiC + Mo^*$	2000		5			
$TiC + W^*$	2000		5			
$TiC + MgO^*$	1800		2			
$TiC + ZrO_2^*$	2200		1			
$TiC_{0.96} +$ $+ ZrO_2$	To 1300	No reaction	1	[728]	1969	See [543]. Phase X has a NaCl type of structure. With temp. rise the lattice constant falls from 4.67 A (1300°) to 4.43 A (1900°) and does not alter subsequently
	1300—1600	$TiC + ZrO_2 +$ $+$ To X				
	1900—2000	X				

Reacting mixture	Temp., °C	Basic reaction products	Reaction time, h	Source	Year	Notes
1	2	3	4	5	6	7
TiC ZrC HfC NbC $\}$ + C TaC Mo$_2$C WC	2200	No reaction	—	[640]	1965	
TiC + Fe	1410	—	—	[421]	1965	This is the eutectic temperature
TiC + Co	1380					
TiC + Ni	1340					
ZrC + Fe	1480					
ZrC + Co	1410					
ZrC + Ni	1350					
HfC + Fe	1440					
HfC + Co	1410					
HfC + Ni	1360					
VC + Fe	1330					
VC + Co	1350					
VC + Ni	1350					
NbC + Fe	1390					
NbC + Co	1400					
NbC + Ni	1330					
TaC + Fe	1360					
TaC + Co	1400					
TaC + Ni	1340					
Cr$_3$C$_2$ + Fe	1280					
Cr$_3$C$_2$ + Co	1285					
Cr$_3$C$_2$ + Ni	1255					
Mo$_2$C + Fe	1310					
Mo$_2$C + Co	1310					
Mo$_2$C + Ni	1330					
WC + Fe	1490					
WC + Co	1390					
WC + Ni	(1453)					
ZrC + Nb*	1400	—	5	[640]	1965	This is the initial-reaction temperature

Reacting mixture	Temp., °C	Basic reaction products	Reaction time, h	Source	Year	Notes
1	2	3	4	5	6	7
$ZrC + Ta^*$ $ZrC + Mo^*$ $ZrC + W^*$ $ZrC + MgO^*$ $ZrC + ZrO_2^*$	2200 2000 2200 2000 2200	—	5 5 5 2 1	[640]	1965	This is the initial-reaction temperature
$ZrC_{0.98} +$ $+ ZrO_2$	To 1800 1900— 2100	No reaction $X +$ tr. ZrO_2	1	[728]	1969	The lattice constant of ZrC diminishes. Phase X has the NaCl type of structure. Its lattice constant increases with temperature rise
$ZrC + Si$	To 1000 1100— 1200 1300 1400— 1500 1600— 1700	No reaction $ZrC + Si$ $ZrC + ZrSi_2 +$ $+ Si$ $ZrC + ZrSi +$ $+$ tr. SiC $ZrC + ZrSi +$ $+ \beta\text{-}SiC$	1	[727]	1969	
$ZrC + Nb^*$	1600 1800 2000 2200	Reacts weakly Same Reacts Same	2	[637]	1962	
$ZrC + Ta^*$	1600 1800 2000 2000 2200	No reaction " " Reacts weakly Same	2 2 2 5 2	[637]	1962	

Reacting mixture	Temp., °C	Basic reaction products	Reaction time, h	Source	Year	Notes
1	2	3	4	5	6	7
ZrC + Mo*	1000—2000	ZrC + Mo	2	[637]	1962	
	2000	Reacts weakly	5			
	2200	Reacts	2			
ZrC + W*	1600—2000	ZrC + W	2—5	[637]	1962	
ZrC (67%) + + B_4C (33%)	1400—2150	ZrB_2 + B_4C + C	—	[725]	1952	
HfC + Nb*	1400	—	5	[640]	1965	This is the initial-reaction temperature
HfC + Ta*	1400		5			
HfC + Mo*	1500		5			
HfC + W*	2000		5			
HfC + MgO*	2200		1			
HfC + ZrO_2^*	2200		1			
HfC + Nb*	1600	Reacts weakly	2	[637]	1962	
	1800	Same				
	2000	Reacts				
	2200	Same				
HfC + Ta*	1400—1600	HfC + Ta	2—5	[637]	1962	
	1800	Reacts weakly	2			
	2000	Reacts	2			
	2200	Same	2			

Reacting mixture	Temp., °C	Basic reaction products	Reaction time, h	Source	Year	Notes
1	2	3	4	5	6	7
HfC + Mo*	1000—1800	HfC + Mo	2—5	[637]	1962	
	1800	Reacts weakly	5			
	2000	Reacts	2			
	2200	Same	2			
HfC + W*	1000—1800	HfC + W	2—5	[637]	1962	
	2000	Reacts weakly	2			
	2200	Same	2			
$VC_{0.82}$ + ZrO_2	1300	No reaction	1	[728]	1969	
	1400	ZrO_2 + VC + tr. ZrC				
	1500	ZrO_2 + VC + ZrC + tr. V_2C				
	1600	ZrO_2 + ZrC + V_2C + tr. VC				
	1700	tr. ZrO_2 + ZrC + V + V_2C				
	1900	tr. ZrO_2 + ZrC + V + tr. V_2C				
	2000	ZrC + V + tr. V_2C				
VC (67%) + B_4C (33%)*	1500—2100	VB_2 + B_4C + C	—	[725]	1952	
NbC + Nb*	1600 1800	NbC + Nb Reacts	2	[637]	1962	
	2000 2200	Same "				

Reacting mixture	Temp., °C	Basic reaction products	Reaction time, h	Source	Year	Notes
1	2	3	4	5	6	7
NbC + Ta*	1600 1800	NbC + Ta Reacts weakly	2—5 2	[637]	1962	
	2000	Reacts	2			
	2200	Same	2			
NbC + Mo*	1000— 1800	NbC + Mo	2—5	[637]	1962	
	1800	Reacts weakly	5			
	2000 2200	Same Reacts	2 2			
NbC + W*	1600— 2000 2200	NbC + W Reacts weakly	2	[637]	1962	
NbC (77%) + + B_4C (23%)	1500— 2100	$NbB_2 + B_4C + + C$	—	[725]	1952	
NbC + ZrO_2	1400	No reaction	1	[543]	1966	Interaction occurs according to chemical-analysis data Lines are widened and new reflexes appear
	1400— 2000	ZrO_2 + NbC				
	2100	ZrO_2 + NbC				
NbC + Nb*	1700	—	5	[640]	1965	This is the initial-reaction temperature
NbC + Ta* NbC + Mo* NbC + W* NbC + MgO* NbC + ZrO_2^*	1700 1700 2200 1800 2200		5 5 5 2 I			

RESISTANCE TO REACTION IN THE SOLID PHASE AND WITH NITROGEN (continued)

Reacting mixture	Temp., °C	Basic reaction products	Reaction time, h	Source	Year	Notes
1	2	3	4	5	6	7
TaC + Mo*	1000—1800	TaC + Mo	2—5	[637]	1962	
	2000	TaC + Mo	2			
	2000	Reaction weak	5			
TaC + W*	1600—2000	TaC + W	2—5	[637]	1962	
	2000	Reaction weak	5			
	2200	Reacts	2			
TaC + Nb*	1600	Reaction weak	2	[637]	1962	
	1800	Same				
	2000	Reacts				
	2200	Same				
TaC (67%) + + B_4C (33%)	1500	$TaB_2 + B_4C + + C$	—	[725]	1952	
Cr_3C_2 (67%) + + B_4C (33%)	1700	$CrB_2 + B_4C + + C$				
Mo_2C (67%) + B_4C (33%)	1800	$MoB_2 + B_4C + + C$				
WC (77%) + + B_4C (23%)	2100	$W_2B_5 + B_4C + + C$				
Fe_3C (71%) + + B_4C (29%)	1500	$FeB + B_4C + + C$				
TaC + Nb*	1400	—	5	[640]	1965	This is the initial-reaction temperature
TaC + Ta*	>2200		5			
TaC + Mo*	2000		5			
TaC + W*	1900		5			
TaC + MgO*	2000		2			
TaC + ZrO_2^*	2200		1			

Reacting mixture	Temp., °C	Basic reaction products	Reaction time, h	Source	Year	Notes
1	2	3	4	5	6	7
$Cr_7C_3 + Ta^*$	1800—2000	$Cr_7C_3 + Ta$	—	[629]	1959	
$Cr_7C_3 + Mo^*$	1800—2000	$Cr_7C_3 + Mo$	—			
$Cr_7C_3 + W^*$	1800—2000	$Cr_7C_3 + W$	—			
$Cr_3C_2 + Cr_2O_3$	1150—1200	$Cr_{23}C_6 + CO$	1	[762]	1971	
$Cr_3C_2 + ZrO_2$	1200	No reaction	1	[728]	1969	See [543]
	1300	$Cr_3C_2 + ZrO_2 + tr. Cr_7C_3$				
	1400	$Cr_3C_2 + ZrO_2 + tr. Cr_7C_3, ZrC$				
	1500	$Cr_3C_2 + Cr_7C_3 + ZrC + tr. ZrO_2$				
	1600	$Cr_7C_3 + ZrC + Cr_3C_2 + tr. ZrO_2$				
	1800	$Cr_3C_2 + ZrC$				
$Mo_2C + ZrO_2$	1400	No reaction	1	[728]	1969	
	1500—1600	$Mo_2C + ZrO_2 + tr. Mo, ZrC$				
	1700—1800	$Mo_2C + ZrC + Mo$				
	1900	$ZrC + Mo + tr. Mo_2C$				
	2000	$ZrC + Mo$				
$Mo_2C + Nb^*$	1900	—	5	[640]	1965	This is the initial-reaction temperature
$Mo_2C + Ta^*$	1800		5			
$Mo_2C + Mo^*$	1700		5			
$Mo_2C + W^*$	1700		5			
$Mo_2C + MgO^*$	1800		2			
$Mo_2C + ZrO_2^*$	2000		2			

Reacting mixture	Temp., °C	Basic reaction products	Reaction time, h	Source	Year	Notes
1	2	3	4	5	6	7
$Mo_2C + Mo$	1300—2000	$Mo_2C + Mo$	—	[629]	1959	
$Mo_2C + W$	1800—2000	$Mo_2C + W$				
$Mo_2C +$ $+ MoSi_2 +$ $+ Mo$	1800—1900	$Mo_2C +$ $+ MoSi_2 + Mo$				
$Mo_2C +$ $+ Mo_2B_5 +$ $+ Mo$	1800—1900	$Mo_2C +$ $+ Mo_2B_5 + Mo$				
$W_2C + Mo$	1800—2000	$W_2C + Mo$	—	[629]	1959	
$W_2C + B_2O_3$	800—1400	$W + B + CO$	—	[729]	1959	
$WC + Mo$	1800—2000	$WC + Mo$	—	[629]	1959	
$WC + 3W +$ $+ B_4C$	2100—2200	$W_2B_5 + (WB)$	—	[726]	1959	
$WC + B_4C +$ $+ C$	2100—2200	$W_2B_5 + (WB)$	—	[726]	1959	
$WC + B_2O_3$	1400	$W_2C + B + CO$	—	[729]	1959	
$WC + Nb*$	1900	—	5	[640]	1965	This is the initial-reaction temperature
$WC + Ta*$	1700		5			
$WC + Mo*$	2000		5			
$WC + W*$	2000		5			
$WC + MgO*$	2000		2			
$WC + ZrO_2^*$	2200		1			
$WC + ZrO_2$	1500	No reaction	1	[728]	1969	
	1600	$ZrC + WC +$ $+ tr. W_2C$				
	1700—1800	$ZrC + W +$ $+ W_2C$				
	1900—2000	$ZrC + W$				

RESISTANCE TO REACTION IN THE SOLID PHASE AND
WITH NITROGEN (continued)

Reacting mixture	Temp., °C	Basic reaction products	Reaction time, h	Source	Year	Notes
1	2	3	4	5	6	7
$5WC + V_2O_5$	900—1400	S. s.* V—W	—	[730]	1965	
$4WC + Nb_2O_5$	1400	S. s. W— Nb, tr. NbC, NbO_2	0,5			
$5WC + Nb_2O_5$	1400	S. s. W— Nb, NbC (little) NbO_2 (little)	0,5			
$6WC + Nb_2O_5$	1400	S. s. W— Nb, NbC (little) tr. NbO_2, tr. W_2C	0.5			
$8WC + Nb_2O_5$	1400	S. s. W— Nb, W_2C, NbC (little)	0.5			
$5WC + Ta_2O_5$	1400	S. s. W— Ta, TaC (little) tr., TaO_2	0,5			
$6WC + Ta_2O_5$	1400	S. s. W— Ta, TaC (little) W_2C (very small)	0,5			
$8WC + Ta_2O_5$	1400	S. s. W— Ta, TaC (little) W_2C	0,5			
$UN + Mo$	2000	γ-U—Mo	1	[799]	1969	Vacuum
$UN + Mo$	2000	No reaction	1	[799]	1969	Nitrogen; $p = 500$ torr
$UN + W$	2400	Same	2	[799]	1969	Nitrogen; $p = 500$ torr
$UN + ThO_2$	2400	"	2	[799]	1969	Nitrogen; $p = 500$ torr
$UN + ThO_2$	2000	Weak reaction	1	[799]	1969	Vacuum
$UN + BeO$	600	No reaction	25	[800]	1968	Argon
	1100	UO_2 (weak reaction)				

*S. s. = Solid solution.

Reacting mixture	Temp., °C	Basic reaction products	Reaction time, h	Source	Year	Notes
1	2	3	4	5	6	7
UN + MgO	600	Strong reaction	10	[800]	1968	Air
	600	UO$_2$, U$_2$N$_3$ (weak reaction)	25			Argon
	1100	UO$_2$, (Mg) (strong reaction)	25			Argon
UN + α-Al$_2$O$_3$	600	UO$_2$, (Al), AlN (weak reaction)	25	[800]	1968	Argon
	1100	UO$_2$, Al, AlN (strong reaction)				
UN + γ-Al$_2$O$_3$	600	Weak reaction	10	[800]	1968	Air
	1100	Strong reaction	10			Air
	600	UO$_2$, U$_2$N$_3$, (Al) (weak reaction)	25			Argon
	1100	UO$_2$, Al, AlN, α-Al$_2$O$_3$ (strong reaction)	25			Argon
UN + SiO$_2$·H$_2$O	600	Strong reaction	10	[800]	1968	Air
UN + SiO$_2$	600	No reaction	25	[800]	1968	Argon
	1100	UO$_2$, (Si) (strong reaction)				
UN + CeO$_2$	600	(CeN)	10	[800]	1968	Air
	1100	UO$_2$, Ce$_2$O$_3$ (strong reaction)				

Reacting mixture	Temp., °C	Basic reaction products	Reaction time, h	Source	Year	Notes
1	2	3	4	5	6	7
$UN + ThO_2$	600—1100	No reaction	25	[800]	1968	Argon
$UN + TiO_2$	600	UO_2, U_2N_3 (weak reaction)	10	[800]	1968	Air
	1100	UO_2, TiN, Ti_2O_3, Ti_3O_5 (strong reaction)				
$UN + ZrO_2$	600	No reaction	10	[800]	1968	Air
	1100	UO_2, ZrN (weak reaction)				
$TiN + Nb*$	1800	—	5	[640]	1965	This is the initial-reaction temperature
$TiN + Ta*$	2000		2			
$TiN + Mo*$	2000		2			
$TiN + W*$	2000		5			
$TiN + MgO*$	1300		2			
$TiN + ZrO_2^*$	1400		2			
$TiN + Nb*$	1800	Weak reaction	5	[637]	1962	
	2000	Reacts	2			
	2000	Same	5			
	2100	Strong reaction	2			
$TiN + Ta*$	1600	$TiN + Ta$	2	[637]	1962	
	1800	$TiN + Ta$	5			
	2000	Reacts	2			
	2100	Strong reaction	5			

Reacting mixture	Temp., °C	Basic reaction products	Reaction time, h	Source	Year	Notes
1	2	3	4	5	6	7
TiN + Mo*	1600—1800	TiN + Mo	2—5	[637]	1962	
	2000	Weak reaction	2			
	2000	Same				
	2100	Reacts	5			
			5			
TiN + W*	1600—1800	TiN + W	2—5	[637]	1962	
	2000	Weak reaction	2			
	2000	Reacts	5			
	2100	Strong reaction	2			
ZrN + Ta*	2000—2100	ZrN + Ta	2—5	[637]	1962	
ZrN + Mo*	1800—2100	ZrN + Mo	2—5	[637]	1962	
	2100	Weak reaction	5			
ZrN + W*	2000—2100	ZrN + W	2—5	[637]	1962	
ZrN + Nb*	2000	—	5	[640]	1965	This is the initial-reaction temperature
ZrN + Ta*	>2100					
ZrN + Mo*	2100					
ZrN + W*	>2100					
ZrN + MgO*	1400					
ZrN + ZrO$_2^*$	1300					
Disilicides of rare-earth metals + Al$_2$O$_3$	1100—1600	No reaction	—	[260]	1959	
	1600	Reacts				

Reacting mixture	Temp., °C	Basic reaction products	Reaction time, h	Source	Year	Notes
1	2	3	4	5	6	7
$TiSi_2 + 2B$ $ZrSi_2 + 2B$ $CrSi_2 + 2B$ $MoSi_2 + 2B$	1450 1650 1550 1750	TiB_2 ZrB_2 CrB, little $CrSi_2$ MoB, little $MoSi_2$	—	[625]	1957	
$Ti + Si + N_2$	1840 1870	TiN $TiN + X$	—	[260]	1959	
$Nb + Si + N_2$	1840 1870	$Nb_5Si_3N_7$ $NbSi_2 + Nb_5Si_3N_7$	—	[260]	1959	
$(Ta, Nb) + Si + Si_3N_4 + N_2$	2080	$(Ta, Nb) Si_2 + (Ta, Nb)_5 \cdot Si_3N_7 + (Ta, Nb) Si_{0,6}$	—	[260]	1959	
$Ta + Si + N_2$	1330 1870	$Ta_5Si_3N_7 + X$ $X + TaSi_2$	—	[260]	1959	
$Ta + Si + Si_3N_4 + N_2$	1380—2080	$TaSi_2 + Ta_5Si_3N_7 + Ta_2N$	—	[260]	1959	
$Ta_2N + Si_3N_4 + N_2$	1840	$TaSi_2 + Ta_5Si_3N_7$	—	[260]	1959	
$MoSi_2 + Mo$	1800—2000	Strong reaction	—	[629]	1959	
$MoSi_2 + Ta$	1900—2050	$Mo_3Si + TaSi$	—	[629]	1959	
$MoSi_2 + ZrO_2$	1700	No reaction	—	[637]	1962	
$B + C + N_2$	1400 1400	BNC + weak lines for B_4C BNC + weak lines for B_4C	0.25 3	[732]	1971	

435

Reacting mixture	Temp., °C	Basic reaction products	Reaction time, h	Source	Year	Notes
1	2	3	4	5	6	7
$B + C + N_2$	1600	BNC + weak lines for B_4C	0,25	[732]	1971	
	1600	BNC + weak lines for B_4C	3			
	1800	BNC	0,25			
	1800	BNC	3			
	2000	BNC	0.25			
	2000	BNC	3			
	2200	BNC	0,25			
	2200	BNC	1			
	2200	BNC + weak lines for B_4C	2			
	2200	B_4C + graphite	3			
$B + C + NH_3$	1600	$C + B_4C + BNC$	0,5	[732]	1971	
	1600	$C + B_4C + BNC$	2			
	1800	$C + BNC$	0,5			
	1800	$C + BNC$	1			
	1800	C base	2			
	1800	C base	3			
	2000	C base + BNC	0.5			
	2000	C base	1			
	2000	C base	3			
$B_4C + N_2$	1600	B_4C	0,25	[732]	1971	
	1600	B_4C	3			
	1800	B_4C	0,25			
	1800	B_4C	3			
	2000	B_4C	0,25			
	2000	B_4C	2			
	2000	B_4C + weak lines of a phase based on BN	3			
$B_4C + NH_3$	1600	B_4C + (BNC or C)	0,25	[732]	1971	
	1600	B_4C + (BNC or C)	3			
	1800	B_4C + (BNC or C)	0,25			
	1800	B_4C + (BNC or C)	3			

Reacting mixture	Temp., °C	Basic reaction products	Reaction time, h	Source	Year	Notes
1	2	3	4	5	6	7
$B_4C + NH_3$	2000	$B_4C +$ (BNC or C)	0.25	[732]	1971	
	2000	$B_4C +$ (BNC or C)	3			
	2100	$B_4C +$ (BNC or C)	1			
	2200	B_4C	1			
$B_4C + BN$	1800—2200	$B_4C + BN + + BNC$	—	[732]	1971	In various gaseous media
$SiC + Ti$	900	Ti_3Si	300	[733]	1973	Reaction of SiC fiber with Ti-matrix
	1000	No reaction	1—10			
	1000—1300	Ti_5Si_3	1—10			
$SiC + W$	900	No reaction	300	[733]	1973	Reaction of SiC fiber with W-substratum
$BN + Ti$	1200—2000	TiB, TiN, TiB_2	10 min	[722]	1973	Hot pressing
$BN + Zr$	1200—2000	ZrB_2, ZrB, ZrB_{12}, tr. BN	10 min	[722]	1973	
$BN + Hf$	1600	HfN, Hf, BN	10 min	[722]	1973	
$BN + ZrN$	2000	ZrN, ZrB_2, BN	10 min	[722]	1973	
$BN + ZrB_2$ $BN + TiB_2$ $BN + HfB_2$	—	—	—	[722]	1973	No reaction up to the hightest possible temperatures
$BN + Mo$	1600	Reacts	—	[731]	1971	New metal-type phase forms

Reacting mixture	Temp., °C	Basic reaction products	Reaction time, h	Source	Year	Notes
1	2	3	4	5	6	7
AlN + Mo	1800	No reaction	—	[731]	1971	
Si_3N_4 + Ti	1200—1600	TiN	5	[318]	1973	Hot pressing
Si_3N_4 + Ti*	1200	Ha Ti_5Si_3 and $TiSi_2$ on $TiSi_2$ On silicon nitride—TiN	—	[318]	1973	Helium p = 0.5 atm

Chapter VIII

EXAMPLES OF THE APPLICATIONS OF REFRACTORY COMPOUNDS

EXAMPLES OF THE APPLICATIONS OF REFRACTORY COMPOUNDS

Compound	Basic properties	Area of use	Source	Year
1	2	3	4	5
UBe_{13}	Nuclear properties; low specific gravity	Nuclear fuel in heat-emitting elements of the dispersion type	[2]	1966
$PuBe_{13}$	Stable neutron yield	Neutron source	[2]	1966
YBe_{13}, $ZrBe_{12}$, $NbBe_{12}$	Nuclear properties	Moderator in nuclear reactors	[2]	1966
All beryllides	Enhanced strength at high temperatures; high oxidation resistance. High thermal conductivity. High strength with low specific gravity	Refractory and other high-temperature alloys with maximum working temperature of up to 1700°C Aviation and rocket techniques		
CaB_6	Refractoriness, low density, satisfactory heat resistance, low neutron yield during thermal emission, high thermal emf	Light heat-resistant alloys, e.g., 10–35% CaB_6, 5–13% B, 60–80% B_4C (density 2.48–2.49), $\sigma_{bend} = 30$–37 kg/mm^2 at 20°C	[254]	1960
		Cathodes in electronics. Thermoelectrodes for high-temperature thermocouples and devices for converting thermal energy into electric energy	[254]	1960

EXAMPLES OF THE APPLICATIONS OF REFRACTORY COMPOUNDS (continued)

Compound	Basic properties	Area of use	Source	Year
1	2	3	4	5
SrB_6	High fusibility, low density, satisfactory heat resistance, low neutron yield during thermal emission	Light heat-resistant alloys, cathodes in electronics	[254]	1960
BaB_6	Good thermal emission properties	Cathodes in electronics	[254]	1960
ScB_2	Low density, infusible	Light heat-resistant alloys	[270]	1960
YB_6	Good thermal emission properties	Cathodes in electronics	[613]	1959
LaB_6	Good thermal emission properties	Cathodes for ionic current sources in cyclotrons and synchophasotrons, electron guns in welding apparatus (for electron beam welding) and furnaces for electron melting of metals and alloys	[613] [614]	1959 1960
		Cathode of high-power microtrons	[254]	1960
CeB_6	Thermal emission properties; high electrical resistance	For making cathodes, in particular for boosting the electrical resistance of lanthanum-boride cathodes	[615]	1959
SmB_6	Semiconductor properties, infusibility	Semiconductor techniques (at high temperatures)	[482]	1961
SmB_6, EuB_6	High neutron absorption capacity, scaling resistance	Nuclear power engineering	[254]	1960

GdB$_6$	High neutron absorption capacity, infusibility	Nuclear power engineering	[254]	1960
TmB$_6$	Low electron work function, γ-radiation	Nuclear power engineering, electronics, devices for converting thermal energy into electrical	[482]	1961
ThB$_4$, ThB$_6$, UB$_2$, UB$_4$, UB$_{12}$	Heat resistance, infusibility, nuclear properties	Nuclear power engineering	[482]	1961
TiB$_2$	Heat resistance, infusibility, scaling resistance	Heat-resistant alloys, e.g., TiB$_2$–CrB$_2$ (4 : 1 mol, parts); density 4.3–4.7; hardness RA 85; bending strength (at 20 °C) 35–50 kg/mm^2; elasticity modulus 32,800 kg/mm^2 (20 °C); thermal expansion factor (20–1200 °C) $8.5 \cdot 10^{-6}$; thermal conductivity 15.7 kcal/(m.h. °C); electrical resistance 32.8 $\mu\Omega \cdot$ cm	[254] [447]	1960 1960
	Great hardness and wear resistance	Cermet hard alloys for cutting metals and drilling rocks. Wear-resistant coatings. Grinding bodies, wear-resistant linings, bearings, sandblasting nozzles	[616] [617] [618] [254]	1960 1961 1962 1960
	Infusibility, resistance to molten metals; linear relationship between electrical resistance and temperature	High-temperature thermocouples for temperature measurements in molten metals and alloys. Tips of metallic immersion thermocouples. Heaters for high-temperature resistance furnaces for use in neutral and reducing atmospheres and in vacuum	[619] [620] [254]	1959 1960 1960

EXAMPLES OF THE APPLICATIONS OF REFRACTORY COMPOUNDS (continued)

Compound	Basic properties	Area of use	Source	Year
1	2	3	4	5
TiB_2		Electrolyzer linings for Al production components for pumps, gutterings, casting funnels in zinc and other nonferrous metals production. Crucibles for precision casting. Pipes for pumping molten metals	[254]	1960
			[62]	1960
	Satisfactory resistance to neutron radiation	TiB_2-Ti cermets for nuclear engineering	[254]	1960
ZrB_2	Heat resistance, infusibility, scaling resistance	Heat-resistant alloys of the "borolitov" type. Basic properties: density 5.2–5.4; hardness RA 88–91; bending strength 48 (16°C), 40 kg/mm² (1200°C); elasticity modulus 22,900 (16°C), 17,400 kg/mm² (1000°C); thermal expansion factor (25–1000°C) $3.2 \cdot 10^{-6}$	[254]	1960
		Refractory, infusible "borolitov" alloys with Mo and Cr bonds	[622]	1960
	Infusibility, high resistance to molten metals, alloys, slags; linear relationship between temperature and electrical resistance over a wide temperature range	High-temperature thermocouples tips for molten steels, cast irons, nonferrous and rare metals and their alloys. Components and reinforcements for furnaces in the ferrous and nonferrous metals industry	[619]	1959
			[620]	1960
		High-temperature electric-resistance furnace heating elements	[254]	1960

Formula	Properties	Applications	Reference	Year
HfB_2, VB_2, NbB_2, TaB_2, CrB, CrB_2	High electrical conductivity, resistance to electric arc	Crucibles for precision metallurgy. Boats for vacuum equipment for spray metallization. Tubes for pumping molten metals	[254]	1960
	High heat resistance, infusibility, scaling resistance	Electrical contacts (e.g., paired with silver) having burning-out resistance	[254]	1960
		Heat-resistant alloys of the "borolitov" type. Basic properties: density 6.77–7.31; hardness RA 77–88; bending strength (1000°C) 87.5–105 kg/mm²; resistivity 27–54 μΩ·cm	[254] [447]	1960
CrB_2	High wear resistance	Wear-resistant weld-on alloys	[254]	1960
Mo_2B_5	Low vapor pressure; infusibility, good alloying action with Mo and W	For soldering W and Mo in radio engineering	[254]	1960
Mo_2B_5, W_2B_5	Resistance to molten metals; thermal-shock resistance, infusibility, thermal conductivity	Crucibles and ingot molds for precision metallurgy	[621]	1960
Borides of Mn, Fe, Co, Ni	Wear resistance, hardness, heat resistance	Wear-resistance and corrosion-resistant coatings	[254]	1960
	High-temperature electric-resistance furnace heating elements	Hydrogenation catalysts	[254]	1960
Bce	Chemical resistance to acids and mixtures of acids in the cold and heated state	Chemical apparatus components	[623]	1959
ThC, ThC_2	Low work function during thermal emission; infusibility	Electrodes for thermoelectronic devices for the direct conversion of thermal to electrical energy	[624]	1961

EXAMPLES OF THE APPLICATIONS OF REFRACTORY COMPOUNDS (continued)

Compound	Basic properties	Area of use	Source	Year
1	2	3	4	5
UC, U₂C₃, UC₂	Infusibility, nuclear properties	Nuclear power engineering (uranium fuel and radiating elements for thermoemitting parts)	[550]	1960
TiC	Infusibility, scaling resistance, heat resistance	1. Heat-resistant-alloys (cermets) for gas-turbine blades, rotors, high-temperature test machine parts (clamps and rollers) Alloys 65% TiC, 15% solid solution TiC–TaC–NbC and 20% Co. Alloys of 73% TiC + TiB₂ and 27% CoSi. Alloys TiC (42.9–63.0) + Cr₃C₂ (5.7–7.1%) + Ni (22.2–50.0%) + Co (7.4–25.9%) + Cr (7.4–11.1%) for temperatures below 1000°C. Cermets based on TiC and other carbides, and also SiC, B₄C, their alloys with oxides for protective coatings of rocket parts, including PRD nozzles and the head components of rockets	[625]	1957
			[625]	1957
		TiC alloys with steel bond (or with carbon steel bond containing up to 2.9% Cr and Mo)	[625]	1957
		2. Wear- and corrosion-resistant coatings on cast irons and steels		
		3. Friction disks for aircraft construction.		
	High wear resistance and hardness	In compositions for cermet hard alloys for cutting steels	[625] [626]	1957 1960

	Properties	Applications		Year
	High electrical conductivity and infusibility, low volatilization rate	Electrodes for arc lamps, electrodes of TiC with stabilizing coatings of silicon nitride or boron nitride for underwater electro-oxygen steel cutting	[627]	1962
	Resistance to reducing gases; linear temperature dependence on electrical resistance; high strength	Thermocouple sheaths for temperatures up to 2500°C in furnaces with reducing and inert atmospheres, and in vacuum; sheaths for metal thermocouples. Metallurgical kiln parts and armatures	[619]	1959
	Resistance to molten metals	Crucibles in nuclear power engineering for making heat exchangers, e.g., from alloy consisting of 80% TiC + 5% WC, or TaC and 15% Co, resistant to molten sodium (900°C for 188 h) and bismuth (1000°C for more than 180 h)	[625]	1957
ZrC	Heat resistance, high oxidation resistance, infusibility	Heat-resistant alloys	[256]	1957
	Low neutron absorption cross section (purified from HfC), infusibility, heat resistance	Nuclear power engineering	[291]	1968
	Thermal-shock resistance, satisfactory thermal emission properties	In compositions of cathodes made of UC–ZrC alloy with high work function in thermoelectronic devices for direct conversion of thermal energy to electrical	[624]	1960
HfC	Resistance to molten metals	Crucibles, boats, tubes	[625]	1957
	Exceptionally high fusion point	Special refractories (e.g., crucible linings for melting refractory metals)	[625]	1957

EXAMPLES OF THE APPLICATIONS OF REFRACTORY COMPOUNDS (continued)

Compound	Basic properties	Area of use	Source	Year
1	2	3	4	5
VC, NbC, TaC	Great hardness and wear resistance	Alloying additives for cermet hard alloys based on WC and TiC (increased resistance, 10–20% for knives)	[625]	1957
NbC, TaC	Resistance to molten metals and metal vapors, satisfactory high-temperature strength, low vapor pressure, good radiation capacity	1. For making heating elements, evaporating elements for Al (service life from TaC 4–7 h, from NbC 1 h at 1500°C).	[625]	1957
		2. Lining crucibles (of TaC) for melting refractory metals 98)	[625]	1957
		3. Heating elements for high-temperature electrical resistance furnaces	[256]	1957
CrC_2	High wear resistance and hardness. High chemical resistance and oxidation resistance	Hard alloys for welding-on	[256]	1957
		Heat-resistant and acid-resistant alloys, e.g., with Ni bond, especially alloy TiC-Cr_3C_2-Ni and Cr_3C_2-WC-Ni (83:2:15)	[628]	1958
			[628]	1958
		Filters in chemical industry and electrodes for electrochemical processes. High-temperature soldering in electronics, e.g., for bonding LaB_6 cathodes to cores of Mo, W, Ta (solder composition Cr_3C_2 + 1% $CaF2$ or NaF)	[625]	1957
Mo_2C		Solder for high-temperature joining in electronics (e.g., ThO_2 to metals)	[629]	1959

Compound	Properties	Applications	Ref.	Year
		High-resistant and hard alloys	[256]	1957
			[625]	1957
	Catalytic properties	Catalysts for hydrogenation of alcohols, cyclohexane, etc.	[630]	1961
WC	High hardness and wear resistance	Cermet hard alloys of the BK type for processing irons, bronzes, brasses, porcelain, earthenware, plastics. Reinforcing elements for crowns used in drilling rocks, diamond substitutes for dressing grinding wheels (alloys $WC + W_2C$ — Likar or Relit)	[626]	1960
			[625]	1957
			[626]	1960
	Catalytic properties	Catalyst in dehydration of alcohols, cyclohexane, etc.	[690]	1961
Be_3N_2	Infusibility	Special refractories	[14]	1969
AlN	Infusibility, thermal-shock resistance, to molten Al, low thermal expansion factor, satisfactory thermal conductivity	Refractories (especially for melting semiconductor alloys)	[14]	1969
UN	Infusibility, chemical resistance to molten metals, nuclear properties	Special refractories (crucibles, boats), nuclear power engineering	[14]	1969
TiN	Heat resistance, infusibility, hardness, wear resistance, resistance to molten metals	1. Special refractories	[14]	1969
		2. Heat-resistant alloys, e.g., MgO + TiN (high thermal-shock resistance, strength at 1090°C is 30% higher than at room temperature)	[625]	1957
		3. Abrasive wheels	[14]	1969

EXAMPLES OF THE APPLICATIONS OF REFRACTORY COMPOUNDS (continued)

Compound	Basic properties	Area of use	Source	Year
1	2	3	4	5
TiN		4. Coatings on titanium parts	[10]	1973
		5. Spraying ingot molds to obtain clean castings	[256]	1957
		6. Nitriding titanium cores for electrical metering devices	[14]	1969
TiN, ZrN	High electrical conductivity, arc resistance	1. Conducting elements for thorium cathodes	[14]	1969
		2. Igniting devices for rectifiers (25% TiN + 75% BeO)		
		3. High-ohmic resistances (TiN + Cr_2N)		
NbN	High electrical conductivity, capable of changing to super-conductivity, at 15 K; heat resistance	1. Detectors	[14]	1969
		2. Bolometers		
		3. Television tubes		
Nitrides of Cr, Fe	High hardness and wear resistance	Wear-resistant coatings on steel	[631]	1960
$BaSi_2$, $LaSi_2$, $CeSi_2$, $DySi_2$	Semiconductor properties, scaling resistance	Semiconductor techniques	[592] [632]	1960 1960
	Heat resistance, can absorb neutrons	Nuclear power engineering (control rods for reactors)	[623] [260]	1960 1959
$TiSi_2$, Ti_5Si_3	Heat resistance, scaling resistance	Heat-resistant alloys Ti_5Si_3–SiC, $TiSi_2$–SiC	[260]	1959

	Properties	Applications	Ref.	Year
V_3Si	Superconductivity ($T_K = 17$ K)	Technical physics and automatics	[260]	1959
Cr_3Si	Heat resistance, scaling resistance	Heat-resistant alloys, e.g., 50% Cr_3Si + 50% Cr	[260]	1959
$MoSi_2$	High reaction capacity in contact with refractory metals and silica, refractoriness, scaling resistance	High-temperature soldering using $MoSi_2$ in electronics, e.g., cathodes of LaB_6 to rods of Mo, Ta, and W	[629]	1959
$MoSi_2$		Heat-resistant alloys for gas turbines, combustion chambers of jet engines and guided missiles, sandblasting nozzles, metallurgical-furnace parts, hot pressing dies, soldering devices	[260]	1959
	High resistance to oxidation and to other chemical agents, and various gases; linear temperature dependence on electrical resistance; high thermal-emf, resistance to molten chlorides	Electrode sheaths for high-temperature thermocouples for temperatures (air) to 1700–1800°C (e.g., thermocouples $MoSi_2$–B_4C, $MoSi_2$–boronized graphite, $MoSi_2$–WSi_2)	[619]	1959
		Electrode sheaths for thermocouples measuring temperatures in reducing atmospheres to 1850°C, and in oxidizing atmospheres to 1700°C	[260]	1959
		Heaters for high-temperature electrical furnaces working in air at up to 1700°C	[291]	1968

EXAMPLES OF THE APPLICATIONS OF REFRACTORY COMPOUNDS (continued)

Compound	Basic properties	Area of use	Source	Year
1	2	3	4	5
$MoSi_2$	High work function during thermo-electron emission; catalytic properties	Protective slag-resistant layers on Mo heaters and other Mo articles	[10]	1973
		High-temperature heaters of complex composition, e.g., 50-90% Mo, 15-50% Si, 1.50% Al	[260]	1959
		Antiemission (grid) coatings in electronics	[260]	1959
		Catalyst for dehydrating alcohols, cyclohexane, etc.	[630]	1961
$MnSi$, $MnSi_2$	Semiconductors	Semiconductor techniques	[632]	1960
		Electrodes of thermal generators with 5-13% efficiency	[633] [634]	1960 1961
$ReSi_2$	Semiconductor properties, scaling resistance to 1600-1700°C	Semiconductor techniques	[632] [633]	1960 1961
Iron silicides	Heat resistance, chemical resistance	Corrosion-resistant and heat-resistant coatings on steel components	[260]	1959
All aluminides	High strength, low density, high thermal conductivity	Modifying additives for various alloys in aviation and automobile industries, engineering	[18]	1965
Aluminides, actinides, and also $ZrAl_2$, $MoAl$, Mo_3Al, $CrAl$, $NiAl$	Nuclear properties in combination with high working temperatures	Nuclear power engineering	[18]	1965

Fe$_2$P	Catalytic properties	Catalysis in chemical industry	[29]	1961
Phases of the Ni–P system	High hardness and wear resistance	Hard and wear-resistant coatings on steels (microhardness up to 950 kg/mm^2)	[29]	1961
CeS, Ce$_2$S$_3$, LaS, La$_2$S$_3$, ThS, Th$_2$S$_3$, Th$_4$S$_7$	Infusibility, low vapor pressure, resistance to molten metals	Refractories for melting infusible metals (Ce, U, Th, Ti, Zr, etc.)	[558]	1961
Ce$_2$S$_3$, La$_2$S$_3$, Th$_2$S$_3$	Semiconductor properties	Semiconductor techniques	[558]	1961
B$_4$C in form of powder	High abrasion capacity, hardness	1. Grinding and polishing hard materials (technical stones, minerals, alloys, glasses, ceramics, quartz); **output** 50–70% of the output of diamond	[254]	1960
		2. Sharpening and reducing cutting plates of hard alloys	[254]	1960
		3. Lapping work		
		4. Metallographic work		
B$_4$C as sintered products with other carbides and metals	High hardness, abrasive capacity, wear resistance, heat resistance, slag resistance, large neutron-absorption cross section	1. Grinding and cutting work	[254]	1960
		2. Heat-resistant and hot strength alloys, e.g., alloy 64% B$_4$C + 36% Fe; σ_i = 24.5 (20°C), 22.8 (870°C); 22.4 (1093°C) and 17.6 kg/mm^2 (1316°C)	[254]	1960
		3. Cutting parts of drilling columns and tools for processing hard materials	[254]	1960

EXAMPLES OF THE APPLICATIONS OF REFRACTORY COMPOUNDS (continued)

Compound	Basic properties	Area of use	Source	Year
1	2	3	4	5
B_4C as sintered products and alloys with other carbides and metals		4. Die tips for welding in protective gases; practically no reaction with metal splash 5. Tools for treating grinding wheels; high finish dressing with 5-10 times output of other diamond substitutes (from hard alloys, silicon carbide, alumina). 6. Sand-blasting nozzles, 300 times more durable than iron 7. Calibers and templates 100-200 times more durable than steel 8. Filters for textile and chemical industry, filament guides in viscose production 9. Matrices for drawing rods from abrasive materials, welding electrodes, etc. 10. Chemical vessels 11. Neutron absorbers, cermets for nuclear technology (e.g., B_4C and Al_2O_3)	[254]	1960
B_4C	High electrical resistance, chemical resistance, semiconductor properties	Borocarbon, nonwire film-type resistances obtained by diffusion treatment of anthracite with boron (nominal resistance from 1 Ω to 300 kΩ, and dis	[254]	1960

452

SiC High electrical resistance, thermal-shock resistance, semiconducting properties, resistance to chemicals	persion power from 0.05 to 300 W; specific dispersion power about 0.4 W/cm^2; temperature coefficient $\leq 5 \cdot 10^{-4}$; in low-ohmic resistances, expansion factor is one order lower		
	Semiconductor sparkers for ignitrons	[254]	1960
	Nonlinear high-ohmic resistances (alloys and chemical compounds B$_4$C and SiC, so-called borosilicocarbides (B$_x$Si$_y$C$_z$)	[635]	1960
	In compositions for contacts based on silver (reduces burnout rate under the effect of an arc)	[631]	1961
	Electrical engineering articles:		
	1. Wave-guide absorbers (in wave guides for transmitting energy at frequencies above 1000 MHz) for finishing parts or complete absorption of input capacity		
	2. Small nonlinear inertia-free resistors (whose resistance critically depends on the direction of the electric field)		
	3. Ignitron sparkers (rectifiers with mercury cathode) in which the cathode spot on the surface of the mercury, being a source of free electrons, develops periodically as the current impulses pass through the semiconductor sparker (in this case, made of SiC)		
	4. Discharger		
	5. Thermocompensators		

EXAMPLES OF THE APPLICATIONS OF REFRACTORY COMPOUNDS (continued)

Compound	Basic properties	Area of use	Source	Year
1	2	3	4	5
SiC	High chemical resistance to acids, alkalis, metallic fusions, and metal vapors	Pump parts for pumping acid, cold and hot solutions; making refrigerators, scrubbers, working with hot gases and causing corrosion, valves of nozzles for spraying chemically active liquids; mixers resistant to corrosion with simultaneous abrasive action of hard constituents in suspension or slurries; diffusers (in particular, those resistant to phosphoric acid), collectors, cyclones subjected markedly to abrasive action of powders or dust-forming products	[623] [611]	1959 1961
	High hardness, abrasive capacity	Grinding, polishing, abrasive products of different shape and size	[631]	1961
	Reducing capacity	Deoxidizing steels, producing ferroalloys	[631]	1961
	Infusibility, thermal shock resistance, slag resistance	Heat-resistant alloys, e.g., SiC + C, SiC + Si + C, SiC + Co, SiC + B, etc.	[635]	1960
α–BN	Infusibility, high electrical resistance, and semiconductor properties	1. Heat insulation for high-frequency vacuum induction furnaces	[636]	1958
		2. Refractory washes in crucibles, heat-resistant holders for automatic welding; crucibles for precision metallurgy, cutoff drains for mixers and converters	[14]	1969
		3. High-temperature wash material		

Material	Properties	Applications		
β-BN (borazon)	Infusibility, high hardness, heat resistance, semiconductor properties	4. Low-duty bearings working in strong and corrosive acid solutions		
		5. Used in compositions of dielectrics		
		6. In compositions of high-temperature semiconductor materials		
		7. Refractories for cryolite-alumina baths in aluminum electrolysis		
		1. As diamond substitute	[254]	1960
		2. In heat-resistant alloys	[260]	1959
		3. High-temperature semiconductors	[260]	1959
Si₃N₄	Infusibility, high resistance to thermal shocks, good heat resistance to 1200–1300°C	In heat-resistant alloys, e.g., SiC + Si₃N₄, B₄C + Si₃N₄, SiC + Si₃N₄ + Fe		
	High chemical resistance to acids, alkali solutions, molten metals, salts and slags	Refractories for pumping molten metals and salts, pumps for melts, linings of baths for making aluminum by electrolysis from cryloite-alumina melts. High refractory properties possessed by articles of SiC bonded with Si₃N₄	[248]	1969
		Nozzle-sheaths of Si₃N₄ and alloys Si₃N₄ with SiC for protecting metallic thermocouples for temperature measurements in molten fluoride in aluminum electrolyzers (resistance 100 h at 940–970°C)	[248]	1969
	High electrical resistance, semiconductor properties	In volume resistors, high-temperature thermistors and thermal resistances of various types	[248]	1969

Appendix

PHASE DIAGRAMS OF SOME BINARY SYSTEMS

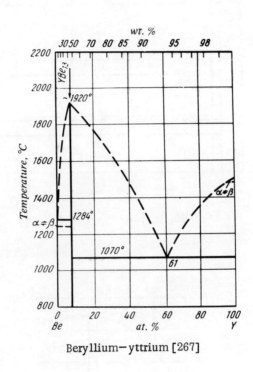

Beryllium—yttrium [267]

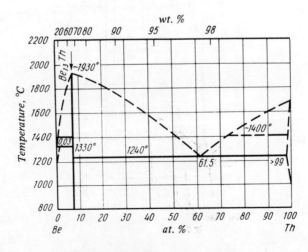

Beryllium—thorium system [267]

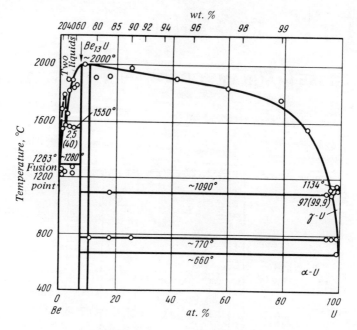

Beryllium—uranium system [2]

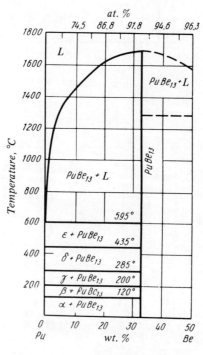

Plutonium—beryllium system [2]

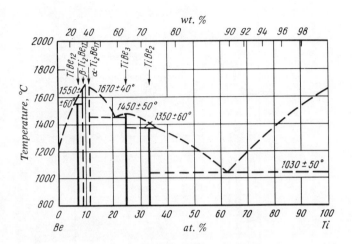

Beryllium–titanium system [267]

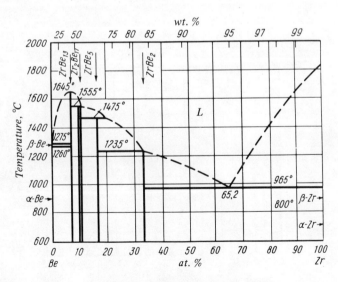

Beryllium–zirconium system [267]

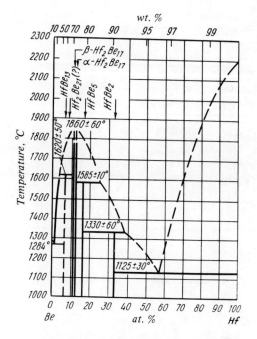

Beryllium—hafnium system [267]

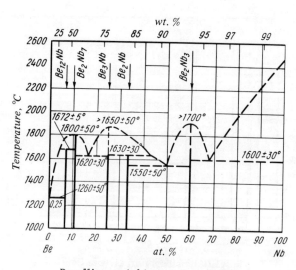

Beryllium—niobium system [268]

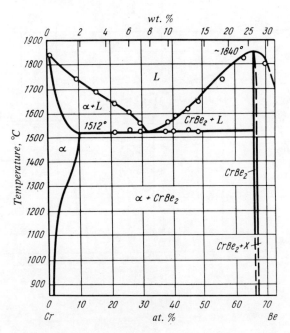

Section of the chromium—beryllium system [2]

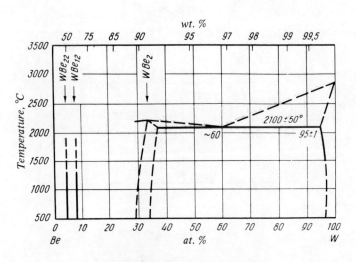

Beryllium—tungsten system [268]

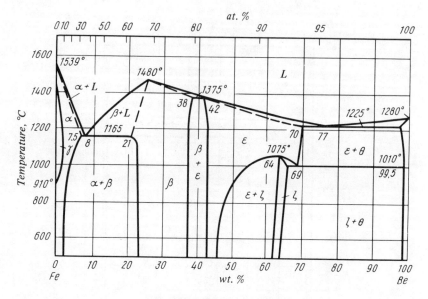

Iron—beryllium system [2]

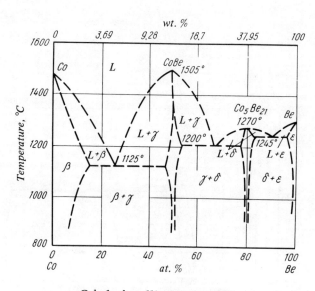

Cobalt—beryllium system [2]

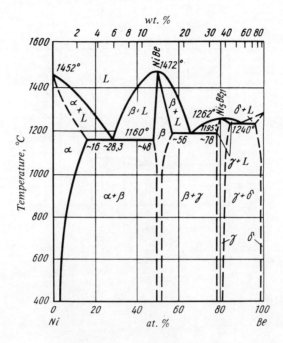

Nickel—beryllium system [2]

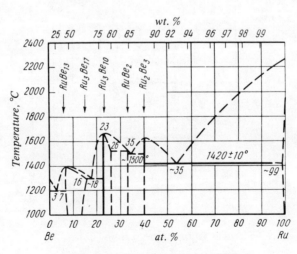

Beryllium—ruthenium system [268]

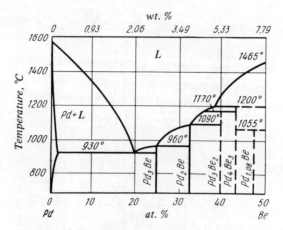

Section of the palladium—beryllium system [2]

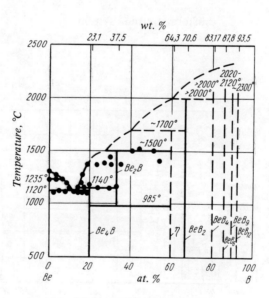

Beryllium—boron system [51]

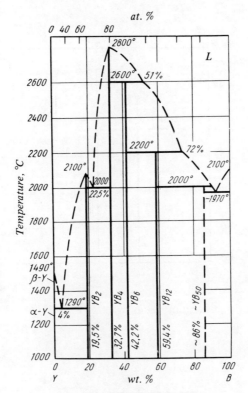

Yttrium—boron system [266]

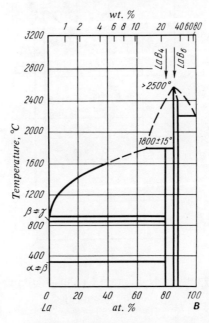

Lanthanum—boron system [267]

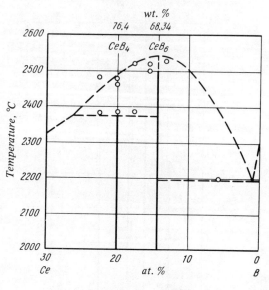

Cerium—boron system [273]

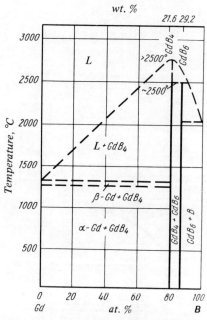

Hypothetical diagram for the gadolinium—
boron system [6]

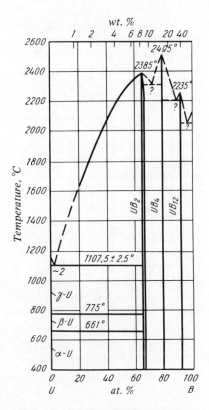

Uranium—boron system [267]

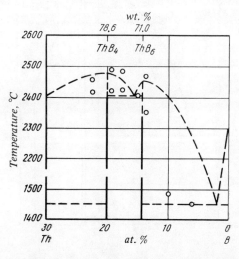

Section of the thorium —boron system [273]

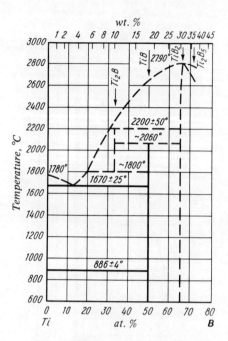

Titanium—boron system [125]

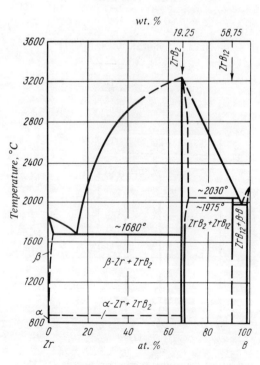

Zirconium—boron system [65]

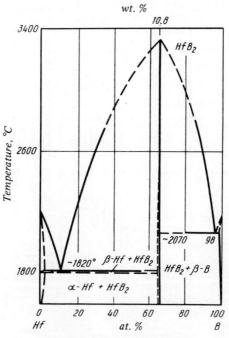

wt. %

Hafnium—boron system [66]

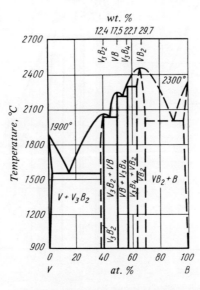

wt. %

Vanadium—boron system [253]

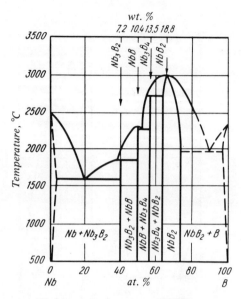

Niobium—boron system [255]

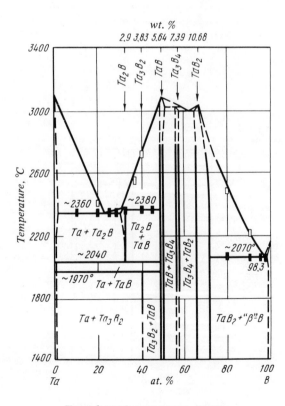

Tantalum—boron system [278]

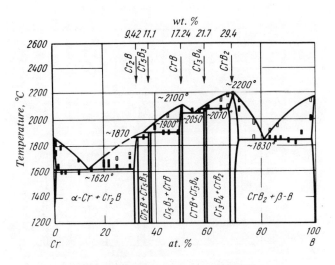

Chromium—boron system [281]

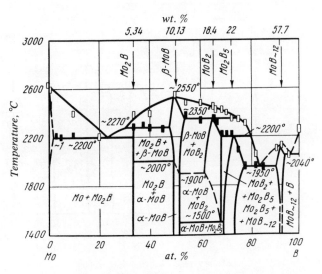

Molybdenum—boron system [279]

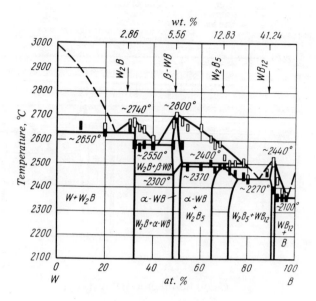

Tungsten—boron system [282]

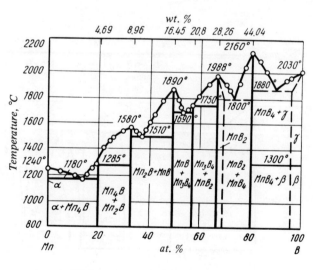

Manganese—boron system [9]

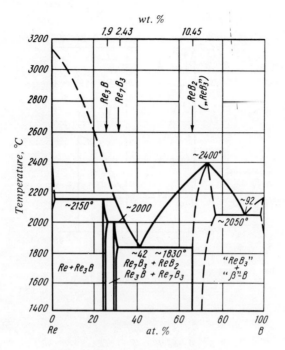

Rhenium—boron system [280]

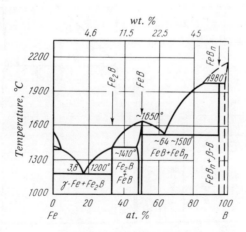

Iron—boron system [283]

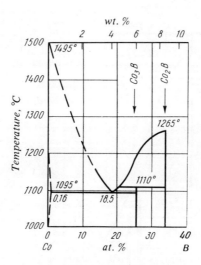

Cobalt—boron system [284]

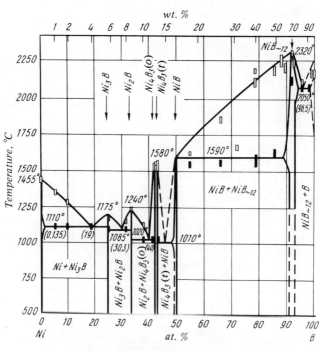

Nickel—boron system [285]

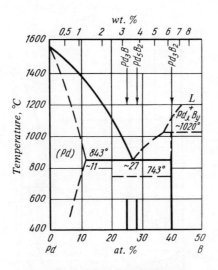

Palladium–boron system [267]

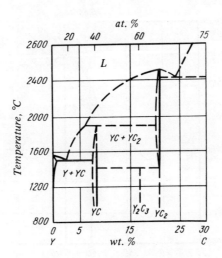

Yttrium–carbon system [266]

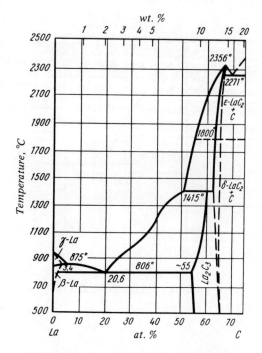

Lanthanum–carbon system [267]

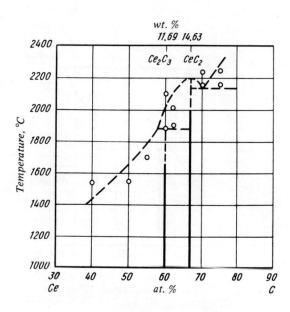

Cerium–carbon system [273]

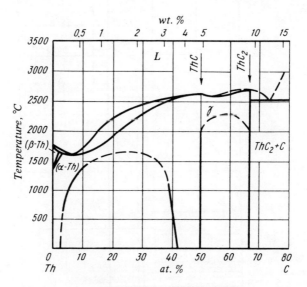

Thorium—carbon system [267]

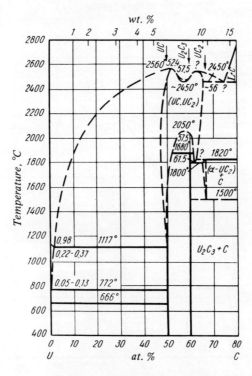

Uranium—carbon system [268]

477

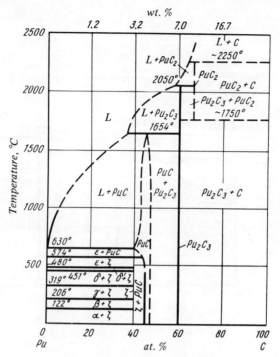

Plutonium—carbon system [11]

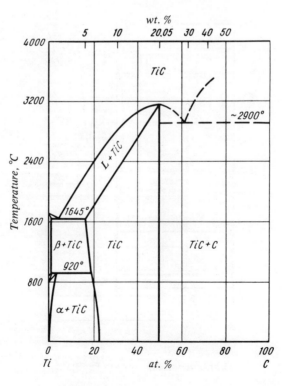

Titanium—carbon system [291]

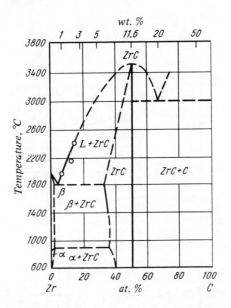

Zirconium—carbon system [291]

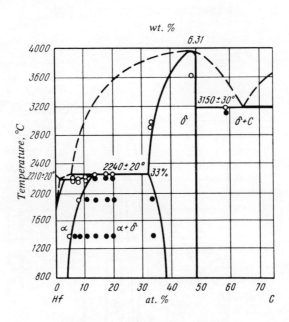

Hafnium—carbon system [357]

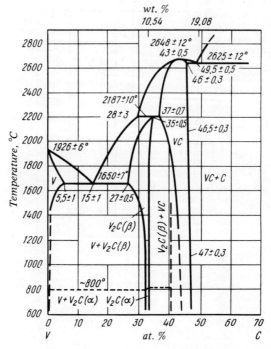

Vanadium—carbon system [289]

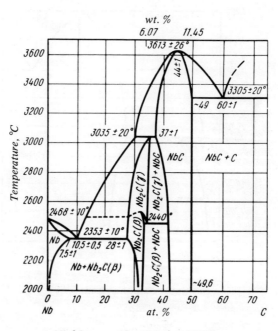

Niobium—carbon system [288]

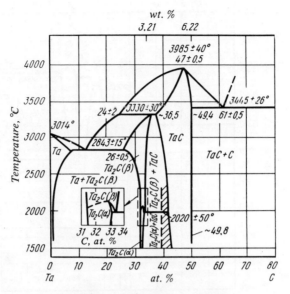

Tantalum—carbon system [289]

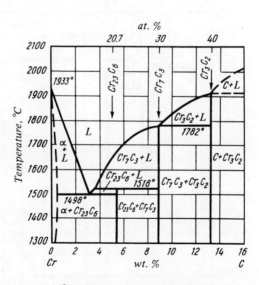

Chromium—carbon system [11]

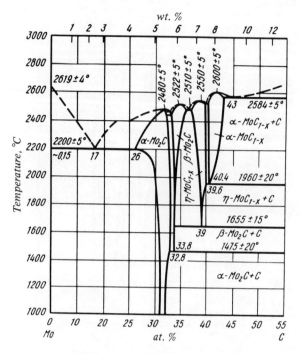

Molbdenum—carbon system [268]

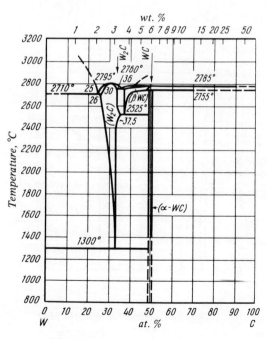

Tungsten—carbon system [268]

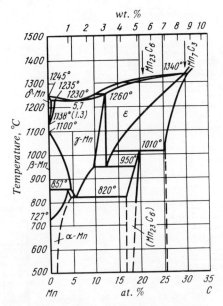

Manganese—carbon system [125]

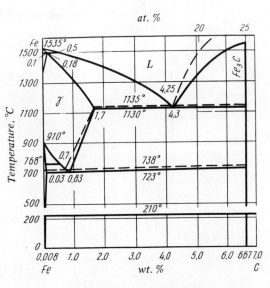

Iron—carbon system [11]

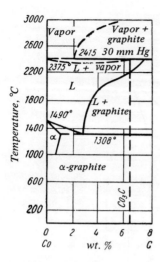

Cobalt—carbon system [11]

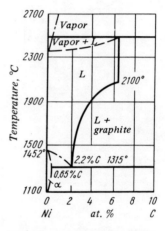

Nickel—carbon system [11]

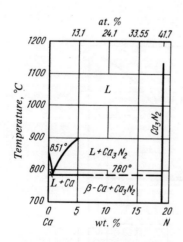

Calcium—nitrogen system [14]

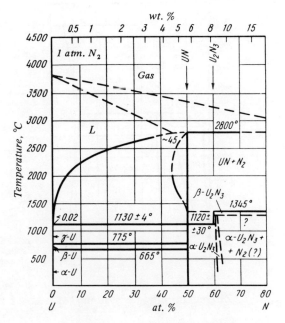

Uranium—nitrogen system [268]

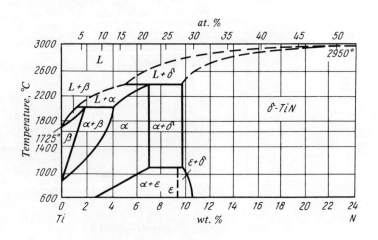

Titanium—nitrogen system [14]

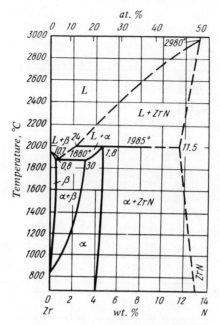

Zirconium—nitrogen system [14]

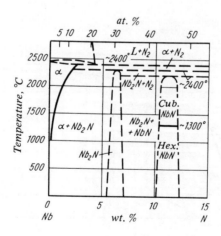

Niobium—nitrogen system [14]

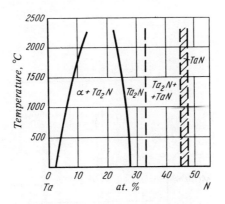

Tantalum−nitrogen system [14]

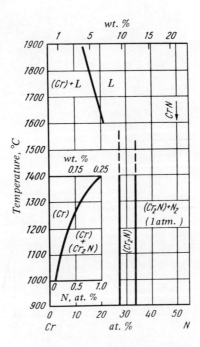

Chromium−nitrogen system [267]

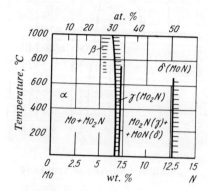

Molybdenum–nitrogen system [14]

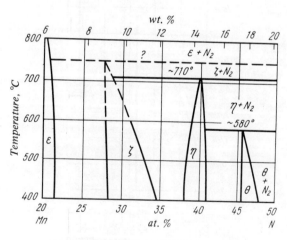

Manganese–nitrogen system [268]

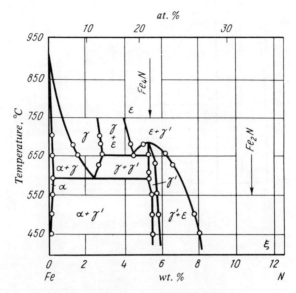

Iron—nitrogen system [14]

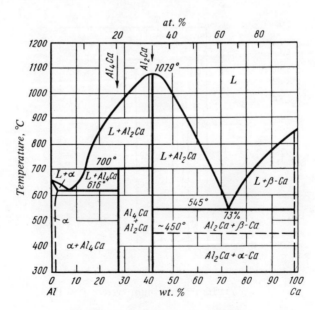

Aluminum—calcium system [18]

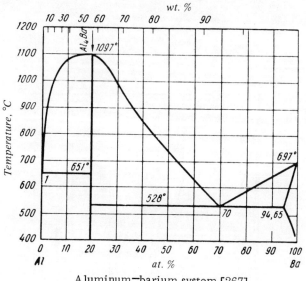

Aluminum—barium system [267]

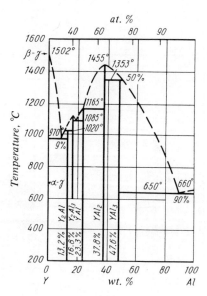

Yttrium—aluminum system [266]

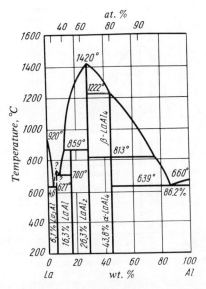

Lanthanum—aluminum system [18]

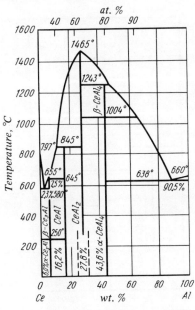

Cerium—aluminum system [18]

491

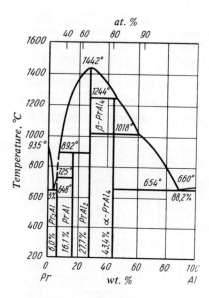

Praseodymium—aluminum system [18]

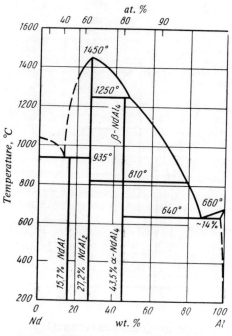

Neodymium—aluminum system [18]

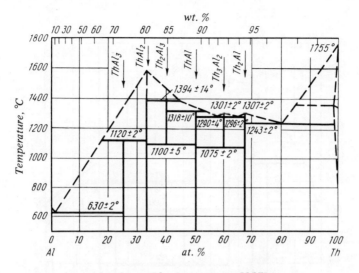

Aluminum—thorium system [267]

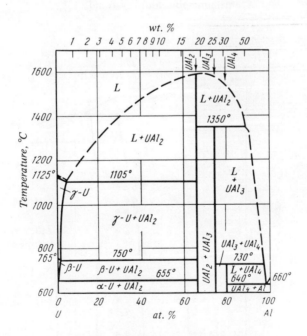

Uranium—aluminum system [18]

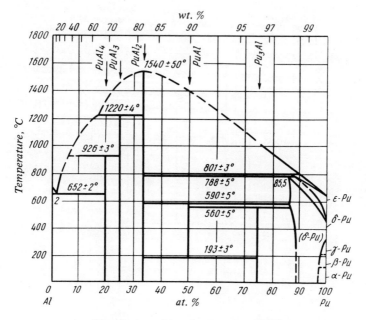

Aluminum—plutonium system [268]

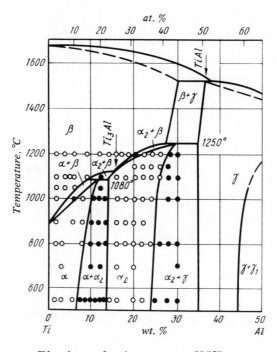

Titanium—aluminum system [359]

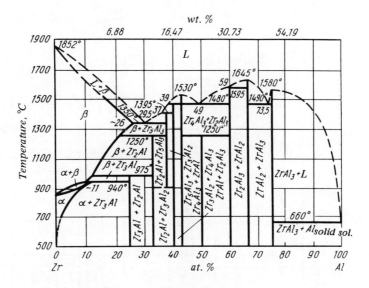

Zirconium—aluminum system [18]

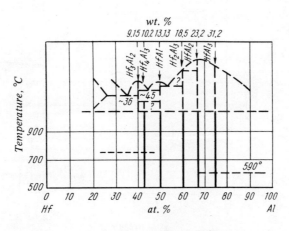

Hafnium—aluminum [18]

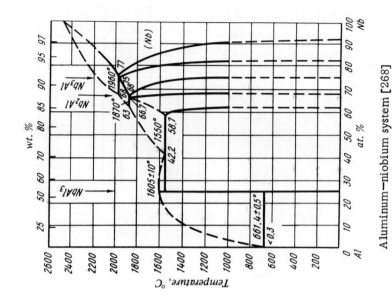

Aluminum—niobium system [268]

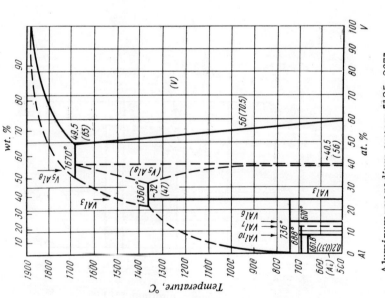

Aluminum—vanadium system [125, 267]

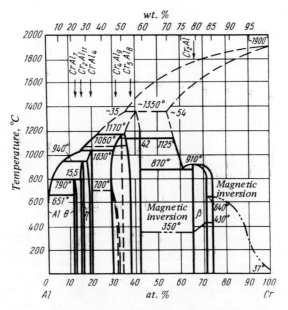

Aluminum–chromium system [268]

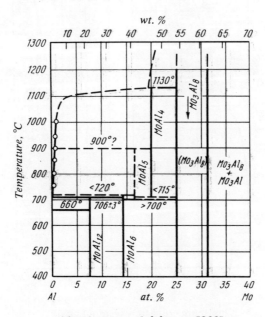

Aluminum–molybdenum [268]

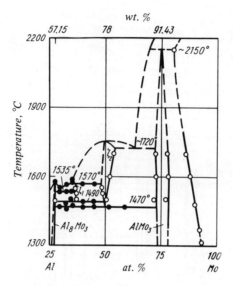

Aluminum—molybdenum system [295]

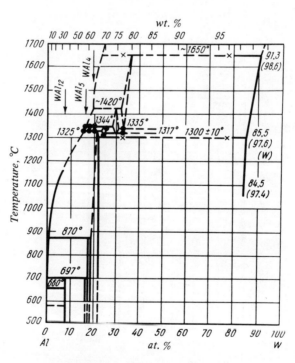

Aluminum—tungsten system [125]

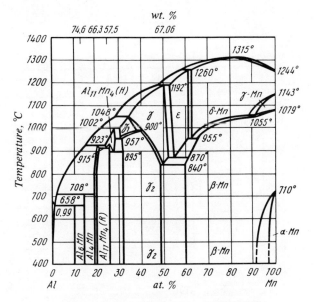

Aluminum—manganese system [296]

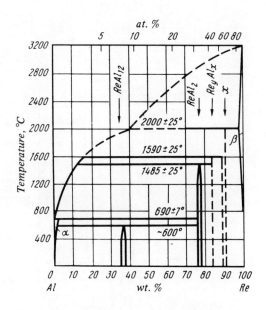

Aluminum—rhenium system [18]

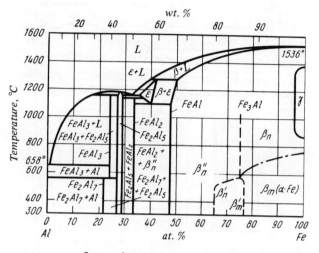

Iron—aluminum system [18]

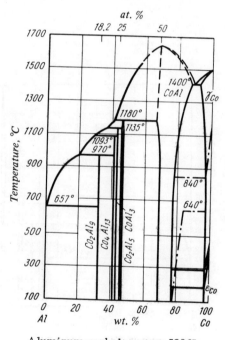

Aluminum—cobalt system [296]·

500

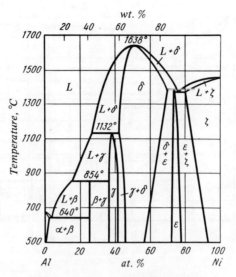

Aluminum—nickel system [18]

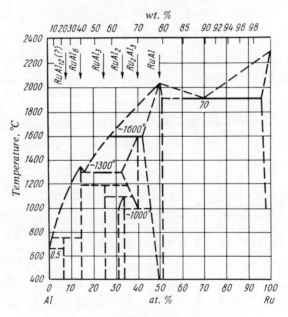

Aluminum—ruthenium system [152]

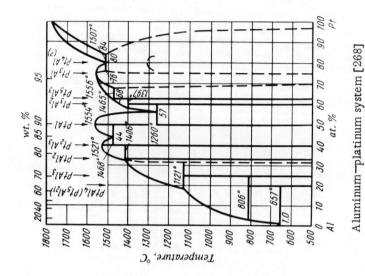

Aluminum—platinum system [268]

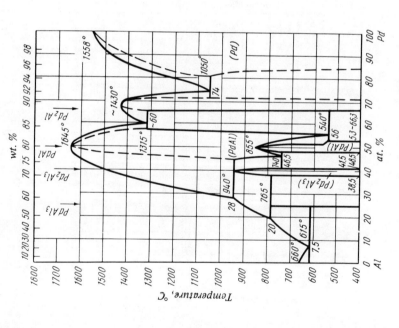

Aluminum—palladium system [268]

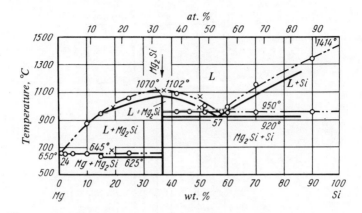

Magnesium—silicon system [125] (two versions)

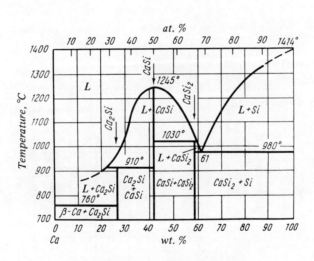

Calcium—silicon system [260]

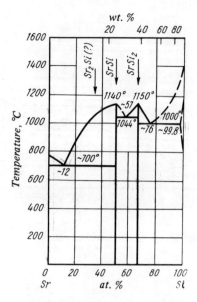

Strontium—silicon system [268]

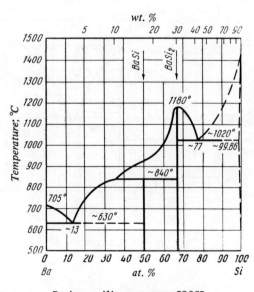

Barium—silicon system [268]

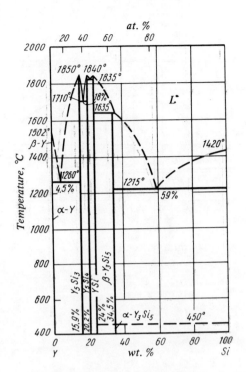

Yttrium—silicon system [266]

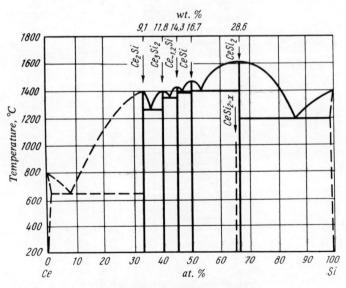

Cerium—silicon system [273]

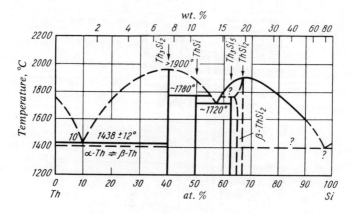

Thorium—silicon system [268]

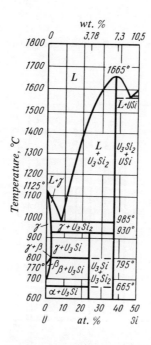

Uranium—silicon system [260]

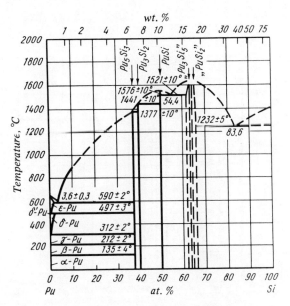

Plutonium—silicon system [268]

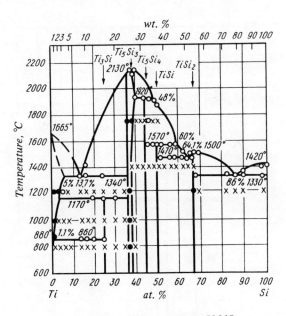

Titanium—silicon system [301]

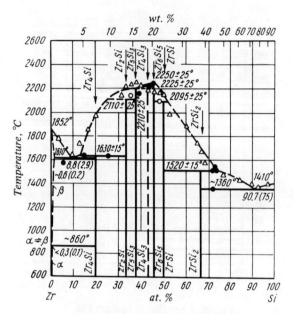

Zirconium—silicon system [125]

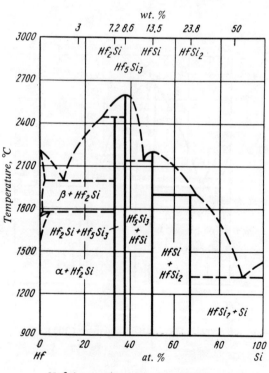

Hafnium—silicon system [267]

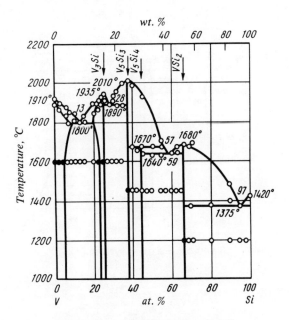

Vanadium–silicon system [166]

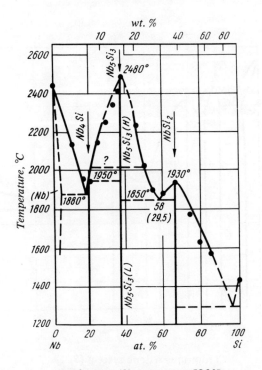

Niobium–silicon system [260]

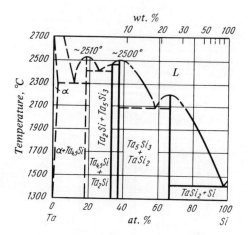

Tantalum—silicon system [125]

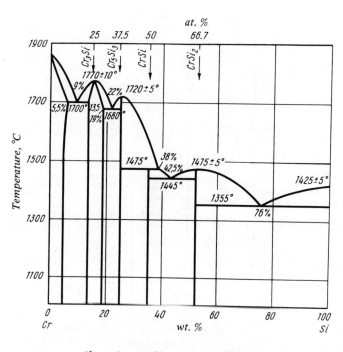

Chromium—silicon system [304]

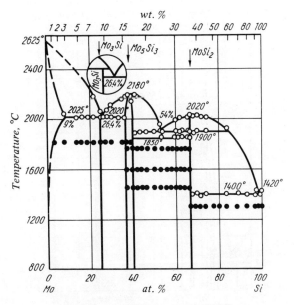

Molybdenum—silicon system [306]

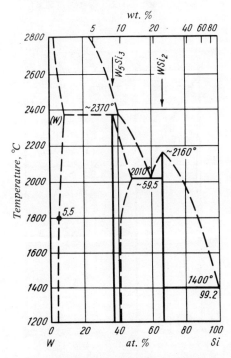

Tungsten—silicon system [267]

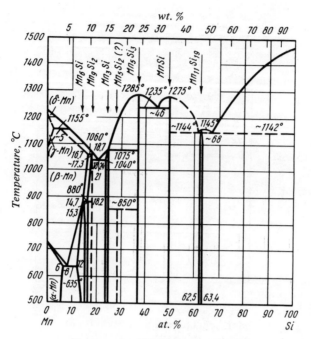

Manganese–silicon system [268]

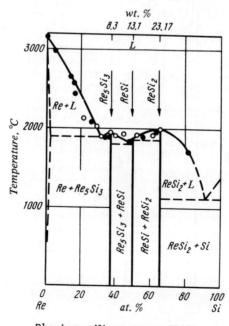

Rhenium–silicon system [267]

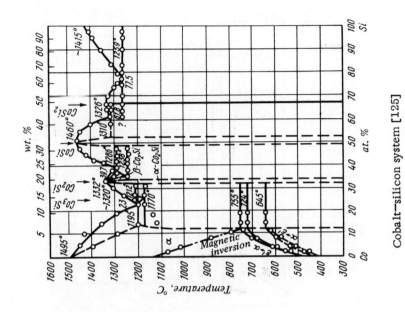

Cobalt—silicon system [125]

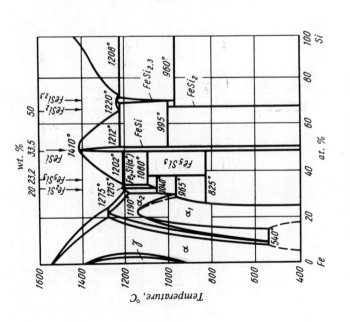

Iron—silicon system [307, 308, 22]

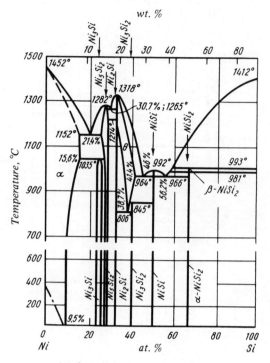

Nickel—silicon system [22]

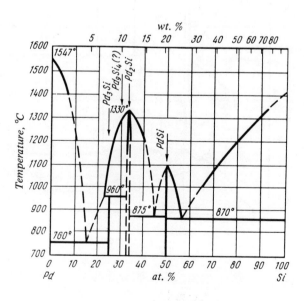

Palladium—silicon system [267]

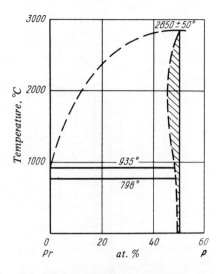

Praseodymium—phosphorus system [27]

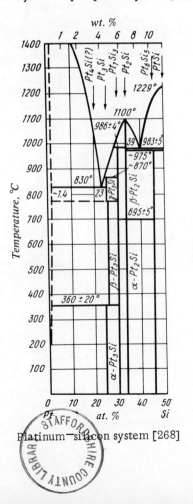

Platinum—silicon system [268]

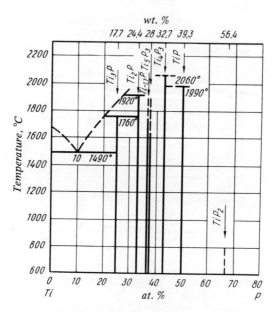

Titanium—phosphorus system [28]

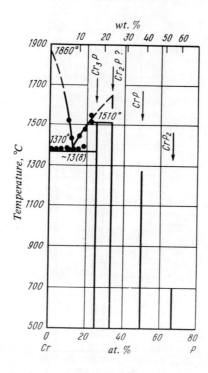

Chromium—phosphorus system [125]

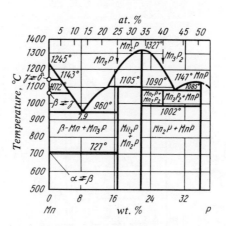

Manganese–phosphorus system [125]

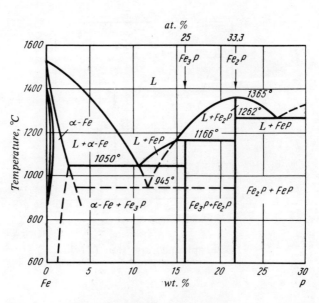

Iron–phosphorus system [125]

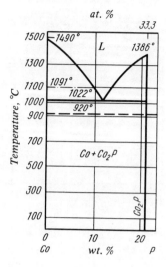

Cobalt–phosphorus system [125]

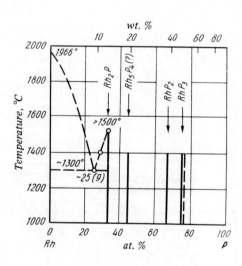

Nickel–phosphorus system [214]

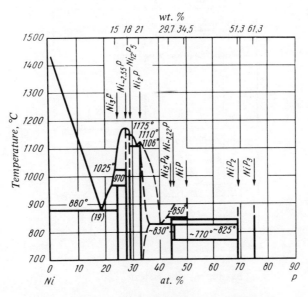

Rhodium–phosphorus system [125]

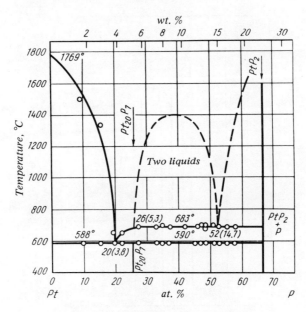

Platinum–phosphorus system [125]

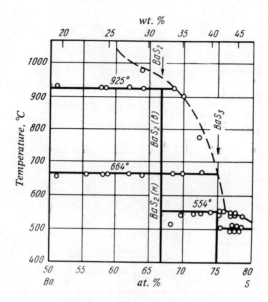

Barium—sulfur system [125]

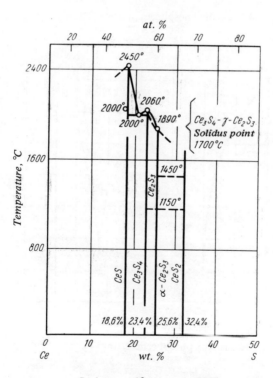

Cerium—sulfur system [33]

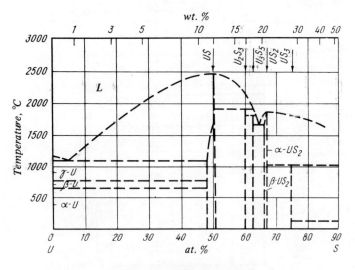

Uranium–sulfur system [267]

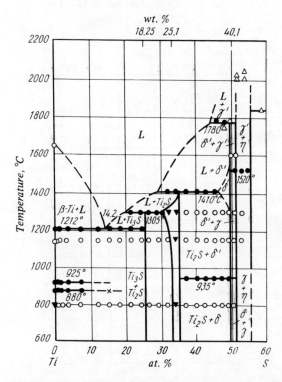

Preliminary version of titanium–sulfur system [33]

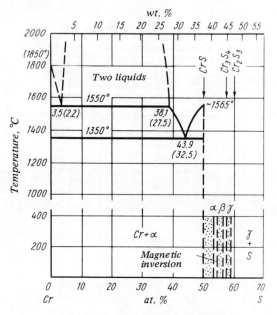

Chromium–sulfur system [125]

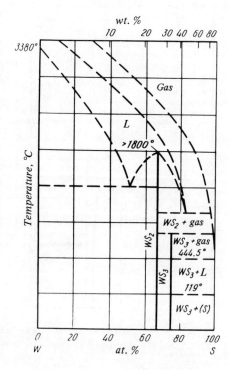

Tungsten–sulfur system [267]

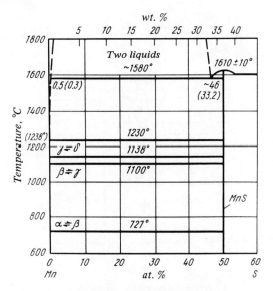

Manganese−sulfur system [125]

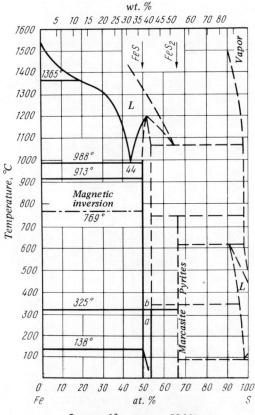

Iron−sulfur system [268]

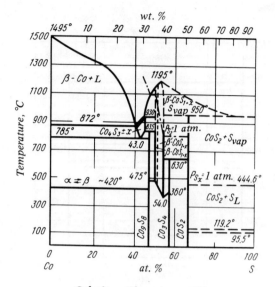

Cobalt—sulfur system [33]

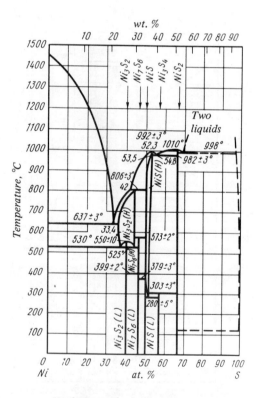

Nickel—sulfur system [267]

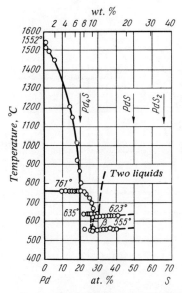

Palladium—sulfur system [125]

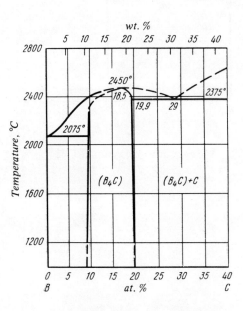

Boron—carbon system [267]

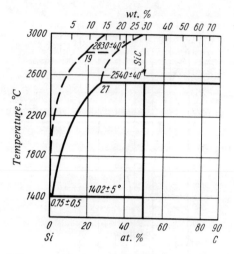

Silicon—carbon system [267] (two versions)

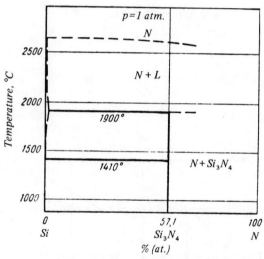

Silicon—nitrogen system [358]

LITERATURE CITED

1. E. Laube and H. Nowotny, Mh. Chemie, 93(3):681-683 (1962).
2. G. V. Samsonov, Beryllides [in Russian], Naukova Dumka, Kiev (1966), p. 108.
3. G. V. Samsonov, T. I. Serebryakov, and V. A. Neronov, Borides [in Russian], Atomizdat, Moscow (1975), p. 386.
4. J. Etourneau, R. Naslain, and S. La Placa, J. Less-Common Metals, 24:183-194 (1971).
5. N. N. Zhuravlev, R. M. Manelis, N. V. Gramm, et al., Poroshk. Metall. No. 2, 95-101 (1967).
6. R. M. Manelis, G. A. Meerson, N. N. Zhuravlev, et al., Poroshk. Metall. No. 11, 77-84 (1966).
7. B. Armas, Rev. Int. Hautes. Temp. Refract., 10(4):231-239 (1973).
8. L. G. Bodrova, M. S. Koval'chenko, and T. I. Serebryakova, Poroshk. Metall., No. 1, 1-4 (1974).
9. L. Ya. Markovskii and E. T. Bezruk, Izv. Akad. Nauk. SSSR Neorg. Mater., 3(12):2165-2169 (1967).
10. G. V. Samsonov and A. P. Épik, Refractory Coatings [in Russian], Metallurgiya, Moscow, (1973), p. 399.
11. T. Ya. Kosolapova, Carbides [in Russian], Metallurgiya, Moscow, (1968).
12. R. Lorenzelli, I. De Dienleveult, and J. Melamed, J. Inorg. Nucl. Chem., 33:3297-3304 (1971).
13. E. Storms, Refractory Carbides, Academic Press, New York, (1967).
14. G. V. Samsonov, Nitrides [in Russian], Naukova Dumka, Kiev (1969), p. 378.
15. R. Kieffer, D. Fisher, and E. Heidler, Metall., 26(2):128-132 (1972).
16. A. Fontbonne and J. C. Gilles, Rev. Int. Hautes Temp. Refract., 6(3):181-192 (1969).
17. R. Colling, Nonstoichiometry [Russian translation], Mir (1974).
18. V. S. Sinel'nikova, V. A. Podergin, and V. N. Rechkin, Aluminides [in Russian], Naukova Dumka, Kiev (1965).
19. H. Nowotny, O. Schob, and F. Benesovsky, Mh. Chemie, 92:1300-1303 (1961).
20. V. S. Sidel'nikova, Tsvetny. Met., No. 10, 69-73 (1964).
21. H. Jacobi and H. J. Engell, Acta Met., 19:701-711 (1971).
22. P. V. Gel'd and F. A. Sidorenko, Silicides of Transition Metals in Period IV [in Russian], Metallurgiya, Moscow (1971).
23. E. Übelacker and P. Lecoco, C. R. Acad. Sci., 262C:793-795 (1966).
24. A. Amamou, P. Bach, and F. Gantier, et al., J. Phys. Chem. Solids, 33(9): 1697-1712 (1972).
25. G. Pilstrom, Acta Chem. Scand., 15(4):893-902 (1961).
26. W. Gold and K. Schubert, Z. Krist., 128:406-413 (1969).
27. K. E. Mironov, I. G. Vasil'eva, and I. G. Pritchina, Rev. Chim. Miner., 10(1-2):383-398 (1973).
28. P. Snell, Acta Chem. Scand., 22(6):1942-1952 (1968).
29. G. V. Samsonov and L. L. Vereikina, Phosphides [in Russian], Akad. Nauk Ukr. SSR, Kiev (1961).

30. B. Sellberg, Acta Chem. Scand., 20:2179-2186 (1966).

31. M. Guittard, C. R. Acad. Sci. Paris, 261:2109-2112 (1965).

32. P. Besancon, C. R. Acad. Sci. Paris, 267C:1130-1132 (1968).

33. G. V. Samsonov and S. V. Drozdova, Sulfides [in Russian], Metallurgiya (1972).

34. E. Trong and M. Huber, C. R. Acad. Sci. Paris, 272C(11):1018-1021 (1971).

35. F. Jellink, Ark. Kemi, 20:447-480 (1963).

36. V. A. Obolonchik, S. V. Radzikovskaya, and V. F. Bukhanevich, Poroshk. Metall., No. 11, 9-14 (1965).

37. V. F. Bukhanevich, S. P. Gordienko, and T. I. Serebryakov, Poroshk. Metall., No. 5, 49-51 (1971).

38. E. Vallet and J. M. Paris, C. R. Acad. Sci. Paris, 264C:203-206 (1967).

39. Y. Jeanin, Bull. Soc. Fr. Mineral. Cristallogr., 10C:528-536 (1967).

40. V. I. Kudryavtsev and G. V. Safronov, Trans. Seminar on Refractory Materials, [in Russian], Akad. Nauk. Ukr. SSR, Kiev, No. 5, 52-55 (1960).

41. T. W. Baker, Acta Cryst., 15(3):175-179 (1962).

42. N. N. Matyushenko, Crystal Structure of Binary Compounds [in Russian], Metallurgiya, Moscow (1969).

43. N. N. Matyushenko, V. N. Karev, and A. P. Svinarenko, Ukr. Fiz. Zhurn., 11:1266 (1963).

44. E. I. Gladyshevskii, P. I. Kripyakevich, and D. P. Frankevich, Kristallografiya, 8(5):788-791 (1963).

45. E. Rudy, F. Benesovsky, H. Nowotny, and L. Toth, Mh. Chem., 92(3):692-700 (1961).

46. Choi Shi Chan and M. V. Mal'tsev, Tsvetn. Metall., No. 5, 133-142 (1959).

47. K. Shubert, Crystal Structures of Bicomponent Phases [in Russian], Metallurgiya, Moscow (1971).

48. N. N. Matyushenko, V. P. Serykh, and Yu. G. Titov, All-Union Conference on Crystal Chemistry of Intermetallic Compounds [in Russian], L'vov State University (1971), pp. 30-31.

49. L. F. Verkhorobin, N. N. Matuyshenko, and A. A. Kruglykh, et al., All-Union Conference on Crystal Chemistry of Intermetallic Compounds [in Russian], L'vov State University (1971), pp. 10-11.

50. N. N. Matyushenko, A. A. Kruglykh, and G. F. Tikhinskii, All-Union Conference on Crystal Chemistry of Intermetallic Compounds [in Russian], L'vov State University (1971), p. 30.

51. J. Stecher and F. Aldinger, Z. Metallkd., 64(10):684-689 (1973).

52. N. V. Vekshina, L. Ya. Markovskii, Yu. D. Kondrashev, et al., Zh. Prikl. Khim., 42(6):1229-1234 (1969).

53. R. Naslain, A. Guette, and M. Barret, J. Solid State Chem., 8:68-85 (1973).

54. N. N. Zhuravlev, A. A. Stepanova, Yu. B. Paderno, and G. V. Samsonov, Kristallografiya, 6(5):791-794 (1961).

55. N. N. Zhuravlev and A. A. Stepanova, Kristallografiya, 3(1):83-85 (1958).

56. P. Peshev, J. Etourneau, and R. Naslain, Mat. Res. Bull., 5:319-328 (1970).

57. J. Etourneau and R. Naslain, C. R. Acad. Sci. Paris, 266C:1452-1455 (1968).

58. Yu. B. Paderno and G. V. Samsonov, Zh. Struk. Khim., 2(2):213-214 (1961).

59. V. V. Odintsov and Yu. B. Paderno, Izv. Akad. Nauk SSSR Neorg. Mater., 7(2):333-334 (1971).

60. A. A. Stepanova and N. N. Zhuravlev, Kristallografiya 3(1):94-95 (1958).

61. V. S. Neshpor and G. V. Samsonov, Zh. Fiz. Khim., 32(6):1328-1332 (1958).

62. K. Schwetz, P. Ettayer, R. Kieffer, et al., J. Less-Common Metals, 26:99-104 (1972).

63. Yu. B. Paderno, At. Énerg. 10(4):396 (1961).

64. B. M. Howlett, J. Inst. Metals, 88:91-92 (1959).

65. K. I. Portnoi, V. M. Romashov, and L. N. Burobina, Poroshk. Metall., No. 7, 68-71 (1970).

66. K. I. Portnoi, V. M. Romashov, I. V. Romanovich, et al., Izv. Akad. Nauk. SSSR Neorg. Mater., 7(11):1987-1991 (1971).

67. E. Rudy, F. Benesovsky, and L. Toth, Z. Metallkd., 54(6):345-353 (1963).

68. M. Elfströmm, Acta Chem. Scand., 15(5):1178 (1961).

69. S. Andersson and T. Lundström, Acta Chem. Scand., 22(10):3103-3110 (1968).

70. T. Lundström, Ark. Kemi, 30(11):115-127 (1968).

71. B. Aronsson, Acta Chem. Scand., 14(6):1414-1418 (1960).

72. S. Andersson and J. O. Karlsson, Acta Chem. Scand., 24(5):1791-1799 (1970).

73. W. Trzebiatowski and W. Rudzinski, J. Less-Common Metals, 6(3):244-247 (1964).

74. B. Aronsson, M. Bäckman, and S. Rundqvist, Acta Chem. Scand., 14(4):1001-1005 (1960).

75. B. Aronsson, E. Stenberg, and I. Aselius, Acta Chem. Scand., 14(3):733-736 (1960).

76. J.-P. Bouchaud, Ann. Chim., 2(6):353-366 (1967).

77. V. S. Neshpor, Yu. B. Paderno, and G. V. Samsonov, Dokl. Akad. Nauk SSSR, 118(3):515-517 (1958).

78. B. Aronsson and S. Rundqvist, Acta Cryst., 15(9):878-887 (1962).

79. C. Mai, M. Ayel, G. Monnier, and R. Riviere, C. R. Acad. Sci. Paris, 267D(11):987-990 (1968).

80. S. Rundqvist, Nature, 181:259-260 (1958).

81. S. Rundqvist and S. Pramatus, Acta Chem. Scand., 21(1):191-194 (1967).

82. S. Rundqvist, Acta Chem. Scand., 13(6):1193-1208 (1959).

83. J. Aselius, Acta Chem. Scand., 14(9):2169-2176 (1960).

84. B. Aronsson, E. Stenberg, and J. Aselius, Nature, 195(4839):377-378 (1962).

85. B. Aronsson, Acta Chem. Scand., 17(7):2036-2050 (1963).

86. B. Aronsson, Ark. Kemi, 16:379-423 (1960).

87. P. Rogl, H. Nowotny, and F. Benesovsky, Mh. Chemie, 102(3):678-686 (1971).

88. F. Aldinger, Metall. 26(7):711-718 (1972).

89. H. Rassaerts, H. Nowotny, G. Vinek, and F. Benesovsky, Mh. Chemie, 98(2):460-488 (1967).

90. H. Jedlicka, H. Nowotny, and F. Benesovsky, Mh. Chemie, 102(2):389-403 (1971).

91. G. V. Samsonov, G. N. Makarenko, and T. Ya. Kosolapova, Dokl. Akad. Nauk SSSR, 144(5):1062-1065 (1962).

92. G. V. Samsonov, T. Ya. Kosolapova, and G. N. Makarenko, Zh. Neorg. Khim. 7(5):975-979 (1962).

93. T. Ya. Kosolapova and G. N. Makarenko, Ukr. Khim. Zh., 30(8):784-787 (1964).

94. F. Spedding, K. Gschneider, and A. Danne, J. Am. Chem. Soc., 80(17):4499-4503 (1958).

95. R. Vikkery, R. Sedlacek, and A. Rusen, J. Chem. Soc., 159:498-501 (1959).

96. G. Bacchella, P. Meriel, M. Pinot, and R. Lallement, Bull. Soc. Fr. Mineral. Cristallogr., 89:226-228 (1966).

97. G. Dean, R. Lallement, et al., C. R. Acad. Sci. Paris, 259:2442-2444 (1964).

98. J. Bauer, J. Less-Common Metals, 37:161-165 (1974).

99. R. Lorenzelli and I. Dieuleveult, J. Nucl. Mat., 29:349-353 (1969).

100. L. Jakesova and D. Jakes, At. Energy Rev., 1(3):3-46 (1963).

101. R. Lorenzelli, C. R. Acad. Sci. Paris, 266C:900-902 (1968).

102. S. Nagakura and S. Oketani, Trans. Iron Steel Inst. Japan, 8:265-268 (1968).

103. H. Rassaerts, F. Benesovsky, and H. Nowotny, Planseeber. Pulvermetall., 14(3):178-183 (1966).

104. N. Terao, C. R. Acad. Sci. Paris, 275C:1165-1168 (1972).

105. Y. Guerin and Ch. Novion, Rev. Int. Hautes Temp. Refract., 8(3-4):311-314 (1971).

106. J. L. Martin, A. Rocher, B. Jouffrey, and P. Costa, Phil. Mag., 24(192):1355-1364 (1971).

107. A. Rouault, P. Herpin, and R. Fruchart, Ann. Chim., 5(6):461-470 (1970).

108. R. Fruchart and A. Rouault, Ann. Chim., 4(3):143-145 (1969).

109. S. Rundqvist and G. Runnsjö, Acta Chem. Scand., 23(4):1191-1199 (1969).

110. K. Yvon, J. Nowotny, and F. Benesovsky, Mh. Chemie, 99(2):726-729 (1968).

111. J. P. Senateur, Ann. Chim., 2(2):103-122 (1967).

112. J. P. Bouchard and R. Fruchart, Bull. Soc. Chim. Fr., Ser. S(275):1579-1583 (1964).

113. S. V. Popova and L. G. Boiko, High Temp. — High Pressures, 3(2):237-238 (1971).

114. M. D. Lyutaya, G. V. Samsonov, and O. P. Kulik, in: Chemistry and Physics of Nitrides [in Russian], Naukova Dumka, Kiev (1968), pp. 21-46.

115. J. Gaude, P. L'Haridon, Y. Laurent, and J. Lang, Bull. Soc. Fr. Mineral. Cristallogr., 95:56-60 (1972).

116. G. Lobier and J. P. Marcon, C. R. Acad. Sci. Paris, 268C:1132-1135 (1969).

117. M. Billy and B. Teyssedre, Bull. Soc. Chim. Fr., No. 5, 1537-1540 (1973).

118. J. C. Gilles, C. R. Acad. Sci. Paris, 266C:546-547 (1968).

119. O. T. Khorpyakov, Yu. B. Paderno, and V. P. Dzeganovskii, X-ray Standards for Hard and Refractory Compounds [in Ukrainian], Akad. Nauk. Ukrainian SSR (1961), p. 63.

120. R. Karam and R. Ward, Inorg. Chem., 9(6):78 (1970).

121. G. V. Samsonov and T. S. Verkhoglyadova, Ukr. Khim. Zh., 30(2):143-146 (1964).

122. V. N. Rechkin, L. K. Lamikhov, and T. N. Samsonova, Kristallografiya, 9(3):405-408 (1964).

123. C. Becle and R. Lemaire, Acta. Cryst., 23(5):840-845 (1967).

124. A. E. Vol, Structure and Properties of Binary Metallic Systems [in Russian], Vols. I, II Fizmatgiz (1959, 1962).

125. M. Khansen and K. Anderko, Structure of Binary Alloys [in Russian], Vols. 1, 2, Metallurgizdat (1962).

126. A. Zalkin, R. Bedford, and D. Sands, Acta Cryst., 12(9):700-702 (1959).

127. D. Sands, A. Zalkin, and O. Krikorian, Acta Cryst., 12(6):461-465 (1959).

128. A. Zalkin, D. Sands, R. Bedford, et al., Acta Cryst., 14(1):63-65 (1961).

129. R. M. Paine and J. A. Carrabine, Acta Cryst., 13(8):680-681 (1960).

130. F. Batchelder and R. Rauechle, Acta Cryst., (10):648-650 (1967).

131. F. Batchelder and R. Rauechle, Acta Cryst., 11(2):122-124 (1958).

132. H. Eich and P. Gilles, J. Am. Chem. Soc., 81(19):5030-5032 (1959).

133. B. McDonald and W. Stuart, Acta Cryst., 13(5):447-448 (1960).

134. W. Klemm and G. Winkelman, Z. Anorg. Chem., 288(87) (1956).

135. K. H. J. Bushchow and J. H. N. van Vucht, Z. Metallkd., 56(1):9-13 (1965).

136. J. H. N. van Vucht and K. H. J. Buschow, Philips Res. Rep., 19(4):319-321 (1964).

137. K. H. J. Buschow, Philips Res. Rep., 20(4):337-339 (1965).

138. J. H. Wernick and S. Geller, Trans. Met. Soc. AIME, 218(5):866-869 (1960).

139. K. H. J. Buschow and J. H. N. van Vucht, Z. Metallkd., 57(2):162-164 (1966).

140. K. H. J. Buschow, J. Less-Common Metals, 8(3):209-212 (1965).

141. K. H. J. Buschow, J. Less-Common Metals, 9(6):452-457 (1965).

142. S. E. Haszlo, Trans. Met. Soc. AIME, 218(5):958-961 (1960).

143. V. I. Kutaitsev, Alloys of Thorium, Uranium, and Plutonium [in Russian], Gosatomizdat, Moscow (1962).

144. P. I. Kripyakevich and E. I. Gladyshevskii, Kristallografiya, 6(1):118-122 (1961).

145. M. V. Nevitt, in: The Electron Structure of Transition Metals and the Chemistry of Their Alloys [in Russian], Metallurgiya, Moscow (1966), pp. 97-165.

146. O. J. C. Runnals and R. R. Boucher, J. Nucl. Mater., 15(1):57-62 (1965).

147. L. E. Edshammar, Acta Chem. Scand., 16(1):20-22 (1962).

148. H. Holleck, F. Benesovsky, and H. Nowotny, Mh. Chemie, 94(2):477-481 (1963).

149. V. S. Sinel'nikova, G. V. Samsonov, and S. N. L'vov, Izv. Akad. Nauk SSSR Metall. i Gorn. Delo, No. 5, 121-127 (1964).

150. V. I. Surikov, A. K. Shtol'ts, et al., Ukr. Fiz. Zh., 15(1):118-119 (1970).

151. M. A. Taylor, Acta Met., 8(4):256-260 (1960).

152. W. Obrowski, Metall., 17(2):108-112 (1963).

153. R. Huch and W. Klemm, Z. Anorg. Allg. Chem., 329:123-126 (1964).

154. E. I. Gladyshevskii, Crystal Chemistry of Silicides and Germanides [in Russian], Metallurgiya, Moscow (1971).

155. E. Parthé, in: Properiétés Thermodynamiques Physiques et Structurales des Derives Semi-metalliques, Ed. du Centre Nat. Rech. Sc., Paris (1967), pp. 195-205.

156. E. I. Gladyshevskii, L. A. Dvorina, et al., Izv. Akad. Nauk SSSR Neorg. Mater., 1(3):321-325 (1965).

157. L. A. Dvorina, V. S. Protasov, et al., Izv. Akad. Nauk SSSR Neorg. Mater., 1(5):711-714 (1965).

158. G. S. Smith, Acta Cryst., 22(6):940-942 (1967).

159. D. Hohnke and E. Parthé, Acta Cryst., 20(4):572-573 (1966).

160. E. I. Gladyshevskii and P. I. Kripyakevich, Izv. Akad. Nauk SSSR Neorg. Mater., 1(5):702-704 (1965).

161. P. L. Blum, J. Silvestre, and H. Vangoyeau, C. R. Acad. Sci. Paris, 260: 5538-5541 (1965).

162. J. Laugier, P. Blum, and R. Tournemine, J. Nucl. Mater., 41:106-108 (1971).

163. V. N. Svechnikov, L. M. Yupko, et al., Dokl. Akad. Nauk SSSR 193(2):393-397 (1970).

164. J. J. Nickel and K. K. Schweitzer, Z. Metallkd., 61:54-57 (1970).

165. P. Spinat, R. Fruchart, and P. Herpin, Bull. Soc. Fr. Mineral. Cristallogr., 93:23-36 (1970).

166. Yu. A. Kocherzhinskii, O. G. Kulik, and E. A. Shishkin, in: The Structure of Phases, Phase Inversions and Phase Diagrams of Metal Systems [in Russian], Nauka, Moscow (1974), pp. 136-139.

167. J. Hallais, Ann. Chim., 6(5):321-330 (1971).

168. C. Dauben, J. Electrochem. Soc., 104:522-524 (1957).

169. O. G. Karpinskii and B. A. Evseev, Izv. Akad. Nauk SSSR Neorg. Mater., 1(3):337-339 (1965).

170. L. F. Verkhorobin and N. N. Matyushenko, Poroshk. Metall., No. 6, 51-53 (1963).

171. N. N. Matyushenko, Poroshk. Metall., No., 20-22 (1964).

172. B. Aronsson, Acta Chem. Scand., 9(7):1107-1110 (1955).

173. R. Duval, R. Pichoir, and B. Roques, Rev. Int. Hautes. Temp. Refract., 5(4): 235-245 (1968).

174. H. Nowotny, The Chemistry of Extended Defects in Nonmetallic Solids, North Holland, Publ. Co., Amsterdam (1970), pp. 223-237.

175. J. P. Piton and M. F. Fay, C. R. Acad. Sci. Paris, 266C:514-516 (1968).

176. K. Khalaff and K. Schubert, J. Less-Common Metals, No. 35, 341-345 (1974).

177. C. Corre and J. M. Genin, Phys. Status Solid, No. 51, K85-K88 (1972).

178. Y. Dusausoy et al., Acta Cryst., 27:1209-1218 (1971).

179. W. Koster and T. Gödecke, Z. Metallkd., 64(6):399-405 (1973).

180. K. Frank and K. Schubert, Acta Cryst., 27:916-920 (1971).

181. L. A. Dvorina and O. I. Popova, Neorg. Mater., 6(11):1969-1972 (1970).

182. I. Engström, Structural Chemistry of Platinum Metal Silicides, Acta Univ. Ups., (1970), p. 54.

183. L. N. Finnie, J. Less-Common Metals, 4(1):24-33 (1962).

184. B. Aronsson and J. Aselius, Acta Chem. Scand., 15(7):1571-1574 (1961).

185. I. Engström, Acta Chem. Scand., 17(3):775-784 (1963).

186. I. Engström and T. Johnsson, Acta Chem. Scand., 19(6):1508-1509 (1965).

187. B. Aronsson and A. Nylund, Acta Chem. Scand., 14(5):1011-1018 (1960).

188. R. Gohle and K. Schubert, Z. Metallkd., 55(9):503-511 (1964).

189. H. Pfisterer and K. Schubert, Z. Metallkd., 41(10):358-367 (1950).

190. S. Rundqvist, Ark. Kemi, 20(7):67-113 (1962).

191. I. M. Novruzly and N. N. Tikhonova, Uch. Zap. Azerb. Gos. Univ. Ser. Khim., No. 3, 19-23 (1965).

192. T. Lundström and P. O. Snell, Acta Chem. Scand., 21(5):1343-1352 (1967).

193. V. N. Eremenko and V. S. Listovnichii, Dokl. Akad. Nauk, Ukr. SSR, No. 9, 1176-1179 (1965).

194. P. C. Nawapong, Acta Chem. Scand., 20(10):2737-2741 (1966).

195. T. Lundström, Acta Chem. Scand., 20(6):1712-1714 (1966).

196. T. Lundström and N. O. Ersson, Acta Chem. Scand., 22(6):1801-1808 (1968).

197. W. Jeitscho and H. Nowotny, Mh. Chemie, 93(5):1107-1109 (1962).

198. T. Lundström, Acta Chem. Scand., 22(7):2191-2199 (1968).

199. H. Jawad, T. Lundström, and S. Rundqvist, Physica Scripta, 3:43-44 (1971).

200. O. Olofsson and E. Ganglberger, Acta Chem. Scand., 24(7):2389-2396 (1970).

201. S. Rundqvist, Nature, 211(5051):847-848 (1966).

202. S. Rundqvist, Acta Chem. Scand., 20(9):2427-2434 (1966).

203. H. Boller and E. Parthé, Acta Cryst., 16(11):1095-1101 (1963).

204. S. Rundqvist, Acta Chem. Scand., 16(2):287-292 (1962).

205. S. Rundqvist and S. Lundström, Acta Chem. Scand., 17(1):37-46 (1963).

206. S. Rundqvist, Acta Chem. Scand., 19(2):393-400 (1965).

207. T. Johnsson, Acta Chem. Scand., 26(1):365-382 (1972).

208. S. Rundqvist, Acta Chem. Scand., 16(4):992-998 (1962).

209. S. Rundqvist, Acta Chem. Scand., 16(1):1-19 (1962).

210. S. Rundqvist, Acta Chem. Scand., 15(2):342-348 (1961).

211. S. Rundqvist, Acta Chem. Scand., 20(8):2075-2080 (1966).

212. E. Dahl, Acta Chem. Scand., 23(8):2677-2684 (1969).

213. S. Rundqvist and N. O. Ersson, Ark. Kemi, 30(10):103-114 (1968).

214. E. Larsson, Ark. Kemi., 23(32):335-365 (1965).

215. S. Rundqvist, Acta Chem. Scand., 15(2):451-453 (1961).

216. S. Rundqvist and L. O. Gullman, Acta Chem. Scand., 14:2246-2247 (1960).

217. C. J. Raub, W. H. Zachariasen, T. H. Geballe, et al., J. Phys. Chem. Solids, 24:1093-1100 (1963).

218. W. H. Zachariasen, Acta Cryst., 16(12):1253-1255 (1963).

219. E. Dahl, Acta Chem. Scand., 21(5):1131-1137 (1967).

220. C. Adolphe, M. Guittard, and P. Laruelle, C. R. Acad. Sci. Paris, 258:4773-4775 (1964).

221. M. Picon and J. Flahaut, C. R. Acad. Sci. Paris, 243:2074-2076 (1965).

222. N. N. Zhuravlev, A. A. Stepanova, and M. P. Shebatinov, Kristallografia, 9(1):116-117 (1964).

223. Vo Van Ten and P. Khodada, Bull. Soc. Chim. Fr., No. 1, 30-39 (1969).

224. P. Besancon, C. Adolphe, and J. Flahaut, C. R. Acad. Sci. Paris, 266C(2): 111-113 (1968).

225. P. Besancon and M. Guittard, C. R. Acad. Sci. Paris, 273C:1348-1351 (1971).

226. A. N. Zelikman, M. V. Teslitskaya, and É. D. Evstigneeva, Izv. Akad. Nauk SSSR Neorg. Mater., 7(2):314-315 (1971).

227. J. P. Marcon and R. Pascard, J. Inorg. Nucl. Chem., 28:2551-2560 (1966).

228. J. P. Marcon and R. Pascard, C. R. Acad. Sci. Paris, 262C:1679-1681 (1966).

229. G. Collin and J. Loriers, C. R. Acad. Sci. Paris, 260(19):5043-5046 (1965).

230. M. Patrie, Bull. Soc. Chim. Fr., No. 5, 1600-1601 (1969).

231. R. Chevalier et al., Bull. Soc. Fr. Mineral. Cristallogr., 10C:564-574 (1967).

232. J. Flahart, L. Domange, and M. P. Pardo, C. R. Acad. Sci. Paris, 258:594-596 (1964).

233. J. P. Marcon, C. R. Acad. Sci. Paris, 265C:235-237 (1967).

234. W. Suski and W. Trzebiatowsky, Bull. Acad. Polo. Sci., Ser. Sci. Chim., 12(5):277-279 (1964).

235. F. Grönvold, H. Haraldsen, et al., J. Inorg. Nucl. Chem., 30:2117-2119 (1968).

236. O. Beckmann, H. Boller, and H. Nowotny, Mh. Chemie., 101(4):945-955 (1970).

237. D. L. Greenaway and R. Nitsche, Propriétés Thermodynamiques Physiques et Structurales des Derives Semi-metalliques, Part 1, Ed. du Centre Nat. Rech. Sc., Paris (1967), pp. 447-457.

238. B. T. Kaminskii et al., Poroshk. Metall., No. 7, 6-10 (1973).

239. M. Chevreton and A. Sapet, C. R. Acad. Sci. Paris, 261, 928-930 (1965).

240. S. Brunie et al., Mat. Res. Bull., 7(4):253-260 (1972).

241. S. Brunie and M. Chevreton, C. R. Acad. Sci. Paris, 258:5847-5850 (1964).

242. D. Hodouin, C. R. Acad. Sci. Paris, 268:1943-1944 (1969).

243. V. Kh. Oganesyan, V. F. Bukhanevich, and S. V. Radzikovskaya, Armen. Khim. Zhurn. 19(3):161-165 (1966).

244. C. F. Bruggen and F. Jellinek, in: Propriétés Thermodynamiques Physiques et Structurales des Derives Semi-metalliques, Part 1, Ed. du Centre Nat. Rech. Sc., Paris (1967), pp. 31-39.

245. B. W. Roberts, NBS Tech. Note No. 482, N. Y. (1969).

246. C. J. Raub, V. B. Compton, et al., J. Phys. Chem. Solids, 26:2051-2057 (1965).

247. G. B. Bokii, Introduction to Crystal Chemistry [in Russian], Moscow State Univ., Moscow (1954).

248. G. V. Samsonov, Nonmetallic Nitrides [in Russian], Metallurgiya (1969).

249. T. V. Andreeva, I. G. Barantseva, et al., TVT 2(5):829-831 (1964).

250. R. Marchand and Y. Laurent, Acta Cryst., 25:2157-2160 (1969).

251. A. Guette, R. Naslain, and J. Galy, C. R. Acad. Sci. Paris, 275C:41-44 (1972).

252. J. P. Mercurio, J. Etoureau, et al., Mat. Res. Bull., 8(7):837-844 (1973).

253. H. Nowotny, F. Benesovsky, and R. Kieffer, Z. Metallkd. 50(5):258-260 (1959).

254. G. V. Samsonov, L. Ya. Markovskii, et al., Boron, Its Compounds and Alloys [in Russian], Akad. Nauk Ukrainian SSR, Kiev (1960).

255. H. Nowotny, F. Benesovsky, and R. Kieffer, Z. Metallkd. 50(7):417-423 (1959).

256. G. V. Samsonov and Ya. S. Umanskii, Hard Compounds of Refractory Metals [in Russian], Metallurgizdat (1957).

257. G. V. Samsonov, G. Sh. Upadkhaya, and V. S. Neshpor, Physical Behavior of Carbides [in Russian], Naukova Dumka, Kiev (1974).

258. Y. Laurent, J. David, and J. Lang, C. R. Acad. Sci. Paris, 259:1132-1134 (1964).

259. O. Whitemore, J. Can. Ceram. Soc., 28:43-48 (1959).

260. G. V. Samsonov, Silicides and Their Use in Technology [in Russian], Akad. Nauk Ukr. SSR, (1959).

261. A. S. Berezhnoi, Silicon and Its Binary Systems [in Russian], Akad. Nauk Ukr. SSR, (1958).

262. G. V. Lashkarev, Yu. B. Paderno, et al., Ukr. Fiz. Zh., 10(5):520-523 (1965).

263. L. Domange, J. Flahaut, and M. Guittard, C. R. Acad. Sci. Paris, 249:697 (1959).

264. G. N. Dubrovskaya and V. Kh. Oganesyan, Izv. Akad. Nauk Armen. SSR, 17(4):387-392 (1964).

265. A. Chretien and G. Odent, C. R. Acad. Sci. Paris, 257:2290-2292 (1963).

266. V. F. Terekhova and E. M. Savitskii, Yttrium [in Russian], Nauka (1967).

267. R. P. Elliott, Constitution of Binary Alloys [Russian translation], Metallurgizdat (1970), Vol. 1, p. 455; Vol. 2, p. 472.

268. F. Shank, Structures of Binary Alloys [in Russian], Metallurgiya (1973).

269. G. V. Samsonov, T. I. Serebryakova, and A. S. Bolgar, Zh. Neorg. Khim., 6(10):2243-2248 (1961).

270. G. V. Samsonov, Dokl. Akad. Nauk, 133(6):1344-1346 (1960).

271. G. A. Kudintseva, M. D. Polyakova, G. V. Samsonov, and B. M. Tsarev, FMM, 6(2):272-275 (1958).

272. G. V. Samsonov, Yu. B. Paderno, and V. S. Forrenko, Poroshk. Metall., No. 6, 24-32 (1963).

273. F. Benesovsky, P. Stechner, H. Nowotny, and W. Rieger, in: Propriétés Thermodynamiques Physiques et Structurales des Derives Semi-metalliques, Part 2, Ed. du Centre Nat. Rech. Sc., Paris (1967), pp. 419-430.

274. G. V. Samsonov and A. E. Grodshtein, Zh. Fiz. Khim., 30(2):379-382 (1956).

275. G. V. Samsonov, N. N. Zhuravlev, et al., Kristallografiya, 4(4):538-542 (1959).

276. J. Etourneau, J. P. Mercurio, et al., C. R. Acad. Sci. Paris, 274C:1688-1691 (1972).

277. G. V. Samsonov and O. N. Zorina, Zh. Neorg. Khim., 1(10):2260-2263 (1956).

278. K. I. Portnoi, V. M. Romashov, and S. E. Salibekov, Poroshk. Metall., No. 11, 89-91 (1971).

279. K. I. Portnoi, Yu. V. Levinskii, et al., Izv. Akad. Nauk SSSR Metall., No. 4, 171-176 (1967).

280. K. I. Portnoi and V. M. Romashov, Poroshk. Metall., No. 2, 41-44 (1968).

281. K. I. Portnoi, V. M. Romashov, et al., Poroshk. Metall., No. 4, 51-57 (1969).

282. K. I. Portnoi, V. M. Romashov, et al., Poroshk. Metall., No. 5, 75-80 (1967).

283. K. I. Portnoi, M. Kh. Levinskaya, and V. M. Romashov, Poroshk. Metall., No. 8, 66-70 (1969).

284. P. T. Kolomytsev, Dokl. Akad. Nauk SSSR, 130(4):767-770 (1960).

285. K. I. Portnoi, V. M. Romashov, et al., Poroshk. Metall., No. 2, 15-21 (1967).

286. P. Rogl, H. Nowotny, and F. Benesovsky, Mh. Chemie, 102(3):678-686 (1971).

287. M. F. Ancey-Moret and M. Y. Deniel, Mem. Sci. Rev. Metall., 70(4):301-317 (1973).

288. N. H. Krikorian, T. C. Wallace, and M. G. Bowman, in: Propriétés Thermo-dynamiques Physiques et Structurales des Derives Semi-metalliques, Part 1, Ed. du Centre Nat. Rech. Sc., Paris (1967), pp. 489-494.

289. G. Hörtz, Metall., 27(7):680-687 (1973).

290. W. S. Williams, High Temp.-High Pressures, 4(6):663-669 (1972).

291. R. Kieffer and F. Benesovskii, Hard Materials [Russian translation], Metallurg-iya (1968).

292. C. P. Kempter, Phys. Status Solidi, 36:K137-K139 (1969).

293. A. Tardif, G. Piquard, and J. Wach, Rev. Int. Hautes Temp. Refract., 8(2):143-148 (1971).

294. W. W. Beaver, Plansee Proceedings 1964, Metallwerk Planser AG, Reutte/Tirol (1965), pp. 682-700.

295. J. Rexer, Z. Metallkd., 62(11):844-848 (1971).

296. T. Gödecke and W. Köster, Z. Metallkd., 63:422-430 (1972).

297. L. A. Dvorina and B. M. Rud', in: Rare-Earth Metals and Their Compounds [in Russian], Naukova Dumka, Kiev, (1970), pp. 142-147.

298. J. A. Perri, J. Phys. Chem., 63(4):616-619 (1959).

299. G. V. Samsonov, Yu. B. Paderno, and T. S. Verkhoglyadova, Dokl. Akad. Nauk Ukr. SSR, No. 2, 224-226 (1963).

300. E. I. Gladyshevskii and P. I. Kripyakevich, Zh. Strukt. Khim., No. 5, 853-859 (1964).

301. H. J. Goldshmidt, Interstitial Alloys [Russian translation], Mir, Moscow, Vol. 1, p. 423; Vol. 2, p. 463 (1971). [Original published by Plenum Press, New York (1968).]

302. G. V. Samsonov, V. S. Neshpor, and V. A. Ermakova, Zh. Neorg. Khim., 3(4):868-878 (1958).

303. V. P. Elyutin and Yu. A. Pavlov, High-Temperature Materials [in Russian], Metallurgiya (1972).

304. V. N. Svechnikov, Yu. A. Kocherzhinskii, and L. M. Yupko, Nauchn. Tr. IMF Akad. Nauk Ukr. SSR, Naukova Dumka, 19:212-218 (1964).

305. L. D. Dudkin and E. S. Kuznetsova, Poroshkov Metall., No. 6, 20-31 (1962).

306. J. A. Kocherzhinsky, Étude des transformations cristallines à haute tempera-ture, No. 205, 47-51 (1971).

307. H. Warlimont, Z. Metallkd. 59:595 (1968).

308. W. Koster and T. Godecke, Z. Metallkd., 59:602 (1968).

309. T. Mager and E. Wachtel, Z. Metallkd., 61(11):853-856 (1970).

310. K. E. Mironov, J. Cryst. Growth, 3, 4:150-152 (1968).

311. U. D. Veryatin, V. P. Mashirev, N. G. Ryabtsev, et al., Thermodynamic Properties of Inorganic Substances [in Russian], Atomizdat, Moscow (1965).

312. R. L. Ripley, J. Less-Common Metals, 4:498-503 (1962).

313. H. Boller and E. Parthé, Acta Cryst., 16:1095-1101 (1963).

314. J. Dismukes and J. White, Inorg. Chem., 3:1220-1228 (1964).

315. K. A. Schneider, Alloys of Rare-Earth Metals [in Russian], Mir, Mowcow (1965).

316. J. P. Marcon and R. Pascard, Rev. Int. Hautes Temp. Refract., 5(1):51-54 (1968).

317. A. Noël, J. Tudo, and G. Tridot, C. R. Acad. Sci. Paris, 264C:443-445 (1967).

318. A. L. Burykina and V. P. Kosteruk, Poroshk. Metall., No. 12, 49-54 (1973).

319. L. Toth, Transition Metal Carbides and Nitrides, Academic Press, New York (1971).

320. R. B. Kotel'nikov, S. N. Bashlykov, et al., Especially Refractory Elements and Compounds [in Russian], Metallurgiya (1969).

321. D. A. Prokoshkin and E. V. Vasil'eva, Niobium Alloys [in Russian], Nauka, Moscow (1964).

322. Nat. Bur. Stand. USA, Tech. Note No. 270-4 (1969).

323. Nat. Bur Stand. USA, Tech. Note No. 270-5 (1971).

324. Nat. Bur. Stand. USA, Tech. Note No. 270-6 (1971).

325. Nat. Bur. Stand. USA, Tech. Note No. 270-7 (1973).

326. P. Peshev, Rev. Int. Hautes Temp. Refract., 4(4):289-296 (1967).

327. G. V. Samsonov and T. I. Serebryakova, Zh. Prikl. Khim., 33(3):563-570 (1960).

328. K. D. Modylevskaya and G. V. Samsonov, Ukr. Khim. Zh., No. 1, 55-61 (1959).

329. G. V. Samsonov, Yu. B. Paderno, and S. U. Kreingol'd, Zh. Prikl. Khim., 34(1):10-15 (1961).

330. R. Meyer and H. Pastor, Bull. Soc. Fr. Ceram., No. 66, 59-80 (1965).

331. A. N. Krestovnikov and M. S Vendrikh, Izv. Vyssh. Uchebn. Zaved. Tsvetn. Metall., No. 2, 54-57 (1959).

332. P. Peshev, L. Leyarovska, and G. Bliznakov, J. Less-Common Metals, 15:259-267 (1968).

333. A. N. Krestovnikov and M. S. Vendrikh, Izv. Vyssh. Uchebn. Zaved. Chern. Metall., No. 3, 13-16 (1960).

334. A. S. Bolgar, A. G. Turchapin, and V. V. Fesenko, Thermodynamic Properties of Carbides [in Russian], Naukova Dumka (1973).

335. A. S. Bolgar, V. V. Fesenko, and S. P. Gordienko, Poroshk. Metall., No. 2, 100-103 (1966).

336. V. V. Fesenko and A. G. Turchanin, in: Refractory Carbides [in Russian], Naukova Dumka (1970), pp. 200-204.

337. V. N. Eremenko, G. M. Lukashenko, and R. I. Polotskaya, in: Refractory Carbides [in Russian], Naukova Dumka, Kiev (1970), pp. 204-207.

338. R. A. Andrievskii, V. V. Khromonozhkin, E. A. Galkin, et al., At. Énerg., 26(6):494-498 (1969).

339. R. Kieffer, H. Nowotny, P. Ettmayer, et al., Mh. Chemie, 101(1):65-82 (1970).

340. A. Neckel and H. Nowotny, 5th Int. Leichtmetalltagung Leoben, 1968, Dusseldorf, Aluminium-Verlag, GmbH, pp. 2-5.

341. G. Jangg, R. Kieffer, and H. Kogler, Z. Metallkd., 59(7):546-552 (1968).
342. G. V. Samsonov, Refractory Compounds of Rare-Earth Metals with Non-metals [in Russian], Metallurgiya (1964).
343. V. S. Neshpor and G. V. Samsonov, Zh. Prikl. Khim., 33(5):993-1001 (1960).
344. Yu. M. Golutvin and É. G. Maslennikova, Izv. Akad. Nauk SSSR Met. No. 2, 193-207 (1968).
345. Yu. M. Golutvin and T. M. Kozlovskaya, Zh. Fiz. Khim., 34(10):2350-2354 (1960).
346. T. G. Chart, High Temp. — High Pressures, 5(3):241-252 (1973).
347. Yu. M. Goryachev and T. G. Kutsenok, High Temp. — High Pressures, 4(6): 663-669 (1972).
348. B. V. Fenochka, S. P. Gordienko, and V. V. Fesenko, in: Rare-Earth Metals and Their Compounds [in Russian], Kiev, Naukova Dumka, pp. 204-209 (1970).
349. B. Degroise and J. Oudar, Bull. Soc. Chim. Fr., No. 5, 1717-1719 (1970).
350. C. Gentaz, G. Bienvenu, and A. Bouissiba, Hautes Temp. Refract., 10(3): 161-172 (1973).
351. É. S. Sarkisov, Zh. Fiz. Khim., 28:627 (1954).
352. G. V. Samsonov, in: Physics and Physicochemical Analysis [in Russian], Metallurgizdat, Moscow, No. 1 (1957), pp. 192-222.
353. O. I. Shulishova, in: High-Temperature Cermets [in Russian], Kiev, Izv. Akad. Nauk Ukr., SSR (1962).
354. B. W. Roberts, Rep. No. 63-RL-3252M, N.Y. (1963).
355. B. W. Roberts, NBS Techn. Note 724, N.Y. (1972).
356. V. S. Fomenko, Emission Properties of Materials [in Russian], Naukova Dumka (1970).
357. V. N. Eremenko, T. Ya. Velikanova, and S. V. Shabanova, in: Structure of Phases, Phase Inversions, and Phase Diagrams of Metal Systems [in Russian], Nauka, Moscow (1974), pp. 129-132.
358. E. Gugel, P. Ettmayer, and A. Schmidt, Ber. Dtsch Keram. Ges., 45(8):395-402 (1968).
359. I. I. Kornilov, M. A. Volkova, E. N. Pylaeva, et al., in: New Research into Titanium Alloys [in Russian], Nauka, Moscow (1965), pp. 48-55.
360. M. Kh. Karapet'yants, Chemical Thermodynamics [in Russian], Goskhimiz-dat, Moscow (1953).
361. O. Kubaschewski and E. L. Evans, Metallurgical Thermochemistry, Perga-mon Press, London (1958).
362. V. E. Ganenko, G. A. Berezovskii, V. S. Neshpor, et al., in: Studies in the Physics of Solids [in Russian], Nauka, Novosibirsk (1969), pp. 127-133.
363. P. V. Gel'd and F. G. Kusenko, Izv. Akad. Nauk SSSR Metall. i Toplivo, No. 2, 79-82 (1960).
364. J. Danan, C. H. Novion, and H. Dallaporta, Solid State Commun., 10(9): 775-778 (1972).
365. A. Kirfel and A. Newhaus, High Temp. — High Pressure, 3(1):81-87 (1971).
366. A. Heiss and M. Djemal, Rev. Int. Hautes Temp. Refract., 8(3-4):287-290 (1971).

367. J. P. Marcon, J. Inorg. Nucl. Chem., 32:2581-2590 (1970).

368. Y. S. Tyan, L. E. Toth, and Y. A. Chang, J. Phys. Chem. Solids, 30:785-792 (1969).

369. P. Costa and R. Lallement, J. Phys. Chem. Solids, 25:559-564 (1964).

370. G. M. Klimashin, V. S. Neshpor, and V. P. Nikitin, et al., Pis'ma Zh. Eksp. Teor. Fiz., 12:147-149 (1970).

371. V. S. Neshpor, V. I. Novikov, et al., Pis'ma Zh. Eksp. Teor. Fiz. 54(1):25-28 (1968).

372. J. J. Veyssie, D. Brochier, A. Nemoz, and J. Blanc, Phys. Lett., 14(4):261-262 (1965).

373. W. Rossteutscher and K. Schubert, Z. Metallkd., 56:813 (1965).

374. J. Bonnerot, J. Hallais, S. Barisic, and J. Labbe, J. de Physique, 30(8-9):701-713 (1969).

375. F. N. Kozlov, L. V. Petrusevich, et al., Izv. Akad. Nauk SSSR Met., No. 6, 88-91 (1969).

376. V. I. Alekseev and L. A. Shvartsman, Dokl. Akad. Nauk. SSSR 141:346 (1961).

377. T. S. Verkhoglyadova and L. A. Dvorina, Zh. Prikl. Khim., 38(8):1716-1725 (1965).

378. E. Trong and M. Huber, C. R. Acad. Sci. Paris, 268C:1771-1774 (1969).

379. V. N. Prilepskii, E. N. Timofeeva, et al., Izv. Akad. Nauk SSSR Neorg. Mater., 6(11):2069-2070 (1970).

380. J. Schmitt and R. Setton, Verres et Refract., 18(4):319-325 (1964).

381. A. N. Krestovnikov and M. S. Vendrikh, Izv. Vyssh. Uchebn. Zaved. Tsvetn. Metall., No. 1, 73-75 (1958).

382. C. Affortit, J. Nucl. Mater., 34:105-107 (1970).

383. V. V. Khromonozhkin and R. A. Andrievskii, in: Thermodynamics, Vol. I, Int. Atomic Energy Agency, Vienna (1966), pp. 359-366.

384. A. G. Turchanin and V. V. Fesenko, Zh. Fiz. Khim., 42:1026 (1968).

385. V. P. Bondarenko, E. N. Fomichev, and V. V. Kandyba, High Temp. — High Pressures, 5(1):5-7 (1973).

386. A. G. Turchanin and V. V. Fesenko, Poroshk. Metall., No. 1, 88-92 (1968).

387. A. S. Bolgar, E. A. Guseva, and V. V. Fesenko, Poroshk. Metall., No. 1, 40-43 (1967).

388. E. A. Guseva, A. G. Turchanin, V. V. Morozov, et al., Zh. Fiz. Khim., 45:2971 (1971).

389. A. G. Turchanin, S. S. Ordan'yan, and V. V. Fesenko, Poroshk. Metall., No. 9, 23-27 (1967).

390. A. S. Bolgar, E. A. Guseva, V. A. Gorbatyuk, et al., Poroshk. Metall., No. 4, 60-63 (1968).

391. V. P. Bondarenko, L. A. Dvorina, et al., Porosh. Metall., No. 11, 47-49 (1973).

392. Yu. M. Golutvin, Zh. Fiz. Khim., 33(8):1798-1805 (1959).

393. Yu. M. Golutvin and T. M. Kozlovskaya, Zh. Fiz. Khim., 36(2):362-364 (1962).

394. Yu. M. Golutvin and Lan Tsin-Kun, Zh. Fiz. Khim., 35(1):129-141 (1961).

395. V. P. Bondarenko, E. N. Fomichev, and A. A. Kolashnik, in: Thermal Physical Properties of Solids [in Russian], Nauka (1971), pp. 167-169.

396. Yu. M. Golutvin, T. M. Kozlovskaya, and É. G. Maslennikova, Zh. Fiz. Khim., 37(6):1362-1368 (1963).

397. I. A. Smirnov, Phys. Status Solidi, 14(2):363-404 (1972).

398. G. Busch, P. Junod, and O. Vogt, in: Propriétés Thermodynamiques Physiques et Structurales des Derives Semi-metalliques, Ed. du Centre Nat. Rech. Sc., Paris (1967), pp. 337-342.

399. E. A. Guseva, A. G. Turchanin, and A. S. Bolgar, Poroshk. Metall., No. 1, 71-76 (1974).

400. H. L. Schick, Thermodynamics of Certain Refractory Compounds, Academic Press, New York (1966).

401. S. P. Gordienko and B. V. Fenochka, Zh. Fiz. Khim., 47(4):1030-1031 (1973).

402. J. F. Lynch, C. G. Ruderer, and W. H. Duckworth, Engineering Properties of Selected Ceramic Materials, Am. Cer. Soc., Columbus (1966).

403. G. Olcese, Atti Acad. Naz. Lincei. Cl. Sci. Fis. Mat. Natur. Rend., 40(4): 629-634 (1966).

404. S. A. Gutov, K. E. Mironov, I. E. Paukov, et al., Zh. Fiz. Khim., 46(2):539-541 (1972).

405. S. P. Gordienko, B. V. Fenochka, and V. V. Fesenko, Rare-Earth Metals and Their Refractory Compounds [in Russian], Naukova Dumka, Kiev (1971).

406. J. Drowart, in: Condensation and Evaporation of Solids, Gordon and Breach, New York (1964), pp. 255-310.

407. V. V. Fesenko, A. S. Volgar, and S. P. Gordienko, Rev. Hautes Temp. Refract., 3(3):261-271 (1966).

408. I. I. Kostetskii, S. N. L'vov, Yu. A. Kunitskii, Izv. Akad. Nauk SSSR Neorg. Mater., 7(6):951-955 (1971).

409. R. J. Fries, J. Chem. Phys., 37(2):320-322 (1962).

410. A. S. Bolgar, V. V. Fesenko, and S. P. Gordienko, Poroshk. Metall., No. 2, 100-103 (1966).

411. M. Hoch, D. P. Dingleby, and H. Johnson, J. Amer. Ceram. Soc., 77:304 (1954).

412. V. V. Fesenko and A. S. Bolgar, Volatilization of Refractory Compounds [in Russian], Metallurgiya, Moscow (1966).

413. G. V. Samsonov, A. S. Bolgar, and T. S. Verkhoglyadova, Izv. Akad. Nauk SSSR Metall. i Toplivo, No. 1, 142-146 (1961).

414. R. F. Flowers and E. G. Rauh, J.Inorg. Nucl. Chem., 28(6/7):1355-1366 (1966).

415. C. A. Stearns and F. J. Kohl, J. Chem. Phys., 54(12):5180-5182 (1971).

416. R. C. Faircloth, R. H. Flowers, and F. C. W. Pummery, J. Inorg. Nucl. Chem., 30(2):499-502 (1968).

417. G. Balducci, A. Capalbi, G. De Maria, and M. Guido, J. Chem. Phys., 43(6): 2136-2139 (1965).

418. C. L. Hoenig, N. D. Stout, and P. C. Nordine, J. Am. Cer. Soc., 50(8):385-388 (1967).

419. G. Balducci, A. Capalbi, and M. Guido, J. Chem. Phys., 51(7):2871-2875 (1969).
420. G. Balducci, G. De Maria, and G. Guido, J. Chem. Phys., 51(7):2876-2879 (1969).
421. L. Ramqvist, Int. J. Powder Metall., 1(4):2-21 (1965).
422. C. Myers and A. Searcy, J. Am. Chem. Soc., 79:527 (1957).
423. V. I. Torbov, L. A. Egorov, V. I. Chuklin, et al., Izv. Akad. Nauk SSSR Neorg. Mater., 7(9):1618-1621 (1971).
424. V. V. Fesenko, Poroshk. Metall., No. 1, 85-88 (1961).
425. G. V. Samsonov, V. S. Fomenko, and Yu. B. Paderno, Ogneupory, No. 1, 40-42 (1962).
426. T. I. Serebryakova, Yu. B. Paderno, and G. V. Samsonov, Optika i Spektrosk., 8:410-412 (1960).
427. B. A. Fridlender, V. S. Neshpor, B. G. Ermakov, et al., Inzh.-Fiz. Zh., 24(2):294-296 (1973).
428. E. Brame, J. Margrave, and V. Meloche, J. Inorg. Nucl. Chem., 5:48 (1957).
429. S. N. L'vov, V. F. Nemchenko, and Yu. B. Paderno, Dokl. Akad. Nauk SSSR, 149(6):1371-1372 (1963).
430. G. V. Samsonov, A. S. Bolgar, E. A. Guseva, et al., High Temps. — High Pressures, 5(1):29-33 (1973).
431. S. N. L'vov, V. F. Nemchenko, P. S. Kislyi, et al., Poroshk. Metall., No. 4, 20-25 (1962).
432. S. N. L'vov, M. I. Lesnaya, I. M. Vinitskii, and V. Ya. Naumenko, TVT 10(6):1327-1329 (1972).
433. E. T. Bezruk and L. Ya. Markovskii, Izv. Akad. Nauk SSSR Neorg. Mater., 4(3):447-449 (1968).
434. Yu. A. Kunitskii and É. V. Marek, Poroshk. Metall., No. 3, 56-59 (1971).
435. I. I. Kostetskii and S. N. L'vov, FMM, 33(4):773-779 (1972).
436. V. S. Neshpor and G. M. Klimashin, Izv. Akad. Nauk SSSR Neorg. Mater., 1(9):1545-1546 (1965).
437. I. G. Korshunov, V. E. Zinov'ev, and P. V. Gel'd, TVT, 11(4):889-891 (1973).
438. V. S. Neshpor, S. S. Ordan'yan, A. I. Avgustinik, et al., Zh. Prikl. Khim., 37:2375-2382 (no date).
439. E. K. Storms and P. Wagner, High Temp. Science, 5:454-462 (1973).
440. B. A. Fridlender and V. S. Neshpor, TVT, 8(4):795-798 (1970).
441. B. H. Morrison and L. L. Sturgess, Rev. Int. Hautes Temp. Refract., 7(4): 351-358 (1970).
442. B. A. Fridlender, V. S. Neshpor, and V. V. Sokolov, TVT, 11(2):424-427 (1973).
443. V. S. Neshpor, S. S. Ordan'yan, et al., Izv. Akad. Nauk SSSR, Neorg. Mater., 6(11):1954-1960 (1970).
444. G. V. Samsonov and T. S. Verkhoglyadova, Dokl. Akad. Nauk SSSR, 142(3): 608-612 (1962).

445. F. Cabannes, High Temp. — High Pressures, 3(2):137-151 (1971).

446. S. N. L'vov, V. F. Nemchenko, and G. V. Samsonov, Poroshk. Metall., No. 6, 70-74 (1961).

447. D. Westbrook, Problem of Modern Metallurgy, No. 4 [Russian translation], (1960), p. 111.

448. E. N. Nikitin, Zh. Tekh. Fiz., 28(1):23-25 (1958).

449. V. S. Neshpor and V. L. Yupko, Zh. Prikl. Khim., 36, 1139-1142 (1963).

450. V. S. Neshpor and I. G. Barantseva, I.F.Zh. 6(1):109-113 (1963).

451. F. I. Ostrovskii, R. P. Krentsis, and P. V. Gel'd, Izv. Vyssh. Uchedn. Zaved. Fiz., No. 7, 130-131 (1970).

452. R. P. Krentsis, F. I. Ostrovoskii, and P. V. Gel'd, FTP, 4(2):403-405 (1970).

453. R. P. Krentsis, F. I. Ostrovskii, et al., Izv. Vyssh. Uchebn. Zaved. Fiz., No. 8, 157-158 (1970).

454. G. V. Samsonov, T. V. Andreeva, and T. V. Dubovik, High Temp. — High Pressures, 4(5):537-539 (1972).

455. N. N. Zhuravlev and A. A. Stepanova, Poroshk. Metall., No. 6, 83-84, (1964).

456. E. N. Severyanina, E. M. Dudnik, and Yu. B. Paderno, Poroshk. Metall., No. 12, 72-74 (1973).

457. G. Beckmann and R. Kiessling, Nature, 178:1341-1343 (1956).

458. V. V. Odintsov and Yu. B. Paderno, At. Énerg., 30(5):453-454 (1971).

459. G. V. Samsonov, B. A. Kovenskaya, and T. I. Serebryakova, Izv. Vyssh. Uchebn. Zaved. Fiz., No. 1, 19-23 (1971).

460. G. V. Samsonov, B. A. Kovenskaya, et al., TVT, No. 1, 195-197 (1971).

461. T. Ya. Kosolapova, L. T. Makarenko, et al., Zh. Prikl. Khim., 44(5):953-958 (1971).

462. G. N. Makarenko, L. T. Pustovoit, et al., Izv. Akad. Nauk SSSR Neorg. Mater., 1(10):1787-1790 (1965).

463. H. Nowotny and E. Laube, Planseeber. Pulvermetall., 9(1/2):54-59 (1961).

464. G. V. Samsonov and V. Ya. Naumenko, TVT, 8(6):1093-1096 (1970).

465. G. V. Samsonov, V. Ya. Naumenko, et al., Izv. Akad. Nauk SSSR Neorg. Mater., 10(1):52-56 (1974).

466. G. V. Samsonov, I. I. Timofeeva, et al., Izv. Akad. Nauk. SSSR Neorg. Mater., 11(1):62-65 (1975).

467. J. Gachon and B. Schmitt, C. R. Acad. Sci. Paris, 272C:428-431 (1971).

468. G. V. Samsonov and E. V. Petrash, Metalloved. i Obrab. Met., No. 4, 19-24 (1955).

469. G. V. Samsonov and T. S. Verkhoglyadova, Zh. Neorg. Khim., 6(12):2732-2737 (1961).

470. G. V. Samsonov and V. S. Sinel'nikova, Tsvetn. Met., No. 11, 92-95 (1962).

471. V. S. Neshpor and M. I. Reznichenko, Ogneupory, No. 3, 134-137 (1963).

472. V. I. Lazorenko, B. M. Rud', Yu. B. Paderno, et al., Izv. Akad. Nauk SSSR, Neorg. Mater., 10(6):1150-1151 (1974).

473. J. Beaudouin, C. R. Acad. Sci. Paris, 263(18):1065-1067 (1966).

474. N. N. Zhuravlev and A. A. Stepanova, At. Énerg., 13(2):183-184 (1962).

475. L. P. Andreeva and P. V. Gel'd, Izv. Vyssh. Uchebn. Zaved. Chern. Metall., No. 2, 111-114 (1965).

476. E. M. Dudnik and V. Kh. Oganesyan, Poroshk. Metall., No. 2, 60-62 (1966).

477. I. G. Kerimov, T. A. Mamedov, et al., in: Thermophysical Properties of Solids [in Russian], Kiev (1971), pp. 19-23.

478. G. Vuillard, Rev. Hautes Temp. et Refract., 2(2):173-174 (1965).

479. E. Baughan, Trans. Faraday Soc., 55:2025 (1959).

480. B. F. Ormont, ed., Compounds of Variable Composition [in Russian], Khimaya, Leningrad (1969).

481. K. B. Yatsimirskii, Zh. Neorg. Khim., 6(3):518-525 (1961).

482. G. V. Samsonov and Yu. B. Paderno, Borides of Rare-Earth Metals [in Russian], Akad. Nauk Ukrainian SSR (1961).

483. J. Castaing, R. Candron, et al., Solid State Commun., 7(20):1453-1456 (1969).

484. J. P. Meriurio, J. Etourneau, et al., C. R. Acad. Sci. Paris, 268B:1766-1769 (1969).

485. V. N. Gurin, Izv. Akad. Nauk SSSR Neorg. Mater., 9(8):1289-1307 (1973).

486. R. Kuentzler, C. R. Acad. Sci. Paris, 270:197-199 (1970).

487. R. Kuentzler, C. R. Acad. Sci. Paris, 266B:1099-1101 (1968).

488. A. Combarieu, P. Costa, et al., C. R. Acad. Sci. Paris, 256:5518-5521 (1963).

489. L. E. Toth, M. Ishikawa, and Y. A. Chang, Acta Metall., 16(9):1183-1185 (1968).

490. J. Mazur and W. Zacharko, Acta Phys., 35:91-99 (1969).

491. A. I. Avgustinik, S. S. Ordan'yan, and V. N. Fishchev, Izv. Akad. Nauk SSSR, Neorg. Mater., 9(7):1169-1171 (1973).

492. J. J. Veyssie, P. Haen, et al., C. R. Acad. Sci. Paris, 260:4980-4983 (1965).

493. B. N. Oshcherin, Izv. Vyssh. Uchedn. Zaved. Fiz., No. 2, 73-78 (1972).

494. E. M. Savitskii, V. V. Baron, et al., (eds.), Metal Physics of Superconducting Materials [in Russian], Nauka, Moscow (1969).

495. É. A. Polonik, I. A. Grikit, et al., Zh. Prikl. Spektrosk., 19(2):202-206 (1973).

496. B. Prevot and C. Carabatos, C. R. Acad. Sci. Paris, 273B(2):93-96 (1971).

497. B. N. Oshcherin, in: Rare-Earth Metals and Their Compounds [in Russian], Kiev (1970), pp. 30-39.

498. V. S. Neshpor and G. V. Samsonov, FMM, 4(1):181-183 (1957).

499. V. N. Paderno, Izv. Akad. Nauk SSSR Metall. i Toplivo, No. 6, 176-178 (1962).

500. G. V. Samsonov and V. P. Latysheva, FMM, 2(2):309-319 (1956).

501. G. V. Samsonov and A. P. Épik, FMM, 14(3):479-480 (1962).

502. A. N. Krasnov and A. L. Burikina, Rev. Int. Hautes Temp. Refract., 8(1):15-24 (1971).

503. G. Hörz and K. Lindenmaier, Z. Metallkd., 63(5):240-247 (1972).

504. I. I. Kovenskii, U.F.Zh., 8(7):757-760 (1963).

505. R. A. Andrievskii, V. N. Zagryazkin, et al., FMM, 21(1):140-143 (1966).

506. L. V. Pavlinov and V. N. Bykov, FMM, 19(3):197-399 (1965).

507. P. Son, M. Miyake, and T. Sano, Technol. Rep. Osaka Univ., 18(828):317-324 (1968).

508. Y. Fujikava, P. Son, et al., Hunnon Kindsoku Hakkaisi (Japan), 34(12): 1259-1263 (1970).

509. L. N. Aleksandrov and V. Ya. Shchelkonogov, Poroshk. Metall., No. 4, 28-32 (1964).

510. V. S. Eremeev, A. S. Panov, et al., Thermodynamics, Vol. 2, Int. Atomic Energy Agency, Vienna (1966), pp. 161-172.

511. R. Ducros and P. Le Goff, C. R. Acad. Sci. Paris, 267C:704-706 (1968).

512. P. L. Gruzin, Yu. A. Polikarpov, and M. A. Shumilov, Zavod. Lab., 21(4): 417-423 (1955).

513. E. Pink, Berg. Hüttenmänn. Monatsh., 112(9):266-269 (1967).

514. H. Jehn, G. Hörz, and E. Fromm, Metall., No. 7, 783-790 (1971).

515. E. Fromm and H. Jehn, High Temp. — High Pressures, 3(5):553-564 (1971).

516. J. Pouliquen, S. Offret, and J. Fouquet, C. R. Acad. Sci. Paris, 274C, 1760-1763 (1972).

517. G. V. Samsonov, M. S. Koval'chenko, and T. S. Verkhoglyadova, Inzh. — Fiz. Zh., No. 3, 62-67 (1959).

518. G. V. Samsonov, M. S. Koval'chenko, and T. S. Verkhoglyadova, Zh. Neorgan. Khim., 4(12):2759-2766 (1959).

519. G. V. Samsonov and L. A. Solonnikova, FMM, 5(3):565-566 (1957).

520. E. Fitzer and F. K. Schmidt, High Temp. — High Pressures, 3(4):445-460 (1971).

521. S. J. Wang and H. Grabke, Z. Metallkd., 61(8):1729-1733 (1970).

522. P. S. Kislyi and G. V. Samsonov, Fiz. Tverd. Tela, 2(8):1729-1733 (1960).

523. G. V. Samsonov and V. M. Sleptsov, Dokl. Akad. Nauk Ukr. SSR, No. 10, 1116-1118 (1959).

524. N. V. Vekshina and L. Ya. Markovskii, Zh. Prikl. Khim., 34(10):2171-2175 (1961).

525. L. N. Kugai and T. N. Nazarchuk, Zh. Anal. Khim., 16(2):205-208 (1961).

526. M. D. Lyutaya, and Z. S. Akinina, Izv. Akad. Nauk SSSR Neorg. Mater., 1(3):1039-1043 (1965).

527. T. Ya. Kosolapova and L. T. Domasevich, Poroshk. Metall., No. 5, 1-5 (1970).

528. L. Ya. Markovskii and G. V. Kaputovskaya, Zh. Prikl. Khim., 33(3):569-577 (1960).

529. G. V. Samsonov, A. T. Pilipenko, and T. N. Nazarchuk (eds.), Analysis of Hard Refractory Compounds [in Ukrainian], Akad. Nauk SkSSR (1961).

530. P. Peshev and G. Bliznakov, J. Less-Common Metals, 14:23-32 (1968).

531. B. Beck, Planseeber. Pulvermetall. 9:96 (1961).

532. M. P. Slavinskii, Physicochemical Properties of Elements [in Russian], Metallurgizdat (1952).

533. B. Hajek, V. Brozek, et al., Collect. Czech. Chem. Commun., 36:3236-3243 (1971).

534. B. Hajek, V. Brozek, et al., Collect. Czech. Chem. Commun., 36:1537-1545 (1971).

535. F. Petru, V. Brozek, and B. Hajek, Rev. Chim. Min.,7:515-537 (1970).

536. V. Brozek, M. Popl, and B. Hajek, Collect. Czech. Chem. Commun., 35:2724-2737 (1970).

537. T. Ya. Kosolapova, L. T. Domasevich, et al., Ukr. Khim. Zh., 39:(1)19-21 (1973).

538. Nuclear Reactors: Material of the Commission on Atomic Energy, USA, Part 3 [Russian translation], IL. M. (1956).

539. J. Besson, P. Blum, et al., C. R. Acad. Sci. Paris, 261(8):1859-1861 (1965).

540. V. P. Kopylova, Zh. Prikl. Khim., 34:1936-1939 (1961).

541. A. P. Brynza, T. Ya. Kosolapova, et al., Poroshk. Metall., No. 8, 67-72 (1971).

542. E. E. Kotlyar and T. N. Nazarchuk, Izv. Akad. Nauk SSSR Neorg. Mater., 2(10):1782 (1966).

543. T. Ya. Kosolapova, V. B. Fedorus, et al., Izv. Akad. Nauk SSSR Neorg. Mater., 2(8):1521-1523 (1966).

544. M. D. Lyutaya and V. F. Bukhanevich, Zh. Neorg. Khim., 7(11):2487-2494 (1962).

545. O. I. Popova and G. T. Kabannik, Zh. Neorg. Khim., 5(4):930-934 (1960).

546. O. I. Popova, Ukr. Khim. Zh., 27(1):22-26 (1961).

547. L. A. Dvorina, O. I. Popova, et al., Poroshk. Metall., No. 5, 29-32 (1969).

548. T. Ya. Kosolapova, L. T. Domasevich, et al., in: Rare-Earth Metals and Their Compounds [in Russian], Kiev, (1970), pp. 230-234.

549. E. Jackobsen, B. Treeman, et al., J. Am. Chem. Soc., 78:4850 (1956).

550. G. Siborg and J. Katz, Chemistry of Actinide Elements [in Russian], Atom-izdat, Moscow, (1960).

551. M. K. Disen and G. F. Hüttig, Planseeber. Pulvermetall., 4(1):10-14 (1956).

552. T. Ya. Kosolapova and E. E. Kotlyar, Zh. Neorg. Khim., 3(5):1241-1244 (1958).

553. O. I. Popova and L. A. Dvorina, Zh. Prikl. Khim., 46(9):1928-1931 (1973).

554. G. V. Samsonov and S. N. Endrzheevskaya, Zh. Org. Khim. 33(9):2803-2804 (1963).

555. E. D. Sinitsyna, I. G. Vasil'eva, and K. E. Mironov, Chemistry of Phosphides with Semiconductor Properties [in Russian], Novosibirsk, Izd. SO Akad. Nauk SSSR (1970).

556. S. N. Endrzheevskaya and G. V. Samsonov, Zh. Org. Khim. 35(11):1983-1984 (1965).

557. L. L. Vereikina and G. V. Samsonov, Ukr. Khim. Zh., 28(4):441-443 (1962).

558. G. V. Samsonov and S. V. Radzikovskaya, Usp. Khim., 30(1):60-91 (1961).

559. G. N. Dubrovskaya, Zh. Neorg. Khim., 16(1):12-15 (1971).

560. T. N. Nazarchuk, Zh. Neorg. Khim., 4(2):2665-2669 (1959).

561. G. V. Samsonov, V. S. Neshpor, et al., FMM, 8(4):622-631 (1959).

562. V. S. Neshpor and G. V. Samsonov, Izv. Akad. Nauk SSSR Neorg. Mater., 7(1):50-55 (1971).

563. Yu. B. Paderno and G. V. Samsonov, Dokl. Akad. Nauk SSSR 137(3):646-647 (1961).

564. Yu. Paderno, Yu. M. Goryachev, et al., Electron Technology, 3(1/2):175-184 (1970).

565. R. Johnson and A. Daane, J. Phys. Chem., 65:909 (1961).

566. A. Menth, E. Beuhler, and T. H. Geballe, Phys. Rev. Lett., 22(7):295-297 (1969).

567. D. Sturgeon, J. P. Mercurio, et al., Mat. Res. Bull., 9(2):117-120 (1974).

568. G. V. Samsonov, Yu. A. Kunitskii, and V. A. Kosenko, FMM, 33(4):884-887 (1972).

569. V. D. Budozhapov, F. A. Sidorenko, et al., Izv. Akad. Nauk SSSR Neorg. Mater., 11(1):173-174 (1975).

570. V. A. Kosenko, B. M. Rud', et al., Izv. Akad. Nauk SSSR Neorg. Mater., 7(8):1455-1456 (1971).

571. G. V. Samsonov, V. A. Kosenko, et al., Izv. Vyssh. Uchebn. Zaved. Fizika, No. 6, 146-148 (1972).

572. G. V. Samsonov, V. A. Kosenko, et al., Izv. Akad. Nauk Neorg. Mater., 8(4):771-772 (1972).

573. V. L. Yupko, G. N. Makarenko, et al., in: Refractory Carbides, G. V. Samsonov, ed. [in Russian], Kiev (1970), pp. 148-150 [Consultants Bureau, New York (1974)].

574. C. P. Kempter and N. H. Krikorian, J. Less-Common Metals, 4(3):244-247 (1961).

575. C. H. Novion and P. Costa, J. de Phys., 33(2-3):257-271 (1972).

576. C. N. Novion, P. Costa, and G. Dean, Phys. Lett., 19, No. 6, 455-456 (1965).

577. V. S. Neshpor, V. P. Nikitin, et al., Poroshk. Metall., No. 8, 54-57 (1973).

578. V. S. Neshpor, S. V. Airapetyants, et al., Izv. Akad. Nauk SSSR Neorg. Mater., 2(5):855-863 (1966).

579. G. V. Samsonov, Yu. V. Paderno, et al., Rev. Int Hautes Temp. Refract., 3(2):180 (1966); 5(4):316 (1968).

580. A. S. Borukhovich, P. V. Geld, et al., Phys. Status Solidi, 45:179-187 (1971).

581. L. Ramqvist, Jernkont. Ann., 152:465-475 (1968).

582. S. N. L'vov, V. F. Nemchenko, et al., Dokl. Akad. Nauk SSSR, 135(3):577-580 (1960).

583. K. Bachmann and W. S. Williams, J. App. Phys., 42(11):4406-4407 (1971).

584. G. V. Samsonov, M. D. Lyutaya, et al., Zh. Prikl. Khim., 36(10):2108-2116 (1963).

585. J. McClure, J. Phys. Chem. Solids, 24:871 (1963).

586. R. Didchenko and E. Gorstema, J. Phys. Chem., 24:863 (1963).

587. F. Anselin, N. Lorenzelli, et al., Phys. Lett., 19(3):174-175 (1965).

588. C. H. De Novion and R. Lorenzelli, J. Phys. Chem. Solids, 29:1901-1905 (1968).

589. R. Juza and A. Rabenau, Z. Anorg. Allg. Chem., 285:212-220 (1956).

590. G. V. Samsonov, V. S. Sinel'nikova, et al., Izv. Vyssh. Uchebn. Zaved. Tsvetn. Metall., No. 1, 145-151 (1964).

591. H. Jacobi, B. Vassos, and H. J. Engell, J. Phys. Chem. Solids, 30:1261-1271 (1969).

592. V. S. Neshpor and G. V. Samsonov, Fiz. Tverd. Tela, 2(9):2202-2210 (1960).

593. B. M. Rdu', V. I. Lazorenko, et al., Dokl. Akad. Nauk Ukr. SSR Series A(9):849-851 (1973).

594. D. Robins, Phil. Mag., 3:313 (1958).

595. V. C. Kieu and P. Lecoco, Propriétés Thermodynamique Physiques et Structurales des Derives Semi-metalliques, Part 1, Ed. du Centre Nat. Rech. Sc., Paris (1967), pp. 145-153.

596. V. S. Neshpor and V. L. Yupko, Poroshk. Metall., No. 2, 55-59 (1963).

597. K. E. Mironov, Mat. Res. Bull., 4:257-264 (1969).

598. I. M. Novruzly and N. N. Tikhonova, in: Rare-Earth Metals and Their Compounds [in Russian], Kiev (1970), pp. 153-160.

599. Z. Henkie and W. Trzebiatowski, Phys. Status. Solidi., 35:827-836 (1969).

600. G. Boda, B. Stenström, et al., Phys. Scr., 4:132-134 (1971).

601. L. G. Keiserukhskaya, N. P. Luzhnaya, et al., Izv. Akad. Nauk SSSR Neorg. Mater., 6(10):1869-1871 (1970).

602. G. V. Lashkarev and Yu. B. Paderno, Izv. Akad. Nauk SSSR Neorg. Mater., 1(10):1791-1802 (1965).

603. P. Peshev, G. Bliznakov, and A. Toshev, J. Less-Common Metals, 14:379-386 (1968).

604. W. Piekarczyk and P. Peshev, J. Cryst. Growth, 6:357-358 (1970).

605. M. Francillon, D. Jerome, et al., J. de Phys. 31(7):709-714 (1970).

606. K. M. Dunaev, N. A. Santalov, et al. (eds.), Sulfur Compounds of Uranium [in Russian], Nauka, Moscow (1974).

607. G. Raphael and C. Novion, J. de Phys. 30(2-3):261-266 (1969).

608. J. B. Goodenough, Propriétés Thermodynamiques et Structurales des Derives Semi-metalliques, Part 2, Ed. du Centre Nat. Rech Sc., Paris (1967), pp. 263-292.

609. A. Gerard, Propriétés Thermodynamiques et Structurales des Derives Semi-metalliques, Part 2, Ed. du Centre Nat. Rech. Sc., Paris (1967), pp. 55-61.

610. N. N. Zhuravlev, G. N. Makarenko, et al., Izv. Akad. Nauk SSSR Metall. i Toplivo, No. 1, 133-141 (1961).

611. R. Dial and G. Mangsen, Corrosion, 17:107 (1961).

612. G. V. Samsonov and G. G. Tsebulya, Ukr. Fiz. Zh., 5(6):615-619 (1960).

613. G. V. Samsonov, Tsvetn. Met., 31(2):58-59 (1959).

614. Yu. B. Paderno and G. V. Samsonov, Élektronika, Seriya 1(7):123-129 (1960).

615. G. V. Samsonova, V. S. Neshpor, et al., Inzh. Fiz. Zh., No. 2, 118-121 (1959).

616. V. F. Funke, S. I. Yudkovskii, et al., Zh. Prikl. Khim., 33(4):831-835 (1960).

617. V. F. Funke, S. I. Yudkovskii, Zh. Prikl. Khim., 34(5):1013-1020 (1961).

618. V. F. Funke, S. I. Yudkovskii, and G. V. Samsonov, in: Hard Alloys, No. 4 [in Russian], Metallurgizdat (1962), pp. 92-104.

619. P. S. Kislyi and G. V. Samsonov, Izv. Akad. Nauk SSSR, Metall. i Toplivo, No. 6, 133-137 (1959).

620. G. V. Samsonov, P. S. Kislyi, et al., in: Machine Construction [in Russian], GNTK, Kiev, UkSSR, No. 6 (1960), pp. 85-91.

621. G. V. Samsonov, G. A. Yasinskaya, et al., Ogneupory, No. 1, 35-39 (1960).

622. M. S. Koval'chenko, G. V. Samsonov, et al., Izv. Akad. Nauk SSSR Metall. i Toplivo, No. 2, 115-120 (1960).

623. S. Ya. Plotkin and G. V. Samsonov, Khim. Mashinstr., No. 4, 37-40 (1959).

624. N. D. Morgulis and Yu. P. Korchevoi, At. Énerg., 9(49) (1960).

625. R. Kieffer and P. Schwartzkopf, Hard Alloys [Russian translation], Metallurgizdat (1957).

626. V. S. Rakovskii, G. V. Samsonov, et al., Principles of Hard-Alloy Production [in Russian], Metallurgizdat (1960).

627. V. V. Pen'kovskii and G. V. Samsonov, Avtom. Svarka, No. 2, 39-44 (1962).

628. V. V. Grifor'eva, V. N. Klimenko, et al., Vopr. Poroshk. Metall. i Prochn. Mater., No. 5, 80-89 (1958).

629. Yu. B. Paderno, G. V. Samsonov, et al., Élektronika, No. 4, 165-172 (1959).

630. G. A. Gaziev, O. V. Krylov, et al., Dokl. Akad. Nauk SSSR, 140(4):863-865 (1961).

631. G. V. Samsonov and K. I. Portnoi, Alloys Based on Refractory Compounds [in Russian], Oborongiz, Moscow (1962).

632. V. S. Neshpor and G. V. Samsonov, Dokl. Akad. Nauk SSSR, 133(4):817-821 (1960).

633. V. S. Neshpor and G. V. Samsonov, Dokl. Akad. Nauk SSSR, 134(6):1337-1339 (1960).

634. G. V. Samsonov and G. V. Lashkarev, Dokl. Akad. Nauk UkSSR, No. 6, 1147-1151 (1961).

635. K. I. Portnoi, G. V. Samsonov, et al., Zh. Neorg. Khim., 5(9):2032-2042 (1960).

636. V. M. Sleptsov and G. V. Samsonov, Vopro. Poroshk. Metall. i Prochn. Mater., No. 5, 65-78 (1958).

637. G. V. Samsonov, G. A. Yasinskaya, et al., in: Collection of Papers on Metal Behavior and Production Technology of Cermet Hard Alloys of Refractory Metals and Compounds Based on Them [in Russian], Part 1, Moscow TsIIN, TsM (1972), pp. 156-167.

638. W. Nodge, R. Evans, and A. Haskins, J. Met., 7:824 (1955).

639. D. Robins and I. Jenkins, in: Refractory and Corrosion-Resistant Cermets [Russian translation], Oborngiz, Moscow (1959), p. 195.

640. A. L. Burykina and L. V. Strashinskaya, Fiz. Khim. Mekh. Mater., No. 5, 557-562 (1965).

641. Bor, Gmelins Handbuch der anorg. Chemie, Ergänzungsband, Verlag Chemie, Weinheim (1954), p. 10.

642. Silizium, Gmelins Handbuch der Anorg. Chemie, Berlin (1954).

643. S. Downgang and A. Herr, Solid State Commun., 13(2):197-199 (1973).

644. S. Downgang, C. R. Acad. Sci. Paris, 276:701-704 (1973).

645. Yu. B. Paderno, S. Pokrzywnicki, and B. Stalinski, Phys. Status Solidi, 24:K73-76 (1967).

646. Yu. B. Paderno and S. Pokrzywnicki, Phys. Status Solidi, 24:K11-K12 (1967).

647. H. Watanabe and T. Shinohara, Bull. Japan. Inst. of Metals, 7(8):433-447 (1968).

648. S. N. L'vov, M. I. Lesnaya, et al., Izv. Akad. Nauk SSSR Neorg. Mater., 10(4):600-603 (1974).

649. M. C. Cadeville, J. Phys. Chem. Solids, 27:667-670 (1966).

650. E. Legrand and S. Neov, Solid State Commun., 10(9):883-885 (1972).

651. D. J. Lam, M. H. Mueller, et al., J. de Phys., No. 2-3, 917-919 (1971).

652. H. Bittner and H. Goretski, Monatsh. Chem., 93:1000-1003 (1962).

653. R. Caudron, F. Ducastelle, et al., J. Phys. Chem. Solids, 31:291-297 (1970).

654. J. J. Veyssie and F. Anselin, Propriétés Thermodynamiques Physiques et Structurales des Derives Semi-metalliques, Part 1, Ed. du Centre Nat. Rech. Sc., Paris (1967), pp. 349-356.

655. G. Busch, P. Junod, and O. Vogt, Propriétés Thermodynamiques Physiques Structurales des Derives Semi-metalliques, Part 2, Ed. du Centre Nat. Rech. Sc., Paris (1967), pp. 325-336.

656. W. Trzebiatowski, T. Palewski, et al., Propriétés Thermodynamiques Physiques Structurales des Derives Semi-metalliques, Part 2, Edu. du Centre Nat. Rech. Sc., Paris (1967), pp. 499-508.

657. W. Trzeviatowski, R. Troc, et al., Bull. Acad. Pol. Sci. Ser. Sci. Chim., 10, (8):395-398 (1962).

658. M. Nasr Eddine, F. Sayetat, and E. F. Bertanut, C. R. Acad. Sci. Paris, 269, 574-577 (1969).

659. K. Taylor and M. Darby, Physics of Rare Earth Solids, Halsted Press, New York (1972).

660. B. Stalinski and S. Pokrzywnicki, Phys. Status Solidi, 14:K157-K160 (1966).

661. B. Barbara, C. Becle, and R. Lemaire, Z. Angew. Phys., 31(2):113-116 (1971).

662. C. Becle, R. Lemaire, R. Pauthenet, C. R. Acad. Sci. Paris, 266(15):994-997 (1968).

663. J. B. Goodenough, Magnetism and the Chemical Bond, Interscience Wiley, New York (1963).

664. V. T. Matthias, H. Corenzwit, and W. H. Zachariasen, Phys. Rev., 112:89 (1958).

665. K. Sekizawa and K. Yasukochi, Propriétés Thermodynamiques Physiques et Structurales des Derives Semi-metalliques, Part 1, Ed. du Centre Nat. Rech. Sc., Paris (1967), pp. 439-445.

666. Nhung van Nguen, A. Balret, and J. Laforest, J. de Phys., 32(2-3):1133-1134 (1971).

667. E. A. Dmitriev, F. A. Sidorenko, et al., Poroshk. Metall., No. 6, 96-100 (1970).

668. E. Wachtel and E. Mager, Z. Metallkd., 61(10):762-766 (1970).

669. W. Trzebiatowski and T. Palewski, Phys. Status Solidi, 34:K51-K54 (1969).

670. R. Fruchart, A. Roger, and J. P. Senateur, J. Appl. Phys., 40(3):1250-1257 (1969).

671. E. F. Bertaut, G. Roult, et al., J. de Phys., 25(5):582-595 (1964).

672. P. Belougne, J. V. Zanchetta, et al., J. de Chimie Physique, 70(4):682-683 (1973).

673. M. B. Khusidman and V. S. Neshpor, Poroshk. Metall., No. 8, 72-77 (1970).

674. J. Zupan, M. Komas, and D. Kolar, J. Appl. Phys., 41(13):5337-5338 (1970).

675. R. Kuentzler and R. Jesser, Solid State Commun., 13:915-917 (1973).

676. R. Kuentzler, A. Herr, and R. Jesser, Solid State Commun., 12(9):873-875 (1973).

677. G. Foex, Constantes Selectionnes, Diamagnetisme et Paramagnetisme, Paris (1957).

678. R. G. Barnes and R. B. Creel, Phys. Lett., A29(4):203-204 (1969).

679. E. Fruchart, A. M. Triquet, et al., Ann. Chim., 9(7-8):323-332 (1964).

680. R. Lallement, P. Costa, et al., J. Phys. Chem. Solids, 26:1255-1260 (1965).

681. Z. Glowacki and J. Baer, Zesz. Nauk. Politech. Poznanskiej Mech., No. 7, 213-218 (1966).

682. A. Rouault and R. Fruchart, Ann. Chim., 5(5):335-340 (1970).

683. J. P. Rebouillat and J. J. Veyssie, C. R. Acad. Sci. Paris, 259:4239-4240 (1964).

684. M. Chabanel, C. Janot, and J. P. Motte, C. R. Acad. Sci. Paris, 266:419-422 (1968).

685. K. D. Modylevskaya, M. D. Lyutaya, et al., Zavod. Lab., 27(11):1345 (1961).

686. A. I. Miklashevskii, Zavod. Lab., 7(2):168 (1938).

687. M. Humenik and W. Kingery, J. Am. Cer. Soc., 37:18 (1954).

688. H. Goretzki and W. Scheuermann, High Temp. — High Pressures, 3(6):649-658 (1971).

689. V. I. Tumanov, V. F. Funke, et al., Zh. Fiz. Khim., 36(7):1574-1577 (1962).

690. V I. Tumanov, V. F. Funke, et al., Porosh. Metall., No. 5, 43-44 (1963).

691. H. Kotsch and G. Putzky, Abhandlungen der Deutschen Academie der Wissenschaften zu Berlin, Math., Phys., und Technik, No. 1, 249-252 (1966).

692. G. V. Samsonov, A. D. Panasyuk, et al., Izv. Vyssh. Uchebn. Zaved. Tsvetn. Metall., No. 4, 81-86 (1974).

693. G. S. Markevich and L. Ya. Markovskii, Zh. Prikl. Khim., 33:1008 (1960).

694. V. S. Neshpor and G. V. Samsonov, Nauchn. Tr. Mintsvetmetzoloto. Technology of Nonferrous Metals [in Russian], No. 29, Metallurgizdat (1958), 349-356.

695. G. A. Meerson, G. V. Samsonov, et al., Nauchn. Tr. Minsvetmetzoloto, Technology of Nonferrous Metals, [in Russian], No. 29, Metallurgizdat, (1958), 323-328.

696. K. I. Portnoi, G. V. Samsonov, et al., Zh. Prik. Khim., 33(3):577-582 (1960).

697. R. Voitovich and É. A. Pugach, Oxidation of Refractory Compounds [in Russian], Kiev (1968).

698. R. Favre and F. Thevenot, Bull. Soc. Chim. Fr., No. 11, 3911-3916 (1971).

699. M. D. Lyutaya and T. I. Serebryakova, Izv. Akad. Nauk SSSR Neorgan. Mater., 1(7):1044-1048 (1965).

700. R. F. Voitovich and É. A. Pugach, Poroshk. Metall., No. 3, 86-92 (1974).

701. R. F. Voitovich and É. A. Pugach, Poroshk. Metall., No. 2, 63-68 (1972).

702. R. F. Voitovich and É. A. Pugach, Poroshk. Metall., No. 11, 67-74 (1973).

703. R. F. Voitovich and É. A. Pugach, Poroshk. Metall., No. 4, 59-64 (1973).

704. Ju. Morgenthal-Uhlmann, Abhandlungen der Deutschen Acad. des Wissenschaften zu Berlin, Math., Phys., und Technik, No. 1 (1966), pp. 253-255.

705. T. Ya. Kosolapova and G. V. Samsonov, Zh. Fiz. Khim., 35(2):363-366 (1961).

706. J. Besson, C. Moreau, and J. Philippot, Bull. Soc. Chim. Fr., No. 189, 1069-1074 (1964).

707. M. D. Lyutaya, O. P. Kulik, et al., Poroshk. Metall., No. 3, 72-75 (1970).

708. J. Binder and R. Steinitz, Planseeber. Pulvermetall., 7, 18 (1959).

709. R. F. Voitovich and É. A. Pugach, Poroshk. Metall., No. 1, 63-70 (1974).

710. R. F. Voitovich and É. A. Pugach, Poroshk. Metall., No. 2, 43-49 (1974).

711. G. V. Samsonov, V. A. Lavrenko, et al., Poroshk. Metall., No. 1, 46-49 (1974).

712. R. Kieffer, F. Benesovsky, and H. Schmidt, Z. Metallkd., 47(4):247-253 (1956).

713. I. G. Vasil'eva, K. E. Mironov, et al., in: Rare-Earth Metals and Their Compounds [in Russian], Kiev, (1970), pp. 160-165.

714. R. Meyer and H. Pastor, Planseeber. Pulvermetall., 17:111-122 (1969).

715. H. Pastor and R. Meyer, Hautes Temp. et Refract., 11(1):41-54 (1974).

716. A. L. Burykina and T. M. Evtushok, Poroshk. Metall., No. 6, 75-78 (1965).

717. G. A. Yasinskaya and M. S. Groisberg, Poroshk. Metall., No. 6, 36-37 (1963).

718. V. F. Funke and S. I. Yudovskii, in: Investigations of Steels and Alloys [in Russian], Nauka, Moscow, (1964), pp. 108-113.

719. É. V. Marek, Yu. B. Kuz'ma, et al., Poroshk. Metall., No. 2, 70-73 (1971).

720. O. A. Medvedeva, Poroshk. Metall., No. 2, 38-45 (1972).

721. R. Kiessling and L. Liu, J. Metals, 3:639 (1951).

722. G. V. Samsonov, A. L. Burykina, et al., Poroshk. Metall., No. 11, 50-57 (1973).

723. K. I. Portnoi, Yu. V. Levinskii, et al., Izv. Akad. Nauk SSSR Metall. i Toplivo, No. 2, 147-149 (1961).

724. A. Betkher and A. Schneider, in: Trans. 2nd Internat. Conf. on Peaceful Use of Atomic Energy: Nuclear Fuels and Nuclear Materials, Geneva (1959), 269.

725. F. Glaser, J. Metals, 4:391 (1952).

726. G. V. Samsonov, Vop. Porosh. Metall. Prochn. Mater., No. 7, 72-98 (1959).

727. T. Ya. Kosolapova and É. V. Kutysheva, Poroshk. Metall., No. 11, 61-65 (1969).

728. V. B. Fedorus, T. Ya. Kosolapova, et al., Rev. Int. Hautes Temp. Refract., 6(3):193-198 (1969).

729. A. Erb, Ann. Chim., 14:713 (1959).

730. F. A. Josien and R. Renaud, C. R. Acad. Sci. Paris, 260:2239-2242 (1965).

731. R. A. Alfintseva, Poroshk. Metall., No. 1, 83-87 (1971).

732. T. Ya. Kosolapova, G. N. Makarenko, et al., Poroshk. Metall., No. 1, 27-33 (1971).

733. A. L. Burykina, Yu. V. Dzyadykevich, et al., Poroshk. Metall., No. 9, 74-76 (1973).

734. A. Hivert and J. Poulignier, Rev. Int. Hautes Temp. Refract, 5, No. 1, 55-61 (1968).

735. E. M. Savitskii, Mechanical Properties of Intermetallic Compounds, New York, J. Wiley (1960), pp. 106-110.

736. V. K. Kharchenko and L. I. Struk, Poroshk. Metall., No. 2, 87-91 (1962).

737. I. Binder, J. Amer. Cer. Soc., 43:287 (1960).

738. G. V. Samsonov, V K. Kazakov, et al., Poroshk. Metall., No. 2, 60-63 (1974).

739. J. J. Gangler, High Temp. — High Pressures, 3(5):487-502 (1971).

740. L. L. Vereikina, V. N. Rudenko, et al., Zavod. Lab., 26(5):620-621 (1960).

741. G. V. Samsonov and V S. Neshpor, Vopr. Porosh. Metall. Prochn. Mater., No. 5, 3-8 (1958).

742. K. J. Matterson and H. Jones, A Study of the Tetraborides of Uranium and Thorium, Trans. Brit. Cer. Soc., 60:475-491 (1961).

743. A. B. Lyashchenko, P. I. Mel'nichuk, et al., Poroshk. Metall., No. 5, 10-14 (1961).

744. K. I. Portnoi, A. A. Mukaseev, et al., Poroshk. Metall., No. 3, 32-33 (1968).

745. C. Novion and R. Lallement, Rev. Int. Hautes Temp. Refract., 9(1):117-122 (1972).

746. J. L. Chermant, Rev. Int. Hautes Temp. Refract., 6(4):299-312 (1969).

747. J. L. Martin and P. Costa, C. R. Acad. Sci. Paris, 273C:89-91 (1971).

748. K. I. Portnoi and V. N. Gribkov, Porosh. Metall., No. 5, 10-14 (1970).

749. M. Cantagrel and M. Marchal, Rev. Int. Hautes Temp. Refract., 9:93-100 (1972).

750. V. S. Neshpor and S. S. Ordan'yan, Porosh. Metall., No. 1, 23-27 (1964).

751. A. A. Ivan'ko, Hardness [in Russian], Naukova Dumka, Kiev (1968).

752. G. V. Samsonov and V. Kh. Oganesyan, Dokl. Akad. Nauk UkSSR, No. 10, 1317-1321 (1965).

753. L. Ramqvist, Jernkonterets Ann., 152:517-523 (1968).

754. L. Ramqvist, Jernkonterets Ann., 153:159-175 (1969).

755. G. V. Samsonov, Dokl. Akad. Nauk SSSR, 86(2):325-329 (1952).

756. G. V. Samsonov, Dokl. Akad. Nauk SSSR, 113(6):1299-1302 (1957).

757. P. T. Kolomyitsev, Dokl. Akad. Nauk SSSR, 124(6):1247-1250 (1959).

758. A. S. Sobolev and A. S. Fedorov, Izv. Akad. Nauk SSSR Neorg. Mater., 3(4):723-727 (1967).

759. G. V. Samsonov, G. N. Makarenko, et al., Zh. Prikl. Khim., 34(7):1444-1448 (1961).

760. J. L. Chermant, P. Delavignette, et al., J. Less-Common Metals, 21:89-101 (1970).

761. M. S. Koval'chenko, V. V. Dzhemelinskii, et al., Poroshk. Metall., No. 8, 87-91 (1971).

762. M. S. Koval'chenko and Yu. I. Rogovoi, Poroshk. Metall., No. 2, 93-99 (1971).

763. F. W. Vahldiek and S. A. Mersol, Anisotropy in Single-Crystal Refractory Compounds, Vol. 2, Plenum Press, New York (1968), p. 199.

764. B. O. Haglund, Prakt. Metallogr., 7:173-182 (1970).

765. G. V. Samsonov and T. S. Verkhoglyadova, Zh. Strukt. Khim. 2(5):617-618 (1961).

766. Yu. D. Chistyakov, G. V. Samsonov, and M. V. Mal'tsev, in: Trudy VNITO Metallurgov, Vol. 2, Metallurgizdat (1954), pp. 169-191.

767. L. A. Dvorina, Izv. Akad. Nauk SSSR Neorg. Mater., 1(10):1722-1777 (1965).

768. N. V. Kolomoets, V. S. Neshpor, et al., ZhTF, 28(11):2382-2389 (1958).

769. Yu. B. Paderno, V. L. Yupko, et al., Izv. Akad. Nauk SSSR Neorg, Mater., 2(4):626-628 (1966).

770. V. G. Grebenkina and E. N. Denbnovetskaya, in: Refractory Carbides [in Russian], Naukova Dumka, Kiev (1970), pp. 160-163.
771. G. V. Samsonov and V. S. Sinel'nikova, in: High-Temperature Cermets [in Russian], Akad. Nauk UkSSR, Kiev (1960), pp. 120-123.
772. V. S. Neshpor, V. P. Nikitin, and N. A. Skaletskaya, Poroshk. Metall., No. 8, 54-57 (1973).
773. P. V. Geld, V. A. Tskhai, A. S. Borukhovich, et al., Phys. Status Solidi, 42:85-88 (1970).
774. G. A. Kudintseva, V. S. Neshpor, et al., in: High-Temperature Cermets [in Russian], Akad. Nauk UkSSR, Kiev (1962), pp. 109-112.
775. G. Bliznakov, I. Tsolovski, and P. Peshev, Rev. Int. Hautes Temp. Refract., 6(3):159-164 (1969).
776. G. V. Samsonov, V. P. Dzeganovskii, and I. A. Semashko, Kristallografiya, 4(1):119-120 (1959).
777. G. A. Kudintseva, G. M. Kuznetsova, V. P. Bondarenko, et al., Poroshk. Metall., No. 2, 45-47 (1968).
778. W. S. Fomenko, Radotekhi. Élektron., 6:1406 (1961).
779. G. V. Samsonov, V. S. Fomenko, and Ju. A. Kunitzkij, Rev. Int. Hautes Temp. Refract., 10(1):11-14 (1973).
780. G. V. Samsonov, V. S. Neshpor, and G. A. Kudintseva, Radiotekh Élektron., 2(5):631-636 (1957).
781. G. G. Gnesin, G. S. Oleinik, et al., TVT, 8(3):663-665 (1970).
782. W. S. Fomenko, W. A. Lawrenko, et al., Rev. Int. Hautes Temp. Refract., 8(3-4):315-318 (1971).
783. L. N. Okhremchuk, I. A. Podchernyaeva, et al., Radiotekh. Élektron., 15(5):1114-1115 (1970).
784. G. V. Samsonov, V. S. Fomenko, et al., Ukr. Fiz. Zh., 3(6):700-703 (1963).
785. G. V. Samsonov, V. S. Fomenko, et al., in: Chemistry and Physics of Nitrides [in Russian], Naukova Dumka, Kiev (1968), pp. 162-167.
786. V. I. Marchenko, G. V. Samsonov, et al., Radiotekh. i Électron., 8(6):1076-1082 (1963).
787. V. I. Marchenko, G. V. Samsonov, et al., ZhTF, 34(1):128-131 (1964).
788. Con van Kieu, C. R. Acad. Sci. Paris, 260:111-113 (1965).
789. E. S. Garf, Yu. B. Paderno, et al., in: Rare-Earth Metals and Their Compounds [in Russian], Naukova Dumka, Kiev (1970), pp. 101-109.
790. L. M. Falicov and J. C. Kimball, Phys. Rev. Lett., 22(19):997-999 (1969).
791. V. M. Bertenev, S. M. Blokhin, and K. E. Mironov, in: Rare-Earth Metals and Their Compounds [in Russian], Naukova Dumka, Kiev (1970), pp. 147-153.
792. L. A. Ivanchenko, G. V. Lashkarev, et al., in: Rare-Earth Metals and Their Compounds [in Russian], Naukova Dumka, Kiev (1970), pp. 186-191.
793. G. V. Samsonov and V. S. Sinel'nikova, Ukr. Fiz. Zh., 6(5):687-690 (1961).
794. V. S. Sinel'nikova, T. Niemyski, J. Panczyk, et al., J. Less-Common Metals, 23:1-6 (1971).
795. G. V. Samsonov, I. F. Pryadko, and L. F. Pryadko, A Configurational Model of Matter [in Russian], Naukova Dumka, Kiev (1971) [Consultants Bureau, New York (1973).]

796. K. D. O'Dell and E. B. Hensley, J. Phys. Chem. Solids, 33(2):443-449 (1972).

797. S. P. Gordienko and B. V. Fenochka, Zh. Fiz. Khim., 47(4):1030-1031 (1973).

798. K. A. Gingerich, J. Chem. Phys., 54(9):3720-3722 (1971).

799. C. Politis, F. Thümmler, and H. Wedemeyer, J. Nucl. Mater., 32:181-192 (1969).

800. H. Kleykamp, P. Korogiannakis, et al., Kernforschungszentrum Karlsruhe Ber., 66:KFK-771 (1968).

801. R. Kuentzler and A. Meyer, C. R. Acad. Sci. Paris, 266:755-758 (1968).

802. V. S. Neshpor, in: Metal Physics [in Russian], Nauka, Moscow (1971), pp. 333-341.

803. R. Caudron, J. Castaing, and P. Costa, Solid State Commun., 8(8):621-625 (1970).

804. A. P. Épik, Poroshk. Metall., No. 5, 21-25 (1963).

805. G. V. Samsonov, ed., Handbook of the Physicochemical Properties of the Elements, IFI/Plenum, New York. (1968).

806. G. V. Samsonov and A. P. Épik, Dokl. Akad. Nauk Ukr. SSR, No. 1, 67-71 (1964).

807. G. Sh. Viksman, S. P. Gordienko, and B. V. Fenochka, Zh. Fiz. Khim. 49(5): 1258-1259 (1975).

808. A. I. Nakopechnikov, L. V. Pavlinov, et al., FMM, 22(2):234-238 (1966).

809. Suzuki Hiroshige, Kimura Shushichi, Hase Teizo, et al., Bull. Tokyo Inst. Technol., No. 90, 105-115 (1969).

810. G. Ya. Meshcheryakov, R. A. Andrievskii, and V. N. Zagryazkin, FMM, 25(1):189-193 (1968).

811. F. A. Schmidt and O. N. Carlson, J. Less-Common Metals, 26(2):247-253 (1972).

812. D. F. Kalinivich, I. I. Kovenskii, and M. D. Smolin, Ukr. Fiz. Zh., 14(2): 339-341 (1969).

813. S. V. Zemskii and M. N. Spasskii, FMM, 21(1):129-131 (1966).

814. P. S. Rudman, Trans. Metall. Soc. AIME, 239(12):1949-1954 (1967).

815. A. Schepela, J. Less-Common Metals, 26(1):33-43 (1972).

816. J. Klein, J. Appl. Phys., 38(1):167-170 (1967).

817. G. V. Grigor'ev and L. V. Pavlinov, FMM, 25(5):836-838 (1968).

818. R. Tate, G. Edwards, and E. Hakkila, J. Nucl. Mater., 29(2):154-157 (1969).

819. G. Siebel, C. R. Acad. Sci. Paris, 256:661-663 (1963).

820. R. A. Krakowski, J. Nucl. Mater., 32(1):120-125 (1969).

821. H. M. Lee and L. K. Barrett, J. Nucl. Mater., 27(3):275-284 (1968).

822. V. Phillippe, Determination du coefficient d'autodiffusion de l'uranium dans son monocarbure, Grenoble, Rapp. CEA (1968), No. 3434, p. 64.

823. A. I. Evstyukhin, G. I. Solov'ev, et al., in: Method of Isotope Indicators in Scientific Research and In Industrial Production [in Russian], Atomizdat, Moscow (1971), pp. 166-171.

824. S. Sarian, J. Appl. Phys., 39(11):5036-5041 (1968).

825. S. Sarian, J. Appl. Phys., 39(7):3305-3310 (1968).

826. D. L. Kohlstedt, W. S. Williams, et al., J. Appl. Phys., 41(11):4476-4484 (1970).
827. S. Sarian, J. Appl. Phys., 40(9):3515-3520 (1969).
828. V. N. Zagryazkin, FMM, 28(2):292-297 (1969).
829. R. A. Andrievskii and K. P Gurov, FMM, 39(1):57-61 (1975).
830. S. Sarian and J. M. Criscione, J. Appl. Phys., 38(4):1794-1798 (1967).
831. Yu. N. Vil'k, S. S. Nikol'skii, et al., TVT, 5(4):607-611 (1967).
832. Yu. N. Vil'k and S. S. Ordan'yan, Zh Fiz. Khim., 48(8):2150-2151.
833. R. A. Andrievskii, V. V. Klimenko, et al., FMM, 28(2):298-303 (1969).
834. S. Sarian, J. Phys. Chem. Solids, 33(8):1637-1643 (1972).
835. T. Ya. Meshcheryakov and V N Zagryazkin, FMM, 32(4):883-885 (1971).
836. W F. Brizes, J. Nucl. Mater., 26(2):227-231 (1968).
837. R Resnik and L Scigle, Trans. Met. Soc. AIME, 236(21):1732-1738 (1966).
838. V. S. Eremeev and A. S. Panov, Izv. Akad. Nauk SSSR Neorg. Mater., 4(9):1507-1510 (1968).
839. V. S. Eremeev, P. L. Gruzin, and A. S. Panov, in: Method of Isotropic Indicators in Scientific Research and Industrial Production [in Russian], Atomizdat, Moscow (1971), pp. 60-65.
840. C. P. Buhsmer and P. H. Crayton, J. Mater. Sci., 6(7):981-988 (1971).
841. I. I. Spivak, Izv. Akad. Nauk SSSR Neorg. Mater., 5(6):1138-1139 (1969).
842. Yu. F. Khromov, V. P. Yanchur, and V. S. Eremeev, FMM, 33(3):642-644 (1972).
843. R. A. Andrievskii, V. S. Eremeev, et al., Izv. Akad. Nauk SSSR Neorg. Mater., 3:2158-2160 (1967).
844. K. Torkar, H. Cel, and A. Illigen, Ber. Dtsch. Keram. Gesl., 43:162-166 (1966).
845. R. Ghoshtagore and R. Coble, Phys. Rev., 145:623-626 (1966).